Biology of the Reptilia

Volume 18, Physiology E

Hormones, Brain, and Behavior

Biology of the Reptilia

Edited by
Carl Gans

Volume 18, Physiology E
Hormones, Brain, and Behavior

Coeditor for this volume
David Crews

The University of Chicago Press
Chicago and London

The University of Chicago Press, Chicago 60637
The University of Chicago Press, Ltd., London

Printed in the United States of America

01 00 99 98 97 96 95 94 93 92 5 4 3 2 1

ISBN (cloth): 0–226–28122–1
ISBN (paper): 0–226–28124–8

Library of Congress Cataloging-in-Publication Data
(Revised for vol. 18)

Gans, Carl, 1923–
Biology of the reptilia.
Vol. 14–15 published: New York: Wiley; vol. 16 published: New York : A. R. Liss.
Vol. 18 published: Chicago : University of Chicago Press
Includes bibliographies and indexes.
1. Reptiles. I. Title.
QL641.G3 597.9 68-9113

⊗ The paper used in this publication meets the minimum requirements of the American National Standard for Information Sciences—Permanence of Paper for Printed Library Materials, ANSI Z39.48–1984.

Contents

Dedication

On 26 September 1990, the herpetological community lost the counsel and company of Angus Bellairs, scholar and raconteur, herpetologist and anatomist, physician and teacher. A most knowledgeable investigator, he also cherished his idiosyncrasies, whether a professed fear of airports or an absolute admiration for large cats (and an unwillingness to accept the reality of their predation on his other friends, the lizards). He was committed to carrying out his own research deliberately and personally, from the collection of specimens to their histological preparation and final illustration. He was loved and admired by his graduate students, and his company was enjoyed by a much wider audience.

It seems appropriate to dedicate this volume of the Biology of the Reptilia *to Angus, as he was critical in the initial planning of the series, although he then saw it as a much more restricted enterprise. He coedited the first volume, provided advice on several others, and collaborated in the writing of two major chapters. Several of his students continue to participate. Beyond this, the dedication provides me the opportunity of acknowledging the aid of a friend who left us much too early.*

When I initially inquired about material for an obituary, I was surprised to hear that Angus had written his own and had sent it to Susan Bryant about the time of his "official" retirement. As the document retains some of the flavor of his conversation, I am publishing it here (with his wife Ruth's kind permission). A couple of amplifications and additions are provided in brackets.

Angus d'Albini Bellairs was born in 1918 and educated at Stowe School, Queens' College, Cambridge, and University College Hospital, London. Interested in natural history, especially in reptiles, for as long as he can remember, he was delighted to find that it was possible to read zoology at Cambridge, in combination with the traditional medical subjects. At Cambridge, his zoological interests became directed to comparative anatomy and he was lucky enough to meet several teachers who later became lifelong friends: in particular Dixon Boyd and Frank Goldby, Hugh Cott and Rex Parrington, who, as Angus later wrote, "showed in his elegant and stimulating lectures on fossil reptiles the best that the formal discipline of a university course can offer."

When Angus graduated from Cambridge in 1939 the clouds of World War II were looming, and, foreseeing that he would be better off as a medical officer than as a private in the infantry, he managed to qualify in medicine. He was called up into the R.A.M.C. in 1942 and found the army surprisingly less unpleasant than he had expected—and in some ways more democratic than university life! He was posted as M.O. to the 4th Divisional Engineers, a unit for which he retains the greatest affection and whose annual reunion dinner he is still happy to attend. His unit went almost immediately to North Africa and took part in the final battles in Tunisia; it went briefly to Egypt and then to Italy, participating in the fighting around Cassino. Angus was then suddenly posted across the world to a branch of the Army Biological Directorate (then headed in London by the distinguished biologists F. A. E. Crew and Lancelot Hogben) at the headquarters of General Slim's 14th ("forgotten") Army in Burma. This transfer, which carried a promotion to Major, may possibly have been in the nature of a "kicking upstairs" (a commendable fate!). Everywhere Angus went, he collected reptiles and there was a general rumour (which he may have encouraged) that anyone in his unit who reported sick with a reptile in a tin was bound to receive a day off duty. Of course this posting to the Far East afforded further opportunities for collection, and also fascinating human experiences in the course of duty. One of Angus's main projects in the 14th Army was a follow-up of non-commissioned soldier patients in Indian hospitals, which took him through many weird and wonderful medical institutions throughout the length and breadth of the sub-continent. He has pleasant memories of a delightfully alcoholic Colonel commanding a hospital who did his ward rounds on a bicycle when he was not shooting imaginary tigers through the window of his hut. Also of a capricious medical documentation of the Indian Army; an aged sepoy diagnosed as suffering from senility might be temporarily downgraded to Category B for further treatment. By the strangest of military chances, Victory Day found Angus in Times Square, New York, where he participated in celebrations possibly reminiscent of Mafeking Night in London—as yet unclouded by forebodings of the future of the Bomb.

Angus had long decided that his career after the war was not to be in clinical medicine, and it would have been logical for him to seek a post in a zoology department of one of our expanding universities. However, J. D. Boyd, then Professor of Anatomy at the London Hospital Medical College, persuaded him that anatomy, which at that time contained a stronger tradition of comparative work than it does today, might be a suitable milieu in which to combine his zoological interests with his medical experience, and offered him a Lectureship in his department. During this period, Angus naturally paid his respects to Professor G. R. (later Sir Gavin) de Beer, the doyen of cranial morphologists, in the Anatomy Department of University College, London. There he met a research student, Ruth Morgan, whom he subsequently married and who is now herself Professor of Embryology in the same department. They have a daughter, Vivien, in her final year as a medical student at Oxford. [Vivien St. Joseph is now a general practitioner in Colchester and has a son.]

In 1953, J. D. Boyd moved to the Chair of Anatomy in Cambridge and managed to take both Angus and his wife with him. Unhappily, the transplantation was not a successful one; at that time the (appropriately named) Anatomy School seemed to be an underprivileged department in a highly privileged university, and the teaching commitments, which extended through part of the long vacation, were ferocious. In 1953 Angus obtained the Readership on Anatomy at St. Mary's Hospital Medical School, where he has remained happily ever since; in 1970 he was awarded a personal Chair in Vertebrate Morphology [created especially for him by the University of London].

Unlike some of his colleagues, Angus has the highest regard for that "vast acephalic monster" (as H. G. Wells termed it), the University of London, which, if properly utilised, offers a range of facilities and talent throughout its various colleges which no other university in Britain can match. Moreover, the capital contains two other venerable scientific institutions to which Angus feels deeply attached: the British Museum (Natural History)—despite its current exhibition policy, which he deplores—and the Zoological Society of London. His relationship with the Zoo has been particularly happy and he derives the greatest pleasure from his honorary appointment as consulting herpetologist to the Society. He has also been a member of the Society's Publications Committee (in effect, the editorial board of the Journal of Zoology) for many years. He was a founder member of the British Herpetological Society and the first editor of its journal.

Angus has made a number of scientific travels in post-war years. In 1955 he obtained a Royal Society Travelling Bursary to visit South Africa and enjoyed the zoology (especially crocodile-watching), though he found the politics depressing. In 1970 he was Visiting Professor at the University of Kuwait, and as a person of the highest reputed moral integrity, was put in charge of the girl students' field excursions, teaching them to handle lizards and harmless snakes without fear—a considerable feat. In 1973 he took part in an expedition to the Galápagos Islands to study the behaviour of giant tortoises, and he is an enthusiastic visitor to the U.S.A., the principal centre of modern herpetology.

Angus's seventy scientific publications have been mainly on reptiles, though he has also studied birds and mammals. Many of them have been concerned with the structure and development of the skull and its associated organs, a once-predominant biological discipline, but today a somewhat recondite field. He is at present examining the skull in a fascinating group of burrowing reptiles, the amphisbaenians, in collaboration with Carl Gans of the University of Michigan. His studies on cranial morphology are embodied in a substantial monograph (1981) in the important serial publication *Biology of the Reptilia*. Other work has been concerned with reptilian teratology; he can claim to have discovered cleft palate of a type very similar to the human malformation in snakes, and has just finished what is probably the first anatomical description of cyclopia in turtles. He has also studied regeneration of the tail in lizards in collaboration with Susan V. Bryant, a former graduate student, now a professor in the University of California; they have recently prepared another large monograph on autotomy and regeneration for a forthcoming volume in the *Biology of the Reptilia* [published in volume 15]. Another former graduate student is Paul Maderson of Brooklyn College, New York, renowned for his studies on the reptilian skin.

Although Angus enjoys scientific collaborations with his friends, he likes to think of himself as something of a loner, pursuing a little-known subject in his own way, and relatively independent of the vagaries of biological fashion and of grant awards. Indeed, he sometimes refers to himself as a "scientific antique collector." He prepares most of his own material, types his own papers and illustrates them with his own drawings. He likes writing and has written or co-authored three books, two of which (*The Life of Reptiles*, Weidenfeld, 1969, and *Reptiles*, Hutchinson's University Library, 1975; with J. Attridge) have been translated into foreign languages. [The first edition of *Reptiles* was actually written by Angus alone and published by Hutchinson's University Library in 1958.] In these books he has tried to provide a synthesis of the various kinds of knowledge available about reptiles, from physiology to behaviour, and likes to paraphrase T. E. Lawrence in saying, "Nothing reptilian is alien to me."

Despite his somewhat unusual (though by no means unique) scientific interests for a teacher in a medical school, Angus likes to feel that he has contributed to medical education and has done something to mitigate its deplorably narrow but congested curriculum. On first coming to St. Mary's, he designed, in collaboration with Professor Goldby, a type of introductory course in which some of the more general aspects of biology could be touched on. He feels that every doctor—indeed every educated person—should be taught something about man's place in the Animal Kingdom, his effect on the world's ecosystem, and about animal behaviour, in particular the sexual and territorial imperatives. He believes that every medical school should contain at least one teacher who is primarily a naturalist.

Angus hopes to continue his scientific research and writing into his retirement, but he has many other interests: all natural history, especially of the domestic cat which seems to show an admirable combination of friendliness and independence, antique collecting, military history and modern fiction. He very much hopes to write a novel, partly based on academic life; however, he realises that writing a novel is very different from writing books on reptiles, and whether he can really do it remains to be seen . . .

There is much that could be added to this auto-obituary, as the account typically does not note the honors that accrued to him, for instance his honorary membership of societies, herpetological and zoological. The Zoological Society of London honored him with a Symposium on reptilian biology that resulted in a festschrift (edited by Professor M. W. J. Ferguson). In 1989, he was the Honorary President of the First World Congress of Herpetology, and the occasion saw the publication of his novel The Isle of Sea Lizards. *Herpetologists from around the world then had a chance to chat with him, and many went home with an inscribed copy of the tale of tropical islands, academic politics, diatribes on research, museum collections and conservation, and tales about the discovery of a colony of mosasaurs in the Pacific. It is sad that the promised second novel could not be completed and sadder still that so many of us will have to miss future occasions for conversations.*

December 1990

Carl Gans

Preface

The present volume of *Biology of the Reptilia* concentrates on the interaction of the hormonal and nervous systems in influencing the behavior of reptiles, primarily that associated with reproduction. As many recent efforts in this area have been stimulated by David Crews, I was delighted to have his association in assembling chapters reviewing these topics.

In editing these chapters we found that the coverage of species was highly uneven, as would be characteristic of a relatively new field. We have encouraged our authors to stress what is known on the basis of experimental studies and to note the species from which that knowledge derives. However, we have also asked them to refer to the many initial field observations that belong to what is sometimes called the Natural History literature. Although such records are no substitute for the experimental approaches, they are often unexcelled in permitting decisions about the probable generality of phenomena, in disclosing exceptions to theory, and in stimulating further investigations. Too rarely are such data noted in initial field reports; this is particularly regrettable at a moment when more and more species are becoming rare and a separate expedition for each species is no longer an acceptable luxury.

What should also be clear is that major taxonomic groups still need to be sampled. For entire orders we lack even preliminary information, and the information that has been gathered suggests that surprising discoveries remain in store for the discerning investigator.

It is the desirability of such future comparison that has led us once again to attempt to standardize nomenclature. I continue to appreciate George Zug's careful check of Latin names, which permits us to refer to species by the currently appropriate terminology. I am very grateful to our contributors for their responsiveness to the comments of multiple reviewers and for their patience while awaiting the completion of the last-arrived manuscripts of the volume. It is also my pleasure to acknowledge the assistance of David Crews and of the large numbers of reviewers. Their efforts represent the key to the generation of volumes intended to remain useful to the herpetological, indeed the zoological, community. At the risk of omitting some names we would like to mention E. Allen, J. T. Bagnara, G. M. Burghardt, E. H. Burtt, Jr., V. M. Cassone, D. A. Chiszar, D. Duvall, N. B. Ford, W. A. Gern, M. T. Mendoça, F. L. Moore, W. C. Sherbrooke, and G. L. Vaughn, as well as the authors of the several chapters. Katherine Vernon aided substantially with the correspondence. Finally, we appreciate the support of our institutions toward the ever increasing costs of postage and copying.

Ann Arbor, Michigan CARL GANS

1

The Interaction of Hormones, Brain, and Behavior: An Emerging Discipline in Herpetology

DAVID CREWS AND CARL GANS

CONTENTS

I. INTRODUCTION

The interaction of the endocrine and nervous systems and the way this interaction affects the behavior of animals have been major areas of study for the past 70 years. A rich tradition of research has demonstrated that these systems are central to, and finely tuned by, complex and critical behavioral interactions. However, most work in what is now termed psychoneuroendocrinology has focused on birds and mammals and, until recently, has generally been ignored by herpetologists.

This is curious, because two herpetologists, Gladwyn Kingsley Noble and Llewellyn T. Evans, were pioneers in the field of behavioral endocrinology. In the mid-1930s both were conducting experimental studies focused on the effects of hormones on sexual behaviors. This was shortly after the first steroid hormones had been characterized and the stimulatory role of pituitary hormones on gonadal growth had been demonstrated. Noble and Evans, working independently and at the same time as mammalian physiologists such as William C. Young and his colleagues, removed the gonads from *Anolis carolinensis*. These early studies demonstrated the dependence of sexual behavior on the secretion of gonadal hormones. They also were among the first scientists to document that administration of pituitary extracts stimulates gonadal activity and hence sexual behavior. Together their results confirmed the then novel concept of feedback control in reproduction.

Noble was a particularly astute observer of behavior. For example, he found that mating inhibits further sexual receptivity in female *Anolis carolinensis* (Greenberg and Noble, 1944); this was the first suggestion of this now widely accepted finding for any vertebrate. The work of Noble and his colleagues also furnishes one of the first references to the multiplicity of factors that regulate successful reproduction. They note that "besides the endocrine factors undoubtedly involved in this fluctuation, there were the less tangible effects of changes which tended to throw established relationships out of balance. The most common variations were: (1) displacement of males because of fights, accidents, or disease; (2) emergence of dominant females; and (3) environmental factors such as light, weather, and food supply" (Greenberg and Noble, 1944, p. 402).

Finally, Noble envisioned the nature of ovarian cycles, particularly as they related to sexual receptivity, by writing "oestrus in the lizard depends upon ovarian endocrine secretions. It would seem that copulation is possible only when ripe ova have developed to the point where the associated follicular or interstitial cells produce sufficient hormone to activate the female mating reflexes" (Greenberg and Noble, 1944, p. 434). Further he demonstrated that "ovariectomy eliminates estrus completely while administration of crystalline sex hormones to spayed or intact females or frog anterior pituitaries to intact females induces full sexual receptivity" (Noble and Greenberg, 1941, p. 469). The reader should keep in mind that this research was conducted at a time that the classic work showing a hormone-behavior relationship in laboratory mammals had just been published (Young, 1941).

However, the impetus that might have made green anoles as common a laboratory animal as small rodents are today was cut short. Noble died prematurely at the height of his research career, whereas

Evans essentially ceased all research activities for want of an academic position. Still, this early literature continues to offer a rich source of ideas and suggestive evidence.

Recently, herpetological contributions to behavioral physiology have flowered, and this volume documents the initial results. Much of what proceeds is associated with reproduction, from the initial recognition and evaluation of potential sex partners to decisions about reproduction, in addition to those associated with survival. The chapters included here synthesize much of the existing literature in this newly emerging area. The contributors also point the way to the many remaining questions in this important area. Here, especially, we see how reptiles offer a number of unique model systems with which to address a variety of significant questions in behavioral physiology (Table 1.1). Finally, we hope that the volume will stimulate established researchers to expand their research programs and will encourage new investigators to enter this exciting area of research.

II. A VIEW OF RESEARCH STRATEGIES

The earliest studies in function, physiology, and behavior inevitably attempted to identify particular components and to treat these under seemingly neutral environmental conditions, independent of the situations in which they arose. Only as the techniques for study improved and as more work from other areas became available did this viewpoint start to change. It then was recognized that, for instance, the effects of behavioral states represented more than a source of noise to be eliminated. We now know that these states represent a critical aspect, as natural selection often has matched processes to expected environmental conditions. This association has resulted in a perspective that may be referred to as the ecology of mechanisms. We believe that it represents a major frontier for modern behavioral neuroscience and that reptiles will be shown to be the ideal animals for addressing this aspect. Basically, the ecology of mechanisms suggests that neuroendocrine mechanisms are correlated with the environment. The results of studies designed to explore this viewpoint promise to resolve questions of selection and historical causation.

This volume will document that reptiles match many of the demands of modern approaches to behavioral endocrinology, which accounts for the excitement, indeed almost the frontier attitude, prevalent in this field. Another reason for the excitement is that the discipline of herpetology is still very much dominated by the natural history perspective; students consciously investigate the ways their animals match the other aspects of the environment. This approach generates an excellent knowledge base for answering questions about the "what," "where," "when," and "why" of topics. It also in turn provides a superb platform for launching investigations into questions about the "how" of processes.

In addition to the natural history tradition of herpetology, the field has always had a strong focus on comparative studies. The comparative approach has generated data sets combining ecological and evolutionary perspectives and points the way to the analysis of a variety

Table 1.1 Some possible comparisons using reptiles as model systems

Questions	Comparison	Example
How does behavior change in speciation? What is the effect of ploidy on neuroendocrine mechanisms?	Ancestor vs. descendant Diploid vs. triploid	*Cnemidophorus inornatus* × *C. uniparens*
Are there functional differences in genotypic sex determining systems (GSD)?	Male heterogamety vs. female heterogamety	*C. inornatus* × *Thamnophis sirtalis parietalis*
How does environmental sex determination (ESD) differ from GSD?	Genotypic sex determination vs. temperature-dependent sex determination	*Tryonyx spiniferus* × *Trachemys scripta*
How does the behavior of different reptiles differ?	Phylogeny	Clades
How can there be different phenotypes of the same sex?	Alternative reproductive patterns	*Thamnophis*
What is the functional relationship between gamete growth, sex steroid hormone secretion, and mating behavior?	Associated vs. dissociated reproduction	*Anolis carolinensis* × *Thamnophis* spp.
What are the similarities and differences between egg-laying and live birth?	Oviparity vs. viviparity	*Sceloporus aeneus*
How can the environment shape behavior and its controlling mechanisms?	Habitat and life-style	Temperate, tropical, xeric, etc., and/or aquatic, terrestrial, etc., species

Note: Some of these comparisons can be found within the same species whereas others require comparisons across species, genera, and clades.

of natural experiments or "experiments of nature." The evolution of pattern is best illustrated by the diversity of organisms. By studying closely related species living in different habitats, we see how each has become adapted. By studying distantly related species that occupy equivalent habitats, we can evaluate whether the solutions to similar problems are different or analogous.

Any phenomenon being investigated must be studied at multiple levels if we are to understand the factors underlying the diversity and evolution of the mechanisms controlling behavior and physiology. Any major study must include analyses of (1) the relevant physical and sociosexual factors imposed by the environment, (2) the perception and integration of these stimuli by the central nervous system, (3) the interaction of the nervous and endocrine systems in regulating the internal state, and (4) the characterization of the way that changes in internal state influence subsequent changes in behavior and physiology.

Such a research program implies combined analyses of animals in the field and the laboratory. Indeed, this research is best done with animals that can be studied easily both in nature and in the laboratory. Such studies can prove to be complementary; the field has always been a valuable testing ground for adaptive functions, whereas the laboratory is the only possible arena for determining the physiological basis of phenomena observed in the field. We will see that reptiles facilitate such conduct of research in both the field and the laboratory.

All of the above requires an interdisciplinary approach that combines a number of levels of biological organization, including the molecular, physiological, morphological, organismal, ecological, and phylogenetic. This lets one approximate the ways by which causal mechanisms and functional outcomes operate at each level of biological organization. At the same time, this process illuminates the relationships among these several levels.

III. WHICH ANIMAL TO STUDY?

Such a broadly based approach to the study of behavioral physiology must start with a decision about an organism suitable for study. Candidate species should satisfy at least six criteria:

1. The species should be obvious, indeed conspicuous, in nature and hence easily observed.

2. The species should reproduce reliably and at frequent intervals in the laboratory and should grow rapidly. If the focus of the research is developmental or genetic, the species must have a short generation time.

3. The species should have an interesting and sufficiently complex

social organization so that parallels may be drawn to mammalian and avian social systems.

4. It should be possible to manipulate the animals experimentally in the field. Animals transferred to seminatural laboratory conditions should continue to exhibit behavioral patterns and maintain the social organization observed in nature.

5. Species of the selected group should occupy a variety of diverse habitats. This should make it likely that members differing in behavior patterns will have encountered distinct environmental constraints and consequently will differ in the physiological mechanisms that underlie their behavior.

6. Finally, it is desirable to search for a group for which a body of basic information exists about the evolution, ecology, behavior, and physiology of the species to be studied. By studying groups for which the phylogenetic relationships have been ascertained, we can learn how behavior evolves as it is shaped by the environment. Experiments may then be designed to address defined questions.

To the above attributes must be added the obvious ones. First, the organisms must be accessible. Convenience of field sites or the availability of a steady supplier is essential. Second, and of course implicit in the first selection, is the need to consider the status of the wild population; this point also applies to the choice of commercial and private collectors. With the possible exception of salvage collection, as in areas that are slated for environmental destruction for commercial purposes, it is a responsibility of the investigator to assure that the removal of individuals proceeds at a rate that has no significant effect on the survival of the species or its component populations. One approach toward assuring this is the monitoring of numbers, body proportions, and mass for members of populations subject to long-term sampling. Finally, an association with breeders is often desirable. However, it is important to keep in mind that domestication in some species not only produces phenotypic changes in behavior but also modifies genotypes, affecting behavior, morphology, and physiology (Marmie et al., 1989).

Ultimately, the questions being studied will change. One shifts from understanding the process to understanding of its role in diverse organisms. First of all, an evolutionary study of diversity demands that each major grouping must be sampled. This approach means that the processes identified in detailed studies of individual species will be searched for more widely. One needs to know whether the process initially characterized applies only to a single species, to a particular group (genus, family, order), to a particular environment (tropical, temperate, xeric), or to a functional association (terrestrial, aquatic). One must test whether the factors observed influence the nature of the process under study. These objectives require sampling across a wider range. Naturally, such sampling should be cost-effective and take into account the probability that differences are likely to be encountered.

One can generate probability statements about the possible rate of success in discovering unknown conditions by using established patterns (based upon the conditions in related forms) and the phylogenetic placement and ecological circumstances of particular animals. Presumably, each major group requires attention; hence, one would sample pleurodires and cryptodires (and again marine turtles, and the residual cryptodires) independently. Similarly, most studies on reptiles would demand some information about *Sphenodon,* a genus that is the last living representative of a group that flourished in the Triassic. Naturally, the search for completeness must be balanced against the status of the surviving populations (Daugherty et al., 1990; May, 1990); happily, at least one species of *Sphenodon* remains extremely common (Gans, 1983).

Turtles separated early from the main line leading to other reptiles and birds (Fig. 1.1). This separation was close to that passing to pelycosaurs, therapsids, and mammals. One might assume that some of the "primitive" conditions retained by turtles represent the pelycosaurian state; however, they could just as well be modifications involved with the drastic shift to a new habitus and mode of life history. Resolution of this matter may benefit from consideration of other groups (the outgroup comparison), such as the lepidosaurians. Similar test designs can be framed for groups of lizards and snakes.

However, for other questions the ecology of the animals would be more important. For instance, an aquatic species differs profoundly from a terrestrial one and presumably more from a desert than from a forest species. Similarly, species that form dense aggregations would

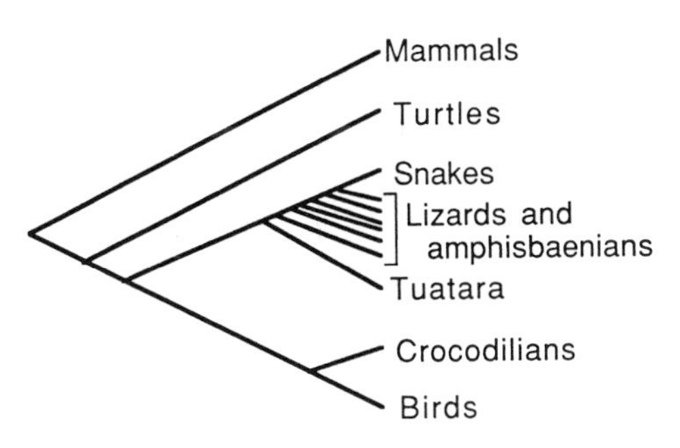

Fig. 1.1. Phylogenetic relationships among vertebrates. Note that crocodilians are more closely related to birds than they are to turtles; that is, they share a much more recent ancestor. The affinities of the lepidosaurians remain unresolved.

be expected to show differences from those that spend much of their life in isolation or at low densities. One would also expect differences between diurnal and nocturnal species, surface-living and subterranean ones, sea level and high-altitude forms.[1]

One of the key decisions in establishing comparisons must, of course, establish that the components (morphological, physiological, or behavioral) being analyzed are indeed equivalent. The kind of biological equivalence referred to as homology implies that elements derive from the same component in an ancestral form common to the two (or more) species being compared. Whereas complex evolutionary considerations underlie the decision, it is generally and appropriately made on operational grounds (Gans, 1985). However, the classic theory of homology is likely to require changes as we learn more about the possible effects of genetic transfers outside of the classic sexual mechanisms (Gray and Fitch, 1983; see also Hillis and Moritz, 1990). Also, it is important to remember that functions or behaviors using homologous structures have profoundly different implications from those using nonhomologous ones. Only the former are likely to document that we are dealing with an evolutionary sequence. Naturally there is the need for care and assurance that the two not be confounded, which would limit the potential utility of the comparison. Hence, these kinds of studies have to be based on modern studies of morphology (including molecular morphology). As important for functional comparison is analogy, which represents a pure and often most direct expression of adaptation (Gans, 1985).

IV. WHY USE REPTILES?

Reptiles are ideal for comparisons, such as those described above, revealing as they do both complex adaptations and multiple solutions for many tasks. Hence, the potential for testing of generalizations, of possibly unitary explanations, proves to be one of their many merits. Indeed, the literature suggests that reptiles represent a source of important new model systems in neuroscience, endocrinology, and ethology. Why is this so?

Reptiles constitute the remnants of a constellation of natural experiments that allow one to investigate still unanswered questions in new ways. Whereas Recent reptiles have had as much time since the origin of amniote vertebrates as have the birds and mammals, some of their processes appear to be closer to those of the ancestral groups. Reptiles can also be used to establish and test analogous conditions to those seen in birds and mammals, as they often solve similar problems in distinct ways. Like all organisms, they incorporate characteristics reflecting past historical influences as well as current adaptation; however, the processes of change may have occurred sufficiently recently so that intermediate states remain, aiding our interpretation.

Their utility is of course based on the fact that reptiles represent a grade rather than a clade; for instance, crocodilians are much more closely related to birds than to squamates. Reptiles share a suite of aspects, such as the amniotic egg, with birds and mammals, but lack endothermy (which appears to have arisen independently in birds and mammals). The situation is complicated further by the fact that the surviving reptiles represent residuals of many different lineages, with the transitional states often missing. The diversity of reptiles has recently been emphasized as a result of the trend to strict cladistic classification in which paraphyletic groups, those that exclude a descendant lineage, are rejected as being defined artificially.

Cladistic schemes would not permit the term *reptiles* (unless it were to include birds) for two reasons. First, the lineage that includes all Recent reptiles would inevitably include birds, which represent a sister group to crocodilians. Second, the reptile-like fossils that gave rise to the mammals presumably left the basic "reptilian" line at or near the origin of turtles. Hence some cladists would place the earliest "premammals" into a group separate from their closest relatives, the earliest "prereptiles," with which they likely shared many "reptilian" characteristics. The basic branching pattern has been known for decades; however, a strict application of phylogenetic rules to taxonomy seems to overemphasize it (Fig. 1.1). At the very least, this approach needs modification whenever the aim of the research is an understanding of analogy and the analysis of past functional patterns. Consequently, there is no reason for absolute rejection of such terms as reptiles in general zoological parlance.

Whenever studying reptiles we must consider the multiple disparate conditions—the terminal states of a branching tree. We must recognize the benefits and the demands of such diversity and use it to our advantage. As biologists, we tend to search for generality among the phenomena we discover, for the unitary principles rather than multiple explanations. However, this may lead to the error of excessive generalization. Perhaps herpetologists should not complain because such generalization has generated attention to our organisms (Gans, 1991). For example, early students looked at turtle muscle and heart because of the assumption that the principles there exemplified applied to all muscles and all hearts, certainly to those of humans. The observation is not totally wrong, because certain aspects of turtle hearts do apply generally. However, each line of turtles had the historical option of modifying its physiology and behavior to local conditions. For instance, the different locomotor patterns of *Geochelone* and *Malacochersus* in East Africa each are associated with major sensory and behavioral correlates (Ireland and Gans, 1973). Similarly, reproductive constraints may show effects on processes such as sexual selection. Fitzgerald (1982) found that mate preference was fundamentally associated with whether females laid a fixed number of eggs per clutch. In species for which there is a positive correlation between egg number and size, males always chose the larger female. In those species in which egg number is fixed and independent of the size of the female, the males exhibit no preference. Thus, the substantial merit of reptiles is precisely their physiological, ecological, and ethological diversity; consequently, they represent a constellation of natural experiments.

V. SOME PRACTICAL BENEFITS AND COSTS OF REPTILES

It may be argued that living reptiles represent the closest we can come to the several ancestral conditions that led to present-day mammals and birds. Information on what is plesiomorphic (fundamental or preexisting) in branches of evolutionary sequences and what is apomorphic (derived) can best be gathered by studying a variety of species and requires analysis of nonmammalian forms. In many ways reptiles remain sufficiently simple in their adaptive patterns so one can see remnants of the ways in which they solved problems posed by the environment, often by avoiding them behaviorally rather than by restructuring morphology or physiology.

Further benefits of reptiles include the fact that many species may be maintained in the laboratory much more easily than birds and mammals. Better than for any other amniote, we have the ability to create a meaningful environment in the laboratory. As a result of generations of zookeepers and terrarium keepers, and more recently of breeders concentrating on reintroduction into the wild, there is substantial information on methods for dealing with reptile maintenance and, more important, with conditions required for breeding reptiles. Careful analysis of this literature again stresses the need to pay attention to diversity; the conditions tolerated and found amenable by multiple reptiles differ on the generic and specific level.[2]

Maintaining a captive reproducing colony in the laboratory, however, is not a simple or straightforward procedure. Establishing a successful breeding program requires a thorough knowledge of the internal and external factors controlling reproduction (Crews and Garrick, 1980). Some of this information can be obtained from studies of the natural history and reproductive cycles of related species. These data sometimes are sufficient to carry out the aims of the investigator. However, the various levels of complexity underlying reproduction often require a more systematic approach.

Two primary levels of complexity are the physiological and the psychological. The physiological level includes the organization and interrelationships of the nervous and endocrine physiological systems (among others), whereas the psychological level centers on the behavior of the organism and its relations to the physical, climatic, and social environments.

In nature, the physiological and psychological levels are usually smoothly integrated. However, displacement of reptiles from habitats in which they have evolved and attempts to reestablish them in the laboratory necessitate an appreciation for and an understanding of this integration. Indeed, most problems encountered will concern one or more aspects of the artificial environment constructed for the organisms.

Particularly important for successful captive reproduction are effects exerted on physiological state, behavior, and social structure. Fig. 1.2 illustrates some of the aspects in this interactive network of environment, behavior, and physiology, aspects shown to be important in effecting reproduction (Crews, 1979a; Crews and Garrick, 1980; Whittier and Crews, 1987). Potential influences may be grouped into several main divisions. For instance, the conditions of the exter-

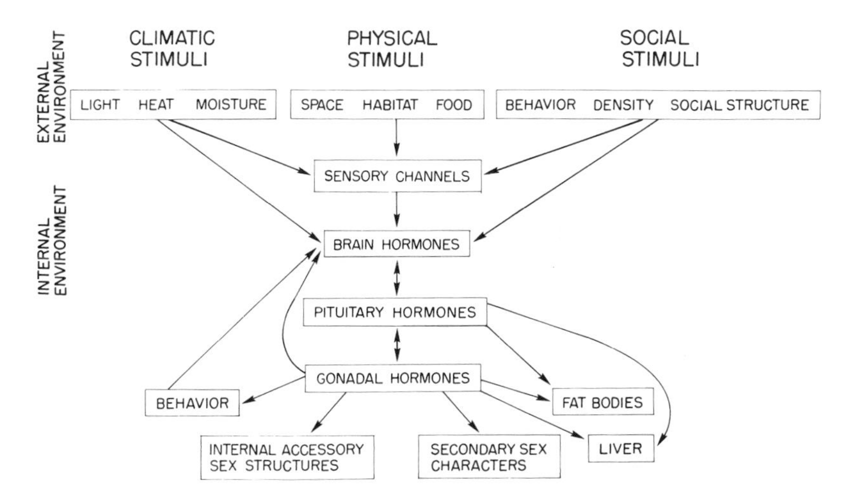

Fig. 1.2. Successful reproduction is a consequence of the reciprocal interaction and coordination of the external and internal environments. The external environment includes climatic, physical, and sociosexual stimuli. These and other stimuli are transduced into neural impulses, many of which are integrated in the hypothalamus. Reproductive processes are in large part regulated by hormones of the hypothalamus-pituitary-gonadal axis, which are themselves controlled by positive and negative feedback relationships. The changing internal state of the organism in turn affects how it interacts with its environment. (From Crews and Garrick, 1980.)

nal environment can be considered to consist of three broad categories of stimuli: climatic stimuli, which include light, heat, and moisture; physical stimuli, which include space, habitat, and food; and social stimuli, which for our purposes include the behavior of conspecifics and the species-typical social organization. All of these environmental stimuli profoundly influence physiology and, to varying degrees, can be manipulated in a captive setting. These caveats, rather than representing a set of difficulties for the use of reptiles, represent a spectrum of opportunities in ways in which the life of reptiles may be modulated to assay their control systems.

VI. REPTILES AS A SOURCE OF HYPOTHESES

Particular reptiles may help deal with specific questions. As such, they provide examples of the Krogh phenomenon, which suggests that nature has given us a best animal for resolving any biological problem. However, studies of reptiles often disclose that these animals do things differently and thus force us to look at phenomena in a unique way or search for the reasons underlying a situation that otherwise appears opaque. Study of atypical organisms can be especially useful whenever their unusual adaptations illustrate alternative solutions to particular problems. They "often force one to abandon standard methods and standard points of view" with the result that, "in trying to comprehend their special and often unusual adaptation, one often serendipitously stumbles on new insights" (Bartholomew, 1982).

Turtles, for example, have long resisted the efforts of behavioral physiologists. We have not yet developed the ability to think like a turtle, to recognize what they are about, indeed to get them to do anything interesting. Yet this ability is critical for establishing what might be an experiment relevant to such an animal and what are the modes with which such an animal might react to the cues appropriately presented. This requires information, not only about the natural history of the organisms, but about their sensory and physiological capacities. With this we might be able to ask whether we would expect the animals to respond by glandular secretion, by change of breathing pattern or heart rate, or again by modifying their behavior (responding or not) to cues that might otherwise elicit mating or feeding.

VII. SOME EXAMPLES

A key to understanding interspecific variation in neuroendocrine controlling mechanisms lies in understanding the constraints that have given rise to diverse reproductive patterns. Species that presumably evolved under different constraints exhibit different patterns of reproduction and therefore are likely to have fundamentally different neuroendocrine mechanisms controlling their reproduction and associated behaviors (see below). Here, diversity can be exploited profitably by those interested in the fundamental nature of behavioral mechanisms.

The comparative approach also promises to increase our understanding of broader intellectual and theoretical issues in behavioral physiology. For example, a main impediment to studies of the evolution of behavior-controlling mechanisms is the absence of ancestors. Fossils do not behave.[3] However, there are means other than fossils to see snapshots of evolutionary stages. For example, currently living unisexual lizards are known to have descended from bisexual species, representatives of which still exist. This permits comparisons of "ancestral" species with known "descendant" species. Furthermore, by the nature of their reproduction, parthenogenetic animals simultaneously remove two of the major confounding factors in research into sexual differences, namely individual genetic variation and gender (Crews, 1989). Genetic variance is removed because the extent of possible change by mutations of individual parthenogens is small, compared to that likely in recombination. Gender is removed because male individuals are absent or extremely rare.

These facts cause us to emphasize the remarkable pseudosexual behaviors seen in some all-female species of lizards. In at least six species, unisexual individuals have been observed to exhibit both mounting and receptive behaviors (Crews and Fitzgerald, 1980; Crews, 1987a) similar to those observed during mating in the sexual ancestral species (Crews, 1980, 1989). Similar pseudosexual behaviors occur in nature in both the mourning gecko *Lepidodactylus lugubris* (Werner, 1980; McCoid and Hensley, 1991) and the unisexual whiptail lizard *Cnemidophorus uniparens* (Crews and Young, 1991; L. Vitt, personal communication). In the parthenogens, these facilitate reproduction just as did courtship and copulatory behavior in the ancestral bisexual species (Crews et al., 1986).

Recent studies indicate complementary sexual dimorphisms in two hypothalamic areas of the brain (the anterior hypothalamus-preoptic area and the ventromedial hypothalamus) of sexual whiptail lizards known to be involved in male-typical and female-typical sexual behaviors, respectively (Crews et al., 1990). It is interesting that the brain of the unisexual whiptail only has the characteristics of female members of the ancestral bisexual species.

Reptilian reproductive patterns differ in other ways. For example, the predictability of the environment has shaped behavior-controlling mechanisms. The green anole, *Anolis carolinensis*, is perhaps that reptile, the physiological and behavioral relationships of which we know most. In this and other species inhabiting predictable environments with prolonged periods supporting reproduction, gonadal activity and mating period are associated. As the breeding season approaches, gonadal activity (gametogenesis and steroidogenesis) increases; this peaks during the breeding season (Fig. 1.3). Many of the hormone-brain-behavior relationships discovered in *A. carolinensis* are similar to those found in animals on which the more conventional models have been based (Crews, 1979b, 1980).

However, not all animals live like anoles. Some reptiles occupy environments that predictably assure only brief periods that provide the opportunity to breed (leading to a dissociated reproductive pattern). Still other species live in areas that have very brief periods that are unpredictable in their occurrence (leading to the constant reproductive pattern) (see Fig. 4.1 in Crews, 1987a).

It stands to reason that the neuroendocrine mechanisms regulating reproduction must reflect the environment in which reproduction takes place (Crews and Moore, 1986). The red-sided garter snake, *Thamnophis sirtalis parietalis,* which exhibits a dissociated reproductive pattern, has become an animal model system with which to study neuroendocrine adaptations. Extensive studies (reviewed in Crews, 1983, 1987b, 1990) have revealed that rather than depending upon the activational effects of sex steroid hormones to regulate sexual behavior, red-sided garter snakes display to physical cues by direct behav-

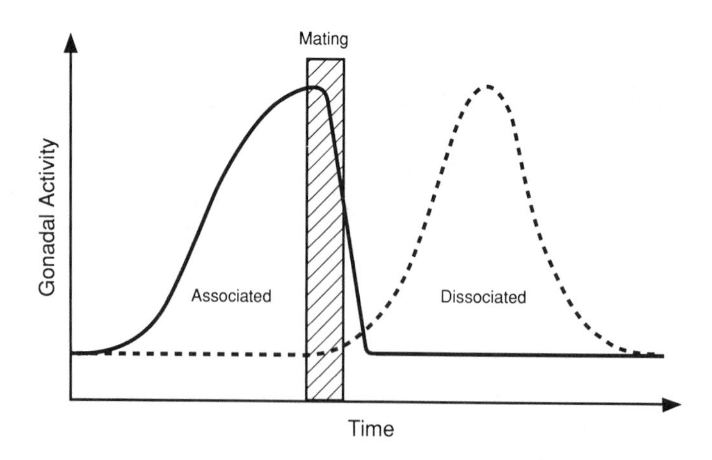

Fig. 1.3. Two reproductive patterns exhibited in reptiles. Gonadal activity is defined in terms of the maturation and shedding of gametes and/or the secretion of sex steroid hormones. In individuals exhibiting the associated reproductive pattern (solid line), gonadal activity increases immediately before mating. In individuals exhibiting the dissociated reproductive pattern (dashed line), gonadal activity is minimal at mating. (Redrawn from Crews, 1987a.)

ioral and physiological responses. The following conclusions may be drawn from this research: (1) Gonadal growth and sexual behavior are not necessarily functionally associated; (2) sexual behavior need not depend on increased levels of gonadal steroid hormones; (3) the initiation, maintenance, and termination of sexual behavior are independent and controlled by different cues; and (4) males and females of a particular species may regulate similar reproductive (behavioral) events by using different proximate cues.

Another realm in which the study of reptiles can contribute significantly concerns the nongenetic aspects of sexual differentiation. In mammals, both the uterine environment and maternal milk are known to contain substantial concentrations of hormones that influence growth and the development of secondary sex structures, including the brain and hence behavior (Vom Saal, 1983; Clark et al., 1991; Kacsoh et al., 1989; Koldovsky, 1980). There is no reason why a similar phenomenon, which we might refer to as *hormonal inheritance*, should not occur in egg-laying animals. The yolk is a rich repository of hormones and their precursors. For example, in teleost fish, thyroid hormones of the maternal circulation are reflected in the yolk and, in turn, influence the development of young fry (Brown et al., 1988). This suggestion that yolky eggs contain hormones and their precursors that influence sex determination and differentiation of the embryo warrants investigation.

Alternative forms of sex determination in reptiles also present important opportunities. As in mammals, we find that experience during embryonic development shapes much of adult sexuality. In the leopard gecko, *Eublepharis macularius*, a low incubation temperature produces only female hatchlings (cold females), whereas relatively high incubation temperatures produce mostly males, with approximately 20% of the hatchlings growing up as gonadal females (hot females) (Crews and Bull, 1987).

Adult *Eublepharis macularius* exhibit marked sexual dimorphisms in phenotype (Crews, 1988). For example, males, as well as hot females, have patent pubic pores, whereas cold females have closed pores. Head size also is sexually dimorphic, with males having wider heads than females. However, with each sex, the higher the incubation temperature, the wider the head of the adult (see Fig. 9 in Crews, 1988).

The endocrine physiology of the adult also is influenced by the temperature experienced during incubation (see Table 2 in Gutzke and Crews, 1988). Radioimmunoassay of circulating concentrations of sex steroid hormones revealed systematic differences in the levels of total androgens and estradiol between males and females and within the adults of known incubation temperatures. Overall, circulating concentrations of total androgens are significantly lower in adult females than in adult males. The differences detected in the hormonal profiles between males from eggs incubated at different temperatures are not statistically significant. However, within the gonadal females, androgen levels are significantly higher, and estrogen levels significantly lower, in hot females compared to cold females. It is significant that hot females have never been observed to mate, to ovulate, or to lay eggs. The effect of male-producing temperature parallels the well-known masculinizing effects of exogenous androgens on neonatal female mammals. It is possible that the primary alteration is at the neuroendocrine level, as it is in androgenized female rodents. Onset of sexual maturation also varies predictably with incubation temperature; hence cold females first yolk ovarian follicles substantially before females that experience warmer incubation temperatures.

Finally, the sexual behaviors of adult *Eublepharis macularius* from different incubation temperatures vary significantly (see Table 1 in Gutzke and Crews, 1988). The responses of females to male courtship differ markedly depending upon incubation temperature. Cold females exhibit sexual receptivity when courted by a male, and tests with either male- or female-stimulus animals show no evidence of heterotypical behavior. On the other hand, all of the hot females commonly respond with heterotypical behavior. Females from eggs incubated at 29° C are intermediate in the frequency of heterotypical behavior displayed. Behavioral differences among individuals within the same treatment group may reflect hormonal differences. To illustrate, one female incubated at 32° C exhibited a much lower frequency of heterotypical behavior compared with other 32° C females; this female also had the lowest level of androgen and the highest level of estrogen compared to the other hot females.

Males from different incubation temperatures do not differ in their homotypical or heterotypical behavior to male- or female-stimulus animals. However, males are produced over a smaller range of temperatures (3° versus 6° C), which may be why less behavioral variation is observed.

It is interesting that in both sexes agonistic behavior depends on incubation temperatures (see Fig. 2 in Gutzke and Crews, 1988). Both males and females are more likely to exhibit aggressive behavior if they had experienced high temperatures during incubation; this relationship is found whether animals are tested in their home cage or in a neutral site. Incubation temperature also affects the probability of aggressive behavior by males toward other males but not toward females. However, hot females are equally aggressive toward either sex.

Studies such as those described here and elsewhere in this volume prove that reptiles can be valuable animal model systems for the study of fundamental questions in behavioral biology. Investigations incorporating the various components discussed above are likely to place reptiles into a well-deserved position as the preferred alternative in the hierarchy of vertebrate model systems.

VIII. HOW REPTILIAN ADAPTATIONS ALLOW ONE TO STUDY THE PATTERNS OF CHANGE

We have noted previously the merits of the reptilian grade as a most diverse assemblage. One might ask why there is so much diversity and proceed from this question to one of expectation. Clearly we do not see a linear sequence or an assembly of sequential modifications in related organisms. Rather one may view things as an assembly of experiments, and this lends it value in the present context.

The diversity observed is generated by the generally rather sloppy nature of adaptation (Gans, 1988). Natural selection means only sufficiency; consequently nonlimiting aspects may vary substantially without lethal effect. Many such variations contribute to what may be referred to as excessive constructions, aspects of the phenotype that are not used at the moment but may allow usage of environmental changes. Many traits furthermore have their major utility during a brief but important phase of the life cycle. Their cost may perhaps be averaged not only over that of a single sex, but over two sexes. This leads to situations in which the functional implications of organisms' characteristics may not appear clear. However, the investigator's innocence regarding function does not mean that a particular component lacks function. Also, once presumed adaptive roles are identified, rigorous tests can be designed to determine whether these are indeed real.

Some of the utility of reptiles to the physiologist is that they represent the only surviving ectothermal amniotes. The first animals to emerge onto land are classed as amphibians, as are the modern gymnophionans, frogs, and salamanders. However, Recent amphibians, and especially salamanders, are highly modified by being adapted for and restricted to a set of environments that hardly mirrors those into which the first terrestrial radiation proceeded. This likely makes them quite distinct from the conditions of their ancestors. The elemental characteristic that unites reptiles with birds and mammals is the cleidotic egg; all three groups are amniotes (Romer and Parsons, 1986). Several other reptilian attributes, such as those associated with locomotion, respiration, and kidney function, probably make better models for fundamental principles for higher vertebrates.

Reptiles are particularly useful in the behavioral neurosciences. Although the "triune brain," sometimes adduced as a sequence of structures with the central area (reptilian brain) being more ancient than more distal areas (neomammalian brain), represents an attractive concept, it is actually more allegory than fact; the vertebrate brain is not organized as a layer cake (Reiner, 1990). A more useful perspective is that instead of continuing complication, there may well be simplifications, as is the rule in many other reptilian systems.

On the other hand, it can be useful to regard reptiles as walking limbic systems. Their complex behaviors and social organization allow particularly "pure" preparations for the study of specific limbic structures (Greenberg et al., 1979). Reptiles may lack the complications of cortical intercalation seen in birds and mammals. However, they still face tasks equivalent to those of birds and mammals—and solve them quite well. The power of this approach is one of focusing on simple systems in order to understand complex systems. Primitive systems may not be simple, but it may be easier to see the relationships therein. The merits of this perspective are especially evident from the many examples seen in the study of invertebrate neurobiology.

To what extent are these complications regarding sex and response to potential predation again to be seen as part of the emergence onto land? Mating, particularly mating involving internal fertilization, involves approach to another individual. This exposure establishes vulnerability because of possible errors with the misidentification of individuals or in the act of mating itself. Other issues may involve the pattern of predations on young, subadults, and small adults. The pronounced structural and behavioral diversity of reptiles, with distinct patterns seen in turtles, lizards, snakes, and crocodilians, again establishes options for study.

The processes obviously proceeded in fishes and marine invertebrates, but the shift from an aquatic to a terrestrial medium imposed new problems. Hence we would suggest that the conditions seen be evaluated in an evolutionary framework. Air differs from water not just in its mass, but also in its carrying capacity for signals. Vision of distant objects became much more important, but also became complicated as the terrestrial environment was diversified with the origin of ever more complex vegetations. With vision came a marked increase in the detection range not exceeded until the special case seen in the development of aquatic mammals. Furthermore, we see the visual system develop toward the detection of patterns. However, this capacity was hardly the initial stage in the origin of vision; instead we see object detectors, contrast detectors, and similar special groupings (Lettvin et al., 1957). The retention of the capacity for chemical signaling demonstrated in the following chapters would have helped to bridge the transition to a more fine-grained visual detecting system.

Also, the tetrapod advent into the dry land environments and the radiation there as assemblages of small- to medium-sized insect predators must have involved major new sensory tasks with the advent of avian predators. Such interacting factors may well provide the reasons for the complexity of detection pattern among reptilian sexes. Questions regarding mate recognition, characterizations of individuals, conspecifics, enemies, and prey shifted with the development of sensory capacities by the other players with which the individual reptile shared the stage.

IX. SUMMARY

There are two motivations for biological study: to improve our appreciation and comprehension of the nature of the world and to help humankind (Lehrman, 1971). In the present context, this suggestion may be paraphrased. One may study modern behavioral physiology for its intrinsic interest, in order to learn about the mechanisms by which reptiles control their reproductive patterns under diverse environmental circumstances. Alternatively, one may proceed in a rather similar fashion in order better to understand the patterns underlying the reproductive systems of birds and mammals. This chapter should have made clear that the steps and cautions in which these two approaches proceed will ultimately be congruent, because both need to disentangle past and present influences shaping organismic diversity.

In presenting this volume, we hope to have characterized the interest of the general area. More than this, we hope to have demonstrated that reptiles represent an untapped resource of material for the resolution of these significant questions and to permit their analysis in an effective manner.

ACKNOWLEDGMENTS

We thank David M. Hillis, Henry Mushinsky, Michael Ryan, and George R. Zug for comments on drafts of this manuscript. The work of DC and CG is supported, respectively, by a Research Scientist Award from the National Institute of Mental Health and the Leo Leeser Foundation.

NOTES

1. A word of caution, however. Whereas it is in our nature to make such comparisons in terms of the two extremes of a single scale, this is fallacious for at least two reasons. The first is that in most cases the organisms do not

occupy only the extremes, but rather many aspects of the spectrum. Actually, these extreme cases often show the adaptive condition in its most extreme state and thus facilitate its recognition (Gans, 1980; Bartholomew, 1982; Crews, 1984). It may still be interesting to examine extreme adaptations if the context is kept in mind. A second and remarkably common error has been called the two-species fallacy (Gans, 1989). It assumes that all phenotypic differences between two organisms are due to their divergent ecology; thus desert-aquatic or low elevation–high elevation comparisons often appear in the literature. In most cases there are many additional differences between the animals being compared, not the least the phyletic one. Hence, the biological solution is analogous to that of resolving simultaneous equations, and one needs at least as many pairs of species for comparisons as there are possible states of difference (such as phylogeny, xeric-mesic, high elevation–low elevation, and forest-savannah). Finally, the difficulty remains in interpreting what one sees.

2. It should be noted that many animal care regulations do not take into account the special requirements of reptiles. For these animals, temperature variation and heat lamp cycles, moisture cycles and a diverse, more than a "clean," substrate are critical. The regulations appear to suit certain "laboratory" mammals, but not natural animals that have not been bred for domestication. This anthropocentrism becomes more critical the further one departs from the mammalian condition. For example, the cleaning of fecal material from cages is part of all Institutional Animal Care regulations. Yet in *Iguana iguana,* ingestion of fecal material is crucial to normal development (Troyer, 1982).

3. Whereas fossils may not behave, the organisms that left us their remnants did do so. Indeed, examination of bone structure, intracranial casts, and so on has revealed a great deal of information about probable behavior and social organization of these ancient animals. Examples can be found in Hotton et al. (1986) and Thomas and Olson (1980).

REFERENCES

Bartholomew, G. A. (1982). Scientific innovation and creativity: a zoologist's point of view. *Am. Zool.* 22, 227–235.

Brown, C. L., Doroshov, W. I., Nunez, J. M., Hadley, C., Vaneenennaam, J., Nishioka, R. S., and Bern, H. (1988). Maternal triiodothyronine injections cause increases in swimbladder inflation and survival rates in larval striped bass, *Morone saxatilis. J. Exp. Zool.* 248, 168–176.

Clark, M. M., Crews, D., and Galef, B. G., Jr. (1991). Circulating concentrations of sex steroid hormones in pregnant and fetal mongolian gerbils. *Physiol. Behav.* 49, 239–43.

Crews, D. (1979a). Endocrine control of reptilian reproductive behavior. In

Endocrine Control of Sexual Behavior (C. Beyer, ed.). Raven Press, New York, pp. 167–222.
Crews, D. (1979b). Neuroendocrinology of lizard reproduction. *Biol. Reprod.* 20, 51–73.
Crews, D. (1980). Interrelationships among ecological, behavioral and neuroendocrine processes in the reproductive cycle of *Anolis carolinensis* and other reptiles. In *Advances in the Study of Behavior,* Vol. 11 (J. S. Rosenblatt, R. A. Hinde, C. G. Beer, and M. C. Busnel, eds.). Academic Press, New York, pp.1–74.
Crews, D. (1983). Control of male sexual behavior in the Canadian red-sided garter snake. In *Hormones and Behavior in Higher Vertebrates* (J. Balthazart, E. Pröve, and R. Gilles, eds.). Springer-Verlag, Berlin, pp. 398–406.
Crews, D. (1984). Gamete production, sex hormone secretion and mating behavior uncoupled. *Horm. Behav.* 18, 22–28.
Crews, D. (1987a). Diversity and evolution of behavioral controlling mechanisms. In *The Psychobiology of Reproductive Behavior: an Evolutionary Perspective* (D. Crews, ed.). Prentice-Hall Inc., Englewood Cliffs, N.J., pp. 88–119.
Crews, D. (1987b). Functional associations in behavioral endocrinology. In *Masculinity/Femininity: Basic Perspectives* (J. M. Reinisch, L. A. Rosenblum, and S. A. Saunders, eds.). Oxford University Press, Oxford, pp. 113–126.
Crews, D. (1988). The problem with gender. *Psychobiology* 16, 321–334.
Crews, D. (1989). Unisexual organisms as model systems for research in the behavioral neurosciences. In *Evolution and Ecology of Unisexual Vertebrates* (R. M. Dawley and J. P. Bogart, eds.). New York State Museum, Albany, N.Y., pp. 132–143.
Crews, D. (1990). Neuroendocrine adaptations. In *Hormones, Brain and Behaviour in Vertebrates* (J. Balthazart, ed.). S. A. Karger, Basel, Switzerland, pp. 1–14.
Crews, D., and Bull, J. J. (1987). Evolutionary insights from reptile sexual differentiation. In *Genetic Markers of Sexual Differentiation* (F. Haseltine, M. McClure, and E. Goldberg, eds.). Plenum Press, New York, pp. 11–26.
Crews, D., and Fitzgerald, K. (1980). "Sexual" behavior in parthenogenetic lizards (*Cnemidophorus*). *Proc. Natl. Acad. Sci. U.S.A.* 77, 499–502.
Crews, D., and Garrick, L. D. (1980). Methods of inducing reproduction in captive reptiles. In *Reproductive Biology of Captive Reptiles* (J. B. Murphy and J. T. Collins, eds.). Society for the Study of Amphibians and Reptiles, pp. 49–70.
Crews, D., and Moore, M. C. (1986). Evolution of mechanisms controlling mating behavior. *Science* (N.Y.) 231, 121–125.
Crews, D., and Young, L. J. (1991). Pseudocopulation in nature in unisexual whiptail lizards. *Anim. Behav. U.S.A.* In press.
Crews, D., Camazine, B., Diamond, R. M. M., Tokarz, R., and Garstka, W. (1984). Hormonal independence of courtship behavior in the male garter snake. *Horm. Behav.* 18, 29–41.
Crews, D., Grassman, M., and Lindzey, J. (1986). Behavioral facilitation of reproduction in sexual and parthenogenetic whiptail (*Cnemidophorus*) lizards. *Proc. Natl. Acad. Sci. U.S.A.* 83, 9547–9550.

Crews, D., Wade, J., and Wilczynski, W. (1990). Sexually dimorphic nuclei and the bisexual brain. *Brain, Behav. Evol.* 36, 262–270.

Daugherty, C. H., Cree, A., Hay, J. M., (London) and Thompson, M. B. (1990). Neglected taxonomy and continuing extinctions of tuatara (*Sphenodon*). *Nature* (London) 347, 177–179.

Fitzgerald, K. (1982). "Mate Selection as a Function of Female Body Size and Male Choice in Several Lizard Species." Ph.D. dissertation, University of Colorado, Boulder.

Gans, C. (1980). *Biomechanics. Approach to Vertebrate Biology* [reissue]. The University of Michigan Press, Ann Arbor.

Gans, C. (1983). Is *Sphenodon punctatus* a maladapted relict? In *Advances in Herpetology and Evolutionary Biology: Essays in Honor of Ernest E. Williams* (A. G. J. Rhodin and K. Miyata, eds.). Museum of Comparative Zoology, Cambridge, Mass., pp. 613–620.

Gans, C. (1985). Differences and similarities: comparative methods in mastication. *Am. Zool.* 25, 291–301.

Gans, C. (1988). Adaptation and the form-function relation. *Am. Zool.* 28, 681–697.

Gans, C. (1989). Concluding remarks: morphology, today and tomorrow. In *Trends in Vertebrate Morphology* (H. Splechtna and H. Hilgers, eds.). *Fortschr. Zool.* 35, 631–637.

Gans, C. (1991). The status of herpetology. In *Herpetology* (K. Adler, ed.). In press.

Gray, G. S., and Fitch, W. M. (1983). Evolution of antibiotic resistance genes: the DNA sequence of a kanamycin resistance gene from *Staphylococcus aureus*. *Mol. Biol. Evol.* 1, 57–66.

Greenberg, B., and Noble, G. K. (1944). Social behavior of the American chameleon (*Anolis carolinensis* Voigt). *Physiol. Zool.* 17, 392–439.

Greenberg, N., MacLean, P. D., and Ferguson, J. L. (1979). Role of the paleostratum in species-typical display behavior of the lizard (*Anolis carolinensis*). *Brain Res.* 172, 229–241.

Gutzke, W. H. N., and Crews, D. (1988). Embryonic temperature determines adult sexuality in a reptile. *Nature* (London) 332, 832–834.

Hillis, D. M., and Moritz, C. (1990). Molecular systematics: context and controversies. In *Molecular Systematics* (D. M. Hillis and C. Moritz, eds.). Sinauer, Sunderland, Mass., pp. 1–10.

Hotton, N., III, MacLean, P. D., Roth, J. J., and Roth, E. C. (1986). *The Ecology and Biology of Mammal-like Reptiles.* Smithsonian Institution Press, Washington, D.C.

Ireland, L. C., and Gans, C. (1973). The adaptive significance of the flexible shell of the tortoise *Malacochersus tornieri*. *Anim. Behav.* 20, 778–781.

Kacsoh, B., Terry, L. C., Meyers, J. S., Crowley, W. R., and Grosvenor, C. E. (1989). Maternal modulation of growth hormone secretion in the neonatal rat. I. Involvement of milk factors. *Endocrinology* 125, 1326–1336.

Koldovsky, O. (1980). Hormones in milk. *Life Sci.* 26, 1833–1836.

Lehrman, D. S. (1971). Behavioral science, engineering and poetry. In *The Biopsychology of Development* (E. Tobach, L. R. Aronson, and E. Shaw, eds.). Academic Press, New York, pp. 459–471.

Lettvin, J. Y., Maturana, H. R., McCulloch, W. S., and Pitts, W. H. (1957). What the frog's eye tells the frog's brain. *Proc. Inst. Radio Engin.* 47, 1940–1951.

Marmie, W., Kuhn, S., and Chiszar, D. (1989). Behavior of captive-raised rattlesnakes (*Crotalus enyo*) as a function of rearing conditions. *Zoo Biol.* 9, 241–246.

May, R. M. (1990). Taxonomy as destiny. *Nature* (London) 347, 129–130.

McCoid, M. J., and Hensley, R. A. (1991). Pseudocopulation in *Lepidodactylus lugubris* (Gekkonidae). *Herpet. Rev.* 22, 8–9.

Noble, G. K., and Greenberg, B. (1941). Effects of seasons, castration and crystalline sex hormones upon the urogenital system and sexual behavior of the lizard (*Anolis carolinensis*). *J. Exp. Zool.* 88, 451–479.

Reiner, A. (1990). An explanation of behavior. *Science* (N.Y.) 250, 303–305.

Romer, A. S., and Parsons, T. S. (1986). *The Vertebrate Body,* 5th ed. W.B. Saunders, Philadelphia.

Thomas, R. D. K., and Olson, E. C. (1980). *A Cold Look at the Hot-blooded Dinosaurs.* West View Press. Boulder, Colo.

Troyer, K. (1982). Transfer of fermentative microbes between generations in a herbivorous lizard. *Science* (N.Y.) 216, 540–542.

Vom Saal, F. S. (1983). The interaction of circulating estrogens and androgens in regulating mammalian sexual differentiation. In *Hormones and Behavior in Higher Vertebrates* (J. Balthazart, E. Pröve, and R. Gilles, eds.). Springer-Verlag, Berlin, pp. 158–177.

Werner, Y. L. (1980). Apparent homosexual behaviour in an all-female population of a lizard, *Lepidodactylus lugularis* and its probable interpretation. *Z. Tierpsychol.* 54, 144–150.

Whittier, J. M., and Crews, D. (1987). Seasonal reproduction: patterns and control. In *Hormones and Reproduction in Fishes, Amphibians and Reptiles* (D. O. Norris and R. E. Jones, eds.). Plenum Press, New York, pp. 385–409.

Young, W. C. (1941). Observations and experiments on mating behavior in female mammals. *Q. Rev. Biol.* 16, 135–156; 311–335.

2

Physiological Regulation of Sexual Behavior in Female Reptiles

Joan M. Whittier and Richard R. Tokarz

CONTENTS

I. INTRODUCTION

The three categories of sexual behavior used in this chapter are those originally defined by Beach (1976). These are attractivity, proceptivity, and receptivity. *Attractivity* refers to the behavior and attributes of females that increase their stimulus value to sexually active males. *Proceptivity* includes the behavior of females that leads to contact with sexually active males. *Receptivity* defines the behavior of females that is required for insemination to occur. In addition to these three categories of sexual behavior of females, this chapter also deals with rejection or nonreceptive behavior of females as a distinct phenomenon. Rejection behavior is defined as active behavior of the female that results in avoidance or rebuffs of sexual advances by males. In any discussion of sexual behavior, the nature of the displays may also be limited by factors such as the phenotype or morphology of the individual or species.

Attractivity in female reptiles involves several sensory modalities, with vomeronasal/olfactory or visual cues being primary. Some reptilian species produce semiochemicals to signal attractivity, and these chemicals may depend on gonadal secretions (see Chapter 4). Communication of the status of females, especially with respect to previous mating activity with other males, may also depend on semiochemicals produced by the male or female during copulation (see Chapter 4). Visual cues, particularly coloration, may signal attractivity of females; in some species these cues appear to depend on gonadal hormones (Cooper and Clarke, 1982; Cooper and Crews, 1987, 1988). Finally, female size is an attribute that may serve as an attractiveness stimulus to males (Garstka et al., 1982).

Proceptive behavior of female reptiles has received little attention. No evidence links physiological state to behavior of females leading to mating activity. Movement of females to male territories in territorial species and increased activity of females that are ready to engage in sexual behavior would be examples of proceptive behavior. It needs to be characterized further.

As for other female vertebrates, sexual receptivity has been the focus of most research. In some vertebrate species, female sexual receptivity can be quantified by observing the intensity or frequency of expression of particular female postures. For example, a sexually receptive female rat will stand in a posture termed *lordosis,* remaining in place with her hindquarters raised and her tail deflected (Bermant and Davidson, 1974). In reptiles, the neck-bending posture of female *Anolis carolinensis* can be cited as a female receptive behavior (see Fig. 2.1). This posture is thought to facilitate the male in obtaining a grip of the female's neck with his jaws before mating.

In snakes, female sexual receptivity is most often defined by two activities: (1) a straightening of the posterior half of the body so that males can more easily intromit and (2) a cloacal gaping behavior in which the anal scute is lifted away from the posterior region of the cloaca (Carpenter and Ferguson, 1977; Schuett and Gillingham, 1988). These behaviors have been observed in viperid, colubrid, and boid snakes (Murphy et al., 1978; Armstrong and Murphy, 1979; Gillingham, 1979; Barker et al., 1979; Gillingham et al., 1983; Secor, 1987). Although posterior straightening of the body may be required for intromission to occur in some species, this behavior does not necessarily indicate that a female is receptive. Rather, it is perhaps more useful to view posterior coiling as a rejection behavior. Some species do not show cloacal gaping (Secor, 1987; Schuett and Gillingham, 1988). Cloacal gaping lasts only a few seconds (Schuett and Gillingham, 1988), and it usually is followed immediately by copulation; hence the behavior cannot be used as a measure of female sexual receptivity in the absence of mating. In studies in which snake sexual receptivity has been quantified, the incidence of mating is used as the index of sexual receptivity (Ross and Crews, 1977, 1978; Whittier et al., 1985).

Some female vertebrates may show no specific behavioral postures reflecting receptivity but only remain stationary as the male approaches and then allow him to mount (Beach, 1976). For example, in the monitor lizard *Varanus bengalensis*, "do nothing" behavior on the part of the female during a courtship bout stimulates subsequent male courtship and thus may indicate female receptivity (Auffenberg, 1983). Furthermore, the behavioral postures that signify receptivity may differ among members of a closely related genus. For example, unlike other anoline species, female *Anolis aeneus* do not exhibit the neck-bend posture when they are sexually receptive, even though a neck grip is obtained by the male during a copulatory sequence (Stamps, 1975). However, active participation of females in discrete behavior associated with sexual receptivity may be more common than is often observed.

The inability to quantify discrete receptive behavior often may reflect the lack of perception of subtle behavioral changes or actions of the females. This is especially true if mating occurs only at certain times of year or under particular conditions. In such situations the occurrence of mating may indicate the degree of female sexual receptivity in a population. However, such a relationship would not apply if sexually unreceptive females could be forced to copulate. Naturally occurring forced copulations have been described in some species of reptiles (Werner, 1978; Devine, 1984), although the frequency of such behaviors in nature is unknown.

The concept of receptivity was originally generated to categorize behavior occurring between two individuals of the opposite sex that culminated in the insemination of the female. It is now recognized that a number of species of reptiles reproduce parthenogenetically (see Cole, 1975; Darevsky et al., 1985; Moritz, 1983, 1984, 1987; Moritz and King, 1985; Moritz et al., 1989). These unisexual species consist of all-female parthenoforms that do not have testes or male sexual organs. Females of some unisexual species from two families (Teiidae and Gekkonidae) engage in behavior that is remarkably similar to the courtship and copulatory behavior of closely related sexual species

Fig. 2.1. *Anolis carolinensis*. Mating sequence. (A) A sexually active male performs a courtship display toward a female. (B) Sexually receptive female stands in place and neck-bends for courting male. After taking a neck grip, the male mounts the female and intromits one of his two hemipenes. (From Crews, 1980, with permission of Academic Press.)

(Crews et al., 1983; Werner, 1978). One female shows malelike behavior in that it mounts another female in what has been termed *pseudocopulation* (Crews and Fitzgerald, 1980; Werner, 1980). Although there are no intromittent organs and intromission does not occur, the sexual behaviors exhibited by females are included in the discussion of sexual receptivity. Thus, sexual behavior is not defined by the sex of the individuals engaging in the behavior.

Another aspect of sexually receptive behavior that needs to be considered is the physiological basis of nonreceptive behavior, or rejection behavior. Rejection or nonreceptive behavior may occur in a particular context, such as nonreceptive behavior following mating. Rejection behavior under these circumstances may be regulated differently from nonreceptive behavior observed in other contexts.

Female sexual behavior is usually expressed under precisely defined conditions. As copulation is associated with fertilization success, sexual behavior leading up to copulation may be especially affected by natural selection and sexual selection. That is, sexual behavior should be shaped by evolutionary constraints to the extent that it affects reproductive success and, ultimately, survival of offspring. In some groups of reptiles, including crocodilians and most species of lizards, mating coincides with fertilization. Thus the timing of sexual behavior might be expected to be closely related to timing of the critical event of offspring production. Other reptiles, including the tuatara, turtles, lizards, and snakes, store sperm in the reproductive tract of females (Rahn, 1940; Fox, 1956; species compiled by Devine, 1984; Moffat, 1985). Sperm viability ranges from 2½ months in *Uta stansburiana* (Cueller, 1966) to 4 years in the diamond-backed terrapin *Malaclemys terrapin* (Barney, 1922), and to 5 and 7 years in the snakes *Leptodeira annulata polysticta* and *Acrochordus javanicus*, respectively (Haines, 1940; Magnusson, 1979). This capability alters the relationship between sexual behavior, copulation, and fertilization (see Parker, 1984; Saint Girons, 1982; Devine, 1984).

Various factors may have led to the evolution of sperm storage; in turtles and snakes, the purported lack of territorial behavior, associated with unpredictable encounters between males and females, has been suggested as a possible selective factor (Devine, 1984). It is likely that the same selective factors that have led to sperm storage may lead to a dissociation between reproductive behavior and gonadal sex steroid secretions, particularly in snakes. This is because sperm storage permits separation of the evolutionary factors that regulate the timing of sexual behavior from those that regulate the timing of the production of young. The differences in the control of sexual behavior between groups that store sperm and those that do not could provide comparative information about the flexibility of behavior-controlling mechanisms.

Reptiles, as other vertebrates, usually reproduce seasonally because conditions in the environment that influence offspring survival vary regularly with time (see Tinkle, 1969; Stearns, 1976; Callard and Ho, 1980; Duvall et al., 1982; Saint Girons, 1982; Licht, 1984; Crews, 1979a; Whittier and Crews, 1987). Proximate environmental cues that are predictably associated with ultimate factors governing offspring survival (*Zeitgebers*) often are coopted as predictors of future favorable environmental conditions. It is of interest, then, to examine how the organism integrates these *Zeitgebers* in influencing sexual behavior. Evolutionary patterns and mechanisms that integrate environmental cues and control seasonal breeding in reptiles can be compared to patterns found in other groups of vertebrates.

II. PATTERNS IN THE REGULATION OF FEMALE SEXUAL BEHAVIOR

Vertebrates as a group show common patterns in the physiological controls with which they regulate the expression of sexual behavior. The information available on these processes in reptiles pertains to only a few species, and more groups need to be sampled. Reptiles show a diversity of evolutionary and life history patterns that provide unique opportunities for investigation of the inherent flexibility underlying mechanisms that control behavior.

Three patterns relate gonadal function and the expression of sexual behavior (Fig. 2.2; Crews, 1984; Crews and Moore, 1986) and form a useful conceptual basis for understanding how ovarian factors influence female sexual behavior. The three patterns include the associated, dissociated, and constant types of reproduction.

In species exhibiting the associated pattern, sexual behavior is associated with maximal follicular size, elevated sex steroid levels, and ovulation. Reptiles that exhibit the associated pattern include most temperate and tropical lizards and the crocodilians (Callard and Ho, 1980; Licht, 1984). This is the pattern of the green anole lizard (*Anolis carolinensis*), one of the reptiles most thoroughly studied with regard to physiological control of sexual behavior (see Crews, 1980, for review).

In species exhibiting the dissociated pattern, sexual behavior is not correlated positively with follicle size, sex steroid secretion, and ovulation. An extensively studied species exhibiting the dissociated pattern of reproduction is *Thamnophis sirtalis parietalis* (see Garstka et al., 1982, and Whittier and Crews, 1987, for a review). The dissociated pattern of reproduction apparently has evolved numerous times and is found in diverse families of reptiles (Crews, 1984). The dissociated pattern is seen in females of several species of lizards, including the skinks *Hemiegris peroni* (Smyth and Smith, 1968), *Eumeces egregius* (Mount, 1963), *Lampropholis guichenoti* and *L. delicata* (Joss and Minard, 1985); the gecko *Phyllodactylus marmoratus* (King, 1977); and the iguanid *Sceloporus grammicus microlepidotus* (Guillette and Casas-Andreu, 1981). Among snakes, it is seen in all species of *Thamnophis* studied thus far (family Colubridae); *Vipera aspis* (family Viperidae; Saint Girons, 1982); *Micrurus fulvius* (family Elapidae; Quinn, 1979); and *Enhydrina schistosa* (family Hydrophiidae; Voris and Jayne, 1979).

As animals exhibiting the dissociated pattern of reproduction have gonads of minimal size when sexual behavior is expressed, we could expect that the role of the ovary in the control of the expression of sexual behavior may be quite different from that in species exhibiting an associated reproductive tactic. Relatively few of the species exhibiting a dissociated reproductive pattern have been studied with respect to this problem, and the factors that regulate sexual behavior in these species remain to be clarified. Study of species with a dissociated pattern of reproduction will provide information about the ability of the neuroendocrine axis to accommodate change in an evo-

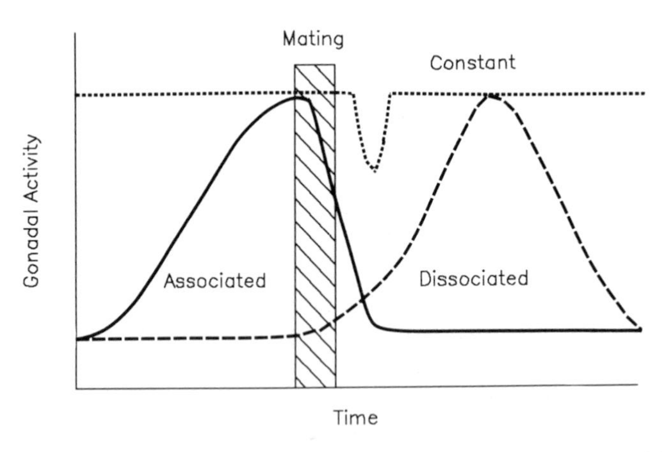

Fig. 2.2. Three reproductive patterns (associated, dissociated, constant) exhibited in vertebrates. Gonadal activity is defined in terms of the maturation and shedding of gametes and/or the secretion of sex steroid hormones. In individuals exhibiting the associated reproductive pattern, gonadal activity increases immediately before mating. In species exhibiting the dissociated reproductive pattern, gonadal activity is minimal at mating. In species exhibiting the constant reproductive pattern, the gonads are maintained at or near maximum activity. The associated and dissociated patterns are common among reptiles; the constant pattern of reproduction has yet to be described in a species of reptile. (From Whittier and Crews, 1987.)

lutionary time frame. Further, it allows analysis of those aspects of neuroendocrine function that have remained relatively unchanged from earlier vertebrate patterns of reproduction.

Species exhibiting the constant pattern of reproduction maintain ovarian follicles and sex steroid secretions at relatively high levels, yet mating activity and ovulation may not occur until precise environmental conditions are met. Whereas the latter pattern has been studied in the zebra finch (*Taeniopygia guttata castanotis*), a species that inhabits unpredictable environments (Sossinka, 1980; Vleck and Priedkalns, 1985), no species of reptiles have yet been shown to exhibit this pattern of reproduction. However, some species of reptiles inhabiting unpredictable environments may prove to show the constant pattern. Study of reptilian species in tropical and arid regions of Australia, where species diversity is high and the environment relatively unpredictable, may provide examples of animals using the constant pattern of reproduction.

III. ENVIRONMENTAL FACTORS AND FEMALE SEXUAL BEHAVIOR

The influence of the environment on sexual reproduction in reptiles has received a great deal of attention (see Crews, 1979a; Crews and Garrick, 1980; Crews and Greenberg, 1981; Duvall et al., 1982; Licht, 1984; Seigel and Ford, 1987; Whittier and Crews, 1987, for reviews). Most of this literature has dealt with the effect of environmental factors on gonadal or hormonal events, rather than on the expression of sexual behavior. In general, temperature appears to be an essential cue that regulates the reproductive cycle of reptiles.

The mechanism by which temperature influences reproductive function is not understood in reptiles. Elevated environmental temperatures or access to a heat source is a minimum requirement for ectotherms to engage in sexual behavior. Perhaps this basic dependency on environmental heat predisposes many ectotherms to use temperature, rather than photoperiod, as an environmental cue for seasonal breeding. Photoperiod is regular and predictable; in contrast, temperature often fluctuates widely. Physiological mechanisms linked to regulation of sexual function must integrate such fluctuations.

Very little is presently known about the physiological control of sexual behavior of female crocodilians and turtles. Environmental factors undoubtedly play a role, and evidence suggests that temperature is a critical factor (see Ferguson, 1985; Mendonça, 1987, for reviews).

More information is available for snakes and lizards. In the European lizard, *Lacerta vivipara*, females appear to initiate the ovarian cycle by assessing both duration and level of cold temperature exposure (Gavaud, 1983). Furthermore, they may regulate the onset of sexual function by responding to fluctuating temperatures.

Recent studies of *Thamnophis sirtalis parietalis* indicate that temperature may be a key environmental factor governing female receptivity. In this species the vernal sexual cycle is easily studied in the field, and receptivity can be induced in captivity by exposure to specific environmental conditions (Aleksiuk and Gregory, 1974; Gregory, 1974, 1977; Hawley and Aleksiuk, 1976; Crews and Garstka, 1982; Garstka et al., 1982; Bona-Gallo and Licht, 1983; Whittier et al., 1987b). Thus, female garter snakes that are exposed to cold temperatures (4° C) for at least 7 weeks, and then warmed to 28° C, initiate vitellogenesis (Fig. 2.3; Bona-Gallo and Licht, 1983; Whittier et al., 1987b). However, females exposed to only 4 weeks of 4° C, or females kept at 28° C, do not undergo ovarian development. Temperature has a stimulatory effect on sexual behavior whenever the ovary is in a regressed condition, and levels of sex steroids are low (Bona-Gallo and Licht, 1983; Garstka et al., 1985; Whittier et al., 1987a). Manipulation of photoperiod before, during, or after the cold-temperature dormancy has no influence on temperature-induced sexual behavior (Whittier et al., 1987b). For example, females maintained in complete darkness become sexually attractive and receptive to males after transfer from 4° C to 28° C (Whittier et al., 1987b). These data suggest that temperature is a primary factor for the induction of sexual behavior in this species. However, the physiological mechanisms involved in the temperature-dependent activation of female sexual behavior remain to be understood in this and other reptiles.

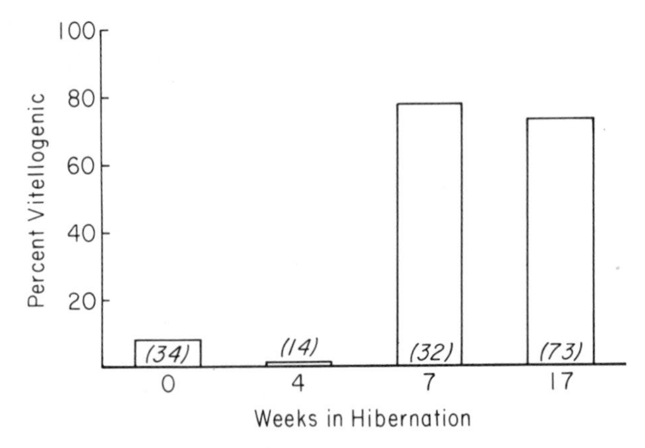

Fig. 2.3. *Thamnophis sirtalis parietalis.* Effect of exposure of female snakes to hibernationlike conditions of 4° C and total darkness on incidence of vitellogenesis on emergence. (From Whittier et al., 1987b, *Can. J. Zool.*)

IV. OVARIAN REGULATION OF FEMALE SEXUAL BEHAVIOR

A. Turtles

The physiological factors controlling sexual behavior of freshwater turtles are difficult to study under field conditions, and there has been limited success in inducing reproduction in captive animals (Mendonça, 1987). Indirect evidence suggests a regulatory role of sex steroids in the control of female sexual behavior. The brain of a freshwater turtle (*Trachemys scripta elegans;* Kim et al., 1978, 1981) concentrates estrogen in specific areas, including the nucleus septi lateralis and bed nucleus of stria terminalis in the septal region, as well as the parvocellular portion of the preoptic area. Magnocellular neurons in both the paraventricular and supraoptic nuclei also concentrate radioactivity. Two nuclei in the central hypothalamus, the nucleus infundibularis and nucleus ventromedialis, and the ventromedial nucleus of the amygdala concentrate radioactive estradiol in this species of turtle. Portions of the hippocampal and piriform cortex also concentrate radioactivity. These estrogen-concentrating areas are similar to those found in other reptiles (Morrell et al., 1979; Demski, 1984) as well as in mammals and birds (Morrell and Pfaff, 1978). This suggests that turtles have an endocrine control mechanism that governs estrogen regulation of reproductive behavior and neuroendocrine function that is similar neuroanatomically to that in other reptiles as well as in birds and mammals.

The control of reproductive behavior in both captive and wild populations of female green sea turtles (*Chelonia mydas*) has been studied in conjunction with observations on nesting and ovulation (Licht et al., 1979, 1980). Both captive and wild animals have low levels of luteinizing hormone (LH) and follicle-stimulating hormone (FSH) before the breeding period but increasing levels through the mating period (Licht et al., 1979, 1980). Levels of progesterone and testosterone are elevated during the mating period, whereas estradiol remains low. The significance of elevated testosterone levels in relation to the control of the expression of reproductive behavior is not established; however, testosterone may act by its conversion to estrogen (aromatization) in the brain (see Callard et al., 1978).

Castration or hormone replacement has not been tried in sea turtles in order to assess how these procedures affect reproductive behavior. Thus it is impossible to conclude that the observed correlation between the timing of sexual behavior and the levels of pituitary and steroid hormones represents a causal mechanism.

B. Crocodilians

The relationship between ovarian state, mating, and fertilization in crocodilians is either unknown or appears to follow no set pattern

(Ferguson, 1985). In *Alligator mississippiensis,* one of the most studied species, preovulatory follicular development is correlated with a seasonal increase in plasma levels of estradiol and testosterone (Lance et al., 1985). However, no experimental studies have examined the role of ovarian hormones in the sexual behavior of crocodilians.

C. Rhynchocephalians

No information is available on the physiological control of sexual behavior in rhynchocephalians. However, the reproductive and population biology of this long-lived species has been studied recently (see Newman et al., 1979; Dawbin, 1982; Newman, 1982, 1987; Newman and Watson, 1985; Castanet et al., 1988). As this group of reptiles is an evolutionary outgroup (Hennig, 1966), information about these animals can be valuable for making evolutionary comparisons.

D. Lizards

In lizards physiological control of reproduction has been studied in some detail (see Crews, 1979a). In *Anolis carolinensis,* females are receptive to a courting male only during the latter half of the follicular cycle, during which the ovary contains a large preovulatory follicle (Fig. 2.4; Noble and Greenberg, 1941; Crews, 1973a; Jones et al., 1983). Female receptivity is thus correlated with the stage of follicular maturation. A female in the first ovarian cycle of the breeding season becomes receptive when the diameter of the largest ovarian follicle is 6 mm or greater. This follicle is ovulated when it reaches a diameter of 8 mm. It is during this time that the sexually receptive neck-bending posture is exhibited by the female in response to male courtship. The period of full receptivity lasts for about 7 days until ovulation occurs. Following ovulation, a mated female becomes unreceptive to male courtship for a period of several days, during which oviposition occurs. Receptive behavior usually reappears once the next largest follicle reaches a diameter of between 3.5 and 6.0 mm. *Anolis aeneus* shows a similar relationship between follicular stage and sexual behavior, although no neck-bending is observed (Stamps, 1975).

The pattern of ovarian sex steroid secretion in intact female *Anolis carolinensis* suggests a causal relationship between actions of these hormones and expression of sexual receptivity. Estradiol 17β (E2) levels increase as the largest ovarian follicle grows and becomes filled with yolk (Crews, 1980). Levels of plasma progesterone (P) are elevated throughout each ovulatory cycle in *A. carolinensis* (Jones et al., 1983). Several lines of evidence further support the hypothesis that the expression of receptive behavior requires ovarian steroid secretion in *A. carolinensis.* First, sexual receptivity declines following re-

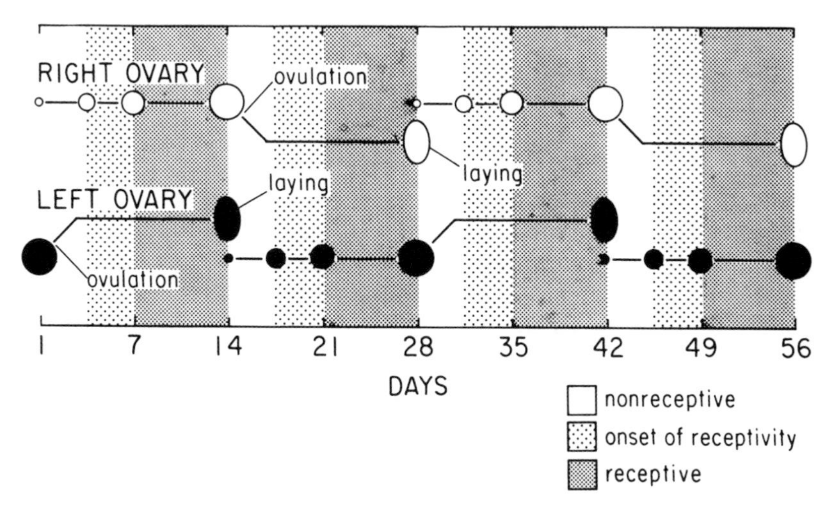

Fig. 2.4. *Anolis carolinensis.* Relationship between maturation and ovulation of the ovarian follicles and sexual receptivity. During the breeding season, females undergo cycles of sexual receptivity that are correlated with the maturation and ovulation of a single follicle alternately between ovaries. The exact onset of estrus varies among females, occurring when the largest ovarian follicle is between 3.5 and 6.0 mm in diameter. See text for details. (From Crews, 1977.)

moval of the ovaries, and ovariectomized females never show receptive behavior (Evans, 1936; Noble and Greenberg, 1941; Crews, 1973a, 1977; Valenstein and Crews, 1977). Second, administration of some kinds of ovarian steroids to ovariectomized female *A. carolinensis* induces sexual receptivity. For example, E2 treatment of ovariectomized females reinstates sexual receptivity (Fig. 2.5; Mason and Adkins, 1976; Valenstein and Crews, 1977; Tokarz and Crews, 1980). Progesterone treatment facilitates receptivity in ovariectomized females primed with a subthreshold dose of E2 (McNicol and Crews, 1979; Wu et al., 1985). The interval between the subthreshold priming dose of E2 and the administration of P must be between 24 and 48 hours in order for P to induce female receptivity (Wu et al., 1985). Thus, high P levels together with elevated levels of E2 help to initiate sexually receptive behavior just before ovulation.

Intraperitoneal administration of ^{3}H-E2 causes radioactivity to concentrate in areas of the brain of *Anolis carolinensis* that in other animals are associated with the expression of sexual behavior (Crews, 1975, 1979b). These include the amygdala, septum, medial preoptic area, anterior hypothalamic area, and the ventromedial and periventricular nuclei of the hypothalamus (Morrell and Pfaff, 1978; Morrell et al., 1979). The brain of *A. carolinensis* also has estrogen-dependent progestin binding sites, particularly in the hypothalamus (Tokarz et al.,

1981). Thus, *A. carolinensis* provides evidence that ovarian steroid secretions are necessary for the expression of sexual receptivity and that ovarian steroids influence sexual receptivity by central binding of estrogen and progesterone.

Estradiol-17β has also been shown by itself to induce sexual receptivity in intact or ovariectomized female broad-headed skinks, *Eumeces laticeps* (Cooper et al., 1986). Estrogen stimulation of receptivity is not limited to lizards having a particular reproductive mode because females of *E. laticeps*, unlike female *Anolis carolinensis*, produce a single large clutch in late spring. Treatment with estrogen of intact

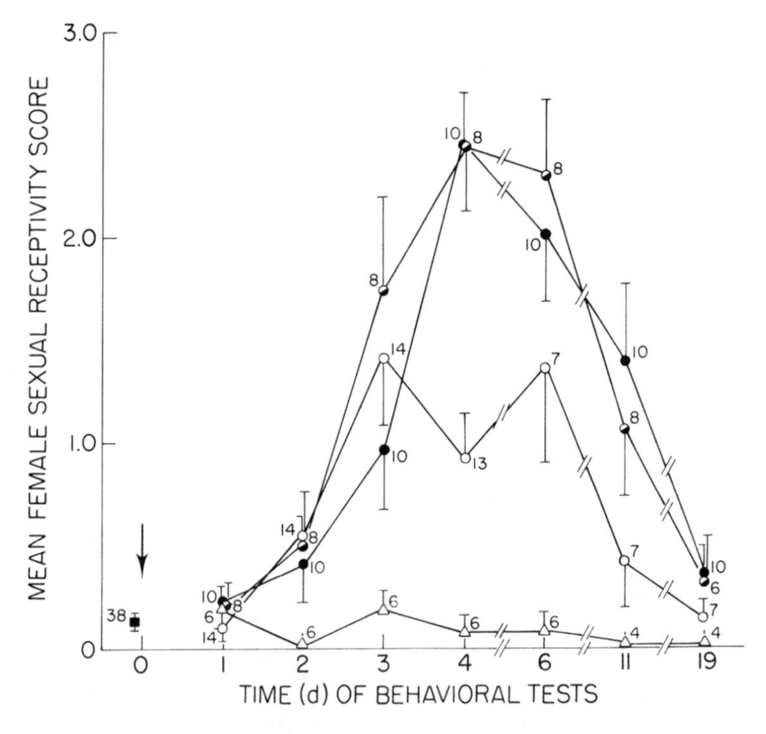

Fig. 2.5. *Anolis carolinensis.* Estrogen induction of sexual receptivity in the female. After a behavioral pretest (closed square), ovariectomized females were treated (arrow) with a subcutaneous injection of steroid-suspending vehicle (open triangles); 0.8 μg estradiol benzoate (EB, open circles); 1.4 μg EB (closed circles); or 4.0 μg EB (half-closed circles). Behavioral tests occurred daily on days 1 to 4. Criteria assigned to females as sexual receptivity score: 0 = ran away from courting male; 1 = initially ran away from courting male but received a neck grip within 10 seconds of his approach; 2 = did not run away; 3 = did not run away and displayed a neck bend. Each point represents the mean (± standard error of the mean). Sample sizes are noted next to treatment symbols. (From Tokarz and Crews, 1980, with permission from Academic Press.)

Eumeces fasciatus and *E. inexpectatus,* the closest relatives of *E. laticeps,* also induces sexual receptivity (Cooper et al., 1986).

Although the relationship between estradiol levels and sexual receptivity observed in female keeled earless lizards, *Holbrookia propinqua,* is inconclusive due to measurement problems, Cooper and Crews (1988) suggest that elevated levels of this steroid may be responsible for the shift from avoidance of males to sexual receptivity. The observed timing of the period of sexual receptivity in this species suggests that there may be an estrogen-progesterone synergism. Brightly colored female *H. propinqua* have elevated levels of progesterone and androgens, leading to the suggestion that increased secretion of these steroids shortly before ovulation may stimulate pigment deposition and later activate various female rejection displays (Cooper and Crews, 1988). This suggestion is supported by earlier findings. Thus exogenous progesterone and androgen induce brightening in plain-colored females; after brightening, the steroids make females more aggressive toward males (Cooper and Clarke, 1982; Cooper and Crews, 1987).

Androgens that can be converted (aromatized) to estrogen also induce sexual receptivity in ovariectomized female *Anolis carolinensis* (Noble and Greenberg, 1941; Mason and Adkins, 1976; Adkins and Schlesinger, 1979). Although intact female *A. carolinensis* have low circulating levels of plasma androgens (testosterone, T, and 5α-dihydrotestosterone, DHT; 0.475 ng/ml, N = 12, Crews, 1980), ^{3}H-T and ^{3}H-DHT-concentrating cells have been identified in the brain of females of this species (Morrell et al., 1979). These androgen-labeled cells are located in the same areas that contain estrogen-labeled cells (Morrell et al., 1979). Exogenously administered DHT, an androgen that cannot be converted to an estrogen by aromatization, does not stimulate sexual receptivity in *A. carolinensis.* Thus, the stimulatory effect of T on sexual receptivity in *A. carolinensis* may be due to its conversion to an estrogen (Crews and Morgentaler, 1979). Aromatase enzymes, which convert T and other aromatizable androgens to estrogens, have been identified and localized in the brains of turtles and snakes (Callard et al., 1977; 1978; Callard, 1985). The endogenous role of androgens in the control of sexual behavior in females remains to be studied in reptiles.

In the parthenogenetic whiptail lizard, *Cnemidophorus uniparens,* females exhibit both malelike and femalelike copulatory behavior, depending on their ovarian stage (Fig. 2.6; Crews and Fitzgerald, 1980; Moore et al., 1985a). The femalelike phase of behavior is correlated with ovarian stage; only preovulatory females with yolking follicles exhibit the receptive posture (Fig. 2.7; Crews and Fitzgerald, 1980;

Fig. 2.6. *Cnemidophorus uniparens*. Pseudosexual behavior sequence in this unisexual parthenogenetic lizard and its sexual ancestor, *C. inornatus*. (From Crews and Fitzgerald, 1980.)

Moore et al., 1985a). Thus, femalelike sexual postures are adopted by intact females at a time when plasma has maximal E2 concentrations (Fig. 2.7; Moore et al., 1985b). Ovariectomy abolishes receptive behavior, and estrogen replacement therapy of ovariectomized females restores sexual receptivity (Moore et al., 1985a; Grassman and Crews, 1986; Crews and Moore, 1991). However, estrogen replacement is only effective when animals are placed in the proper context; receptive behavior of estrogen-treated ovariectomized females occurs when the treated females are housed with females capable of exhibiting malelike behavioral postures (Grassman and Crews, 1986). The pattern of estrogen and progesterone secretion during the reproductive cycle of female *Cnemidophorus inornatus* is nearly identical to that found in female *C. uniparens; C. inornatus* is one of the sexually reproducing ancestors of the triploid *C. uniparens* (see Crews et al., 1986; Moore and Crews, 1986). This suggests that the estrogen dependence of sexual receptivity in the two species may be similar. The central action of hormones in mediating sexual behavior in the sexual and parthenogenetic cnemidophorine species remains to be investigated.

E. Amphisbaenians

No information is available on the reproductive behavior and physiological mechanisms of the amphisbaenians. A few studies have documented reproductive cycles in this group of reptiles (Loveridge, 1941; Bons and Saint Girons, 1963).

F. Snakes

The physiological control of sexual behavior in snakes has been studied most thoroughly in female *Thamnophis sirtalis parietalis* (Fig. 2.8). Female mating behavior is easily studied in the spring both in the field or in the laboratory; hence, most of the studies of the physiological regulation of sexual behavior have focused on this mating period. However, in some years, two-thirds of females returning to hibernacula in the autumn had recently deposited sperm in their oviducts (Halpert et al., 1982; Whittier and Crews, 1986a). These data suggest

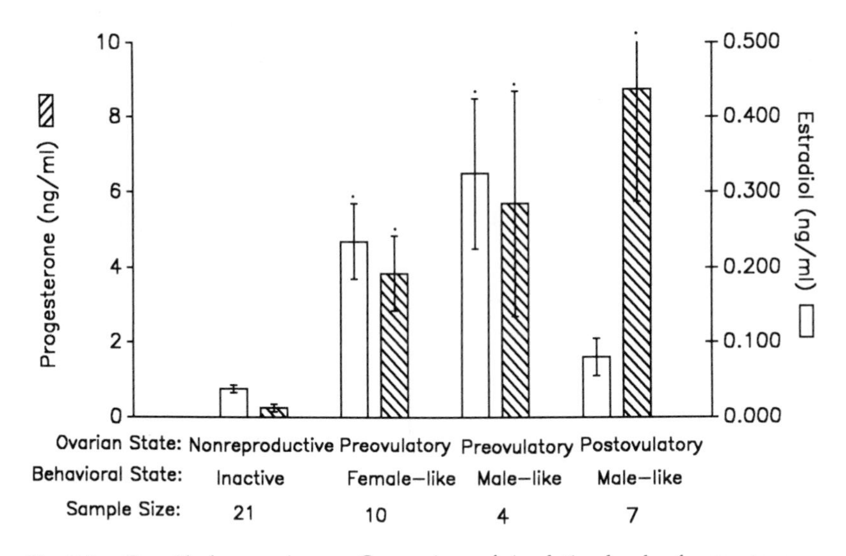

Fig. 2.7. *Cnemidophorus uniparens.* Comparison of circulating levels of progesterone and estradiol immediately after completion of a pseudocopulation. Asterisks indicate circulating levels that are significantly elevated over those of nonreproductive animals. (From Moore et al., 1985b, with permission from Academic Press.)

Fig. 2.8. *Thamnophis sirtalis parietalis.* Mating aggregation in the red-sided garter snake. The newly emerged attractive female (the large individual) is being courted by a group of males.

the possibility of a significant period of mating in late summer that has yet to be studied in detail.

Attractivity is associated with the secretion of an attractiveness pheromone on the dorsal surface of female *Thamnophis sirtalis parietalis* (see Chapter 4). Although females show cloacal gaping (Ross and Crews, 1977; 1978), this behavior has not been quantified as a measure of receptivity. Mating usually occurs within seconds of the cloacal gape, so that the behavior has not been used as a test of receptivity in the absence of mating. Thus, receptivity is assumed only when actual mating occurs (Ross and Crews, 1977, 1978; Whittier et al., 1985).

The role of the ovary in the control of sexual behavior in *Thamnophis sirtalis parietalis* is unclear, as no traditional ovariectomy and hormone replacement has been attempted. However, sexual attractivity and sexual receptivity are regulated differentially in *T. sirtalis parietalis* because some attractive females are unreceptive to male courtship (Ross and Crews, 1977, 1978; Devine, 1977; Bona-Gallo and Licht, 1983; Whittier et al., 1985). Possible mechanisms for the differential regulation of sexual attractivity and receptivity in this species include (1) independent regulation of the two behaviors or (2) sensitivity of the two behaviors to different levels of the same regulator.

Most evidence suggests that neither the ovary nor ovarian sex steroid secretions play an activational role in the control of sexual receptivity in female *Thamnophis sirtalis parietalis.* Sexual receptivity is negatively correlated with follicular size, in that females mate when their ovarian follicles are of minimal size. Thus, both during the spring and the proposed late summer period of mating, follicular size is minimal at 3 to 5 mm (Gregory, 1977; Garstka et al., 1982). Although pharmacological dosages of estrogen (40 μg/70 g BW, daily, for 7 days; Camazine et al., 1980) can induce both sexual attractivity and receptivity at almost any time of year, endogenous levels of E2 are basal at the time sexual behavior is expressed under natural and nature-simulating conditions (see Fig. 2.9, time 0 and 6 hours, respectively; Garstka et al., 1985; Whittier et al., 1987a). Testosterone and P cannot be detected in newly emerged females in the field and under nature-simulating laboratory conditions, times when sexual behavior is maximally expressed (Whittier et al., 1987a). However, these observations do not rule out the possibility that other ovarian secretions, including some other estrogen, aromatizable androgen, or other steroidal or nonsteroidal hormones, may stimulate the vernal period of sexual receptivity.

Specific regions of the brain of *Thamnophis sirtalis parietalis* concentrate ^{3}H-E2, including the ventral amygdaloid nucleus, medial nucleus sphericus, septum, paleostriatum, accessory olfactory tract, and the bed nucleus of the stria terminalis, all in the lateral telencephalon

(Halpern et al., 1982). In the medial telencephalon, radioactivity concentrates in the retrobulbar pallium, dorsolateral septum, and medial preoptic areas. Several hypothalamic areas also concentrate ^{3}H-E2, including the anterior, periventricular, ventromedial, and arcuate nuclei. Central grey and tegmental areas, as well as a region in the obex,

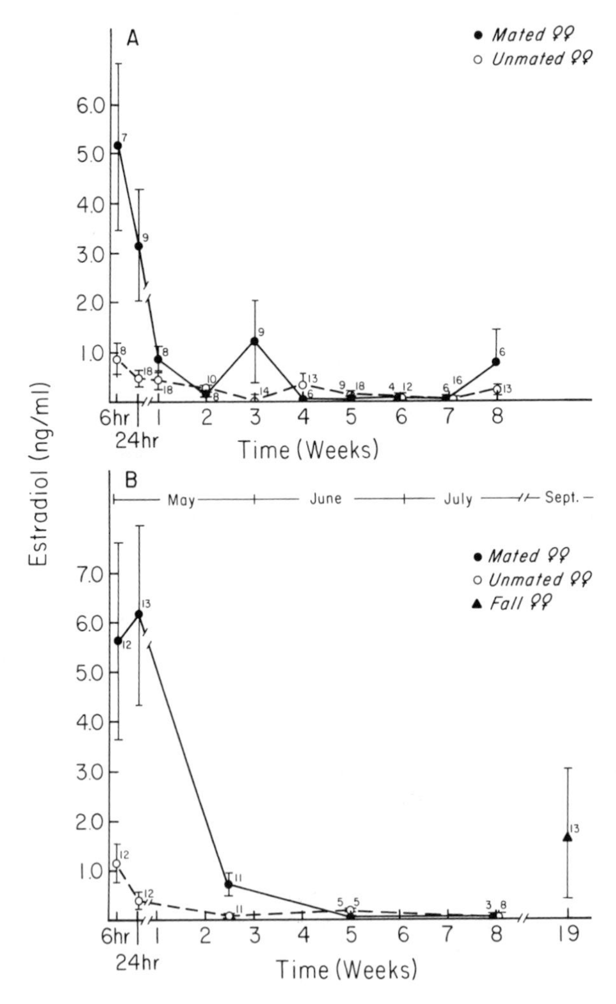

Fig. 2.9. *Thamnophis sirtalis parietalis.* Plasma estradiol 17β concentrations (ng/ml; mean + SD) in females hibernated in the laboratory (A) or in the field (B). Females either mated (solid circles) or did not mate (open circles) after exposure to male courtship. Sample sizes are indicated by the small numbers near each observation. In (B), a final sample from postreproductive females was obtained in the field during the fall. (From Whittier et al., 1987a, with permission from Academic Press.)

concentrate radioactivity in *T. sirtalis parietalis* (Halpern et al., 1982). These areas in mammals have been shown to be related to the control of sexual behavior (Halpern et al., 1982). The significance of these estrogen-concentrating sites to the control of sexual behavior in snakes is not known.

The expression of sexual behavior and ovarian development appears to be regulated by different mechanisms in *Thamnophis sirtalis parietalis*. For example, females that are unlikely to produce offspring because of their small size or low body weight often are sexually attractive and receptive. In one year 50% of females <50 cm snout-vent length, a size class that rarely delivers offspring, returned to the hibernacula in the fall with recently deposited sperm in their oviducts (Whittier and Crews, 1986a). In another study, females at spring emergence that had relatively low body weights for their size were shown to be less likely to undergo ovarian development than heavier females, but body weight did not affect the incidence of mating (Whittier and Crews, 1990). Furthermore, more than 95% of females have been shown to mate on emergence in the spring (Garstka et al., 1982), even in years such as 1983, when reproductive success was low for unknown reasons (Whittier et al., 1985; Whittier and Crews, 1990).

Because garter snakes can store sperm for long periods, a successful strategy may involve mating in seasons when production of offspring is unlikely. The sperm can be stored and used later during more favorable environmental and internal conditions for ovarian and embryonic development. Interestingly, extremely large females have been observed to leave the hibernaculum without mating (Garstka et al., 1982; Garstka, 1982); these observations suggest that large females are not always sexually receptive, although they attract the most males (Garstka et al., 1982).

Mating of *Thamnophis sirtalis parietalis* in the spring of the year results in elevation of plasma levels of E2 within 6 hours (see Fig. 2.9; Garstka et al., 1982; Whittier et al., 1987a). Following mating, females immediately become sexually unreceptive and within 24 hours become unattractive to males (Fig. 2.10; Ross and Crews, 1977, 1978; Devine, 1977; Whittier et al., 1985). Thus females become sexually unattractive to males, although endogenous levels of E2 are maximal. The surge in estradiol may facilitate the process of vitellogenin production by the liver and foster ovarian development (Garstka et al., 1982); however, normal ovarian development can occur in the absence of a postmating E2 surge (Whittier and Crews, 1986b; Whittier et al., 1987a; Mendonça and Crews, 1989). The postmating increase in E2 may also evacuate stored sperm in the anterior regions of the oviduct (Halpert et al., 1982). Elaboration of E2 after mating in female garter

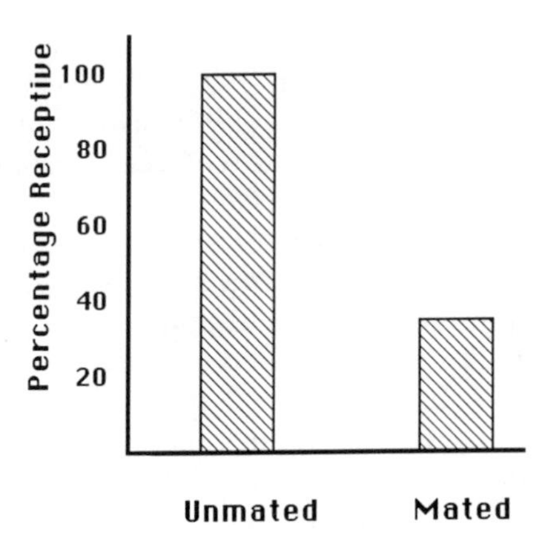

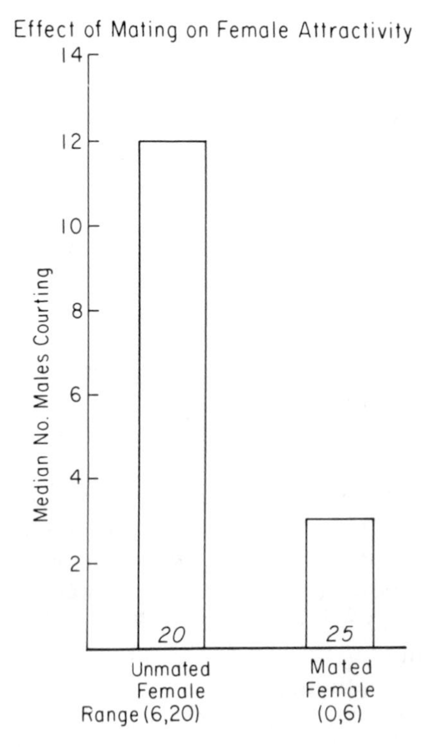

Fig. 2.10. *Thamnophis sirtalis parietalis.* Influence of mating on subsequent sexual behavior of females. (A) Incidence of receptivity (defined by mating) within 2 hr of a previous mating. (B) Incidence of attractivity (defined by number of a total of twenty males courting) within 12 to 24 hr after mating. Both measures were determined in outdoor enclosures in the field. The number of females tested is indicated on the lower portion of each bar. (A, adapted from Whittier et al., 1985, with permission of Springer-Verlag; B, from Whittier and Crews, 1986b, with permission from Williams and Wilkins.)

snakes may involve a neural reflex; spinal transection of females before mating abolishes the postmating E2 surge and subsequent ovarian development (Mendonça and Crews, 1990).

V. NEUROENDOCRINE REGULATION OF FEMALE SEXUAL BEHAVIOR

A. Gonadotropin-Releasing Hormone

The importance of nonsteroidal central mediators in the control of sexual behavior is becoming increasingly evident. One of the most active mediators of vertebrate sexual behavior is gonadotropin-releasing hormone (GnRH: reviewed by Jackson, 1978; King and Millar, 1979, 1980; Kalra and Kalra, 1983; Demski, 1984; Licht and Porter, 1987; Sherwood, 1987a,b). This peptide is thought to act centrally as a neurotransmitter that induces sexual receptivity. Gonadotropin-releasing hormone also regulates the expression of steroid-induced sexual receptivity by stimulating gonadotropin release, which in turn stimulates steroid synthesis by the ovary (see Conn et al., 1987). In rats, GnRH effectively stimulates sexual receptivity only when administered to estrogen-primed females. Thus, both the neurotransmitting and neuroendocrine roles of GnRH need to be considered in the regulation of sexual receptivity (reviewed by McEwen et al., 1987). For example, GnRH facilitates lordosis in ovariectomized, estrogen-primed rats that have been hypophysectomized (Pfaff, 1973; Moss and McCann, 1975). This effect of GnRH on sexual receptivity in rats does not require pituitary hormones or adrenal steroids; however, estrogen replacement therapy is needed. Infusion of GnRH into the hypothalamus or mesencephalic central grey area of female rats also potentiates lordosis (Foreman and Moss, 1977; Sakuma and Pfaff, 1980). Gonadotropin-releasing hormone has been shown to affect the incidence of sexual receptivity and lead to release of gonadotropin in most classes of vertebrates for which these problems have been examined (for reviews see Cheng, 1977; Morali and Beyer, 1979; Moss, 1977, 1979; Moss and Dudley, 1979; Stacey, 1981, 1983; Moore, 1987; Sherwood, 1987a).

Gonadotropin-releasing hormone occurs as a family of related decapeptides that have been found in all classes of vertebrates. Multiple forms of the peptide have been found in some species (see King and Millar, 1979; 1980). In reptiles, the amino acid sequences of GnRH(s) have not been determined, but information is available on chromatographic and immunoreactive characteristics of the peptides (summarized in Table 2.1). Based on the similarity of structure to mammalian GnRH, these peptides are referred to as GnRH, even though the gonadotropin-releasing activity of the peptides in some taxa remain to be determined (reviewed by Sherwood, 1987a,b). In the lizards *Chal-*

Table 2.1 Survey of GnRH in reptiles

Taxon species	Form of GnRH	Concentration of GnRH	Reference
Turtles			
Chersina angulata	?[a]	5.2–8.9 ng/ hypothalamus	King & Millar, 1979 King & Millar, 1980
Chrysemys picta	Avian I & II	7.8 ng/brain	Sherwood & Whittier, 1988
Crocodilians			
Alligator mississippiensis	Avian form	0.816–0.278 ng/mg protein	Lance et al., 1985
	Avian I & II	14.6 ng/brain	Powell et al., 1985, 1986
Lizards			
Chalcides ocellatus	Avian II[b]	0.1 ng/brain	Powell et al., 1985, 1986
Podarcis sicula sicula	Avian II[b]	0.1 ng/brain	Powell et al., 1985, 1986
Cordylus cordylus niger	Avian II[b]	3.7 ng/brain	Powell et al., 1985, 1986
Snakes			
Thamnophis sirtalis parietalis	Avian I[b]	0.3–0.4 ng/brain	Sherwood & Whittier, 1988

[a]Unidentified form of GnRH differing from mammalian GnRH.
[b]Several other forms of GnRH unlike mammalian or avian forms were found in these species.

cides ocellatus, *Podarcis sicula sicula*, and *Cordylus cordylus*, the major form of GnRH is avian II-like, but salmonlike and several unusual forms are also present in lesser amounts (Table 2.1; Powell et al., 1985, 1986).

No studies have examined the influence of GnRH on the expression of sexual receptivity in turtles, crocodilians, the tuatara, or snakes. The role of GnRH in the control of sexual behavior has been tested for two species of iguanid lizards (Alderete et al., 1980; Phillips et al., 1985; Phillips and Lasley, 1987). Treatment of estrogen-primed ovariectomized *Anolis carolinensis* with mammalian GnRH induces sexual receptivity in a dose-dependent manner (Alderete et al., 1980). However, the specificity of the behavioral response of female *A. carolinensis* to mammalian GnRH is not clear; another peptide, thyrotropin-releasing hormone (TRH), also stimulates sexual receptivity (Alderete et al., 1980). In female *Iguana iguana* pulsatile administration of GnRH (avian II) using intraperitoneal osmotic pumps increases male courtship and intermale aggressive behavior, suggesting increased female attractivity in response to the peptide treatment (Phillips et al., 1985; Phillips and Lasley, 1987). These responses follow after approximately

3 days of GnRH administration. In addition, copulation occurs from 5 to 10 days after implantation (Phillips and Lasley, 1987). Furthermore, both avian II GnRH and mammalian GnRH stimulate sexual behavior in *I. iguana* when administered in a pulsatile fashion. An equivalent dosage of avian II GnRH, administered in a constant rather than pulsatile fashion, is equally effective in stimulating behavioral responses (Phillips and Lasley, 1987). Some female *I. iguana* to which GnRH is administered exhibit a subsequent out-of-season period of sexual behavior that is followed by nest building and oviposition (Phillips and Lasley, 1987). This response suggests that GnRH has some persistent long-term effects on the neuroendocrine axis.

Because GnRH influences sexual behavior in a few species of lizards, it is appropriate to examine the distribution of this peptide in the brain. Immunoreactive GnRH (irGnRH) in the reptilian brain has been mapped in a turtle, *Chinemys reevesii;* two lizards, *Gekko gecko* and *Eumeces okadae;* and four species of snake, *Elaphe climacophora, E. quadrivirgata, E. conspicillata,* and *Rhabdophis tigrina* (Fig. 2.11; Nozaki and Kobayashi, 1979; Nozaki et al., 1984). In the species examined, irGnRH-containing perikarya are concentrated in the preoptic and septal regions of the hypothalamus. Projections of fibers from these areas in all species terminate in the median eminence (Nozaki et al., 1984). These findings are consistent with the hypothesis that GnRH in these reptiles is a hypothalamic releasing factor.

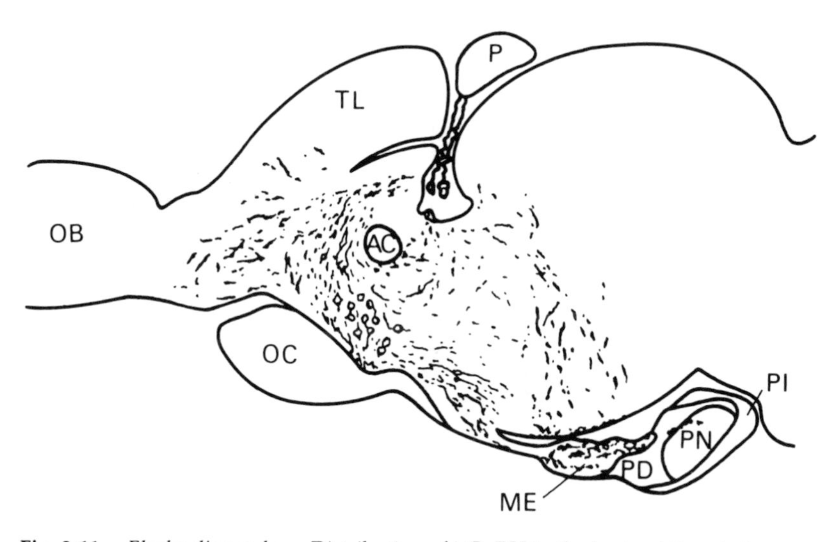

Fig. 2.11. *Elaphe climacophora.* Distribution of irGnRH in the brain. AC, anterior commissure; OB, olfactory bulb; OC, optic chiasm; ME, median eminence; P, pineal; PD, pars distalis; PI, pars intermedia; PN, pars nervosa; TL, telencephalon. (From Nozaki et al., 1984, with permission of Japanese Scientific Society Press.)

Diffuse extrahypothalamic distribution of GnRH-staining fibers is characteristic of vertebrates. Extrahypothalamic irGnRH fibers are found in all species of reptiles examined (Nozaki et al., 1984). Most of the irGnRH fibers project either dorsally or caudally from the preoptic and septal regions of the hypothalamus to the region of the anterior commissure. Additional fibers project anteriorly to the cerebral hemispheres and medial and lateral septal nuclei. The functional significance of extrahypothalamic fibers is not known; however, some of these diffuse fibers may influence the expression of sexual behavior (Nozaki et al., 1984).

B. Prostaglandins

Prostaglandins (PGs) are another class of substances that have been shown to activate or facilitate sexual behavior in diverse vertebrates. Prostaglandins induce spawning behavior in female goldfish and paradise fish (Stacey, 1976; Stacey and Peter, 1979; Villars et al., 1985). They also inhibit the expression of the release call, a call produced by females that are unreceptive, in female leopard frogs (Diakow and Nemiroff, 1981). In mammals, prostaglandins have been shown to stimulate the expression of sexual receptivity in rats and hamsters (Dudley and Moss, 1976; Hall et al., 1975; Rodriguez-Sierra and Komisaruk, 1977; Hall and Luttge, 1977; Marrone et al., 1979; Buntin and Lisk, 1979). Furthermore, prostaglandins have a central role in the control of GnRH-stimulated gonadotropin release in mammals (Ojeda et al., 1977a, b; Kalra and Kalra, 1983). Maximal levels of prostaglandins in the brain are localized in the median eminence, medial-basal hypothalamus, anterior pituitary, and pineal gland of the rat (Ojeda et al., 1977a, b). All of these structures have a role in neuroendocrine secretion. In rats, areas of the brain that are sensitive to PGE_2, GnRH, estradiol and progesterone overlap (Rodriguez-Sierra and Komisaruk, 1982). However, the role of PGE_2 in stimulating sexual behavior appears not to be mediated solely by its effect on GnRH release. Thus, when administered to certain areas of the brain, PGE_2 stimulates sexual receptivity in ovariectomized female rats, whereas GnRH has this effect only in those that are ovariectomized and implanted with estrogen (Rodriguez-Sierra and Komisaruk, 1977).

Prostaglandins inhibit the expression of sexual behavior in some mammals, such as guinea pigs (Marrone et al., 1979). The behavioral response to prostaglandins appears to be related to whether vaginal or mating stimuli have a stimulatory or inhibitory effect on subsequent sexual receptivity (Tokarz and Crews, 1981). As in guinea pigs, sexual receptivity in *Anolis carolinensis* is inhibited by mating (Greenberg and Noble, 1944; Crews, 1973b; Valenstein and Crews, 1977).

Table 2.2 Effects of intraperitoneal (IP) and intracranial (IC) injection of prostaglandin $F_{2\alpha}$ on sexual receptivity in ovariectomized, estrogen-treated[a] *Anolis carolinensis*

Treatment	N	Percentage receptive	
IP		Pretest	5 min postinjection
Vehicle	7	100	100
$PGF_{2\alpha}$ (15 μg)	7	100	0[b]
IC		Pretest	1 hr postinjection
Vehicle	8	100	88
$PGF_{2\alpha}$ (3.8 μg)	14	100	36[b]

Source: Adapted from Tokarz, R. R., and Crews, D., Effects of prostaglandins on sexual receptivity in the female lizard, *Anolis carolinensis, Endocrinology* 109, 451–457, 1981.
[a]Ovariectomized females were pretreated with 1.4 μg of estradiol benzoate to stimulate sexual receptivity before injections.
[b]Significantly fewer females were sexually receptive compared to vehicle-injected controls and pretreatment behavior (Fischer exact probability test, $p < 0.05$).

After mating, females are sexually unreceptive for the remainder of their 14-day follicular cycle. Females become sexually receptive in the subsequent cycle. Females exposed only to male courtship or to courtship and copulation with hemipenectomized males remain sexually receptive, suggesting that stimuli associated with intromission are required to inhibit female receptivity (Crews, 1973b).

Treatment of sexually receptive female *Anolis carolinensis* with $PGF_{2\alpha}$ blocks the expression of receptivity in a dose-dependent manner (Table 2.2; Tokarz and Crews, 1981). Both PGE_2 and PGE_1 also reduce receptivity in this species. The inhibitory effects of prostaglandins on receptivity appear within 5 minutes of injection and persist for approximately 3 hours; thus, females regain receptivity at 6 and 24 hours post treatment. The rapid onset of PG inhibition of behavior is similar to the time course of female unreceptivity after mating (Table 2.2; Tokarz and Crews, 1981).

Female *Thamnophis sirtalis parietalis* exhibit a dramatic postmating decline in sexual behavior (Fig. 2.10; Ross and Crews, 1977, 1978; Devine, 1977, Whittier et al., 1985). Prostaglandins also appear to be involved in the postmating termination of receptive behavior in these females (Figs. 2.12 and 2.13; Whittier and Crews, 1986b). Intraperitoneal injections of $PGF_{2\alpha}$ at a dosage of 0.5 μg/g BW significantly inhibit sexual receptivity in unmated, newly emerged females (Fig. 2.12; Whittier and Crews, 1986b). Lower doses do not significantly inhibit mating but do increase the latency of females to mate. After mating, the mated females show significantly elevated $PGF_{2\alpha}$ levels in the plasma in contrast with the levels in the plasma of females that received the same amount of male courtship but were prevented from mating (Whittier and Crews, 1989; Fig. 2.13). Levels of plasma $PGF_{2\alpha}$ are significantly elevated within minutes of termination of copulation (copulation lasts 15 to 30 minutes in this species) and continue to rise up to 6 hours post coitus, remaining elevated for at least 24 hours (Fig. 2.13; Whittier and Crews, 1989).

To understand how prostaglandins are affecting sexual behavior it is necessary to establish the source of prostaglandins and their site of action. Prostaglandins are produced by many tissues, including the oviduct and uterus, the ovary, the vasculature, blood cells, and nervous tissue (Kelley, 1981). Tissues including the ovary, oviduct, liver, testis, and epididymidis of *Sceloporus jarrovi* have been shown to produce both $PGF_{2\alpha}$ and PGE_2 (Guillette et al., 1988). Many effects of prostaglandins are attributed to localized actions, such as their role in inflammatory responses or as neuromodulators. Some effects of prostaglandins only occur after transport via the circulation. The release and transport of prostaglandins by the uterus to the ovary in the control of the corpus luteum are an example of a localized hormonal effect of prostaglandins (Milvae and Hansel, 1985). In addition, circulating levels of prostaglandins can be elevated during specific events, such as oviposition in chickens (Olson et al., 1986; Saito et al., 1987). Elevation of the level of prostaglandins in the circulation suggests that these substances act in a classic hormone fashion, although the site of action and the effects in chickens are unknown.

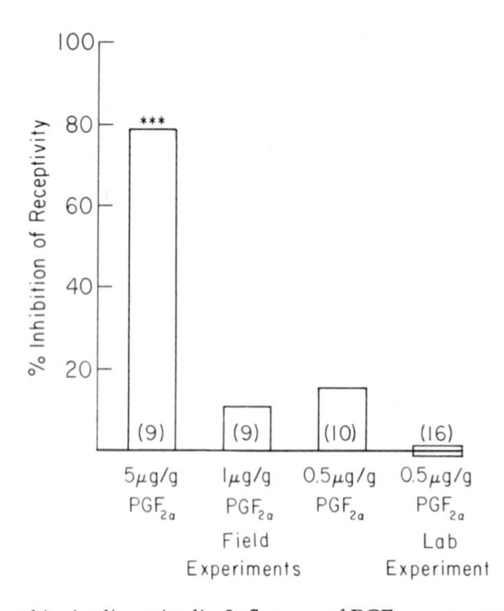

Fig. 2.12. *Thamnophis sirtalis parietalis.* Influence of $PGF_{2\alpha}$ on receptivity in females. Females injected with the highest dosage were significantly less receptive. Sample sizes are indicated in parentheses. (From Whittier and Crews, 1986b.)

Prostaglandins induce oviductal contractions in birds and mammals (Wechsung and Houvenaghel, 1976; Hammond et al., 1981; Spilman and Harper, 1975). Prostaglandins are produced in response to mechanical probing of the reproductive tract (Poyser et al., 1971; Kloeck and Jung, 1973; Kelley, 1981). This suggests that in some species the oviduct may have an important role in PG-induced termination of sexual receptivity. The neurohypophysial octapeptide hormones arginine-8 vasotocin (AVT) and oxytocin can stimulate oviductal contractions in reptiles (LaPointe, 1969; Guillette and Jones, 1980). However, AVT treatment does not inhibit sexual receptivity in *Anolis*

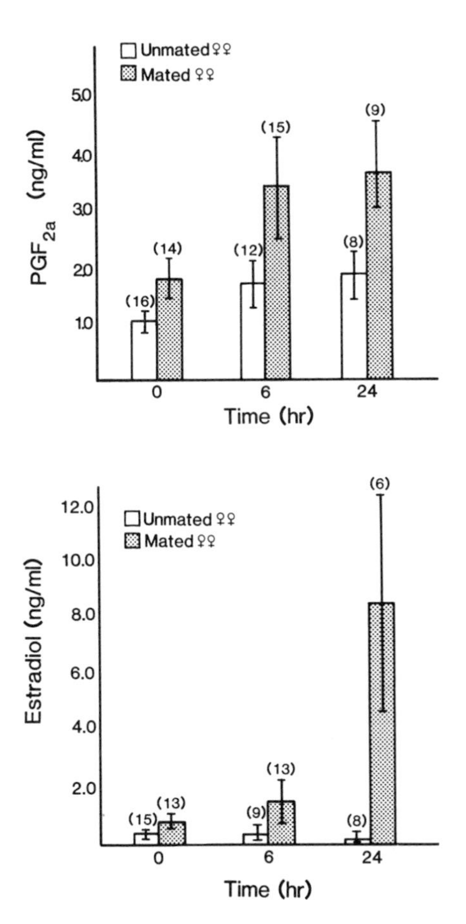

Fig. 2.13. *Thamnophis sirtalis parietalis.* Plasma levels of $PGF_{2\alpha}$ in females that have mated or have been exposed to males, but did not mate. Plasma $PGF_{2\alpha}$ levels of mated females were significantly elevated over those of unmated females at each time sampled. Sample sizes are indicated at the base of each bar. (From Whittier and Crews, 1989.)

carolinensis (Tokarz and Crews, 1981). Thus, AVT does not mimic $PGF_{2\alpha}$-stimulated inhibition of receptivity, although it elicits oviductal contractions.

Prostaglandin $PGF_{2\alpha}$ treatment inhibits sexual receptivity in estrogen-primed females that have been previously ovariectomized and oviductectomized (Tokarz and Crews, 1981). Thus, ovaries or oviducts do not have to be present for inhibition of receptivity during administration of prostaglandin. One interpretation of this outcome is that the exogenously administered prostaglandins may act at other peripheral sites or centrally in the brain. Intracranial injection of a lower dose of $PGF_{2\alpha}$ than that administered systemically depresses receptivity in *Anolis carolinensis* 1 hour after injection (Tokarz and Crews, 1981). The site of action of prostaglandins in the brain has not been investigated. Furthermore, no information is available on plasma or tissue levels of $PGF_{2\alpha}$ in mated and unmated females.

Whereas it is evident that prostaglandins play a role in the postmating termination of sexual behavior in *Thamnophis sirtalis parietalis,* the mechanism of PG action, the source of PG production, and the role of different prostaglandins or substances related to prostaglandins are not known. In *Thamnophis,* one possibility is that prostaglandins are involved in a neuroendocrine reflex that regulates the termination of sexual behavior. A second possibility is that prostaglandins directly influence secretions of the uterus, oviduct, or ovary and that these products in turn influence sexual behavior.

Seminal fluid of male mammals is rich in prostaglandins (Kelley, 1981). It is conceivable that male *Anolis carolinensis* and *Thamnophis sirtalis parietalis* may pass prostaglandins to females in seminal fluids, thereby rendering them unreceptive. Experimental evidence for the effect of seminal prostaglandins on sexual behavior of females is lacking among vertebrates. However, in male crickets, *Teleogryllus commodus* (Orthoptera), males pass prostaglandin synthetase to females; the enzyme then produces prostaglandins, rendering the females unreceptive to other males (Loher and Edson, 1973; Loher, 1979).

Some male snakes, including *Nerodia taxispilota, Thamnophis butleri, T. radix, T. sauritus,* and *T. sirtalis,* as well as *Virginia striatula* and *Coluber constrictor priapus,* deposit a hard gelatinous copulatory plug in the cloaca of females at the end of copulation (Devine, 1975, 1984; Ross and Crews, 1977, 1978). The possible roles of this plug in inducing changes in female receptive behavior are not clear. For example, although 70% of female *Thamnophis sirtalis parietalis* become unreceptive to males after a single mating under natural conditions, 30% of all females may mate two or more times, even though a copulatory plug is present from the earlier mating (Whittier and Crews, 1986a).

Thus, the copulatory plug has limited effectiveness in blocking subsequent mating; the concept of "enforced chastity" imposed by males on females seems an overstatement (Devine, 1975, 1977).

Devine (1984) suggests that gelatinous plugs evolved in snakes that mate in aggregations because in such situations, male mate guarding or male combat behavior would be less effective at ensuring sole access to the female. Although most species that form gelatinous plugs do mate in aggregations, this need not preclude the use of plug formation by species that employ behavioral methods for reducing intermale competition. In the European *Vipera berus,* products of the male renal sex segment cause the posterior oviduct to constrict following mating (Nilson and Andrén, 1982). This oviductal constriction is also referred to as a copulatory plug (Nilson and Andrén, 1982; Andrén and Nilson, 1987), but no gelatinous plug forms (Nilson and Andrén, 1982). *V. berus* exhibits mate guarding and male combat, and mating occurs in small aggregations. The physiological basis of the constrictive effect of male seminal secretions on the oviduct of the adder remains unknown. Also unclear is the function of this oviductal blockage, although it may allow retention of sperm or prevent sperm competition from subsequent matings in the same season. However, there are disputes about the natural occurrence of multiple matings and the incidence of multiple paternity resulting thereby (see Nilson and Andrén, 1982; Stille et al., 1986; Andrén and Nilson, 1987; Stille and Niklasson, 1987). Some snakes, such as *Crotalus viridis viridis* and *Lampropeltis mexicana alterna,* do exhibit mate-guarding behavior but do not form gelatinous plugs (Ludwig and Rahn, 1943; Murphy et al., 1978), although the incidence of other means of cloacal blocking has not been studied. Further research of this type of interaction between males and females might examine potential implications in intrasexual selection.

C. Other Neuropeptides

Other neuropeptides may affect female sexual behavior in reptiles. Thyrotropin-releasing hormone (TRH), a tripeptide named for its activity in mammals, stimulates sexual receptivity in *Anolis carolinensis,* whereas a biologically inactive form of TRH (deamido TRH) does not affect sexual behavior (Alderete et al., 1980). In rats, TRH does not have a stimulatory effect on sexual receptivity; in fact, it has been found to inhibit GnRH induction of lordosis (Moss and McCann, 1973; Moss, 1977). Additional studies are needed to determine possible interactions between GnRH and TRH and other substances that may be involved in the control of sexual behavior. Neuropeptide Y (NPY) has also recently been found to inhibit sexual behavior and concurrently stimulate feeding behavior in rats (Clark et al., 1985). Recent work indicates similar effects in male *Thamnophis sirtalis parietalis* (Crews, 1990). Peripheral or central administration of another peptide, 8-cholecystokinin, has been recently shown to inhibit lordosis in rats (Mendelson and Gorzalka, 1984; Bloch et al., 1987; Babcock et al., 1988), but the role of this peptide in reptiles has not been examined.

VI. CONCLUSIONS

The physiological regulation of female sexual behavior in reptiles has been examined in only a few species. Indeed, most of our knowledge on this subject is based on studies of a single lizard, *Anolis carolinensis,* and a single snake, *Thamnophis sirtalis parietalis.* A few species of turtles and crocodilians have also been studied, but our knowledge of the physiological regulation of female sexual behavior in these reptiles is relatively less comprehensive. Unfortunately, no information on the physiological control of female sexual behavior is available for the rhynchocephalians or amphisbaenians.

In the species of reptiles examined thus far, female sexual behavior appears to be governed by two contrasting patterns of regulation. In the first pattern, females become sexually receptive during the periovulatory period. Thus, female sexual behavior is positively correlated with maximum ovarian follicular stages and with circulating levels of ovarian steroids such as E2 and P. Circulating E2 levels in these species increase around the time of ovulation. Preventing increased E2 by removal of the ovaries prevents sexual receptivity; estrogen replacement therapy restores it. Estradiol-concentrating neurons have been identified in the brains of reptiles in which estradiol mediates female sexual receptivity. The anatomical areas of the brain that bind E2 are analogous to areas known to mediate sexual behavior in other vertebrates, such as mammals. Some evidence suggests that progesterone may also facilitate sexual receptivity in estrogen-primed females. Progesterone is believed to be synthesized by the ovary, and the levels of this steroid are elevated around the time of ovulation. High levels of E2 and P may interact to influence sexual receptivity; the brain areas that bind E2 also have P receptors. Thus, in the associated pattern of regulation, the gonadal steroids produced during follicular maturation and ovulation have most likely been coopted for initiation of the central mechanisms controlling the expression of female sexual receptivity. Interestingly, this pattern of control has apparently been retained in some all-female unisexual species of lizards. The associated control pattern exhibited by some reptiles is similar to that in many other vertebrates, including most domestic and laboratory mammals and birds. The common features of this control pattern allow reptiles to be used as research models for studies of the physiological control of female sexual behavior.

In the second pattern of physiological regulation, the periods of female sexual behavior are not associated with ovarian maturation and ovulation. In this, the dissociated pattern, female sexual behavior is not mediated by elevated levels of gonadal steroids. Such dissociation between the expression of female sexual behavior and ovulation may occur in species for which environmental constraints preclude mating at the time of ovulation. Experimental studies of *Thamnophis sirtalis parietalis* indicate that temperature is an important proximal factor initiating female sexual behavior. The manner in which this mediation is achieved remains unknown, but it is reasonable to suggest that temperature may exert its effect by modulating neuroendocrine function. Although the dissociated pattern of reproduction has been studied in fewer species of vertebrates, animals that exhibit this pattern can be used to investigate the central regulators that control sexual behavior.

Although it is evident that sexual behavior is estrogen-dependent in some species but estrogen-independent in others, the neural basis of these differences remains to be described. It is possible that the same neural circuits responsible for integration of sensory information and activation of sexual behavior occur in both associated and dissociated breeders. The difference in the sensitivity to estrogen may lie in the mechanisms that prime these neural circuits on a seasonal basis. Support for this hypothesis is found in data for males, where behavior in dissociated breeders is controlled by the same neural circuits found in associated breeders (Friedman and Crews, 1985). However, these neural circuits are activated by different mechanisms (Krohmer and Crews, 1987). It may be likely that a similar situation exists for females.

Much work remains to be done on the role of recently discovered natural forms of GnRH in the neuroendocrine control of sexual behavior. Exogenously administered GnRH induces sexual behavior in estrogen-primed female lizards that exhibit the associated reproductive pattern. Gonadotropin-releasing hormone also regulates the release of gonadotropins in some turtles, crocodilians, and lizards. The gonadotropin-releasing activity in turn influences ovarian steroid secretion. This effect of GnRH must be considered as a concerted action in species exhibiting the associated reproductive pattern. The role of GnRH has not been evaluated for control of sexual behavior of reptiles exhibiting the dissociated pattern of reproduction. As we learn more about the control of behavior by bioactive peptides, we can assay the role of these peptides in reptiles. Additional study of functional correlates of neuroanatomical distribution of GnRH peptides in reptiles is also needed.

The role of prostaglandins and other related metabolites in the control of sexual behavior in reptiles needs to be further evaluated. Prostaglandin treatment has been shown to inhibit female sexual behavior in a lizard, *Anolis carolinensis,* and a snake, *Thamnophis sirtalis parietalis.* In *T. sirtalis parietalis,* plasma $PGF_{2\alpha}$ levels are elevated after mating, suggesting the role of the PG is a hormonal one. The behavioral effects of PG treatment in reptiles are consistent with those found in other vertebrates. However, it is unknown whether these responses involve the modulation of GnRH neurons in the central nervous system of reptiles, as occurs in the neuroendocrine pathways of rats. The source of prostaglandin production, the mechanism of action, and the specific identity of the prostaglandins produced in reptiles are areas that deserve further investigation.

Sexual behavior is essentially a social phenomenon, that is, it involves interactions among individuals, most commonly males and females. Sexual receptivity is defined by the culmination of the behavior in copulation or copulatory postures. Although reptiles are not the only group in which internal fertilization has evolved, they are the first entire class of terrestrial vertebrates that rely on internal transfer of sperm as a mechanism for fertilization to occur. The development of internal fertilization, coupled with the recently developed amniotic egg in the early reptiles, must have been linked to the organization of specific patterns of behavior that allow intromissive and copulatory interactions between males and females. The basic patterns of sexual behavior associated with internal fertilization that developed in these early reptiles may still be present. If so, they may constrain the behavior patterns and underlying physiological mechanisms now observed in their reptilian, avian, and mammalian descendants.

Reptiles rely on the capacity to store sperm as a mechanism of internal fertilization, a reproductive mechanism less common in mammals and birds. Sperm storage complicates consideration of environmental and physiological factors that regulate female receptive behavior. The evolutionary role of sperm storage has been analyzed both theoretically and experimentally in insects, another group that relies heavily on sperm storage in sexual tactics (Parker, 1984). Critical points are mate choice, multiple mating with one or more partners, and potential for sperm competition and the influence of these factors on sexual behavior. Both the male and female perspective must be addressed when considering the evolution of physiological control of sexual behavior. These perspectives may not be in the same direction, leading to equilibria or oscillations in traits.

Most information on reptilian reproductive biology has been gathered on temperate zone animals, although temperate species comprise only a fraction of known species of reptiles. Little is known about reproductive behavior and controls among closely related reptiles inhabiting more tropical environments. Comparison of ecological and evolutionary constraints on the mechanisms affecting the expression of behavior are possible. Future strategies for the study of the reproductive behavior and physiology of these animals should examine the patterns in evolutionary outgroups (Hennig, 1966) and in closely related species (or populations) inhabiting different environments. Another strategy involves careful consideration of groups of reptiles exhibiting dissimilar reproductive patterns in different environments. The historical focus on temperate reptiles has also limited the scope of comparisons because species diversity is low in temperate regions, particularly for lizards in North American deserts (Pianka, 1987). As a result of the historical focus on these groups, reproductive biologists studying reptiles have a restricted impression of the flexibility of neuroendocrine- and endocrine-controlling mechanisms. Future studies should provide more information about the natural diversity present in reptiles.

Most information about the physiological control of sexual behavior has been collected in a few snakes and lizards. Presently we do not know whether these findings are generally applicable to other reptiles. Although comparisons often span vertebrate classes, conclusions about the basic neuroendocrine patterns present in distantly related groups must proceed cautiously. Without a more complete picture of the behavioral control patterns occurring in major groups we cannot distinguish between patterns that are truly homologous and those that have evolved many times in parallel.

The physiological control of sexual behavior in reptiles offers basic information to researchers interested in the biological control of motivated behavior. Systematic study of more species of reptiles may facilitate conclusions about the evolution of behavior-controlling mechanisms and the constraints of the environment on these processes. Reptiles are well suited as model organisms for these types of studies, partly because as a group they exhibit a diversity of reproductive mechanisms. For example, within-genera variation of parity type, ranging from oviparity to viviparity, is known in squamates. True placentation has evolved more than once in some of these groups. Yet other groups, such as the turtles and crocodilians, have obligatory oviparity in the entire taxon. Other modes of reproduction, including parthenogenesis, and a wide variety of sex-determining mechanisms in sexual species, such as environmental sex determination, merit further study. As ectotherms, reptiles provide an alternative and more sensitive model in which to evaluate environmental control of reproduction. Finally, information about the control of mating in reptiles may facilitate efforts directed at conservation of rare and endangered species.

ACKNOWLEDGMENTS

We thank D. Crews, R. Mason, M. Mendonça, and M. Moore for encouraging research on this topic. We also thank J. Martin for editorial assistance. Supported in part by a Whitehall Foundation Grant-in-Aid to JMW.

REFERENCES

Adkins, E. L., and Schlesinger, L. (1979). Androgens and the social behavior of male and female lizards (*Anolis carolinensis*). *Horm. Behav.* 13, 139–152.

Alderete, M. R., Tokarz, R. R., and Crews, D. (1980). Luteinizing hormone-releasing hormone and thyrotropin-releasing hormone induction of female sexual receptivity in the lizard, *Anolis carolinensis. Neuroendocrinology* 30, 200–205.

Aleksiuk, M., and Gregory, P. T. (1974). Regulation of seasonal mating behavior in *Thamnophis sirtalis parietalis. Copeia* 1974, 681–689.

Andrén, C., and Nilson, G. (1987). The copulatory plug of the adder, *Vipera berus:* does it keep sperm in or out? *Oikos* 49, 230–231.

Armstrong, B. L., and Murphy, J. B. (1979). The natural history of Mexican rattlesnakes. *Univ. Kansas Mus. Nat. Hist. Special Publ.*, 1–37.

Auffenberg, W. (1983). Courtship behavior in *Varanus bengalensis* (Sauria: Varanidae). In *Advances in Herpetology and Evolutionary Biology* (A. G. J. Rhodin and K. Miyata, eds.). Museum Comparative Zoology, Harvard University, Cambridge, Mass., pp. 535–551.

Babcock, A. M., Bloch, G. J., and Micevych, P. E. (1988). Injections of cholecystokinin into the ventromedial hypothalamic nucleus inhibit lordosis behavior in the rat. *Physiol. Behav.* 43, 195–199.

Barker, D. G., Murphy, J. B., and Smith, K. W. (1979). Social behavior in a captive group of Indian pythons, *Python molurus* (Serpentes, Boidae) with formation of a linear social hierarchy. *Copeia* 1979, 466–471.

Barney, R. L. (1922). Further notes on the natural history and artificial propagation of the diamond-backed terrapin. *Bull. U.S. Bur. Fish.* 38, 91–111.

Beach, F. A. (1976). Sexual attractivity, proceptivity, and receptivity in female mammals. *Horm. Behav.* 7, 105–138.

Bermant, G., and Davidson, J. M. (1974). *Biological Bases of Sexual Behavior.* Harper & Row, Publishers, New York.

Bloch, G. J., Babcock, A. M., Gorski, R. A., and Micevych, P. E. (1987). Cho-

lecystokinin stimulates and inhibits lordosis behavior in female rats. *Physiol. Behav.* 39, 217–224.
Bona-Gallo, A., and Licht, P. (1983). Effect of temperature on sexual receptivity and ovarian recrudescence in the garter snake, *Thamnophis sirtalis parietalis. Herpetologica* 39, 173–182.
Bons, J., and Saint Girons, H. (1963). Écologie et cycle sexuel des amphisbéniens du Maroc. *Bull. Soc. Sci. Nat. Phys. Maroc* 43, 117–158.
Buntin, J. D., and Lisk, R. D. (1979). Prostaglandin E_2-induced lordosis in estrogen-primed female hamsters: relationship to progesterone action. *Physiol. Behav.* 23, 569–575.
Callard, G. V. (1985). Estrogen synthesis and other androgen-converting pathways in the vertebrate brain and pituitary. In *Current Trends in Comparative Endocrinology* (B. Lofts and W. N. Holmes, eds.). Hong Kong University Press, Hong Kong, pp. 1179–1184.
Callard, G. V., Petro, Z., and Ryan, K. J. (1977). Identification of aromatase in the reptilian brain. *Endocrinology* 100, 1214–1218.
Callard, G. V., Petro, Z., and Ryan, K. J. (1978). Conversion of androgen to estrogen and other steroids in the vertebrate brain. *Am. Zool.* 18, 511–523.
Callard, I. P., and Ho, S. M. (1980). Seasonal reproductive cycles in reptiles. *Prog. Reprod. Biol.* 5, 5–38.
Camazine, B., Garstka, W. R., Tokarz, R. R., and Crews, D. (1980). Effects of castration and androgen replacement on male courtship behavior in the red-sided garter snake (*Thamnophis sirtalis parietalis*). *Horm. Behav.* 14, 358–372.
Carpenter, C. C., and Ferguson, G. W. (1977). Variation and evolution of stereotyped behavior in reptiles. In *Biology of the Reptilia,* Vol. 7 (C. Gans and D. W. Tinkle, eds.). Academic Press, New York, pp. 335–554.
Castanet, J., Newman, D. G., and Saint Girons, H. (1988). Skeletochronological data on the growth, age, and population structure of the tuatara, *Sphenodon punctatus,* on Stephens and Lady Alice Islands, New Zealand. *Herpetologica* 44, 25–37.
Cheng, M. F. (1977). Role of gonadotropin releasing hormones in the reproductive behaviour of female ring doves (*Streptopelia risoria*). *J. Endocrinol.* 74, 37–45.
Clark, J. T., Kalra, P. S., and Kalra, S. P. (1985). Neuropeptide Y stimulates feeding but inhibits sexual behavior in rats. *Endocrinology* 117, 2435–2442.
Cole, C. J. (1975). Evolution of parthenogenetic species of reptiles. In *Intersexuality in the Animal Kingdom* (R. Reinboth, ed.). New York, Springer-Verlag, pp. 340–355.
Conn, P. M., McArdle, C. A., Andrews, W. V., and Huckle, W. R. (1987). The molecular basis of gonadotropin-releasing hormone (GnRH) action in the pituitary gonadotrope. *Biol. Reprod.* 36, 17–35.
Cooper, W. E., Jr., and Clarke, R. F. (1982). Steroidal induction of female reproductive coloration in the keeled earless lizard, *Holbrookia propinqua. Herpetologica* 38, 425–429.
Cooper, W. E., Jr., and Crews, D. (1987). Hormonal induction of secondary sexual coloration and rejection behaviour in female keeled earless lizards (*Holbrookia propinqua*). *Anim. Behav.* 35, 1177–1187.

Cooper, W. E., Jr., and Crews, D. (1988). Sexual coloration, plasma concentrations of sex steroid hormones, and responses to courtship in the female keeled earless lizard (*Holbrookia propinqua*). *Horm. Behav.* 22, 12–25.

Cooper, W. E., Jr., Mendonça, M. T., and Vitt, L. J. (1986). Induction of sexual receptivity in the female broad-headed skink, *Eumeces laticeps*, by estradiol-17β. *Horm. Behav.* 20, 235–242.

Crews, D. (1973a). Behavioral correlates to gonadal state in the lizard, *Anolis carolinensis*. *Horm. Behav.* 4, 307–313.

Crews, D. (1973b). Coition-induced inhibition of sexual receptivity in female lizards (*Anolis carolinensis*). *Physiol. Behav.* 11, 463–468.

Crews, D. (1975). Psychobiology of reptilian reproduction. *Science* (N.Y.) 189, 1059–1065.

Crews, D. (1977). The annotated anole: Studies on the control of lizard reproduction. *Am. Sci.* 65, 428–434.

Crews, D. (1979a). Endocrine control of reptilian reproductive behavior. In *Endocrine Control of Sexual Behavior* (C. Beyer, ed.). Raven Press, New York, pp. 167–222.

Crews, D. (1979b). Neuroendocrinology of lizard reproduction. *Biol. Reprod.* 20, 51–73.

Crews, D. (1980). Interrelationships among ecological, behavioral, and neuroendocrine processes in the reproductive cycle of *Anolis carolinensis* and other reptiles. *Adv. Study Behav.* 11, 1–74.

Crews, D. (1984). Gamete production, sex hormone secretion, and mating behavior uncoupled. *Horm. Behav.* 18, 22–28.

Crews, D. (1990). Neuroendocrine adaptations. In *Hormones, Brain and Behaviour in Vertebrates* (J. Balthazar, ed.). S. Karger AG, Basel, pp. 1–14.

Crews, D., and Fitzgerald, K. T. (1980). "Sexual" behavior in parthenogenetic lizards (*Cnemidophorus*). *Proc. Natl. Acad. Sci. USA* 77, 499–502.

Crews, D., and Garrick, L. D. (1980). Methods of inducing reproduction in captive reptiles. In *The Reproductive Biology and Diseases of Captive Reptiles* (J. Murphy and J. T. Collins, eds.). Society for the Study of Amphibians and Reptiles, Lawrence, Kan., pp. 49–70.

Crews, D., and Garstka, W. R. (1982). The ecological physiology of a garter snake. *Sci. Am.* 247, 158–168.

Crews, D., and Greenberg, N. (1981). Function and causation of social signals in lizards. *Am. Zool.* 21, 273–294.

Crews, D., and Moore, M. (1986). Evolution of mechanisms controlling mating behavior. *Science* (N.Y.) 231, 121–125.

Crews, D., and Moore, M. (1991). Psychobiology of reproduction of unisexual lizards. In *Biology of Cnemidophorus Lizards* (J. W. Wright, ed.). University of Washington Press, Seattle. In press.

Crews, D., and Morgentaler, A. (1979). Effects of intracranial implantation of oestradiol and dihydrotestosterone on the sexual behavior of the lizard, *Anolis carolinensis*. *J. Endocrinol.* 82, 373–381.

Crews, D., Gustafson, J. E., and Tokarz, R. R. (1983). Psychobiology of parthenogenesis. In *Lizard Ecology* (R. B. Huey, E. R. Pianka, and T. W. Schoener, eds.). Harvard University Press, Cambridge, Mass., pp. 205–232.

Crews, D., Grassman, M., and Lindzey, J. (1986). Behavioral facilitation of reproduction in sexual and unisexual whiptail lizards. *Proc. Natl. Acad. Sci. USA* 83, 9547–9550.

Cueller, O. (1966). Delayed fertilization in the lizard, *Uta stansburiana. Copeia* 1966, 549–552.

Darevsky, I. S., Kupriyanova, L. A., and Uzzell, T. (1985). Parthenogenesis in reptiles. In *Biology of the Reptilia,* Vol. 15 (C. Gans and F. Billett, eds.). John Wiley & Sons, New York, pp. 411–526.

Dawbin, W. H. (1982). The tuatara *Sphenodon punctatus:* aspects of life history, growth and longevity. In *New Zealand Herpetology* (D. G. Newman, ed.). New Zealand Wildl. Serv., Dept. Intern. Aff., Occas. Publ. 2, 237–250.

Demski, L. S. (1984). The evolution of neuroanatomical substrates of reproductive behavior: sex steroid and LHRH-specific pathways including the terminal nerve. *Am. Zool.* 24, 809–830.

Devine, M. C. (1975). Copulatory plugs in snakes, enforced chastity. *Science* (N.Y.) 187, 844–845.

Devine, M. C. (1977). Copulatory plugs, restricted mating opportunities, and reproductive competition among male garter snakes. *Nature* (London) 267, 345–346.

Devine, M. C. (1984). Potential for sperm competition in reptiles: behavioral and physiological consequences. In *Sperm Competition and the Evolution of Animal Mating Systems* (R. L. Smith, ed.). Academic Press, New York, pp. 509–522.

Diakow, C., and Nemiroff, A. (1981). Vasotocin, prostaglandin, and female reproductive behavior in the frog, *Rana pipiens. Horm. Behav.* 15, 86–93.

Dudley, C. A., and Moss, R. L. (1976). Facilitation of lordosis in the rat by prostaglandin E_2. *J. Endocrinol.* 71, 457–458.

Duvall, D., Guillette, L. J., Jr., and Jones, R. E. (1982). Environmental control of reptilian reproductive cycles. In *Biology of the Reptilia* (C. Gans and F. H. Pough, eds.). Academic Press, London, 13, 201–231.

Evans, L. T. (1936). Social behavior of the normal and castrated lizard, *Anolis carolinensis. Science* (N.Y.) 83, 104.

Ferguson, M. W. J. (1985). Reproductive biology and embryology of the crocodilians. In *Biology of the Reptilia,* Vol. 14 (C. Gans, F. Billett, and P. F. A. Maderson, eds.). John Wiley & Sons, New York, pp. 329–492.

Foreman, M. M., and Moss, R. L. (1977). Effects of subcutaneous injection and intrahypothalamic infusion of releasing hormones upon lordotic response to repetitive coital stimulation. *Horm. Behav.* 8, 219–234.

Fox, W. (1956). Seminal receptacles of snakes. *Anat. Rec.* 124, 519–539.

Friedman, D., and Crews, D. (1985). Role of anterior hypothalamus-preoptic area in the regulation of courtship behavior in the male Canadian red-sided garter snake (*Thamnophis sirtalis parietalis*): lesion experiments. *Behav. Neurosci.* 99, 942–949.

Garstka, W. R. (1982). "Chemical Communication and the Control of Garter Snake Reproductive Cycles." Ph.D. dissertation, Harvard University, Cambridge, Mass.

Garstka, W. R., Camazine, B., and Crews, D. (1982). Interactions of behavior and physiology during the annual reproductive cycle of the red-sided garter snake (*Thamnophis sirtalis parietalis*). *Herpetologica* 38, 104–123.

Garstka, W. R., Tokarz, R. R., Diamond, M., Halpert, A. P., and Crews, D. (1985). Behavioral and physiological control of yolk synthesis and deposition in the female red-sided garter snake (*Thamnophis sirtalis parietalis*). *Horm. Behav.* 19, 137–153.

Gavaud, J. (1983). Obligatory hibernation for completion of vitellogenesis in the lizard *Lacerta vivipara*. *J. Exp. Zool.* 225, 397–405.

Gillingham, J. C. (1979). Reproductive behavior of the rat snakes of eastern North America, genus *Elaphe*. *Copeia* 1979, 319–331.

Gillingham, J. C., Carpenter, C. C., and Murphy, J. B. (1983). Courtship, male combat and dominance in the western diamondback rattlesnake, *Crotalus atrox*. *J. Herpetol.* 17, 265–270.

Grassman, M., and Crews, D. (1986). Progesterone induction of pseudocopulatory behavior and stimulus-response complementarity in an all-female lizard species. *Horm. Behav.* 20, 327–335.

Greenberg, B., and Noble, G. K. (1944). Social behavior of the American chameleon (*Anolis carolinensis* Voigt). *Physiol. Zool.* 17, 392–439.

Gregory, P. T. (1974). Patterns of spring emergence of the red-sided garter snake (*Thamnophis sirtalis parietalis*) in the Interlake region of Manitoba. *Can. J. Zool.* 52, 1063–1069.

Gregory, P. T. (1977). Life history parameters of the red-sided garter snake (*Thamnophis sirtalis parietalis*) in an extreme environment, the Interlake region of Manitoba. *Natl. Mus. Nat. Sci. (Ottawa) Publ. Zool.* 13, 1–44.

Guillette, L. J., Jr., and Casas-Andreu, G. (1981). Fall reproductive activity in the high altitude Mexican lizard *Sceloporus grammicus microlepidotus*. *Herpetologica* 38, 94–104.

Guillette, L. J., Jr., Herman, C. A., and Dickey, D. A. (1988). Synthesis of prostaglandins by tissues of the viviparous lizard, *Sceloporus jarrovi*. *Herpetologica* 22, 180–185.

Guillette, L. J., Jr., and Jones, R. E. (1980). Arginine vasotocin-induced *in vitro* oviductal contractions in *Anolis carolinensis:* effects of steroid hormone pretreatment *in vivo*. *J. Exp. Zool.* 212, 147–152.

Haines, T. P. (1940). Delayed fertilization in *Leptodeira annulata polysticta*. *Copeia* 1940, 116–118.

Hall, N. R., and Luttge, W. G. (1977). Diencephalic sites responsive to prostaglandin E_2 facilitation of sexual receptivity in estrogen-primed ovariectomized rats. *Brain Res. Bull.* 2, 203–207.

Hall, N. R., Luttge, W. G., and Berry, R. B. (1975). Intracerebral prostaglandin E_2: effects upon sexual behavior, open field activity and body temperature in ovariectomized female rats. *Prostaglandins* 10, 877–888.

Halpern, M., Morrell, J. I., and Pfaff, D. W. (1982). Cellular [^{3}H]estradiol and [^{3}H]testosterone localization in the brains of garter snakes: an autoradiographic study. *Gen. Comp. Endocrinol.* 46, 211–224.

Halpert, A. P., Garstka, W. R., and Crews, D. (1982). Sperm transport and storage and its relation to the annual sexual cycle of the female red-sided garter snake, *Thamnophis sirtalis parietalis*. *J. Morphol.* 174, 149–159.

Hammond, R. W., Koelkebeck, K. W., Scanes, C. G., Biellier, H. V., and Hertelendy, F. (1981). Plasma prostaglandin, LH, and progesterone levels dur-

ing the ovulation cycle of the turkey (*Meleagris gallopavo*). *Gen. Comp. Endocrinol.* 44, 400–403.
Hawley, A. W. L., and Aleksiuk, J. (1976). Sexual receptivity in the female red-sided garter snake (*Thamnophis sirtalis parietalis*). *Copeia* 1976, 401–404.
Hennig, W. (1966). *Phylogenetic Systematics.* University of Illinois Press, Urbana.
Jackson, I. M. D. (1978). Phylogenetic distribution and function of the hypophysiotropic hormones of the hypothalamus. *Am. Zool.* 18, 385–399.
Jones, R. E., Guillette, L. J., Jr., Summers, C. H., Tokarz, R. R., and Crews, D. (1983). The relationship among ovarian condition, steroid hormones, and estrous behavior in *Anolis carolinensis. J. Exp. Zool.* 227, 145–154.
Joss, J. M. P., and Minard, J. A. (1985). On the reproductive cycles of *Lampropholis guichenoti* and *L. delicata* (Squamata: Scincidae) in the Sydney region. *Aust. J. Zool.* 33, 699–704.
Kalra, S. P., and Kalra, P. S. (1983). Neural regulation of luteinizing hormone secretion in the rat. *Endoc. Reviews* 4, 311–351.
Kelley, R. W. (1981). Prostaglandin synthesis in the male and female reproductive tract. *J. Reprod. Fert.* 62, 293–304.
Kim, Y. S., Stumpf, W. E., Sar, M., and Martinez-Vargas, M. C. (1978). Estrogen and androgen target cells in the brain of fishes, reptiles, and birds: phylogeny and ontogeny. *Am. Zool.* 18, 425–433.
Kim, Y. S., Stumpf, W. E., and Sar, M. (1981). Anatomical distribution of estrogen target neurons in turtle brain. *Brain Res.* 230, 195–204.
King, J. A., and Millar, R. P. (1979). Heterogeneity of vertebrate luteinizing hormone-releasing hormone. *Science* (N.Y.) 206, 67–69.
King, J. A., and Millar, R. P. (1980). Comparative aspects of luteinizing hormone-releasing hormone structure and function in vertebrate phylogeny. *Endocrinology* 106, 707–717.
King, M. (1977). Reproduction in the Australian gekko *Phyllodactylus marmoratus* (Gray). *Herpetologica* 33, 7–13.
Kloeck, F. K., and Jung, H. (1973). *In vitro* release of prostaglandins from the human myometrium under the influence of stretching. *Am. J. Obstet. Gynecol.* 115, 1066–1069.
Krohmer, R. W., and Crews, D. (1987). Temperature activation of courtship behavior in the male red-sided garter snake (*Thamnophis sirtalis parietalis*): role of the anterior hypothalamus preoptic area. *Behav. Neurosci.* 101, 228–236.
Lance, V. A., Vliet, K. A., and Bolaffi, J. L. (1985). Effect of mammalian luteinizing hormone-releasing hormone on plasma testosterone in male alligators, with observations on the nature of alligator hypothalamic gonadotropin-releasing hormone. *Gen. Comp. Endocrinol.* 60, 138–143.
LaPointe, J. L. (1969). Effect of ovarian steroids and neurohypophysial hormones on the oviduct of the viviparous lizard, *Klauberina riversiana. J. Endocrinol.* 43, 197–205.
Licht, P. (1984). Seasonal cycles in reptilian reproductive physiology. In *Marshall's Physiology of Reproduction* (E. Lamming, ed.). Churchill Livingstone, Edinburgh, pp. 206–282.

Licht, P., and Porter, D. (1987). Role of gonadotropin-releasing hormone in regulation of gonadotropin secretion from amphibian and reptilian pituitaries. In *Hormones and Reproduction in Fishes, Amphibians, and Reptiles* (D. O. Norris and R. E. Jones, eds.). Plenum Press, New York, pp. 61–85.
Licht, P., Wood, J., Owens, D. W., and Wood, F. (1979). Serum gonadotropins and steroids associated with breeding activities in the green sea turtle *Chelonia mydas*. I. Captive animals. *Gen. Comp. Endocrinol.* 39, 274–289.
Licht, P., Rainey, W., and Cliffton, K. (1980). Serum gonadotropin and steroids associated with breeding activities in the green sea turtle, *Chelonia mydas*. II. Mating and nesting in natural populations. *Gen. Comp. Endocrinol.* 41, 116–122.
Loher, W. (1979). The influence of prostaglandin E_2 on oviposition in *Teleogryllus commodus*. *Entomol. Exp. Appl.* 25, 107–119.
Loher, W., and Edson, K. (1973). The effect of mating on egg production and release in the cricket, *Teleogryllus commodus*. *Entomol. Exp. Appl.* 16, 483–490.
Loveridge, A. (1941). Revision of the African lizards of the family Amphisbaenidae. *Bull. Mus. Comp. Zool. Harvard* 87, 353–451.
Ludwig, M., and Rahn, H. (1943). Sperm storage and copulatory adjustment in the prairie rattlesnake. *Copeia* 1943, 15–18.
Magnusson, W. E. (1979). Production of an embryo by an *Acrochordus javanicus* isolated for seven years. *Copeia* 1979, 744–745.
Marrone, B. L., Rodriguez-Sierra, J. F., and Feder, H. H. (1979). Differential effects of prostaglandins on lordosis behavior in female guinea pigs and rats. *Biol. Reprod.* 20, 853–861.
Mason, P., and Adkins, E. K. (1976). Hormones and sexual behavior in the lizard *Anolis carolinensis*. *Horm. Behav.* 7, 75–86.
McEwen, B. S., Jones, K. J., and Pfaff, D. W. (1987). Hormonal control of sexual behavior in the female rat: molecular, cellular, and neurochemical studies. *Biol. Reprod.* 36, 37–45.
McNicol, D., Jr., and Crews, D. (1979). Estrogen/progesterone synergy in the control of female sexual receptivity in the lizard, *Anolis carolinensis*. *Gen. Comp. Endocrinol.* 38, 68–74.
Mendelson, S. D., and Gorzalka, B. B. (1984). Cholecystokinin-octapeptide produces inhibition of lordosis in the female rat. *Pharmacol. Biochem. Behav.* 21, 755–759.
Mendonça, M. T. (1987). Photothermal effects on the ovarian cycle of the musk turtle *Sternotherus odoratus*. *Herpetologica* 43, 82–90.
Mendonça, M. T., and Crews, D. (1989). Effects of fall mating on ovarian development in the red-sided garter snake (*Thamnophis sirtalis parietalis*). *Am. J. Physiol.* 257 (Regulatory, Integrative and Comparative Physiology) 26, R1548-R1550.
Mendonça, M. T., and Crews, D. (1990). Mating-induced neuroendocrine reflex in the gartersnake: faculative ovarian development. *J. Comp. Physiol.* A166, 629–632.
Milvae, R. A., and Hansel, W. (1985). Inhibition of bovine luteal function by indomethacin. *J. Anim. Sci.* 60, 528–531.
Moffat, L. A. (1985). Embryonic development and aspects of reproductive

biology in the tuatara, *Sphenodon punctatus*. In *Biology of the Reptilia*, Vol. 14 (C. Gans, F. Billett, and P. A. Maderson, eds.). John Wiley & Sons, New York, pp. 493–522.

Moore, F. L. (1987). Regulation of reproductive behaviors. In *Hormones and Reproduction in Fishes, Amphibians, and Reptiles* (D. O. Norris and R. E. Jones, eds.). Plenum Press, New York, pp. 505–522.

Moore, M. C., and Crews, D. (1986). Sex steroid hormones in natural populations of a sexual whiptail lizard *Cnemidophorus inornatus*, a direct evolutionary ancestor of a unisexual parthenogen. *Gen. Comp. Endocrinol.* 63, 424–430.

Moore, M. C., Whittier, J. M., Billy, A. J., and Crews, D. (1985a). Male-like behaviour in an all-female lizard: relationship to ovarian cycle. *Anim. Behav.* 33, 284–289.

Moore, M. C., Whittier, J. M., and Crews, D. (1985b). Sex steroid hormones during the ovarian cycle of an all-female parthenogenetic lizard and their correlation with pseudosexual behavior. *Gen. Comp. Endocrinol.* 60, 144–153.

Morali, G., and Beyer, C. (1979). Neuroendocrine control of mammalian estrous behavior. In *Endocrine Control of Sexual Behavior* (C. Beyer, ed.). Raven Press, New York, pp. 33–75.

Moritz, C. (1983). Parthenogenesis in the endemic Australian lizard *Heteronotia binoei* (Gekkonidae). *Science* (N.Y.) 220, 735–737.

Moritz, C. (1984). The origin and evolution of parthenogenesis in *Heteronotia binoei* (Gekkonidae). I. Chromosome banding studies. *Chromosoma* (Berlin) 89, 151–162.

Moritz, C. (1987). Parthenogenesis in the tropical gekkonid lizard *Nactus arnouxii* (Sauria: Gekkonidae). *Evolution* 41, 1252–1266.

Moritz, C., and King, D. (1985). Cytogenetic perspectives on parthenogenesis in the Gekkonidae. In *Biology of Australasian Frogs and Reptiles* (G. Grigg, R. Shine, and H. Ehmann, eds.). Royal Zoological Society of New South Wales, Sydney, pp. 327–337.

Moritz, C., Brown, W. M., Densmore, L. D., Wright, J. W., Vyas, D., Donnellan, S., Adams, M., and Baverstock, P. (1989). Genetic diversity and the dynamics of hybrid parthenogenesis in *Cnemidophorus* (Teiidae) and *Heteronotia* (Gekkonidae). In *The Biology of Unisexual Vertebrates* (R. M. Dawley and J. P. Bogart, eds.). Bulletin 466, New York State Museum, Albany, N.Y., pp. 87–112.

Morrell, J. I., and Pfaff, D. W. (1978). A neuroendocrine approach to brain function: localization of sex steroid concentrating cells in vertebrate brains. *Am. Zool.* 18, 447–460.

Morrell, J. I., Crews, D., Ballin, A., Morgentaler, A., and Pfaff, D. W. (1979). (^{3}H)-estradiol, (^{3}H)-testosterone, and (PH)-dihydro-testosterone localization in the brain of the lizard *Anolis carolinensis:* an autoradiographic study. *J. Comp. Neurol.* 188, 201–224.

Moss, R. L. (1977). Effects of hypothalamic peptides in sex behavior in animals and man. In *Psychopharmacology: a Generation of Progress* (M. A. Lipton, A. DiMascio, and K. F. Killan, eds.). Raven Press, New York, pp. 431–440.

Moss, R. L. (1979). Actions of hypothalamic-hypophysiotropic hormones on the brain. *Ann. Rev. Physiol.* 41, 617–631.

Moss, R. L., and Dudley, C. A. (1979). Sexual function and brain peptides. In *Brain Peptides: a New Endocrinology* (A. M. Motto, Jr., E. J. Peck, Jr., and A. E. Boyd, III, eds.). Elsevier/North Holland Biomedical Press, Amsterdam, pp. 325–346.

Moss, R. L., and McCann, S. M. (1973). Induction of mating behavior in rats by luteinizing hormone-releasing factor. *Science* (N.Y.) 181, 177–179.

Moss, R. L., and McCann, S. M. (1975). Action of luteinizing hormone-releasing factor (LRF) in the initiation of lordosis behavior in the estrone-primed ovariectomized female rat. *Neuroendocrinology* 17, 309–318.

Mount, R. H. (1963). The natural history of the red-tailed skink, *Eumeces egregius* Baird. *Am. Midl. Nat.* 70, 356–385.

Murphy, J. B., Tryon, B. W., and Brecke, B. J. (1978). An inventory of reproduction and social behavior in captive gray-banded kingsnakes, *Lampropeltis mexicana alterna* (Brown). *Herpetologica* 34, 84–93.

Newman, D. G. (1982). Current distribution of the tuatara. In *New Zealand Herpetology* (D. G. Newman, ed.). New Zealand Wildl. Serv. Dept. Intern. Aff., Occas. Publ. 2, 145–147.

Newman, D. G. (1987). Burrow use and population densities of tuatara (*Sphenodon punctatus*) and how they are influenced by fairy prions (*Pachyptila turtur*) on Stephens Island, New Zealand. *Herpetologica* 43, 336–344.

Newman, D. G., and Watson, P. R. (1985). The contribution of radiography to the study of the reproductive ecology of the tuatara, *Sphenodon punctatus*. In *The Biology of Australasian Frogs and Reptiles* (G. Grigg, R. Shine, and H. Ehmann, eds.). Surrey Beatty & Sons Pty Limited, Chipping Norton, Australia, pp. 7–10.

Newman, D. G., Crook, I. G., and Moran, L. R. (1979). Some recommendations on the captive maintenance of tuataras *Sphenodon punctatus* based on observations in the field. *Int. Zoo Yearb.* 19, 68–74.

Nilson, G., and Andrén, C. (1982). Function of renal sex secretions and male hierarchy in the adder, *Vipera berus*, during reproduction. *Horm. Behav.* 16, 404–413.

Noble, G. K., and Greenberg, B. (1941). Effects of seasons, castration and crystalline sex hormones upon the urogenital system and sexual behavior of the lizard *Anolis carolinensis*. I. The adult female. *J. Exp. Zool.* 88, 451–479.

Nozaki, M., and Kobayashi, H. (1979). Distribution of LHRH-like substance in the vertebrate brain as revealed by immunohistochemistry. *Arch. Hist. Jap.* 42, 201–219.

Nozaki, M., Tsukahara, T., and Kobayashi, H. (1984). Neuronal systems producing LHRH in vertebrates. In *Endocrine Correlates of Reproduction* (K. Ochiai, Y. Arai, T. Shioda, and M. Takahashi, eds.). Japan Scientific Society Press, Tokyo/Springer-Verlag, Berlin, pp. 3–27.

Ojeda, S. R., Jameson, H. E., and McCann, S. M. (1977a). Hypothalamic areas involved in prostaglandin (PG)-induced gonadotropin release. I: Effect of PGE_2 and $PGF_{2\alpha}$ implants on luteinizing hormone release. *Endocrinology* 100, 1585–1594.

Ojeda, S. R., Jameson, H. E., and McCann, S. M. (1977b). Hypothalamic areas involved in prostaglandin (PG)-induced gonadotropin release. II: Effect of PGE_2 and $PGF_{2\alpha}$ implants on follicle stimulating hormone release. *Endocrinology* 100, 1595–1603.

Olson, D. M., Shimada, K., and Etches, R. J. (1986). Prostaglandin concentrations in peripheral plasma and ovarian and uterine plasma and tissue in relation to oviposition in hens. *Biol. Reprod.* 35, 1140–1146.

Parker, G. A. (1984). Sperm competition and the evolution of animal mating strategies. In *Sperm Competition and the Evolution of Animal Mating Systems* (R. L. Smith, ed.). Academic Press, New York, pp. 2–61.

Pfaff, D. W. (1973). Luteinizing hormone-releasing factor (LRF) potentiates lordosis behavior in hypophysectomized ovariectomized female rats. *Science* (N.Y.) 182, 1148–1149.

Phillips, J. A., and Lasley, B. L. (1987). Modification of reproductive rhythm in lizards via GnRH therapy. *Ann. N.Y. Acad. Sci.* 519, 128–136.

Phillips, J. A., Karesh, A. W. B., Millar, R., and Lasley, B. L. (1985). Stimulating male sexual behavior with repetitive pulses of GnRH in female green iguanas, *Iguana iguana*. *J. Exp. Zool.* 234, 481–484.

Pianka, E. (1987). *Ecology and Natural History of Desert Lizards.* Princeton University Press, Princeton, N.J.

Powell, R. C., King, J. A., and Millar, R. P. (1985). $[Trp^7.Leu^8]$LH-RH in reptilian brain. *Peptides* 6, 223–227.

Powell, R. C., Ciarcia, G., Lance, V. A., Millar, R. P., and King, J. A. (1986). Identification of diverse molecular forms of GnRH in reptile brain. *Peptides* 7, 1101–1108.

Poyser, N. L., Horton, E. W., Thompson, C. J., and Los, M. (1971). Identification of prostaglandin $F_{2\alpha}$ released by distension of guinea-pig uterus *in vitro*. *Nature* (London) 230, 526–528.

Quinn, H. R. (1979). Reproduction and growth of the Texas coral snake. *Copeia* 1979, 453–463.

Rahn, H. (1940). Sperm viability in the uterus of the garter snake, *Thamnophis*. *Copeia* 1940, 109–115.

Rodriguez-Sierra, J. F., and Komisaruk, B. R. (1977). Effects of prostaglandin E_2 and indomethacin on sexual behavior in the female rat. *Horm. Behav.* 9, 281–289.

Rodriguez-Sierra, J. F., and Komisaruk, B. R. (1982). Common hypothalamic sites for activation of sexual receptivity in female rats by LHRH, PGE_2, and progesterone. *Neuroendocrinology* 35, 363–369.

Ross, P., Jr., and Crews, D. (1977). Influence of the seminal plug on mating behaviour in the garter snake. *Nature* (London) 267, 344–345.

Ross, P., Jr., and Crews, D. (1978). Stimuli influencing mating behavior in the garter snake, *Thamnophis radix*. *Behav. Ecol. Sociobiol.* 4, 133–142.

Saint Girons, H. (1982). Reproductive cycles of male snakes and their relationships with climate and female reproductive cycles. *Herpetologica* 38, 5–16.

Saito, N., Sato, K., and Shimada, K. (1987). Prostaglandin levels in peripheral and follicular plasma, the isolated theca and granulosa layers of pre- and postovulatory follicles, and the myometrium and mucosa of the shell gland (uterus) during a midsequence-oviposition of the hen (*Gallus domesticus*). *Biol. Reprod.* 36, 89–96.

Sakuma, Y., and Pfaff, D. W. (1980). LH-RH in the mesencephalic central grey can potentiate lordosis reflex in female rats. *Nature* (London) 283, 566–567.

Schuett, G. W., and Gillingham, J. C. (1988). Courtship and mating of the copperhead, *Agkistrodon contortrix. Copeia* 1988, 374–381.

Secor, S. M. (1987). Courtship and mating behavior of the speckled kingsnake, *Lampropeltis getulus holbrooki. Herpetologica* 43, 15–28.

Seigel, R. A., and Ford, N. B. (1987). Reproductive Ecology. In *Snakes: Ecology and Evolutionary Biology* (R. A. Seigel, J. T. Collins, and S. S. Novak, eds.). Macmillan Co., New York, pp. 210–252.

Sherwood, N. (1987a). Gonadotropin-releasing hormones in fishes. In *Hormones and Reproduction in Fishes, Amphibians, and Reptiles* (D. O. Norris and R. E. Jones, eds.). Plenum Press, New York, pp. 31–60.

Sherwood, N. (1987b). The GnRH family of peptides. *Trends Neurosci.* 10, 129–132.

Sherwood, N. M., and Whittier, J. M. (1988). Gonadotropin-releasing hormone (GnRH) from the brains of reptiles: turtles (*Pseudemys scripta*) and snakes (*Thamnophis sirtalis parietalis*). *Gen. Comp. Endocrinol.* 69, 319–327.

Smyth, M., and Smith, M. J. (1968). Obligatory sperm storage in the skink *Hemiegris peroni. Science* (N.Y.) 161, 575–576.

Sossinka, R. (1980). Ovarian development in the opportunistic breeder, the zebra finch *Poephila guttata castanotis. J. Exp. Zool.* 211, 225–230.

Spilman, C. H., and Harper, M. J. K. (1975). Effects of prostaglandins on oviductal motility and egg transport. *Gynecol. Invest.* 6, 186–205.

Stacey, N. E. (1976). Effects of indomethacin and prostaglandin on the spawning behavior of female goldfish. *Prostaglandins* 12, 113–126.

Stacey, N. E. (1981). Hormonal regulation of female reproductive behavior in fish. *Am. Zool.* 21, 305–316.

Stacey, N. E. (1983). Hormones and pheromones in fish sexual behavior. *Bioscience* 33, 552–556.

Stacey, N. E., and Peter, R. E. (1979). Central action of prostaglandins in spawning behavior of female goldfish. *Physiol. Behav.* 22, 1191–1196.

Stamps, J. A. (1975). Courtship patterns, estrus periods and reproductive condition in a lizard, *Anolis aeneus. Physiol. Behav.* 14, 531–535.

Stearns, S. C. (1976). Life-history tactics: a review of the ideas. *Q. Rev. Biol.* 51, 3–47.

Stille, B., and Niklasson, M. (1987). Within season multiple paternity in the adder, *Vipera berus:* a reply. *Oikos* 49, 232–233.

Stille, B., Madsen, T., and Niklasson, M. (1986). Multiple paternity in the adder, *Vipera berus. Oikos* 47, 173–175.

Tinkle, D. W. (1969). The concept of reproductive effort and its relation to the evolution of life histories of lizards. *Am. Nat.* 103, 501–516.

Tokarz, R. R., and Crews, D. (1980). Induction of sexual receptivity in the female lizard, *Anolis carolinensis:* effects of estrogen and the antiestrogen CI-628. *Horm. Behav.* 14, 33–45.

Tokarz, R. R., and Crews, D. (1981). Effects of prostaglandins on sexual receptivity in the female lizard, *Anolis carolinensis. Endocrinology* 109, 451–457.

Tokarz, R. R., Crews, D., and McEwen, B. S. (1981). Estrogen-sensitive progestin binding sites in the brain of the lizard, *Anolis carolinensis*. *Brain Res.* 220, 95–105.

Valenstein, P., and Crews, D. (1977). Mating-induced termination of behavioral estrus in the female lizard, *Anolis carolinensis*. *Horm. Behav.* 9, 362–370.

Villars, T. A., Hale, N., and Chapnick, D. (1985). Prostaglandin $F_{2\alpha}$ stimulates reproductive behavior of female paradise fish (*Macropodus opercularis*). *Horm. Behav.* 19, 21–35.

Vleck, C. M., and Priedkalns, J. (1985). Reproduction in zebra finches: hormone levels and effects of dehydration. *Condor* 87, 37–46.

Voris, H. K., and Jayne, B. C. (1979). Growth, reproduction and population structure of a marine snake, *Enhydrina schistosa* (Hydrophiidae). *Copeia* 1979, 307–318.

Wechsung, E., and Houvenaghel, A. (1976). A possible role of prostaglandins in the regulation of ovum transport and oviposition in the domestic hen. *Prostaglandins* 12, 599–608.

Werner, D. I. (1978). On the biology of *Tropidurus delanonis*. *Z. Tierpsychol.* 47, 337–395.

Werner, Y. L. (1980). Apparent homosexual behavior in an all-female population of a lizard, *Lepidodactylus lugubris* and its probable interpretation. *Z. Tierpsychol.* 54, 144–150.

Whittier, J. M., and Crews, D. (1986a). Ovarian development in red-sided garter snakes, *Thamnophis sirtalis parietalis:* relationship to mating. *Gen. Comp. Endocrinol.* 61, 5–12.

Whittier, J. M., and Crews, D. (1986b). Effects of prostaglandin $F_{2\alpha}$ on sexual behavior and ovarian function in female garter snakes (*Thamnophis sirtalis parietalis*). *Endocrinology* 119, 787–792.

Whittier, J. M., and Crews, D. (1987). Seasonal reproduction: patterns and control. In *Hormones and Reproduction in Fishes, Amphibians, and Reptiles* (D. O. Norris and R. E. Jones, eds.). Plenum Press, New York, pp. 385–409.

Whittier, J. M., and Crews, D. (1989). Mating increases plasma levels of prostaglandin $F_{2\alpha}$ in female garter snakes. *Prostaglandins* 37, 359–366.

Whittier, J. M., and Crews, D. (1990). Body mass and reproduction in female red-sided garter snakes (*Thamnophis sirtalis parietalis*). *Herpetologica* 46, 215–222.

Whittier, J. M., Mason, R. T., and Crews, D. (1985). Mating in the red-sided garter snake, *Thamnophis sirtalis parietalis:* differential effects on male and female sexual behavior. *Behav. Ecol. Sociobiol.* 16, 257–261.

Whittier, J. M., Mason, R. T., and Crews, D. (1987a). Plasma steroid hormone levels of female red-sided garter snakes, *Thamnophis sirtalis parietalis:* relationship to mating and gestation. *Gen. Comp. Endocrinol.* 67, 33–43.

Whittier, J. M., Mason, R. T., Crews, D., and Licht, P. (1987b). Role of light and temperature in the regulation of reproduction in the red-sided garter snake, *Thamnophis sirtalis parietalis*. *Can. J. Zool.* 65, 2090–2096.

Wu, J., Whittier, J. M., and Crews, D. (1985). Role of progesterone in the control of female sexual receptivity in *Anolis carolinensis*. *Gen. Comp. Endocrinol.* 58, 402–406.

3

The Physiological Basis of Sexual Behavior in Male Reptiles

MICHAEL C. MOORE AND JONATHAN LINDZEY

CONTENTS

I. INTRODUCTION

A. Central Paradigms

For organisms to reproduce successfully, they must coordinate reproductive physiology and behavior with events in the external environment. In many animals, the need to coordinate the expression of reproductive behavior and reproductive physiology has led to the evolution of very strong links between these. This coordination is most frequently accomplished through the action of reproductive hormones on the areas of the brain that directly mediate behavior.

Extensive studies on mammals and birds disclose that the sex steroid hormones secreted by the gonads are those most directly involved in the control of reproductive behavior.

Steroid hormones influence behavior by both organizational and activational effects (Phoenix et al., 1959; Arnold and Breedlove, 1985; Adkins-Regan, 1987). The classic formulation of the organizational-activational model proposed that these two effects occur sequentially. Organizational effects are permanent and occur early in development. During organization, reproductive hormones act on the developing nervous system to alter the structure of hormone-sensitive regions, typically influencing whether they develop in a male-typical or female-typical pattern. Activational effects are transitory and occur during adult life. During activation, hormones work on the previously organized neural substrates to activate the expression of adult reproductive behaviors. This activational effect is temporary and lasts only as long as elevated circulating levels of sex steroids are present. Originally it was thought that all hormone-mediated effects on behavior required both organization and activation. Recently the demonstration of more or less permanent effects of hormones during adult life has caused the extent of this dichotomy to be questioned. Arnold and Breedlove (1985) argue convincingly that the classic organizational-activational dichotomy is perhaps more appropriately viewed as extremes on a continuum of possible effects of hormones on behavior.

Despite the recent rethinking of the extent of the dichotomy between organization and activation, this model has been the central paradigm around which the study of hormone effects on behavior in mammals and birds has been focused. Much of the existing information on hormonal control of behavior in reptiles fits neatly within this model as well, and we will use it to structure the material presented in this review. However, a great diversity of reproductive strategies has evolved in reptiles in response to the challenges of reproducing successfully in a diversity of extreme environments. This has led to the evolution of a diversity of behavior-controlling mechanisms that do not always fit into the standard models and paradigms (Crews, 1984; Crews and Moore, 1986; Crews and Bull, 1987; Moore and Marler, 1988). In this review, we hope to document the importance of current research on behavior-controlling mechanisms in reptiles in expanding contemporary views of the general scope of such mechanisms.

Recent research on reptiles has focused attention on the evolutionary plasticity of behavior-controlling mechanisms. Hence, in this review we have decided to depart from the usual phylogenetic format. Generalizations about evolutionarily plastic traits are more successfully derived from common ecological and social situations, that is, from the existence of common selective pressures, than they are from the common evolutionary ancestry that is the basis of the phylogenetic approach. The latter is more useful for traits that are evolutionarily conserved. These traits are more likely to reflect ancestral constraints rather than recent selective pressures. In addition, hormonal control of behavior has been studied in so few species of reptiles that any phylogenetic generalizations would be tentative and premature. Instead, our approach will be to focus on the relatively few case studies for which there are good experimental data on hormonal and neural bases of behavior in reptiles. We will put these case studies in the framework of the central ideas developed in the more extensively studied mammals and birds.

B. Evidence for Hormone Dependence of Sexual Behavior

Because recent research on reptiles has challenged the universality of gonadal steroid hormone dependence of reproductive behaviors, it is crucial to consider the types of evidence that are necessary to establish that a particular behavior is dependent on particular hormones.

It is neither simple nor straightforward to demonstrate that a behavior depends on a particular hormone. The neuroendocrine system is a complex web of interdependent entities. The possibilities for spurious correlations and even misleading experimental results are manifold. No single test of hormone dependence can be considered absolute. It is necessary instead to assemble several different lines of mutually consistent evidence. The biggest danger arises from the fact that most hormones are regulated by negative feedback loops with other hormones. As a consequence, whenever the level of one hormone is low, that of the other is high and vice versa. For example, testosterone has negative feedback effects on the pituitary gonadotropin, luteinizing hormone (LH). The demonstration that removal of testosterone abolishes the expression of a behavior and replacement of testosterone restores it, shows *either* that testosterone stimulates the behavior *or* that LH inhibits it. The possibility for such indirect, "downstream" effects of hormone manipulations necessitates the reliance on multiple lines of evidence.

The following lines of evidence jointly would present a strong case that a particular behavior depended on a particular hormone:

1. The expression of the behavior is at least loosely correlated with elevated circulating levels of the hormone.

2. Expression of the behavior is abolished by chemical or surgical ablation of the source of the hormone and restored by replacing the hormone (Fig. 3.1).

3. The hormone binds in regions of the brain that can be shown by lesion studies, by electrical stimulation, or by electrical recording to be involved in the control of the behavior.

4. Very localized implants of the hormone in this same region of the brain restore the behavior whenever other sources of the hormone have been ablated.

Absence or failure of any of these lines of evidence does not necessarily negate the possibility that the hormone influences the behavior but might suggest that its effect is less direct or requires synergism with other factors. In particular, correlations between elevated circulating levels of the hormone and expression of the behavior can be very misleading. Expression of behavior also may reflect the presence of releasing stimuli, which can vary independently of circulating levels of hormones. Thus an animal may not express a hormone-dependent behavior when hormone levels are high because no releasing stimuli are present or may express it when levels are low because an extremely strong releaser may override the inhibition due to low hormone levels. For the study of sexual behavior in males, it has been particularly difficult to control for the presence of releasing stimuli (usually sexually receptive females) in correlational studies. For this reason we will avoid lengthy discussion of correlational studies (re-

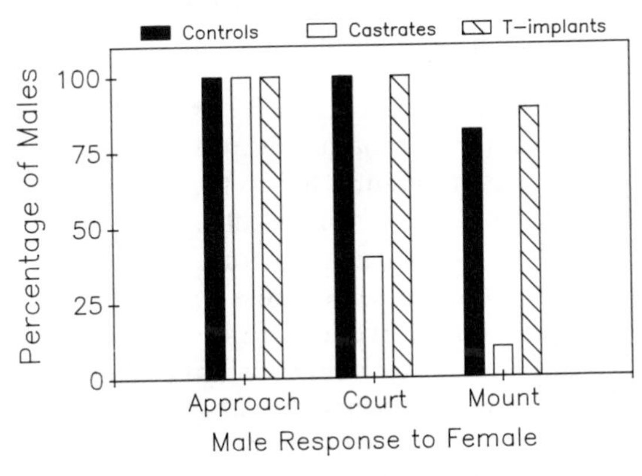

Fig. 3.1. *Sceloporus jarrovi.* Typical results of a castration replacement experiment illustrated with data on the response of free-living males to the presentation of a tethered sexually receptive female in their territory. Note in particular that (1) not all components of the response are affected by hormone manipulations and (2) not all males are equally affected by hormone manipulation, e.g., one castrated male copulated with a female. Behavioral tests in these studies were conducted 2 weeks after surgery. Testosterone implants used produced circulating levels of testosterone that were nearly identical to the levels in the sham-operated controls. (Data from Moore, 1987a, 1988.)

cent reviews by Callard and Ho, 1980; Duvall et al., 1982; Licht, 1982, 1984; Lance, 1984; Crews, 1984; Whittier and Crews, 1987) and instead concentrate on manipulative studies.

C. Ways in which Hormones Can Affect Behavior

Before the last 10 or 15 years, behavioral endocrinology has relied on the study of a relatively limited number of laboratory or domestic stocks of rodents and birds. These studies have led to a rich and detailed view of the actions of hormones in these few species (e.g., Pfaff, 1980; Feder, 1984; Arnold and Breedlove, 1985; Sachs and Meisel, 1988). However, for two reasons these studies have led to an overly narrow view of the diversity and complexity of hormonal control mechanisms.

First, concentration on a few species with a limited range of reproductive strategies led to the dogma that all reproductive behaviors depend directly on elevated levels of sex steroid hormones in the circulation. Recent research on naturally occurring fishes, amphibians, birds, and reptiles (Crews, 1984; Crews and Moore, 1986; Crews, 1987; Moore and Marler, 1988) has changed this view. It now seems more likely that hormones will be used whenever they are an appropriate signal. However, under the constraints of the complex reproductive strategies of naturally occurring species, the onset of mating behavior may be triggered by many different types of signals. Thus, mating behavior may be expressed independently of hormonal signals from the gonads.

Second, the thorough demonstration, using the above-discussed types of evidence from many mammals and birds, that behaviors depend on hormones has frequently led to the temptation of assuming cause and effect. However, few behavioral endocrinologists would actively argue that hormones cause behaviors. Instead, hormones affect the probability that an animal will express a particular behavior under a given set of circumstances (Feder, 1984). In traditional laboratory studies these circumstances are held constant and the hormonal treatment is varied. This produces what appears to be a simple cause-and-effect relationship. However, it is becoming increasingly apparent that internal factors and external environmental variables interact in complex ways to determine the probability of an animal expressing a behavior.

Of particular importance to this point are field studies involving the manipulation of the levels of hormones in animals exposed to the complex and continuously varying natural environments (Wingfield and Moore, 1987; Moore and Marler, 1988; Marler and Moore, 1988, 1989, 1991). Such studies have shown that the expression of behaviors is strongly influenced by hormone levels but that behaviors do not

Table 3.1 Major contrasts available for studies of reptiles

Associated/dissociated strategies	Male/female heterogamety
Viviparous/oviparous	Bisexual/unisexual reproduction
Genotypic/environmental sex determination	Between-sex/within-sex polymorphisms
Endothermy/ectothermy	Laboratory/field endocrinology

Note: Reptiles are useful model systems for studies of behavioral physiology because they provide a large diversity of contrasts that can be utilized as natural experiments to test ideas generated in the less diverse mammals and birds. Perhaps only the teleost fish provide a greater variety of contrasts among the vertebrates.

depend on hormones in simple, direct ways. Although these types of field studies were pioneered in birds (Wingfield et al., 1987; Wingfield and Moore, 1987), several recent studies have used reptiles (e.g., Fig. 3.1) and, as will be pointed out below, there is tremendous potential for future studies of reptilian field endocrinology.

D. Reptiles as Model Systems for Behavioral Endocrinology

Research on reptiles already has played a key role in expanding our view of the diversity of behavior-controlling mechanisms; further studies of reptiles promise to enrich this perspective. In particular, research on reptiles is critically important from a theoretical standpoint because of the tremendous diversity of reproductive patterns displayed within the class (Table 3.1; Callard and Ho, 1980; Duvall et al., 1982; Licht, 1984; Crews, 1984; Whittier and Crews, 1987).

Like their mammalian and avian relatives, many reptiles mate at a time of year when their gonads are enlarged and actively producing gametes. This has been called the *associated reproductive pattern* because mating and gametogenesis are temporally associated (Fig. 3.2). However, there are constraints on this process. For example, organisms inhabiting extreme environments may have insufficient time to produce functional gametes. In these kinds of circumstances, many animals employ a *dissociated reproductive pattern* (e.g., Crews, 1984). In this pattern, gametogenesis occurs long before mating, and gametes are stored until the mating period. Thus, mating occurs when gonads are regressed and circulating levels of steroid hormones are low. The dissociated pattern may be more common in reptiles than in any other class of tetrapod vertebrates. As we will discuss, studies of species with dissociated patterns have led to important insights into the evolutionary plasticity of mechanisms that control behavior.

Reptiles are diverse in the degree of temporal association between gametogenesis and mating behavior; they are even more diverse in other factors that can be profitably exploited in studies of the hormonal control of behavior. Many reptiles are either oviparous or viviparous. (We use these terms in the same sense as Guillette [1987] and consider ovoviviparity to be a form of viviparity. See Angelini and Ghiara [1984] and Shine [1985] for further discussion of the controversy regarding these terms.) Furthermore, even closely related forms may differ in this aspect of reproduction (Guillette, 1982, 1987). There are also considerable differences in the mode of sex determination (Bull, 1983; Crews and Bull, 1987). Many reptiles use genotypic sex determination, with some having female heterogamety, as do birds, and some having male heterogamety, as do mammals. Other reptiles lack sex chromosomes, and sex is determined by a response to environmental conditions (environmental sex determination). Finally, although most reptiles are gonochoristic, having separate males and females, several species consist of only females and reproduce by parthenogenesis (Cole, 1984; Darevsky et al., 1985; Crews and Moore, 1991). As will be developed in the following sections, this diversity provides a rich substrate of natural experiments in which to test prevailing theories about the role that hormones play in establishing sexual differences in behavior and morphology.

From a practical standpoint, reptiles also display a variety of additional advantages over the more frequently studied mammals and birds. Reptilian reproductive behaviors tend to be highly stereotyped, to vary little among individuals, and to be selected from relatively small repertoires. These factors combine to make reptilian behavior very easy to quantify. In the laboratory, many species can be maintained cheaply in small cages and will reliably exhibit the full natural

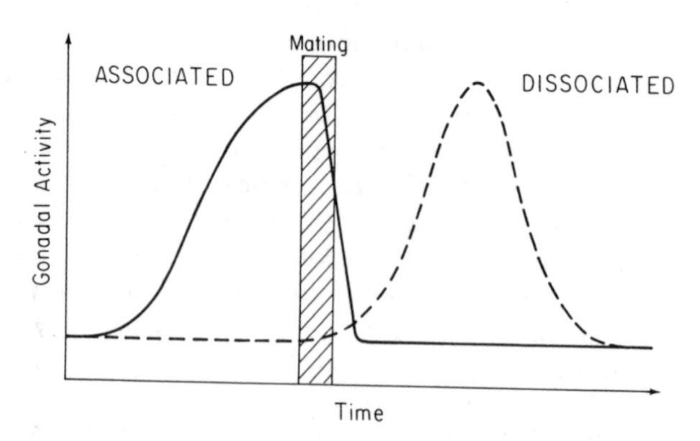

Fig. 3.2. Schematic depiction of the differences between the associated and dissociated reproductive tactics. Individuals exhibiting the associated tactic mate when gonadal activity is high; individuals exhibiting the dissociated tactic mate when gonadal activity is low. The latter usually rely on stored gametes produced at another time of the year. Gonadal activity refers either to the development of eggs and sperm or the occurrence of elevated sex steroid hormone levels. (From Crews, 1984.)

repertoire of behaviors. In the field, reptiles, especially lizards, are often abundant, may be sedentary, and often occupy small home ranges. They are frequently diurnal, and for social communication they rely heavily on sight and hearing, the same sense modalities as their human investigators. Although some reptiles do not share all of these characteristics, the commonalities are sufficient to make many reptiles attractive model organisms. This is particularly true if reptiles are compared to birds and mammals. The latter are often highly mobile, difficult to observe, relatively uncooperative in the laboratory, heavily reliant on sensory modalities not readily discernible by human observers (e.g., olfaction in mammals), and have large individual variation in behavior that makes it difficult to quantify.

However, certain limitations apply to the use of reptiles as model systems. First, only a few species show any kind of parental care. Obviously, reptiles are unlikely to teach us much about the hormonal control of parental care. This is unfortunate because the presence or absence of male parental care increasingly appears to be a major evolutionary determinant of the type of male behavior-controlling mechanisms (Wingfield and Moore, 1987; Wingfield et al., 1987; Moore, 1987b). Although this hypothesis cannot be readily tested within the class Reptilia, studies of reptiles can provide an important contrast to those on mammals and birds (Moore, 1987b). Second, the type of mating system may be an important evolutionary determinant of hormone control mechanisms (Wingfield and Moore, 1987). Because reptiles show little diversity in type of mating system, with most species being promiscuous or polygynous, this idea will be difficult to investigate in reptiles.

II. ORGANIZATION OF MALE REPRODUCTIVE BEHAVIOR

A. Overview

As described above, typical organizational actions of sex steroid hormones differ from typical activational actions in that the former are permanent and occur early in life, usually during a well-defined critical period. The study of organizational actions of hormones has been dominated by the *dominant/neutral model* of sexual differentiation. In this model, one sex, the neutral sex, develops spontaneously without the requirement of the intervention of other factors. The other sex, the dominant sex, develops only when the embryo is actively stimulated by the appropriate factors.

In all the mammals that have been studied so far, the female is the neutral sex and the male is the dominant sex. If an embryo is exposed to appropriate sex steroid hormones during the critical period, it will develop a behavioral phenotype typical of males; if it is not exposed to hormones, it will develop a behavioral phenotype typical of females. Male-typical and female-typical behaviors appear to be controlled by different regions of the brain (Dorner et al., 1968; Pfaff, 1980; Crews and Bull, 1987; Sachs and Meisel, 1988). The anterior hypothalamus-preoptic area (AH-POA) plays a central role in the control of male-typical behavior, whereas the ventromedial nucleus (VMN) is more important in the control of female-typical behavior. Stimulation of the embryo by appropriate sex steroid hormones causes the male-typical behavior-controlling regions of the AH-POA to develop further (masculinization) and the female-typical behavior controlling regions of the VMN to partially regress (defeminization). Lack of stimulation from hormones causes the opposite effect (demasculinization and feminization). Although they may be caused by a single cue, the processes of behavioral masculinization and feminization are independent of one another. For example, because male-typical and female-typical behaviors are controlled by different regions of the brain, a normal male embryo must be both actively masculinized and actively defeminized.

In the very few species of birds that have been studied, the male is the neutral sex and the female is the dominant sex (Adkins-Regan, 1987). Most workers to date have emphasized that a conspicuous parallel difference between mammals and birds is that males are the heterogametic sex in mammals, whereas females are the heterogametic sex in birds. This has led to the hypothesis that the heterogametic sex is always the dominant sex.

Although the dominant/neutral model and the hypothesis of a link between heterogametic sex and dominant sex have been around for many years, neither has been carefully explored in any vertebrate class besides mammals and birds. In particular, it is frequently overlooked that mammals and birds differ in many other ways besides the heterogametic sex and that these characteristics could account for differences in mode of sexual differentiation. As originally pointed out by Mason and Adkins (1976) and further emphasized by Crews and Bull (1987), reptiles provide abundant natural experiments in which to test ideas generated in studies of mammals and birds.

In viviparous mammals, the embryo develops with an intimate attachment to the maternal circulation through the placenta. To the extent that the hormones can penetrate the placental barrier, the developing embryo is constantly exposed to the maternal hormone environment. This has to place constraints on the type of sexual differentiation mechanisms that could evolve. Indeed, it has been argued that alpha-fetoprotein in the fetal circulation protects the embryo from maternal hormones. Several hypotheses could be derived to explain male dominance and female neutrality in this type of environment. For example, the exposure of all embryos to female (maternal) hormones generates an environment in which only a male-dominant mode, and not a female-dominant mode, of sexual differentiation is feasible.

In the oviparous birds, the embryo develops in an egg that is in less contact with the maternal environment than in placental vertebrates. Superficially, it might appear that the hormonal milieu to which the embryo is exposed is largely controlled by the secretions of the endocrine glands of the embryo. However, it has been suggested that maternal steroid hormones may be sequestered in the yolk and thereby affect subsequent development of the embryo (Crews et al., 1989; Brown and Bern, 1989). The only indirect evidence supporting the possibility that behavioral differentiation is affected by maternal hormones is the observation that the injection of estradiol into female Japanese quail (*Coturnix coturnix*) demasculinized their male offspring (E. Adkins-Regan and M. A. Ottinger, unpublished). This result, however, does not prove that the estradiol was sequestered in the yolk nor that endogenous levels of steroids are sufficient to elevate yolk steroids even if the injections had that effect. It is possible that the effect was pharmacological or that the influence was exerted directly on the very early embryo before the egg was shelled. The likelihood of these possibilities is affected by the absence of the influence of the maternal environment in oviparous species once the egg is shelled or laid. Perhaps the yolk contains enzymes or other defense mechanisms that could quickly deactivate maternal hormones. The possibility of maternal influence on behavioral development through yolk constituents has yet to be explored in reptiles. One study in fish suggests that steroids are present in the yolk throughout development and are used up as yolk stores are depleted (Brown and Bern, 1989).

Although many issues remain to be resolved, it appears the embryos of oviparous forms are not independent of maternal influences; however, they still have the potential of being more independent than are those of viviparous species. If we accept the assumption of greater independence in oviparous forms, the type of sexual differentiation mechanism would appear to be much less constrained. Indeed, Adkins-Regan and Ascenzi (1987) have recently provided evidence suggesting that the dominant/neutral model does not apply to birds. Instead, it appears that both feminization and masculinization of behavior may require active stimulation by steroid hormones. Such a complex mechanism of dual stimulation could only work in a situation in which the embryo develops in a self-contained environment. It is therefore possible that the dominant/neutral mode of sexual differentiation of mammals may have developed in response to the constraints imposed by continuous contact with the maternal circulation.

In addition to the possibility that evolution of viviparity required changes in the mechanisms of behavioral differentiation, the evolution of extensive parental regulation of embryonic temperature in birds, viviparous reptiles, and placental mammals may explain the absence of environmental sex determination in these groups. Thus, it appears that several changes in sex-determining mechanisms evolved as the degree of parent-embryo contact increased.

Reptiles provide opportunities to test many of these ideas. First, the existence of several independently evolved taxa of viviparous forms (Guillette, 1987) combined with considerable variation in which sex is heterogametic (Bull, 1983; Adkins-Regan, 1987; Crews and Bull, 1987) provides an opportunity to sort out the relative importance of each as well as opportunities to further our understanding of the basic mechanisms and evolution of sex differentiation. Second, the concept of dominant/neutral sex differentiation developed explicitly in the context of genetic sex determination. Many reptiles lack sex chromosomes, and further understanding of the mechanisms of sex differentiation in these species will advance our knowledge of the fundamental nature of this process. Third, reptiles permit the duality and independence of the masculinization and feminization process to be addressed in two separate situations: species with environmental sex determination and species that reproduce by parthenogenesis. Finally, studies of reptiles may contribute to a newly opened area of the mechanisms of differentiation. Although most vertebrates have different sex-specific phenotypes, it has recently been discovered that many vertebrates, including reptiles, also have multiple within-sex phenotypes (Gross, 1984; Mason and Crews, 1985; Crews and Bull, 1987). Very little is known about the physiological control of differentiation of within-sex phenotypes (Moore, 1991), which in some vertebrates can differ as dramatically in their behavior as do different sexes.

Despite the abundance of opportunities to expand our knowledge of behavioral differentiation by studying reptiles, very little is known about them. In the remainder of this section we will review the limited evidence and attempt to relate this to the views detailed above.

B. Case Studies

1. COMMENTS This discussion will document that we know little about the ontogeny of reptilian behavior. The evidence will be reviewed, but little synthesis can yet be offered (compare Adkins-Regan, 1987; Crews and Bull, 1987; Crews et al., 1987).

The mechanism by which sexual dimorphisms are established remains an important question in organizational theory (Kelly, 1988). There are at least two possibilities. One is that early actions of steroid hormones affect the organization of the brain substrates subserving behavior and produce different neural substrates in males and females. These organizational effects would cause male and female brains to respond differently to the pattern of secretion of adult hormones. The other possibility is that the brain substrates are identical in males and females and that behavioral differences in adults reflect only hormone differences. For example, both sexes may possess identical neural substrates, but testosterone activates male-typical behavior and estrogen activates female-typical behavior. Both these mechanisms have been documented in mammals and birds (Adkins-Regan, 1981, 1987; Adkins-Regan and Ascenzi, 1987; Baum, 1987; Kelly, 1988). They may be distinguished experimentally by castrating adults and treating them with hormonal levels typical of the opposite sex. If the animals show heterotypical sexual behavior (behavior typical of the opposite sex), then the dimorphism results from the difference in adult hormones. If the animals do not show heterotypical sexual behavior, the dimorphism results from the differences in the way the brain centers of the two sexes were organized during early development. However, it should be emphasized that these mechanisms are not mutually exclusive. Both may operate simultaneously, and each may operate separately on different behaviors in the same animal. Furthermore, masculinization or feminization of behaviors is relative. The phenomena simply reflect altered probabilities of expressing homotypical (behavior typical of the same sex) or heterotypical behaviors in each sex.

2. Associated Reproductive Patterns

The only species of reptile with an associated reproductive pattern for which anything is known about differentiation of sexual behavior is the green anole, *Anolis carolinensis*. Very young hatchling *A. carolinensis* exhibit adult display patterns (Cooper, 1971), but it is not known if these are hormone dependent. In gonadectomized adult *A. carolinensis*, both sexes respond to androgens with male-typical sexual behavior (Mason and Adkins, 1976; Adkins and Schlesinger, 1979). This observation suggests that sexual dimorphism in male-typical behaviors simply results from gonadal and hormonal differences rather than organizational effects in the embryo. However, it is much more difficult to induce female-typical behavior with estrogen in males than it is in females (Mason and Adkins, 1976; Adkins and Schlesinger, 1979). This observation suggests that sexual dimorphism in female-typical behavior is not the result of simple activational differences, but rather reflects some organizational effect. Although no further evidence is available, these observations jointly suggest that sex differences in the brain of *A. carolinensis* lie in the ventromedial nucleus (VMN), not the anterior hypothalamus-preoptic area (AH-POA).

3. Dissociated Reproductive Patterns

Somewhat more is known about the organization of behavior in reptiles with dissociated reproductive patterns because of a single study on the red-sided garter snake, *Thamnophis sirtalis parietalis* (Crews, 1985). In this species, sexual behavior of adult males is not activated by androgen; this raises the question of whether androgen plays any role at all in the development of this behavior. Superficially, the sex steroids would be expected to have an organizational role because (1) mating behavior is sexually dimorphic and (2) this sexual dimorphism is unlikely to result from adult hormonal differences because mating behavior is activated by mechanisms independent of steroid hormones.

Male *Thamnophis sirtalis parietalis* show spermatogenesis in late summer, sperm is stored over winter, and mating occurs immediately after emergence from hibernation. During mating, males have small gonads and low circulating levels of sex steroids (Camazine et al., 1980; Crews, 1980a, 1983a, b; Garstka et al., 1982; Crews et al., 1984; Krohmer et al., 1987). Mating behavior is triggered by increased temperature rather than by elevated levels of sex steroid hormones.

Neither castration nor androgen treatment of adult males has any effect on their expression of mating behavior (Camazine et al., 1980; Crews et al., 1984). However, newborn males and females treated with testosterone will court sexually attractive adult females, whereas intact and castrated newborns will not (Table 3.2; Crews, 1985). Whenever these newborns (intacts, castrates, and testosterone-treated males and females) are subsequently exposed to low temperatures for at least 7 weeks and then returned to warm temperatures (a treatment that mimics hibernation and that activates courtship behavior of adult males), only males previously treated with androgen will court females (Table 3.2).

These results are difficult to reconcile with prevailing concepts of sexual differentiation. Because prior treatment of castrated juvenile males with testosterone will cause them to show mating behavior after exposure to increased temperature, their behavior appears to be organized by sex steroid hormones but not activated by them. Behaviors that require hormonal organization but not hormonal activation occur in a few species of other vertebrate classes (Goy and McEwen, 1980; Crews, 1985; Arnold and Breedlove, 1985; Adkins-Regan, 1987; Crews and Bull, 1987) and are thought to be quite rare. This organizational effect of androgen also appears to be unusual in that there is no well-defined perinatal critical period (Arnold and Breedlove, 1985; Adkins-Regan, 1987). Testosterone treatments either before the first hibernation or before the second hibernation were equally effective in eliciting courtship behavior following hibernation (Table 3.2).

Table 3.2 Effect of hormone treatment on neonate garter snakes (*Thamnophis sirtalis parietalis*) and effects of subsequent hormone treatments and hibernations

Treatment as neonates	% Courtship as neonates	% Courtship after first hibernation	Treatment as yearlings	% Courtship during yearling treatment	% Courtship after second hibernation
Male					
Sham	2	0	Testosterone	22	79
Castrate	5	8	Testosterone	61	79
Castrate + T	38	37	Testosterone	31	86
Female					
Sham	1	0	Testosterone	48	0
Castrate + T	29	–	–	–	–
Intact + T	–	0	Testosterone	12	0

Source: Data from Crews, 1985.
Note: Successive treatments and results from a single group of animals are presented on a single line and are arranged in temporal order of occurrence from left to right in the table. Percent courtship indicates the number of total courtship tests on which courtship was observed. A dash (–) indicates that this experiment was not performed.

Thus, the behavioral response of males to increasing temperatures requires organization by early exposure to sex steroid hormones. However, this hormone treatment cannot presently account for the sexual dimorphism of mating behaviors. Contrary to the condition in yearling males, ovariectomy and treatment with androgen before hibernation does not induce male-typical courtship behavior in yearling females when they later emerge from hibernation (Table 3.2; Crews, 1985). Earlier events during development must have sensitized males but not females to this androgen regimen. Furthermore, both males and females express courtship behavior when given testosterone as newborns or when given testosterone after emerging from hibernation as yearlings (Table 3.2). It is not clear what causes both sexes to lose their responsiveness to exogenous androgen as adults. However, it is tempting to use delayed loss of sensitivity to androgen during ontogeny as an argument that temperature activation of sexual behaviors in this species may have evolved subsequent to steroid activation.

Thus, studies to date have demonstrated a limited role for sex steroid hormones in organizing sexual behavior in a species that does not activate this behavior with sex steroid hormones. However, little is known about the mechanisms that initially establish the sexual dimorphism in behavior. In *Thamnophis sirtalis parietalis,* it does seem likely that the organizational effects of steroids represent evolutionarily conserved traits, in short, phylogenetic constraints. In contrast, the lack of activational effects of sex steroids is probably a newly evolved response to environmental constraints (see discussion below).

4. ENVIRONMENTAL SEX DETERMINATION

As discussed above, the development of male-typical and female-typical patterns of behavior (masculinization and feminization) are currently viewed as independent processes that are caused by differences in sex-specific hormones produced from the genetically induced differentiation of the gonads. Thus, sexual differentiation is viewed as a causal cascade starting at the sex chromosomes, leading through the inducer of gonadal differentiation, the differentiation of the gonads, differential hormone production of maturing testes and ovaries, and action of these gonadal hormones on the brain to influence its development in a male-typical or female-typical pattern.

However, this chain of events can only be triggered in species with genetic sex determination. Species with environmental sex determination must employ at least a different initial trigger and may use a substantially different mechanism. Thus, these species permit one to address a minimum of three important questions relevant to general theory of sexual differentiation: (1) To what extent are the series of causal events obligately linked? (2) How independent are the processes of masculinization and feminization from each other? (3) Are sex steroids always involved in sexual differentiation?

Recent work with the leopard gecko (*Eublepharis macularius*) is beginning to provide some preliminary answers to these questions (Crews and Bull, 1987; Gutzke and Crews, 1988). In this species, sexual differentiation is controlled by incubation temperature. At high incubation temperatures, most lizards develop as gonadal males. At low incubation temperatures, all lizards develop as gonadal females. If behavioral differentiation is a direct consequence of gonadal differentiation, then the behavioral differentiation of these animals should be influenced only by the state of their gonads and not by incubation temperature. Alternatively, if some of the behavioral differentiation events are independent of gonadal differentiation, then incubation temperature could affect behavioral differentiation independent of gonadal state. The results provided some support for the latter hypothesis. Although males consistently exhibit homotypic sexual behavior regardless of incubation temperature, females from eggs incubated at warmer temperatures had apparently normal ovaries, but showed increasing masculinization of sexual behaviors, (Table 3.3). High-temperature males and females were more aggressive, whereas low-temperature males and females were more submissive. Thus differentiation of aggressive behavior also appears to have been affected

Table 3.3 Effect of incubation temperature on psychosexual differentiation in the leopard gecko (*Eublepharis macularius*)

Sex of test animal	FEMALE						MALE			
Sex of stimulus animal	Male (%)			Female (%)			Male (%)		Female (%)	
Incubation temperature (°C)	26	29	32	26	29	32	29	32	29	32
Only homotypic response	100	18	0	100	91	67	90	100	100	100
Both responses	0	73	17	0	9	33	10	0	0	0
Only heterotypic response	0	9	83	0	0	0	0	0	0	0

Source: Data from Gutzke and Crews, 1988.
Note: Percentages indicate the fraction of animals in each treatment group exhibiting that response. Note especially the increasing heterotypic responses of warmer-temperature females to both stimulus males and stimulus females.

by incubation temperature and may be at least partially independent of gonadal sex.

However, several important questions remain unanswered that weaken this study. First, the study did not control adult hormone levels and, in fact, adult warm-temperature females had higher androgen and lower estrogen levels than did adult cold-temperature females (Gutzke and Crews, 1988). Similarly, high-temperature males had higher androgen levels. The differences observed therefore may have been downstream organizational or activational consequences of temperature effects on the hormone production of the adult or the embryonic gonads. Experiments with hormone-treated castrates are needed before we can draw definitive conclusions about independent effects on the brain. Second, the hormone profiles of the developing embryos are unknown. Third, the study simply recorded whether the animals showed homotypic or heterotypic sexual behaviors (Table 3.3) without carefully quantifying or defining these. Thus, from the results reported it is difficult to parcel out individual masculinizing and feminizing effects. Indeed, female sexual behavior was apparently only masculinized in the way that females responded to courting males, not in the way that females responded to receptive females. The absence of more detailed quantitative information on the behaviors exhibited makes it impossible to confirm the interpretation of the authors that this behavior was heterotypical.

Although much more work is clearly necessary to settle the issues raised above, the preliminary results from this study suggest (1) that sexual differentiation is not necessarily an obligately linked consequence of gonadal differentiation and (2) that masculinization and feminization are independent processes. Thus, these studies of a species with environmental sex determination support one paradigm derived from work on mammals (the independence of masculinization and feminization) but not the other (the obligate linkage of all sexual differentiation events downstream of gonadal differentiation). This is an excellent example of how studies of novel species with diverse reproductive strategies can improve our understanding of the mechanisms underlying sexual behavior.

5. Parthenogens

Recent work on parthenogenetic, all-female lizards of species of *Cnemidophorus* has provided information supporting the concept that male-typical and female-typical behaviors are controlled by independent brain regions and develop independently. In several species of parthenogenetic *Cnemidophorus,* individuals display both male-typical and female-typical courtship and copulatory behaviors (Crews and Fitzgerald, 1980; Gustafson and Crews, 1981; Crews et al., 1983; Moore et al., 1985a; Crews et al., 1986; Crews and Moore, 1991). Patterns of adult gonadal hormone secretion in the parthenogens are similar to those of gonochoristic females (Moore et al., 1985b; Moore and Crews, 1986). Thus, these individuals have been simultaneously masculinized and feminized by naturally occurring mechanisms. Unfortunately, our current knowledge of the mechanisms of sexual differentiation in these species is limited to a rudimentary understanding of the effects of sex steroid hormones on morphology (Crews et al., 1983; Billy and Crews, 1986; Crews and Bull, 1987).

6. Multiple Within-Sex Phenotypes

Several species of reptiles show multiple within-sex phenotypes. Most commonly these are males that exhibit different reproductive strategies. In garter snakes, some males release the female attractiveness pheromone causing them to be courted by other males (Mason and Crews, 1985). These female mimics, called *she-males,* apparently generate general confusion in other male members of the mating aggregation and increase the probability that the she-male will successfully mate with the female that is being courted in the aggregation. Otherwise, the mating behavior of she-males is identical to that of other males.

Adult she-males have much higher testosterone levels than do other males, but their estrogen levels are identical. It has been proposed that pheromone production, which is estrogen-dependent in females (Garstka et al., 1982), is stimulated in she-males by aromatization of this testosterone to estrogen at target sites that produce this pheromone (Mason et al., 1987). Although this hypothesis has not been tested, it is clear that she-males are morphologically demasculinized because the skin of neonatal males and females transports exogenously induced pheromone, whereas the skin of other adult males does not do so. This demasculinization might occur as a result of intrauterine position effects (Mason and Crews, 1985). In some rodents, females that develop between two males are more masculinized and defeminized than females that develop between a male and a female or between two females (Vom Saal, 1983). It is an intriguing, but untested, possibility that she-male garter snakes are males that develop between two females. If this hypothesis proves correct it would also be significant in providing evidence that females might be the dominant sex in garter snakes. Because female garter snakes are heterogametic, this outcome is consistent with predictions of the hypothesis that heterogametic sex determines dominance, but not with the hypothesis that viviparity causes males to be the dominant sex.

Another situation of multiple male phenotypes occurs in tree lizards, *Urosaurus ornatus*. There are different male types in this species that differ in both morphology and behavior. This species is polymorphic for dewlap color, with at least six distinct types occurring in some populations (Hover, 1985; Thompson and Moore, 1991a). In the two populations that have been studied, males have dewlaps that consist of orange backgrounds with variable-sized central blue spots. In the laboratory, males with large blue spots consistently dominate males with smaller spots (Hover, 1985; Thompson and Moore, 1987, 1991b; Moore and Thompson, 1990). In the field, males with small blue spots are less sedentary than males with larger spots and are much less active in displaying and patrolling behaviors (Thompson and Moore, 1987). This suggests that males with less blue are nonterritorial floater males, whereas males with more blue are territorial (compare Rand, 1987). Although males with more blue are seen courting females significantly more often than males with less blue, both types of males have been observed courting females. This suggests that males with more blue are practicing a different reproductive tactic compared to males with less blue. However, there are no differences in the circulating levels of testosterone and corticosterone in adult males of both types throughout the breeding season (Thompson and Moore, 1987; Moore and Thompson, 1990; Moore, 1991). Because these alternative male reproductive behaviors are expressed only during the breeding season and only by males, it seems likely that they are sex steroid hormone–dependent (Moore and Marler, 1987; Kelly, 1988). As adult steroid hormone levels are identical in males of both morphs, it seems clear that the different behaviors shown by these males result from organizational, not from activational, differences (Moore and Thompson, 1990; Moore, 1991). However, the mechanisms involved remain to be identified.

III. ACTIVATION OF MALE REPRODUCTIVE BEHAVIOR

A. Overview

The activation of sexual behavior in male vertebrates is known mainly from laboratory studies of a few mammalian species. In these, the activation of sexual behavior is largely dependent on testicular androgens (Feder, 1984; Sachs and Meisel, 1988). This narrow approach has underestimated much diversity in the control of sexual behaviors (Crews, 1984; Crews and Moore, 1986); however, the wealth of detail available about control of sexual behaviors in mammals provides convenient points of reference.

The mammals that have been studied most include rats, mice, hamsters, and sheep. In these, sexual behavior of males follows testicular maturation and is temporally associated with high concentrations of testicular androgens in the circulation. Castration abolishes sexual behaviors, whereas exogenous testosterone restores them (Feder, 1984; Sachs and Meisel, 1988).

Aromatization of testosterone to estrogens plays a role in activating components of male sexual behavior in Japanese quail (*Coturnix coturnix*), rats, mice, and red deer (*Cervus elaphus*) (Naftolin et al., 1975; Adkins-Regan, 1981; Feder, 1984; but see Butera and Czaja, 1989, and Kaplan and McGinnis, 1989, for recent studies that may force a reexamination of the aromatization hypothesis). Aromatase activity has been detected in the brains of a wide variety of vertebrates, including several species of lizards, snakes, and turtles (Callard et al., 1978). Thus, it is important to consider the possibility that aromatization of androgens may affect both organization and activation of male sexual behaviors in reptiles. Studies of mammals have also established quite clearly that testicular steroids are not the only factors controlling sexual behaviors. Numerous neurotransmitters and neurohormones affect the expression of male sexual behavior (Sachs and Meisel, 1988). Unfortunately, little is known about these compounds outside of mammals (but see F. L. Moore, 1987a, b).

Two major reproductive strategies are exhibited by male reptiles: dissociated and associated patterns (see Fig. 3.2). Males that undergo prenuptial gametogenesis and have high circulating concentrations of androgens during mating are considered associated breeders. Males exhibiting an associated pattern most closely resemble the generalized models developed from mammalian work. The best-studied example of a reptile with an associated pattern is *Anolis carolinensis* (see Crews, 1979a, 1980a, b, 1987). A dissociated breeder mates during periods characterized by regressed testes and low circulating concentrations of androgens; testicular recrudescence and androgen synthesis take place following mating (postnuptial gametogenesis). The best-studied example of a reptile with a dissociated pattern is the snake, *Thamnophis sirtalis parietalis* (Garstka et al., 1982; Crews, 1982, 1983a, b, 1984). However, these species are strongly associated and strongly dissociated breeders and probably represent extremes on a continuum of breeding strategies. Some species undergo almost constant testicular activity or have two waves of spermatogenesis during a single year (Licht, 1984).

B. Case Studies

1. Lizards with Associated Reproductive Patterns Males of most species of lizards, especially those inhabiting the temperate zone, experience seasonal testicular recrudescence and regression. In most of these species, the sexual behavior of the male lizards is closely associated with high circulating concentrations of androgen and increased testicular androgen secretion (Crews, 1975b, 1977a, b, 1979a, b, c, 1980b, 1987; Crews and Greenberg, 1981; Stamps and Crews, 1976; Licht, 1984).

Males of these species exhibit an associated reproductive pattern that is similar to that of the mammalian species that have been studied thus far. This ecological similarity should be a powerful predictor of similarities in behavior-controlling mechanisms (Crews and Moore, 1986; Moore, 1987b; Moore and Marler, 1988). Consequently, male lizards with associated reproductive patterns should have behavior-controlling mechanisms that are similar to those of the mammals that have been studied. Early studies on *Anolis carolinensis* in the laboratories of G. K. Noble (Noble and Bradley, 1933; Noble and Greenberg, 1940, 1941a, b; Greenberg and Noble, 1944) and L. T. Evans (Evans 1935 a, b, c; 1936 a, b, c, d; 1937; 1938a, b; 1953, 1957) established in this lizard that male sexual behavior is abolished by castration and restored by exogenous administration of androgens. Thus, it had been clear quite early that control of sexual behavior in this lizard with an associated pattern closely resembles that of mammals with a similar pattern, although it remained for Crews (1984) to point out the significance of this similarity.

Many more recent experiments on lizards with associated strategies are consistent with this expectation (Table 3.4). Castration abolishes sexual behaviors in male *Anolis carolinensis* (Crews, 1974a, b, 1975a, b; Crews et al., 1974; Mason and Adkins, 1976; Crews et al., 1978), *Anolis sagrei* (Tokarz, 1986), *Cnemidophorus inornatus* (Lindzey and Crews, 1986), *Sceloporus jarrovi* (see Fig. 3.1; Moore, 1987a), *Eumeces laticeps* (Cooper et al., 1987) and *Uta stansburiana* (Ferguson,

Table 3.4 Effects of exogenous steroids and antisteroids on activation of male sexual behavior in various taxa of reptiles

	Hormone				
	T	DHT	E	P	CPA
Anolis	+ +	+/0	0/+	?	−
Sceloporus jarrovi	+ +	?	?	?	?
Cnemidophorus	+ +	+	?	+	?
Sternotherus odoratus	0/+	?	?	?	?
Thamnophis sirtalis parietalis	0	0	0	0	0

Note: Stimulatory, inhibitory, and neutral effects are designated by +, −, and 0 respectively. A ? denotes lack of data. Abbreviations for hormones are T, testosterone; DHT, dihydrotestosterone; E, estradiol; P, progesterone; CPA, cyproterone acetate (an antiandrogen).

1966). Also, systemic administration of the antiandrogen cyproterone acetate (CPA) inhibits sexual behaviors in intact male *A. sagrei* (Tokarz, 1987), further supporting a stimulatory role for androgens. Other complementary studies show that exogenous testosterone (and in some cases dihydrotestosterone as well) restores sexual behaviors in castrated male *Anolis carolinensis* (Crews, 1973, 1974a, b; Mason and Adkins, 1976; Crews et al., 1978; Adkins and Schlesinger, 1979; Crews, 1979a), *Uta stansburiana* (Ferguson, 1966), *Sphenomorphus kosciuskoi* (Done and Heatwole, 1977), *Cnemidophorus inornatus* (Lindzey and Crews, 1986), *Eumeces laticeps* (Cooper et al., 1987), and *Sceloporus jarrovi* (see Fig. 3.1; Moore, 1988).

Tests of the requirement for aromatization of androgen to estrogen in the brain as part of the mechanisms activating male reproductive behavior have produced conflicting results. Early work by Noble and Greenberg (1940, 1941a, b) on *Anolis carolinensis* showed that testosterone treatments of either immature males or adult females stimulates both male- and female-like sexual behaviors. This result suggests the possibility of conversion of androgen to estrogen, although Noble and Greenberg (1940) themselves only refer to the "estrogenic effects of testosterone." However, the purity of their hormone preparations remains uncertain, and their estrogen treatments unfortunately produced too many toxic side effects. More recent attempts with more controlled doses to avoid the toxic effects of estrogen demonstrate that systemic administration of estradiol alone is ineffective in restoring sexual behavior in male *Anolis* spp. (Mason and Adkins, 1976; Crews et al., 1978; Adkins and Schlesinger, 1979; Tokarz, 1986), although Crews et al. (1978) report a synergistic effect between systemic administrations of estrogen and dihydrotestosterone. However, intracranial implants of estrogen in castrated male *A. carolinensis* do restore sexual behaviors (Crews and Morgentaler, 1979). Several explanations may account for the differences in effects of systemic and intracranial hormone treatments. First, intracranial implants deliver a relatively large dose of hormone to a small area of the brain resulting in a very localized high concentration. Second, sex steroid–binding proteins (SBP) present in the plasma have moderate affinity but high capacity for androgens and estrogens (Wingfield, 1980; Callard and Callard, 1987). These proteins may sequester much of the systemically administered hormones. This problem of sequestration by plasma proteins may be especially acute for the estrogen used in these studies because it is too toxic to reptiles to be administered in high doses. Therefore, because intracranial implants may represent the most sensitive test, their effectiveness in inducing male sexual behavior is significant.

The behavioral effects of systemic treatments with dihydrotestosterone (DHT) are less clear than those with estrogen. In one study (Crews et al., 1978), systemic administration of DHT to castrated male *Anolis carolinensis* did not restore sexual behaviors. Another study in the same laboratory, however, restored sexual behaviors with intracranial implants of DHT (Morgentaler and Crews, 1978). As in the studies with estrogen, the disparity is probably due to the increased effectiveness of intracranial hormone implants (see preceding paragraph). Other studies report that systemic DHT treatments are effective in activating sexual behaviors in castrated male *A. carolinensis* (Adkins and Schlesinger, 1979), *A. sagrei* (Tokarz, 1986), and *Cnemidophorus inornatus* (Lindzey and Crews, 1986). DHT also effectively promotes malelike pseudocopulatory behavior in females of the parthenogenetic *C. uniparens* (Gustafson and Crews, 1981). These conflicting reports are best explained by different experimental designs and, possibly, interspecific differences. Crews et al. (1978) applied hormone treatment 2 weeks after castration, whereas it was applied 1 week after castration in the other studies of *Anolis*. In addition, Crews et al. (1978) began behavioral testing the day following the start of hormone treatment, whereas the researchers in the other *Anolis* studies only started behavioral testing 2 weeks after the start of hormone treatments. Increased length of time after castration and decreased length of DHT treatments would be expected to reduce activational effects of DHT.

The available data clearly indicate that DHT, a nonaromatizable androgen, activates male sexual behaviors under the appropriate conditions (Table 3.4). This argues against a significant role for aromatization of testosterone. Nonetheless, the effectiveness of estrogen whenever it is administered intracranially by itself or systemically with DHT suggests that aromatization may play some role in activation of male sexual behaviors. Further insight into the importance of androgen aromatization in control of male sexual behavior would be gained by examining in a single species (1) the effectiveness of DHT, estrogen, and testosterone in activating sexual behaviors if applied alone and in combination and (2) the ability of inhibitors of aromatase and 5-α-reductase to suppress the behavioral effects of testosterone.

Although activation of sexual behavior of male lizards appears to be largely androgen-dependent, activation may also be influenced independent of testicular androgens (Moore and Marler, 1988; Lindzey and Crews, 1988a). Other factors such as neurohormones and neurotransmitters may play a more proximate role in the immediate activation of sexual behaviors (F. L. Moore, 1987a, b). The presence of other proximate factors is suggested by the fact that (1) sexual behaviors can be expressed independently of testicular androgens and (2) a certain number of males fail to respond to either endogenous or exogenous androgens. These results imply that androgens are not always sufficient by themselves to activate sexual behaviors.

In castrated male *Cnemidophorus inornatus*, sexual behavior is restored partially by exposing castrated males to cycling conspecific females; however, full restoration requires androgen treatment (Lindzey and Crews, 1988a). Castrated male *Anolis carolinensis* maintain sexual behavior if they are left in their home cages (Crews et al., 1974). Similarly, a field study shows that free-living male *Sceloporus jarrovi* do not completely cease sexual behaviors if they are castrated and left on their territories (see Fig. 3.1; Moore, 1987a). Demonstration of such effects in both laboratory and field studies strongly supports the hypothesis that control of sexual behavior involves androgen-independent mechanisms even in species that otherwise appear to have androgen-dependent behaviors. It is becoming increasingly apparent in nearly all species of animals that the social and physical environment affects behaviors as strongly as do sex steroid hormones (Moore and Marler, 1988). More species need to be studied in nature to assess fully the relative importance of hormonal and nonhormonal factors in the regulation of reproductive behavior.

The plasticity of behavioral control mechanisms is also demonstrated by the stimulatory effects of progesterone on sexual behavior in male *Cnemidophorus inornatus* (Lindzey and Crews, 1986). Treatment with either exogenous androgens or exogenous progestins maintains sexual behavior in intact male *C. inornatus* and restores sexual behavior in castrated males (Lindzey and Crews, 1986, 1988b). This finding is curious because progesterone inhibits sexual behaviors in all other male vertebrates examined. Further studies with synthetic progestins indicate that progesterone acts as a progestin, rather than through conversion to an androgen. These studies show that the synthetic progestin R5020, which cannot be converted to an androgen, stimulates sexual behaviors. As nothing similar has been found in any other vertebrate, exploitation of the species differences in response to a class of steroid hormones could help us gain insights into mechanisms that control specificity of hormone-behavior relationships.

2. Dissociated Reproductive Patterns

Species with dissociated reproductive patterns permit an important test of the hypothesis that ecological constraints and reproductive patterns are more important than phylogeny in determining the mechanisms controlling behavior. If this hypothesis is correct, then the behavior-controlling mechanisms of such species should differ from those of both the closely related lizards and the distantly related mammals with associated reproductive patterns. Alternatively, if phylogeny is more important than ecology, the behavior-controlling mechanisms of reptiles with dissociated reproductive patterns should be more similar to those of reptiles with associated reproductive patterns and less similar to those of mammals. Recent work on the red-sided garter snake (*Thamnophis sirtalis parietalis*) has confirmed that this strongly dissociated species does not rely on the androgen-dependent mechanisms typical of reptiles, birds, and mammals with associated reproductive patterns.

Male *Thamnophis sirtalis parietalis* express intense sexual behavior well before seasonal spermatogenesis and androgen synthesis begin; circulating concentrations of androgens are very low at this time (Fig. 3.3; Garstka et al., 1982; Crews, 1983a, b). An early experiment by Crews (1976) suggested that sexual behavior in male garter snakes (*T. s. sirtalis*) is androgen dependent, but this study did not include the appropriate control groups. Males received androgen implants immediately after emergence from hibernation and soon they began to court. Three weeks later the implants were removed, and the courtship soon ceased. However, no unimplanted animals were used as controls. More recent studies (see below) indicate that males in this study had been treated at precisely the time when they would normally begin and cease courting, even in the absence of androgen. It is now very clear that, in keeping with the temporal separation of high circulating concentrations of androgens and high levels of sexual behavior, courtship in adult male *T. s. parietalis* shows no dependency on activational effects of androgens. Neither long-term nor short-term castration causes adult males to cease courtship and neither systemic nor brain implants of androgen stimulate it (Camazine et al., 1980; Garstka et al., 1982; Crews, 1983a, b; Crews et al., 1984; Friedman and Crews, 1985b). Consequently, the androgen-independent sexual be-

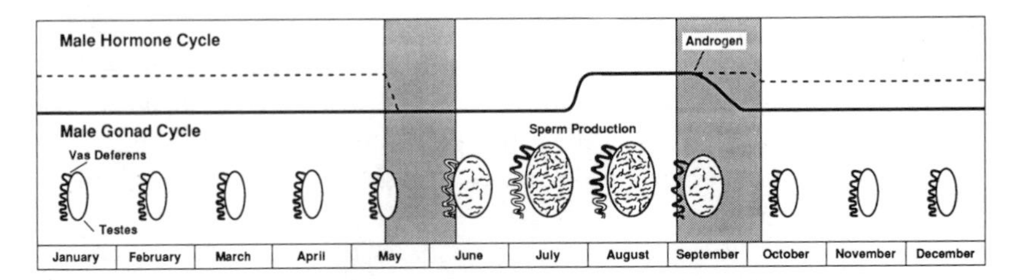

Fig. 3.3. *Thamnophis sirtalis parietalis.* Illustration of the dissociated reproductive cycle of males. The snakes are in their dens from January through early May. The male's testes are small, androgen levels are low, and the vas deferens is filled with sperm stored from the previous summer's spermatogenesis. The snakes emerge and mate in late May while the male's gonads are still small and circulating levels of testosterone are minimal. During the post-mating summer phase the male's testes enlarge, androgen levels rise, and sperm are produced to use in the following season's mating period.

havior is best explained as an adaptation to the environment in which this snake lives. The geographic range of *T. s. parietalis* is characterized by extremely short warm periods, lasting only 3 to 4 months. If males waited for testicular recrudescence and steroidogenesis before initiating sexual behavior, baby snakes would be born close to the onset of hibernation and their survival would be unlikely. It seems reasonable to hypothesize that male *T. s. parietalis* evolved a dissociated reproductive pattern and consequent androgen independence of mating behavior in response to these environmentally imposed time constraints. This dissociated strategy appears to have been evolutionarily retained even in those species that have subsequently invaded more temperate and tropical climates, such as Mexican *T. melanogaster* (Garstka and Crews, 1982).

Sexual behaviors in male *Thamnophis sirtalis parietalis* can only be stimulated by prolonged exposure to cold temperatures (hibernation); males that do not hibernate fail to court females the following spring. It appears that these temperature effects are mediated at least partially via the pineal gland. Males that are pinealectomized before hibernation fail to court on their return to warm temperatures (Nelson et al., 1987; Crews et al., 1988). It is possible that the pineal directly regulates sexual behavior in male red-sided garter snakes. However, because the pineal is very important in time measurement (Underwood, 1988), it seems more likely that the pinealectomized snakes are simply incapable of measuring the duration of hibernation. The critical experiment would involve pinealectomizing males immediately after their return to warm temperatures and determining the effect of this on subsequent courtship behavior. This experiment has not been performed.

These studies of activation of male sexual behavior in red-sided garter snakes have been very important in demonstrating the possible diversity of behavior-controlling mechanisms (Crews, 1984; Crews and Moore, 1986). Furthermore, these studies resulted in more than merely an exercise in understanding an exception to what had been proposed as a universal rule. Rather, they emphasize the plurality and complexity of behavior-controlling mechanisms in all organisms (e.g., Moore and Marler, 1988) and hopefully will serve to move us away from the exclusive reliance on the paradigm that male sexual behavior is always controlled by androgen-dependent mechanisms.

3. Turtles: Intermediate between Associated and Dissociated Reproductive Patterns?

Turtles have not proven to be very amenable to experimental studies of behavioral endocrinology, and only limited information is available about them. Mating behavior is difficult to observe in the field and difficult to reliably elicit in the laboratory. Detailed data on annual testicular cycles are available only for four species of turtles: *Chelonia mydas, Chrysemys picta, Sternotherus odoratus,* and *Trionyx sinensis* (Licht, 1982). There is little detailed information on seasonal patterns of mating activity of turtles in the wild. Both spring and fall matings have been reported in many turtles, but it has not been possible to determine accurately the relationship of sexual activity to patterns of gonadal activity and circulating concentrations of sex steroid hormones. Thus, it is still unclear whether the reproductive patterns of temperate turtles are associated or dissociated. It seems increasingly likely that turtles exhibit an intermediate pattern, but it is unfortunate that so little is known about them because they might represent an evolutionary intermediate between strongly associated and strongly dissociated breeding strategies. In this case, they would be expected to have some kind of intermediate behavior-controlling mechanism.

Sternotherus odoratus is the only species of turtle in which manipulations have been performed to examine steroid dependency of male sexual behaviors (Mendonça, 1987). Male *S. odoratus* exhibit sexual behaviors during the spring and the fall. In this study, the highest levels of courtship were observed during the fall; the testes then are enlarged, and circulating concentrations of androgens are maximal. This seems to demonstrate both associated and dissociated tendencies. Unfortunately, the results of the manipulative experiments have been inconsistent and difficult to interpret (Table 3.5). Because the animals were studied in captivity, it was difficult to elicit behavior in controls and the reproductive condition of stimulus females was not controlled. Nevertheless, the evidence partially supports the concept that

Table 3.5 Effect of season, castration, and testosterone replacement on expression of sexual behavior in male musk turtles (*Sternotherus odoratus*)

			Behavior classes (number of males)				
Month	Treatment	T Level (ng/ml)	1	2	3	4	5
April	Fresh	18.9			1	1	2
	Captive	2.4			1	2	3
	CAS + EM	0.3		1		4	3
	CAS + T	25.0			1	4	5
June	Fresh	6.6				1	4
	Captive	3.5				2	4
	CAS + EM	0.8					6
	CAS + T	41.4				3	5
Aug.	Fresh	28.7	5	1	3		
	Captive	8.4				1	4
	CAS + EM	0.2			4	2	
	CAS + T	41.6		2	2	5	
Sept.	Fresh	38.0	2	6	1	2	
	Captive	13.4				5	
	CAS + EM	0.4				1	4
	CAS + T	30.8		1	3	1	
Oct.	Fresh	68.5	3	1	1	1	
	Captive	10.7		1	1	4	
	CAS + EM	0.7					5
	CAS + T	17.9		1	3	1	

Source: Data from Mendonça, 1987.
Note: Table indicates number of males in each treatment that fall into each of five behavior classes. Note lack of consistent effect of castration or testosterone replacement on male sexual behavior. Abbreviations are CAS, castrated; T, testosterone implant; and EM, empty implant. Behavior classes vary from 1, indicating no courtship behavior, to 5, indicating mounting and copulating.

male courtship behavior is stimulated both by androgens and short days under some conditions. Under other conditions androgen proved to be ineffective. However, as controls did not court either, this result could have been due to stress or unreceptive females (Table 3.5). It seems possible that this species initiates courtship through either androgen-dependent or androgen-independent mechanisms, depending on the season (compare Moore and Marler, 1987, 1988; Moore, 1988, for similar results with lizard territorial behavior). There was also some evidence that male sexual behaviors are expressed in the absence of androgen as long as the animals have been previously exposed to elevated levels of androgens. These results are reminiscent of the studies on neonatal snakes (Crews, 1985). Indeed, the reproductive pattern of *S. odoratus* appears to be intermediate between the associated pattern of species like *Anolis carolinensis* and the dissociated reproductive pattern of species like *Thamnophis sirtalis parietalis*. This makes it less than surprising that the mechanisms controlling mating behavior also are intermediate and difficult to define. The facultative mechanism described above might be advantageous in species that occupy a large geographic range and encounter wide variation in environmental conditions.

IV. NEURAL SUBSTRATES OF MALE SEXUAL BEHAVIOR

A. Overview

The brain is responsible for receiving, transmitting, and processing environmental cues essential to successful reproduction. In addition, the brain serves an integrative function by making decisions regarding appropriate motor behaviors. Neural structures of major importance in controlling male sexual behavior include the hypothalamus, preoptic area, septum, nucleus sphericus, hippocampus, and amygdala. These regions receive a wide range of sensory input and function largely as processing centers. It is certainly no coincidence that these regions of the brain also appear to be the most important sites of action of sex steroid hormones in regulating behavior.

A complete understanding of the action of sex steroid hormones on the neural substrates of male behavior can only follow an understanding of (1) the anatomical structures in which sex steroid hormones bind in the central nervous system, (2) the specific behaviors controlled by each specific anatomical structure, and (3) the neurophysiological changes that are caused by sex steroid hormone binding and the way in which these changes ultimately cause the expression of a particular behavior. For male reptiles, some progress has been made in the first two areas, although almost exclusively with *Anolis carolinensis* and species of *Thamnophis*. Very little, if anything, is known about the third.

Several lines of evidence are needed to establish the involvement of particular neural structures in the neuroendocrine regulation of behavior. First, autoradiography of brain slices from animals that have been injected with radioactive hormones can identify sites in the brain in which sex steroid hormones are concentrated. Second, electrical stimulation or lesions of these sites can help identify the particular behaviors that are controlled by them. Third, microimplants of steroid hormones into localized regions of the brain can identify the hormone and its site specificity of action. Fourth, neurophysiological studies of these brain regions can help identify the changes caused by hormone binding. We will review each of the first three areas for reptiles. However, we will not include the fourth area because virtually nothing is known about this topic in reptiles.

B. Case Studies

1. AUTORADIOGRAPHY STUDIES Three studies have examined the uptake of radioactive hormones in the brain of reptiles: two for *Anolis carolinensis* (Martinez-Vargas et al., 1978; Morrell et al., 1979) and one for an unspecified species of *Thamnophis* (Halpern et al., 1982). These studies demonstrate that the spatial pattern of steroid hormone uptake in the brains of these reptiles is very similar to that in the brains of other vertebrates that have been studied (Kim et al., 1978; Morrell and Pfaff, 1978). As in other vertebrates, the patterns also barely differ between the sexes or for the various hormones studied. These studies reinforce the general conclusion that the patterns of steroid hormone uptake in the vertebrate brain are highly conserved evolutionarily.

The two studies of *Anolis carolinensis* (Martinez-Vargas et al., 1978; Morrell et al., 1979) show that both estrogen and androgen are concentrated by cells in the anterior hypothalamus, medial preoptic area, ventromedial nucleus, periventricular nucleus, tuberal hypothalamus, anterior pituitary, amygdala, and septum. Androgen is bound in only two regions (caudal pallium and mesencephalic tegmental area) in which there is no binding of estrogen. In the study of *Thamnophis* sp. (Halpern et al., 1982), sex steroids are heavily concentrated in several forebrain areas (the ventral amygdaloid nucleus, medial nucleus sphericus, septum, paleostriatum, preoptic area [POA], and retrobulbar pallium), several hypothalamic areas (anterior, periventricular, ventromedial, and arcuate nuclei), and the midbrain tegmentum.

The patterns for these two species are generally similar to one another and to other vertebrates. In particular they emphasize the importance of the preoptic area, of the hypothalamus and of several limbic structures as targets of sex steroid hormones. However, some differences between *Anolis* and *Thamnophis* may be of functional significance. Snakes have a well-developed vomeronasal system that is crucial in male detection of female sex pheromone. In contrast, male *Anolis* lack a well-developed vomeronasal system. In a study of *Thamnophis* sp., nearly every major component of the vomeronasal system contained sex steroid concentrating cells. *Anolis* lacks several of these brain structures, but several other homologous ones are well labeled. Some of the motor nuclei are also labeled differently in the two species. In *Anolis,* the motor nucleus of the trigeminal nerve is labeled; this may relate to the male's copulatory neck grip or to the dewlap display. The same nucleus is unlabeled in garter snakes that do not use a neck grip. Conversely, a small nucleus at the level of the obex is labeled in garter snakes that may be related to the tongue-flicking that males use to transfer the female sex pheromone to the vomeronasal system. Male *Anolis* do not show this behavior, and this nucleus is unlabeled. However, the physiological significance of these differences remains speculative. It would be interesting to examine hormone uptake patterns in lizards of the genus *Cnemidophorus*, which have (1) well-developed vomeronasal organs, (2) tongue-flicking behaviors, and (3) neck-gripping in their repertoire of sexual behaviors.

2. Effects of Brain Lesions and Electrical Stimulation

Several studies have examined the effects of brain lesions in *Anolis carolinensis* (Greenberg, 1977; Greenberg et al., 1979, 1984; Wheeler and Crews, 1978; Farragher and Crews, 1979), *Thamnophis sirtalis parietalis* (Friedman and Crews, 1985a; Krohmer and Crews, 1987a, b), *Caiman crocodilus* (Keating et al., 1970), and *Sceloporus occidentalis* (Tarr, 1977). Fewer studies have investigated the effects of electrical stimulation (Keating et al., 1970; Distel, 1976, 1978; Kennedy, 1975; Sugarman and Demski, 1978), and these investigated nothing beyond dewlap displays that could be related to male sexual behavior. In general these studies agree that the POA and hypothalamus are major sites regulating male sexual behavior. Lesions in the amygdala have fewer effects on sexual behavior and seem to have more effects on male aggressive behavior.

In the studies of *Anolis*, lesions in the ventromedial nucleus (Greenberg et al., 1984), anterior hypothalamus-POA (AH-POA; Wheeler and Crews, 1978) and the basal hypothalamus (Farragher and Crews, 1979) all abolish male courtship and copulatory neck grip behaviors (Fig. 3.4). In the study of the AH-POA, some lesions rostral to the AH-POA also abolish these behaviors and cause testicular regression, but behavior was subsequently restored by systemic testosterone administration. This indicates that the effects of these lesions were an indirect result of interfering with neural control of testicular function (Lisk, 1967).

Studies of *Thamnophis* have also shown that male sexual behavior is abolished by lesions of the AH-POA both before entering hibernation (Krohmer and Crews, 1987a) and after emerging from hibernation (Friedman and Crews, 1985a; Fig. 3.5). Although the AH-POA concentrates sex steroid hormones (Halpern et al., 1982), male *Thamnophis sirtalis parietalis* have a dissociated reproductive pattern and sexual behavior is not dependent on sex steroid hormones (see discussion above). Temperature changes seem to be more important in triggering male sexual behavior than sex steroid hormones. Lesioned males showing deficits in male sexual behavior also show deficits in thermoregulatory behavior (Krohmer and Crews, 1987a). This association makes it likely that lesions in the AH-POA interfere with the

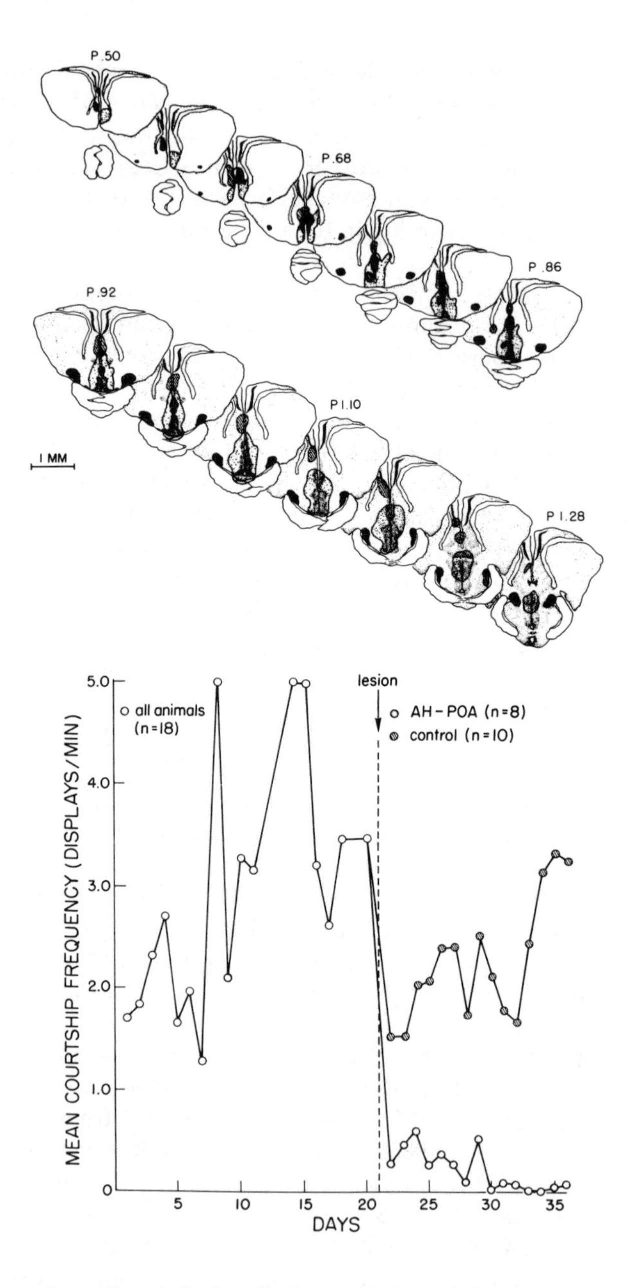

Fig. 3.4. *Anolis carolinensis.* Lesions in the anterior hypothalamus-preoptic area (AH-POA) abolish courtship behavior in castrated, androgen-treated males. Lesions dorsal to the AH-POA lack influence on male behavior. Top: Extent of lesions in areas dorsal to AH-POA (hatched lines) or in the AH-POA (stipple). Bottom: Frequency of male courtship behavior when tested with a sexually receptive female during the course of the experiment. (From Wheeler and Crews, 1978.)

integration of thermal information and sexual behavior rather than with the integration of sex steroid hormone information and sexual behavior, as they do in forms with associated reproductive tactics.

In another study of *Thamnophis sirtalis parietalis*, lesions in the septum and the nucleus sphericus of males before entering hibernation significantly enhanced courtship behavior when these males emerged from hibernation (Krohmer and Crews, 1987b). These two areas concentrate sex steroid hormones (Halpern et al., 1982), have projections to the preoptic area and ventromedial nucleus (Halpern, 1980), and are important components of the vomeronasal system (see discussion above). Therefore, their involvement in the regulation of sexual behavior is not surprising, even though their precise function remains unknown. It is possible that these two areas inhibit expression of sex-

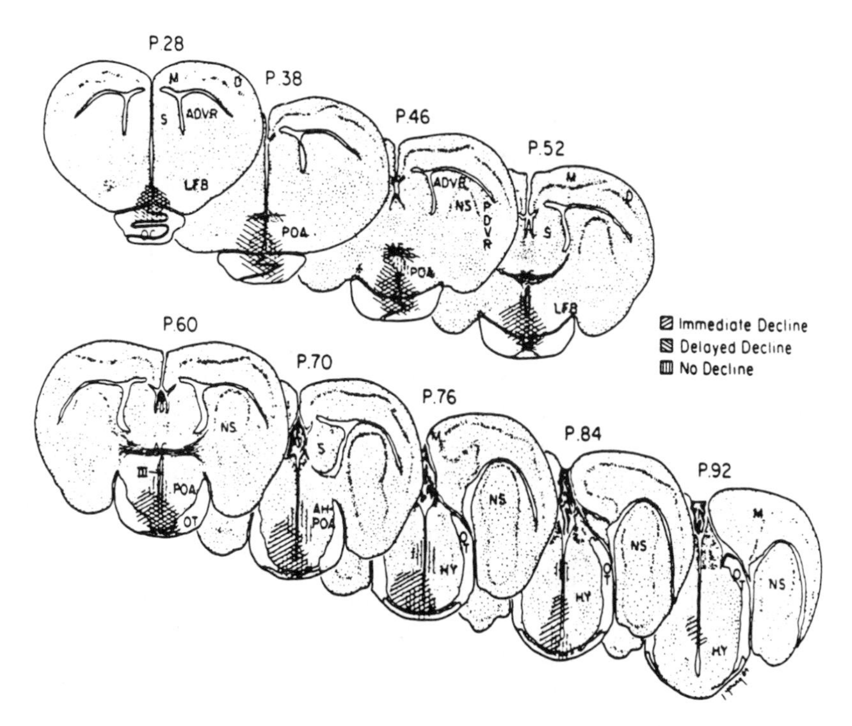

Fig. 3.5. *Thamnophis sirtalis parietalis*. Frontal sections through the male brain illustrating the location of lesions in groups of animals exhibiting different patterns of decline in courtship behavior. As in other species, lesions in the anterior hypothalamus-preoptic area have the greatest effect on courtship behavior. AC, anterior commissure; ADVR, anterior dorsal ventral ridge; AH, anterior hypothalamus; D, dorsal cortex; HY, hypothalamus; LFB, lateral forebrain bundle; M, medial cortex; NS, nucleus sphericus; OC, optic chiasm; OT, optic tract; PDVR, posterior dorsal ventricular ridge; POA, preoptic area; S, septum; III, third ventricle. (From Friedman and Crews, 1985a.)

ual behavior until they are disinhibited by the proper pheromonal stimuli.

3. EFFECTS OF BRAIN IMPLANTS OF STEROID HORMONES

Further evidence that sex steroid hormones mediate a particular behavior in a particular brain structure can be generated with localized microimplants in the brain. A positive response can be interpreted as good evidence for a direct effect. However, one must check for leakage into the cerebrospinal fluid or the peripheral circulation that can be detected through histological examination of peripheral target tissues and radioimmunoassays of plasma samples. In contrast, negative responses to brain implants must be interpreted more cautiously. Lack of an effect could reflect the need for simultaneous hormone stimulation of several brain sites, the need for simultaneous stimulation of additional peripheral targets, or the need for synergism with other hormones.

Anolis carolinensis brain implant studies are consistent with the autoradiographic and lesion studies in identifying the preoptic area and hypothalamus as major sites of sex steroid hormone action. Implants of estradiol, dihydrotestosterone, and testosterone in these regions all restore sexual behaviors in castrated males, whereas implants in adjacent regions do not do so (Fig. 3.6; Morgentaler and Crews, 1978; Crews and Morgentaler, 1979). Further support for the role of AH-POA as the major integrating site for male-typical sexual behavior comes from studies of a parthenogenetic whiptail (Mayo and Crews, 1987) and one of its ancestral species (Rozendaal and Crews, 1989). Implants of dihydrotestosterone into the AH-POA of female *Cnemidophorus uniparens* or male *C. inornatus* caused them to exhibit male-typical sexual behavior. Similar implants into the ventral medial hypothalamus, the usual neural site of regulation of female-typical behavior, did not produce behavioral change.

Consistent with the studies showing that administration of systemic sex steroid hormones does not affect male sexual behavior in *Thamnophis sirtalis parietalis* (Crews et al., 1984; see above discussion), implants of testosterone into the AH-POA, medial and basal hypothalamus, thalamus, medial cortex, and third ventricle all failed to stimulate courtship in noncourting males (Friedman and Crews, 1985b). However, these negative results suffer from the general above-mentioned problems of interpreting negative data.

V. CONCLUDING REMARKS

The overview presented here should indicate that we have reduced our ignorance about the physiological basis of male sexual behavior of reptiles. Nevertheless, the studies are restricted to a very few species,

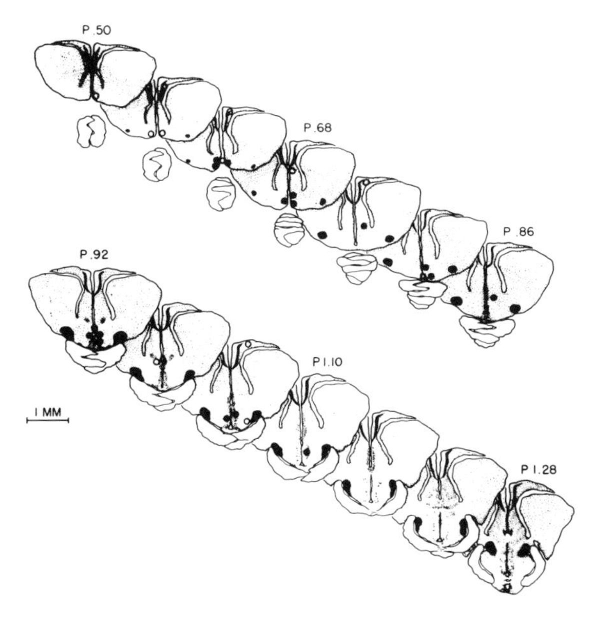

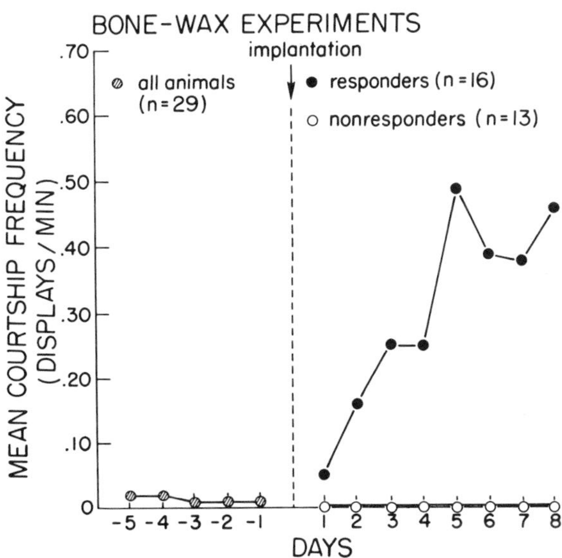

Fig. 3.6. *Anolis carolinensis.* Implants of testosterone into the anterior hypothalamus-preoptic area (AH-POA) of brain of castrated males restore the expression of courtship behavior. Top: Frontal sections through the brains of males indicating location of implants. Bottom: Frequency of courtship behavior of castrated males before and after receiving implants. In both top and bottom figures, filled circles depict males that responded to implantation of testosterone whereas unfilled circles indicate males that did not respond. Note that effective implants (filled circles) are located in the AH-POA. (From Morgentaler and Crews, 1978.)

and we are not even close to a comprehensive understanding of this topic. Fortunately, reptiles are attracting increasing attention as model systems for the investigation of the physiological basis of behavior. The outlined advantages for the use of reptiles for studies of evolutionary and ecological aspects of behavioral physiology are being recognized by a wider and wider audience. Reptiles present many natural experiments permitting tests of the ideas generated in the less polyphyletic mammals and birds. Indeed, many of the hypotheses generated in studies of mammals and birds cannot be tested in these taxa since the proposed causal variables show little variation. For example, reptiles show genotypic sex determination with both male and female heterogamety, environmental sex determination, viviparity and oviparity, and associated and dissociated reproductive tactics. These provide a rich milieu of possible combinations of variables, which facilitates tests of the generality of hypotheses generated in studies of mammals and birds. These tests of universally applicable principles provide the greatest promise for the future of reptilian behavioral physiology.

There is little hope that a comprehensive survey of behavior-controlling mechanisms in the reptiles will be completed any time in the foreseeable future. However, workers who capitalize on the unique advantages of reptiles as model systems are able to challenge and expand the prevailing dogma and disclose general principles underlying the study of behavioral physiology in all organisms. If biology is indeed the science of the search for unity in diversity, then the study of the remarkable diversity presented by the reptiles holds great promise for contributing to the uncovering of fundamental unifying principles.

ACKNOWLEDGMENTS

Preparation of this review was in part supported by a Presidential Young Investigator Award DCB-8451641 from the National Science Foundation to MCM. We thank the editors for this volume, David Crews and Carl Gans, for many useful comments and suggestions. We also thank Maryann Ottinger and Elizabeth Adkins-Regan for useful suggestions.

REFERENCES

Adkins, E., and Schlesinger, L. (1979). Androgens and the social behavior of male and female lizards (*Anolis carolinensis*). *Horm. Behav.* 13, 139–152.

Adkins-Regan, E. (1981). Hormone specificity, androgen metabolism, and social behavior. *Am. Zool.* 21, 257–271.
Adkins-Regan, E. (1987). Hormones and sexual differentiation. In *Hormones and Reproduction in Fishes, Amphibians and Reptiles* (D. O. Norris and R. E. Jones, eds.). Plenum, New York, pp. 1–30.
Adkins-Regan, E., and Ascenzi, M. (1987). Social and sexual behaviour of male and female zebra finches treated with oestradiol during the nestling period. *Anim. Behav.* 35, 1100–1112.
Angelini, F., and Ghiara, G. (1984). Reproductive modes and strategies in vertebrate evolution. *Boll. Zool.* 51, 121–203.
Arnold, A. P., and Breedlove, S. M. (1985). Organizational and activational effects of sex steroids on brain and behavior: a reanalysis. *Horm. Behav.* 19, 469–498.
Baum, J. J. (1987). Hormonal control of sex differences in the brain and behavior of mammals. In *Psychobiology of Reproductive Behavior: an Evolutionary Perspective* (D. Crews, ed.). Prentice-Hall, Englewood Cliffs, N.J., pp. 231–257.
Billy, A., and Crews, D. (1986). The effects of sex steroid treatments on sexual differentiation in a unisexual lizard, *Cnemidophorus uniparens* (Teiidae). *J. Morphol.* 18, 129–142.
Brown, C. L., and Bern, H. A. (1989). Thyroid hormones in early development, with special reference to teleost fishes. In *Development, Maturation and Senescence of Neuroendocrine Systems* (M. P. Schreibman and C. G. Scanes, eds.). Academic Press, San Diego, pp. 289–306.
Bull, J. J. (1983). *Evolution of Sex Determining Mechanisms.* Benjamin/Cummings, Menlo Park, Calif.
Butera, P. C., and Czaja, J. A. (1989). Activation of sexual behavior in male rats by combined subcutaneous and intracranial treatments of 5α-dihydrotestosterone. *Horm. Behav.* 23, 92–105.
Callard, G. V., Petro, Z., and Ryan, K. J. (1978). Conversion of androgen to estrogen and other steroids in the vertebrate brain. *Am. Zool.* 18, 511–523.
Callard, I. P., and Callard, G. V. (1987). Sex steroid receptors and nonreceptor mechanisms. In *Hormones and Reproduction in Fishes, Amphibians, and Reptiles* (D. O. Norris and R. E. Jones, eds.). Plenum Press, New York, pp. 355–384.
Callard, I. P., and Ho, S.-M. (1980). Seasonal reproductive cycles in reptiles. *Prog. Reprod. Biol.* 5, 5–38.
Camazine, B., Garstka, W., Tokarz, R. R., and Crews, D. (1980). The effects of castration and androgen replacement therapy on male courtship behavior in the red-sided garter snake (*Thamnophis sirtalis parietalis*). *Horm. Behav.* 14, 358–372.
Cole, C. J. (1984). Unisexual lizards. *Sci. Am.* 250, 94–100.
Cooper, W. E. (1971). Display behavior of hatchling *Anolis carolinensis*. *Herpetologica* 27, 498–500.
Cooper, W. E., Mendonça, M. T., and Vitt, L. J. (1987). Induction of orange head coloration and activation of courtship and aggression by testosterone in the male broad-headed skink (*Eumeces laticeps*). *J. Herpetol.* 21, 96–101.
Crews, D. (1973). Behavioral correlates to gonadal state in the lizard, *Anolis carolinensis*. *Horm. Behav.* 4, 307–313.

Crews, D. (1974a). Effects of group stability and male aggressive and sexual behavior on environmentally-induced ovarian recrudescence in the lizard, *Anolis carolinensis*. *J. Zool.* (London) 172, 419–441.
Crews, D. (1974b). Effects of castration and androgen replacement on male facilitation of ovarian activity in the lizard, *Anolis carolinensis*. *J. Comp. Physiol. Psychol.* 87, 963–969.
Crews, D. (1975a). Relative effects of different components of the male's courtship display on environmentally-induced ovarian recrudescence and mate selection in the lizard *Anolis carolinensis*. *Anim. Behav.* 23, 349–356.
Crews, D. (1975b). Psychobiology of reptilian reproduction. *Science* (N.Y.) 189, 1059–1065.
Crews, D. (1976). Hormonal control of male and female sexual behavior in the garter snake (*Thamnophis sirtalis sirtalis*). *Horm. Behav.* 7, 451–460.
Crews, D. (1977a). The annotated anole: studies on the control of lizard reproduction. *Am. Scient.* 65, 428–434.
Crews, D. (1977b). Integration of internal and external factors in the regulation of lizard reproduction. In *The Behavior and Neurology of Lizards* (N. Greenberg and P. MacLean, eds.). National Institute of Mental Health, Rockville, Md., pp. 149–171.
Crews, D. (1979a). The neuroendocrinology of reproduction in reptiles. *Biol. Reprod.* 20, 51–73.
Crews, D. (1979b). Endocrine control of reptilian reproductive behavior. In *Endocrine Control of Sexual Behavior* (C. Beyer, ed.). Raven Press, New York, pp. 167–222.
Crews, D. (1979c). The reproductive cycle of the American chameleon. *Sci. Am.* 241, 180–187.
Crews, D. (1980a). Studies in squamate sexuality. *Bioscience* 30, 835–838.
Crews, D. (1980b). Interrelationships among ecological, behavioral, and neuroendocrine processes in the reproductive cycle of *Anolis carolinensis* and other reptiles. In *Advances in the Study of Behavior* (J. S. Rosenblatt, R. A. Hinde, C. G. Beer, and M. C. Busnel, eds.). Academic Press, New York, pp. 1–74.
Crews, D. (1982). The ecological physiology of a garter snake. *Sci. Am.* 244, 158–168.
Crews, D. (1983a). Alternative reproductive tactics in reptiles. *Bioscience* 33, 562–566.
Crews, D. (1983b). Control of male sexual behavior in the Canadian red-sided garter snake. In *Hormones and Behavior in Higher Vertebrates* (J. Balthazart, E. Prove, and R. Gilles, eds.). Springer-Verlag, Berlin, pp. 398–406.
Crews, D. (1984). Gamete production, sex hormone secretion and mating behavior uncoupled. *Horm. Behav.* 18, 22–28.
Crews, D. (1985). Effects of early sex steroid hormone treatment on courtship behavior and sexual attractivity in the red-sided garter snake, *Thamnophis sirtalis parietalis*. *Physiol. Behav.* 35, 569–576.
Crews, D. (1987). *Psychobiology of Reproductive Behavior: an Evolutionary Perspective*. Prentice-Hall, Englewood Cliffs, N.J.
Crews, D., and Bull, J. J. (1987). Evolutionary insights from reptilian sexual

differentiation. In *Genetic Markers of Sex Differentiation* (F. P. Haseltine, M. E. McClure, and E. H. Goldberg, eds.). Plenum, New York, pp. 11–26.

Crews, D., Bull, J. J., and Billy, A. J. (1987). Sex determination and sexual differentiation in reptiles. In *Handbook of Sexology: the Pharmacology of Sexual Function* (J. M. A. Sitsen, ed.). Elsevier Science Publishers, Amsterdam, pp. 1–24.

Crews, D., Camazine, B., Diamond, M., Mason, R., Tokarz, R. R., and Garstka, W. R. (1984). Hormonal independence of courtship behavior in the male garter snake. *Horm. Behav.* 18, 29–41.

Crews, D., and Fitzgerald, K. T. (1980). Sexual behavior in parthenogenetic lizards (*Cnemidophorus*). *Proc. Natl. Acad. Sci. USA* 77, 499–502.

Crews, D., Grassman, M., and Lindzey, J. (1986). Behavioral facilitation of reproduction in gonochoristic and parthenogenetic *Cnemidophorus* lizards. *Proc. Natl. Acad. Sci. USA* 83, 9547–9550.

Crews, D., and Greenberg, N. (1981). Function and causation of social signals in lizards. *Am. Zool.* 21, 273–294.

Crews, D., Gustafson, J. E., and Tokarz, R. (1983). Psychobiology of parthenogenesis in reptiles. In *Lizard Ecology* (R. B. Huey, E. R. Pianka, and T. W. Schoener, eds.). Harvard University Press, Cambridge, Mass., pp. 205–231.

Crews, D., Hingorani, V., and Nelson, R. J. (1988). Role of the pineal gland in the control of annual reproduction behavioral and physiological cycles in the redsided garter snake (*Thamnophis sirtalis parietalis*). *J. Biol. Rhythms* 3, 293–302.

Crews, D., and Moore, M. C. (1986). Evolution of mechanisms controlling mating behavior. *Science* (N.Y.) 231, 121–125.

Crews, D., and Moore, M. C. (1991). Reproductive psychobiology of parthenogenetic whiptail lizards. In *Biology of the Cnemidophorus* (J. Wright, ed.). Allen Press, Lawrence, Kans. In press.

Crews, D., and Morgentaler, A. (1979). Effects of intracranial implantation of oestradiol and dihydrotestosterone on the sexual behavior of the lizard, *Anolis carolinensis*. *J. Endocrinol.* 82, 373–381.

Crews, D., Rosenblatt, J. S., and Lehrman, D. S. (1974). Effects of unseasonal environmental regime, group presence, group composition and mate's physiological state on ovarian recrudescence in the lizard, *Anolis carolinensis*. *Endocrinology* 94, 541–547.

Crews, D., Traina, V., Wetzel, T. F., and Muller, C. (1978). Hormonal control of male reproductive behavior in the lizard, *Anolis carolinensis:* role of testosterone, dihydrotestosterone, and estrogen. *Endocrinology* 103, 1814–1821.

Crews, D., Wibbels, T., and Gutzke, W. H. N. (1989). Action of sex steroid hormones on temperature-induced sex determination in the snapping turtle (*Chelydra serpentina*). *Gen. Comp. Endocrinol.* 76, 159–166.

Darevsky, I. S., Kupriyanova, L. A., and Uzzell, T. (1985). Parthenogenesis in reptiles. In *Biology of the Reptilia: Development B*, Vol. 15 (C. Gans and F. Billett, eds.). Wiley and Sons, New York, pp. 411–526.

Distel, H. (1976). Behavior and electrical brain stimulation in the green iguana, *Iguana iguana*. I. Schematic brain atlas and stimulation device. *Brain Behav. Evol.* 13, 421–450.

Distel, H. (1978). Behavior and electrical brain stimulation in the green iguana, *Iguana iguana*. II. Stimulation effects. *Exper. Brain Res.* 31, 353–367.

Done, B. S., and Heatwole, H. (1977). Effects of hormones on the aggressive behaviour and social organization of the scincid lizard, *Sphenomorphus kosciuskoi*. *Z. Tierpsych.* 44, 1–12.

Dorner, G., Dorke, F., and Moustafa, S. (1968). Differential localization of a male and a female hypothalamic mating center. *J. Reprod. Fert.* 17, 583–586.

Duvall, D., Guillette, L. J., and Jones, R. E. (1982). Environmental control of reptilian reproductive cycles. In *Biology of the Reptilia* (C. Gans and F. H. Pough, eds.). Academic Press, New York, 13, 201–231.

Evans, L. T. (1935a). Effects of Antuitrin S on the male lizard, *Anolis carolinensis*. *Anat. Rec.* 62, 213–218.

Evans, L. T. (1935b). Winter mating and fighting behavior of *Anolis carolinensis* as induced by pituitary injections. *Copeia* 1935, 3–6.

Evans, L. T. (1935c). The effects of pituitary implants and extracts on the genital system of the lizard. *Science* (N.Y.) 81, 468–469.

Evans, L. T. (1936a). Behavior of castrated lizards. *J. Gen. Psychol.* 48, 217–227.

Evans, L. T. (1936b). A study of a social hierarchy in the lizard, *Anolis carolinensis*. *J. Gen. Psychol.* 48, 88–111.

Evans, L. T. (1936c). Territorial behavior of normal and castrated females of *Anolis carolinensis*. *J. Gen. Psychol.* 49, 49–60.

Evans, L. T. (1936d). Social behavior of the normal and castrated lizards, *Anolis carolinensis*. *Science* (N.Y.) 83, 104.

Evans, L. T. (1937). Differential effects of the ovarian hormones on the territorial reaction time of female *Anolis carolinensis*. *Physiol. Zool.* 10, 456–463.

Evans, L. T. (1938a). Cuban field studies of territorial behavior of *Anolis sagrei*. *J. Comp. Psychol.* 25, 97–125.

Evans, L. T. (1938b). Courtship behavior and sexual selection of *Anolis*. *J. Comp. Psychol.* 26, 475–497.

Evans, L. T. (1953). New facts bearing upon the behavior of the male lizard, *Anolis carolinensis*. *Anat. Rec.* 117, 606.

Evans, L. T. (1957). The effects of hormones upon juvenile lizards, *Anolis carolinensis*. *Anat. Rec.* 138, 545.

Farragher, K., and Crews, D. (1979). The role of the basal hypothalamus in the regulation of reproductive behavior in the lizard *Anolis carolinensis*. *Horm. Behav.* 13, 185–206.

Feder, H. H. (1984). Hormones and sexual behavior. *Ann. Rev. Psychol.* 34, 165–200.

Ferguson, G. W. (1966). Effect of follicle-stimulating hormone and testosterone propionate on reproduction of the side-blotched lizard, *Uta stansburiana*. *Copeia* 1966, 495–498.

Friedman, D., and Crews, D. (1985a). Role of the anterior hypothalamus-preoptic area in the regulation of courtship behavior in the male Canadian red-sided garter snake (*Thamnophis sirtalis parietalis*): lesion experiments. *Behav. Neurosci.* 99, 942–49.

Friedman, D., and Crews, D. (1985b). Role of the anterior hypothalamus-preoptic area in regulation of courtship behavior in the male Canadian red-sided garter snake (*Thamnophis sirtalis parietalis*): intracranial implantation experiments. *Horm. Behav.* 19, 122–136.

Garstka, W. R., Camazine, B., and Crews, D. (1982). Interactions of behavior and physiology during the annual reproductive cycle of the red-sided garter snake (*Thamnophis sirtalis parietalis*). *Herpetologica* 38, 104–123.

Garstka, W. R., and Crews, D. (1982). Female control of male reproductive function in a Mexican snake. *Science* (N.Y.) 217, 1159–1160.

Goy, R. W., and McEwen, B. S. (1980). *Sexual Differentiation of the Brain.* MIT Press, Cambridge, Mass.

Greenberg, B., and Noble, G. K. (1944). Social behavior of the American chameleon, *Anolis carolinensis. Physiol. Zool.* 17, 392–439.

Greenberg, N. (1977). A neuroethological investigation of display behavior in a lizard, *Anolis carolinensis. Am. Zool.* 17, 191–201.

Greenberg, N., McClean, P.D., and Ferguson, J. L. (1979). Role of the paleostriatum in species-typical display behavior of the lizard (*Anolis carolinensis*). *Brain Res.* 172, 229–241.

Greenberg, N., Scott, M., and Crews, D. (1984). Role of the amygdala in the reproduction and aggressive behavior of the lizard, *Anolis carolinensis. Physiol. Behav.* 32, 147–151.

Gross, M. R. (1984). Sunfish, salmon, and the evolution of alternative reproductive strategies and tactics in fishes. In *Fish Reproduction: Strategies and Tactics* (G. W. Potts and R. J. Wooten, eds.). Academic Press, New York, pp. 55–75.

Guillette, L. J., Jr. (1982). The evolution of viviparity and placentation in the high elevation Mexican lizard *Sceloporus aeneus. Herpetologica* 38, 94–103.

Guillette, L. J., Jr. (1987). The evolution of viviparity in fishes, amphibians and reptiles: an endocrine approach. In *Hormones and Reproduction in Fishes, Amphibians and Reptiles* (D. O. Norris and R. E. Jones, eds.). Plenum Press, New York, pp. 523–562.

Gustafson, J. E., and Crews, D. (1981). Effect of group size and physiological state of a cagemate on reproduction in the parthenogenetic lizard *Cnemidophorus uniparens* (Teiidae). *Behav. Ecol. Sociobiol.* 8, 267–272.

Gutzke, W. H. N., and Crews, D. (1988). Embryonic temperature determines adult sexuality in a reptile. *Nature* (London) 332, 832–834.

Halpern, M. (1980). The telencephalon of snakes. In *Comparative Neurology of the Telencephalon* (S. O. E. Ebbesson, ed.). Plenum Press, New York, pp. 257–295.

Halpern, M., Morrell, J. I., and Pfaff, D. W. (1982). Cellular [3H] testosterone localization in the brains of garter snakes: an autoradiographic study. *Gen. Comp. Endocrinol.* 46, 211–224.

Hover, E. L. (1985). Differences in aggressive behavior between two throat color morphs in a lizard, *Urosaurus ornatus. Copeia* 1985, 933–940.

Kaplan, M. E., and McGinnis, M. Y. (1989). Effects of ATD on male sexual behavior and androgen receptor binding: a reexamination of the aromatization hypothesis. *Horm. Behav.* 23, 10–26.

Keating, E. G., Kormann, L. A., and Horel, J. A. (1970). The behavioral effects of stimulating and ablating the reptilian amygdala (*Caiman sklerops*). *Physiol. Behav.* 5, 55–59.

Kelly, D. B. (1988). Sexually dimorphic behaviors. *Ann. Rev. Neurosci.* 11, 225–251.

Kennedy, M. C. (1975). Vocalization elicited in a lizard by electrical stimulation of the midbrain. *Brain Res.* 91, 321–325.

Kim, S. Y., Stumpf, W. E., Sar, M., and Martinez-Vargas, C. M. (1978). Estrogen and androgen target cells in the brain of fishes, reptiles and birds: phylogeny and ontogeny. *Am. Zool.* 18, 425–433.

Krohmer, R. W., and Crews, D. (1987a). Temperature activation of courtship behavior in the male red-sided garter snake (*Thamnophis sirtalis parietalis*): role of the anterior hypothalamus-preoptic area. *Behav. Neurosci.* 101, 228–236.

Krohmer, R. W., and Crews, D. (1987b). Facilitation of courtship behavior in the male red-sided garter snake (*Thamnophis sirtalis parietalis*) following lesions of the septum or nucleus sphericus. *Physiol. Behav.* 40, 750–765.

Krohmer, R. W., Grassman, M., and Crews, D. (1987). Annual reproductive cycle in the male red-sided garter snake, *Thamnophis sirtalis parietalis:* field and laboratory studies. *Gen. Comp. Endocrinol.* 68, 64–75.

Lance, V. (1984). Endocrinology of reproduction in male reptiles. *Symp. Zool. Soc.* (London) 52, 357–383.

Licht, P. (1982). Endocrine patterns in the reproductive cycle of turtles. *Herpetologica* 38, 51–61.

Licht, P. (1984). Reptiles. In *Marshall's Physiology of Reproduction* (G. E. Lamming, ed.). Churchill Livingstone, Edinburgh, pp. 206–282.

Lindzey, J., and Crews, D. (1986). Hormonal control of courtship and copulatory behavior in male *Cnemidophorus inornatus,* a direct sexual ancestor of a unisexual, parthenogenetic lizard. *Gen. Comp. Endocrinol.* 64, 411–418.

Lindzey, J., and Crews, D. (1988a). Psychobiology of sexual behavior in the whiptail lizard, *Cnemidophorus inornatus. Horm. Behav.* 22, 279–293.

Lindzey, J., and Crews, D. (1988b). Effects of progestins on sexual behavior in castrated lizards (*Cnemidophorus inornatus*). *J. Endocrinol.* 119, 265–273.

Lisk, R. D. (1967). Neural control of gonad size by hormone feedback in the desert iguana *Dipsosaurus dorsalis dorsalis. Gen. Comp. Endocrinol.* 8, 258–266.

Marler, C. A., and Moore, M. C. (1988). Evolutionary costs of aggression revealed by testosterone manipulations in free-living male lizards. *Behav. Ecol. Sociobiol.* 23, 21–26.

Marler, C. A., and Moore, M. C. (1989). Time and energy costs of aggression in testosterone-implanted free-living male mountain spiny lizards (*Sceloporus jarrovi*). *Physiol. Zool.* 62, 1334–1350.

Marler, C. A., and Moore, M. C. (1991). Energetic costs of aggression compensated for by food supplementation in testosterone-implanted male mountain spiny lizards. *Anim. Behav.* 42, 209–220.

Martinez-Vargas, M. C., Keefer, D. A., and Stumpf, W. E. (1978). Estrogen localization in the brain of the lizard, *Anolis carolinensis. J. Exper. Zool.* 205, 141–147.

Mason, P., and Adkins, E. K. (1976). Hormones and social behavior in the lizard, *Anolis carolinensis*. *Horm. Behav.* 7, 75–86.
Mason, R. T., Chinn, J. W., and Crews, D. (1987). Sex and seasonal differences in the skin lipids in the red-sided garter snake, *Thamnophis sirtalis parietalis*. *Comp. Biochem. Physiol.* 87B, 999–1003.
Mason, R. T. and Crews, D. (1985). Female mimicry in garter snakes. *Nature* (London) 316, 59–60.
Mayo, M., and Crews, D. (1987). Neural control of male-like pseudocopulatory behavior in the all-female lizard, *Cnemidophorus uniparens*: effects of intracranial implantation of dihydrotestosterone. *Horm. Behav.* 21, 181–192.
Mendonça, M. T. (1987). Timing of reproductive behaviour in male musk turtles, *Sternotherus odoratus*: effects of photoperiod, temperature and testosterone. *Anim. Behav.* 35, 1002–1014.
Moore, F. L. (1987a). Regulation of reproductive behavior. In *Hormones and Reproduction in Fishes, Amphibians and Reptiles* (D. O. Norris and R. E. Jones, eds.). Plenum Press, New York, pp. 505–522.
Moore, F. L. (1987b). Behavioral actions of neurohypophysial peptides. In *Psychobiology of Reproductive Behavior: an Evolutionary Perspective* (D. Crews, ed.). Prentice-Hall, Englewood Cliffs, N.J., pp. 61–87.
Moore, M. C. (1987a). Castration affects territorial and sexual behaviour of free-living male lizards, *Sceloporus jarrovi*. *Anim. Behav.* 35, 1193–1199.
Moore, M. C. (1987b). Circulating steroid hormones during rapid aggressive responses of territorial male mountain spiny lizards, *Sceloporus jarrovi*. *Horm. Behav.* 21, 511–521.
Moore, M. C. (1988). Testosterone control of territorial behavior: tonic-release implants fully restore seasonal and short-term aggressive responses in free-living castrated male lizards. *Gen. Comp. Endocrinol.* 70, 450–459.
Moore, M. C. (1991). Application of organization-activation theory to alternative male reproductive strategies: a review. *Horm. Behav.* 25, 154–79.
Moore, M. C., and Crews, D. (1986). Sex steroid hormones in natural populations of a sexual whiptail lizard *Cnemidophorus inornatus*, a direct evolutionary ancestor of a unisexual parthenogen. *Gen. Comp. Endocrinol.* 63, 424–430.
Moore, M. C., and Marler, C. A. (1987). Effects of testosterone manipulations on nonbreeding season territorial aggression in free-living male lizards, *Sceloporus jarrovi*. *Gen. Comp. Endocrinol.* 65, 225–232.
Moore, M. C., and Marler, C. A. (1988). Hormones, behavior and the environment: an evolutionary perspective. In *Processing of Environmental Information in Vertebrates* (M. H. Stetson, ed.). Springer-Verlag, New York, pp. 71–84.
Moore, M. C., and Thompson, C. W. (1990). Field endocrinology of reptiles: hormonal control of alternative male reproductive tactics. In *Progress in Comparative Endocrinology* (A. Epple, C. G. Scanes, and M. H. Stetson, eds.). Wiley-Liss, New York, pp. 685–690.
Moore, M. C., Whittier, J. M., Billy, A. J., and Crews, D. (1985a). Male-like behaviour in an all-female lizard: relationship to ovarian cycle. *Anim. Behav.* 33, 284–289.
Moore, M. C., Whittier, J. M., and Crews, D. (1985b). Sex steroid hormones

during the ovarian cycle of an all-female, parthenogenetic lizard and their correlation with pseudosexual behavior. *Gen. Comp. Endocrinol.* 60, 144–153.

Morgentaler, A., and Crews, D. (1978). Role of the anterior hypothalamic-preoptic area in the regulation of reproductive behavior in the lizard, *Anolis carolinensis:* implantation studies. *Horm. Behav.* 11, 61–73.

Morrell, J. I., Crews, D., Ballin, A., Morgentaler, A., and Pfaff, D. W. (1979). ^{3}H-estradiol, ^{3}H-testosterone, and ^{3}H-dihydrotestosterone localization in the brain of the lizard *Anolis carolinensis:* an autoradiographic study. *J. Comp. Neurol.* 188, 201–224.

Morrell, J. I., and Pfaff, D. W. (1978). A neuroendocrine approach to brain function: localization of sex steroid concentrating cells in vertebrate brains. *Am. Zool.* 18, 447–460.

Naftolin, F., Ryan, K. J., Davis, I. J., Reddy, V. V., Flores, F., Petro, Z., and Kuhn, M. (1975). The formation of estrogens by central neuroendocrine tissues. *Rec. Prog. Horm. Res.* 31, 295–319.

Nelson, R. J., Mason, R. T., Krohmer, R. W., and Crews, D. (1987). Pinealectomy blocks vernal courtship behavior in red-sided garter snakes. *Physiol. Behav.* 39, 231–233.

Noble, G. K., and Bradley, H. T. (1933). The mating behavior of lizards: its bearing on the theory of sexual selection. *Ann. N.Y. Acad. Sci.* 35, 25–100.

Noble, G. K., and Greenberg, B. (1940). Testosterone propionate, a bisexual hormone in the American chameleon. *Proc. Soc. Exper. Biol. Med.* 44, 460–462.

Noble, G. K., and Greenberg, B. (1941a). Effects of seasons, castration and crystalline sex hormones upon the urogenital system and sexual behavior of the lizard *Anolis carolinensis. J. Exper. Zool.* 88, 451–479.

Noble, G. K., and Greenberg, B. (1941b). Induction of female behavior in male *Anolis carolinensis* with testosterone propionate. *Proc. Soc. Exper. Biol. Med.* 47, 32–37.

Pfaff, D. W. (1980). *Estrogens and Brain Function.* Springer-Verlag, New York.

Phoenix, C. H., Goy, R. W., Gerall, A. A., and Young, W. C. (1959). Organizational action of prenatally administered testosterone propionate on the tissues mediating behavior in the female guinea pig. *Endocrinology* 65, 369–382.

Rand, M. S. (1987). The distribution of two color morphs of the male saxicolous lizard, *Sceloporus undulatus erythrocheilus,* and their relationship with habitat in two different Colorado populations. *Am. Zool.* 27, 45A.

Rozendaal, J. C., and Crews, D. (1989). Effects of intracranial implants of dihydrotestosterone on sexual behavior in male *Cnemidophorus inornatus,* a direct sexual ancestor of a parthenogenetic lizard. *Horm. Behav.* 23, 194–202.

Sachs, B. D., and Meisel, R. L. (1988). The physiology of male sexual behavior. In *The Physiology of Reproduction* (E. Knobil and J. D. Neil, eds.). Raven Press, New York, pp. 1393–1486.

Shine, R. (1985). The evolution of viviparity in reptiles: an ecological analysis. In *Biology of the Reptilia: Development B,* Vol. 15 (C. Gans and F. Billett, eds.). Wiley and Sons, New York, pp. 605–694.

Stamps, J., and Crews, D. (1976). Seasonal changes in reproduction and social behavior in the lizard, *Anolis aenus*. *Copeia* 1976, 467–476.

Sugarman, R. A., and Demski, L. S. (1978). Agonistic behavior elicited by electrical stimulation of the brain in western collard lizards, *Crotaphytus collaris*. *Brain Behav. Evol.* 15, 446–469.

Tarr, R. S. (1977). The role of the amygdala in the intraspecific aggressive behavior of the iguanid lizard, *Sceloporus occidentalis*. *Physiol. Behav.* 18, 1153–1158.

Thompson, C. W., and Moore, M. C. (1987). Status signalling in male tree lizards. *Am. Zool.* 27, 50A.

Thompson, C. W., and Moore, M. C. (1991a). Syntopic presence of multiple dewlap color morphs in male tree lizards, *Urosaurus ornatus*. *Copeia* 1991, 493–503.

Thompson, C. W., and Moore, M. C. (1991b). Throat colour reliably signals status in male tree lizards, *Urosaurus ornatus*. *Anim. Behav.* In press.

Tokarz, R. R. (1986). Hormonal regulation of male reproductive behavior in the lizard *Anolis sagrei*: a test of the aromatization hypothesis. *Horm. Behav.* 20, 364–377.

Tokarz, R. R. (1987). Effects of the antiandrogens cyproterone acetate and flutamide on male reproductive behavior in a lizard (*Anolis sagrei*). *Horm. Behav.* 21, 1–16.

Underwood, H. (1988). Circadian organization in lizards: perception, translation, and transduction of photic and thermal information. In *Processing of Environmental Information in Vertebrates* (M. H. Stetson, ed.). Springer-Verlag, New York, pp. 47–70.

Vom Saal, F. (1983). The interaction of circulating oestrogens and androgens in regulating mammalian sexual differentiation. In *Hormones and Behavior in Higher Vertebrates* (J. Balthazart, E. Prove, and R. Gilles, eds.). Springer-Verlag, Berlin, pp. 159–177.

Wheeler, J., and Crews, D. (1978). The role of anterior hypothalamic-preoptic area in the regulation of male reproductive behavior in the lizard, *Anolis carolinensis*: lesion studies. *Horm. Behav.* 11, 42–60.

Whittier, J. M., and Crews, D. (1987). Seasonal reproduction: patterns and control. In *Hormones and Reproduction in Fishes, Amphibians and Reptiles* (D. O. Norris and R. E. Jones, eds.). Plenum Press, New York, pp. 385–409.

Wingfield, J. C. (1980). Sex steroid-binding proteins in vertebrate blood. In *Hormones, Adaptation and Evolution* (S. Ishii, T. Hirano, and M. Wada, eds.). Springer-Verlag, Berlin, pp. 135–144.

Wingfield, J. C., Ball, G. F., Dufty, A. M., Hegner, R. E., and Ramenofsky, M. (1987). Testosterone and aggression in birds. *Am. Scient.* 75, 602–608.

Wingfield, J. C., and Moore, M. C. (1987). Hormonal, social and environmental factors in the reproductive biology of free-living male birds. In *Psychobiology of Reproductive Behavior: an Evolutionary Perspective* (D. Crews, ed.). Prentice-Hall, Englewood Cliffs, N.J., pp. 149–175.

4

Reptilian Pheromones

Robert T. Mason

CONTENTS

I. INTRODUCTION

A. Background

Reptiles may arguably be described as primarily visually oriented creatures. It is by now well known that they exhibit a wide variety of visual signals or behaviors that are important in all aspects of their life cycle (see Carpenter and Ferguson, 1977). Only relatively recently has chemical communication been recognized as an important aspect of reptilian biology. Baumann and Noble conducted the first experimental investigations on the importance of chemical communication in reptiles in the 1920s and 1930s. Baumann (1927, 1929) investigated the courtship behavior of *Vipera aspis*, whereas Noble (1937) investigated courtship, trailing, and aggregation in *Thamnophis* and *Storeria*. It is now clear that the daily problems reptiles face in survival, such as prey detection, predator avoidance, and mate recognition, in many cases require the ability to perceive chemical cues from the environment.

The study of chemical cues, semiochemicals (natural products with signal function), and/or pheromones is now recognized as an integral part of research on reptilian social behavior. For the purposes of this review I will use the original definition of a pheromone, which states that a pheromone is a chemical or semiochemical produced by one individual that effects a change in the physiology or behavior of conspecifics (Karlson and Luscher, 1959). Behavior mediated by chemical cues was observed well into the last century. However, until recently, few experimental studies have sought to document the intricate and complex processes involved in the production and perception of pheromones in reptiles. Accounts of behaviors elicited by pheromones are prevalent in the literature, as are excellent reviews (see Evans, 1961; Carpenter and Ferguson, 1977; Madison, 1977; Burghardt, 1970, 1980; Simon, 1983). The purpose of this review is to document the many behavioral accounts of pheromonal communication in the various reptilian taxa. For each order I have attempted to categorize the behaviors in terms of responses elicited in signal recipients such as sex attraction, aggregation, and the like. In places, I have also included a brief description of those glands and other structures that have been described as producing semiochemicals. More complete descriptions of the histology and histochemistry of these glands may be obtained from the original references. Finally, I have attempted to review critically the literature dealing with the chemical nature of the pheromones themselves. This field of study has only recently received the attention of chemical ecologists; however, even the limited number of available studies has demonstrated dramatically the great potential of reptilian models for investigation of the chemical senses among the vertebrates. Because more work has been done on snakes than any other reptiles, this chapter reverses the usual order of groups and starts with these.

B. Evolution of Semiochemicals in Reptiles

Reptiles possess a remarkable diversity of glands and glandular secretions (Quay, 1972). Indeed, interest in squamate glands began well over a hundred years ago and even included studies by Darwin (1874). But most of these early reports were superficial with little or no investigation of morphology or histology. Gabe and Saint Girons (1965) conducted an extensive study of the cloacal anatomy of 23 families of lepidosaurs. However, the most thorough study of squamate glands is the doctoral dissertation of Anne Whiting (1969), which comprises an excellent survey of the morphology and histology of all the known cloacal glands in squamates. Because of the vast number of synonymous terms for cloacal glands in squamates I will follow Whiting's (1969) terminology.

Gadow (1887) divided the cloaca into three regions: the coprodaeum, the urodaeum, and the proctodaeum. The coprodaeum is really an extension of the large intestine and comprises the largest and most anterior section of the cloaca. The urodaeum receives both the urinary and genital ducts. The most posterior portion, or proctodaeum, is continuous with the urodaeum and opens to the environment through the cloacal vent.

Chemical cues are very efficient energetically in that they are cheap to produce, they relay a message long after the producer is gone, and they work in the dark and over very great distances. In his study of the evolution of social behavior and communication in mammal-like reptiles, Duvall (1986) proposes a mechanism for the evolution of semiochemical communication. It is widely accepted that virtually any exudate can serve as a chemical signal. Duvall states that chemical exudates such as feces or urine, skin lipids, or other metabolic by-products are inexpensive in the sense that they are continually available for use as chemical signals. That these exudates may have been coopted or exapted to serve a semiochemical function has been suggested (Graves et al., 1986; Maderson, 1986) and specifically investigated in reptiles (Duvall et al., 1987). Duvall discusses two ways used by both early and present-day reptiles to mark the substrate. An individual may "passively mark" or deposit secretions as it moves through the habitat. For example, the animal leaves cloacal cues as it rests or drags the cloaca across the ground while walking. These behaviors are distinct from "active marking," in which the animal performs a behavior specifically to leave a chemical cue on the substrate by means such as chin wiping or cloacal rubbing. Finally, specialized exocrine glands are known to elaborate behaviorally active semiochemicals; examples are chin glands in tortoises and cloacal scent glands in snakes.

II. SERPENTES

A. Glands

Reptiles possess a remarkable diversity of glands and glandular secretions (Quay, 1972). In snakes, by far the best-known glands are the paired cloacal scent glands or anal sacs (Fig. 4.1). Interest in these glands began at least several hundred years ago, when Tyson (1683) described the glands as appearing in both sexes of rattlesnakes. These glands appear to be uniquely ophidian and are located in the tail, dorsal to the hemipenes in males and in the corresponding position in females (Whiting, 1969). A duct from the anterior end of each gland

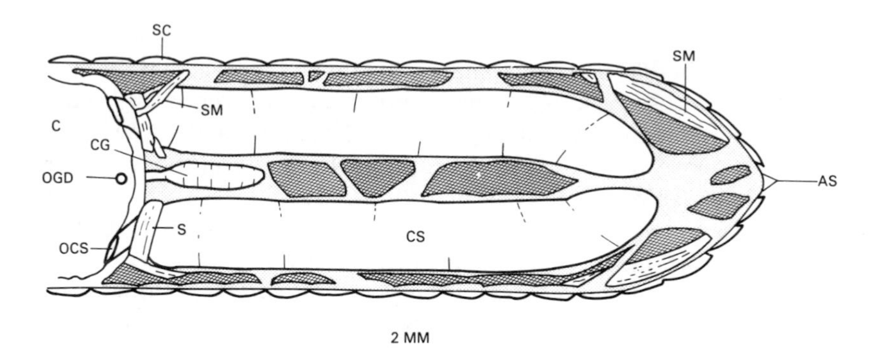

Fig. 4.1 *Leptotyphlops dulcis* ventral view (full section) of the tail region of illustrating the large cloacal sacs and median, unpaired cloacal gland. AS, Apical spine; C, cloaca; CG, cloacal gland; CS, cloacal sac; OCS, opening of cloacal sac duct; OGD, opening of cloacal gland duct; S, voluntary muscle sphincter; SC, scale; SM, skeletal muscle. (From Kroll and Reno, 1971.)

curves laterally and enters the proctodaeum close to the hemipenal opening. The glands are holocrine and produce primarily lipids (Oldak, 1976). The cloacal glands occur in male and female snakes but have been described as being best developed in females (Guibé, 1970). However, this may reflect the greater relative size of the glands in females rather than greater activity. In some snakes the size of the glands varies throughout the year (Kroll and Reno, 1971). There are numerous descriptions of the anatomy and histology of these glands (Gadow, 1887; Portier, 1894; Volsøe, 1944; Gabe and Saint Girons, 1965; Fox, 1965; Whiting, 1969; Kroll and Reno, 1971; Price and La-Pointe, 1981). Whiting (1969) and Kroll and Reno (1971) note the vast number of synonyms for the paired cloacal glands that produce the scent in snakes. In this review, the paired glands will be referred to as cloacal scent glands or scent glands following Whiting's suggestion.

The function of these glands has been the object of controversy for many years. Historically, the suggested functions can be associated with four aspects of ophidian behavior, namely sex attractant, defensive or repellant, aggregation cue, and alarm pheromone. As will become evident in this section, clearly there are multiple functions to the secretions at least in some species of snakes.

An early suggestion is that the glands play a role in sexual behavior. Rathke (1848) suggested that the glands probably aid the sexes in finding one another. Baumann (1929) believed that the male European asp, *Vipera aspis,* is able to follow conspecific females by means of secretions from the scent glands. Volsøe (1944) was also of the opinion that the scent gland secretions of females serve to attract males in European vipers (*Vipera*). However, Noble (1937) demonstrated that the source of attractant signals from female *Thamnophis sirtalis* was located in the dorsal skin and not the cloacal glands. Noble and Clausen (1936) reached the same conclusions in studies of *Storeria dekayi.*

Anyone who has handled a snake can attest to the fact that the cloacal scent gland secretions are malodorous. Cloacal scent gland secretions are often expressed during handling, suggesting that they comprise a defense mechanism, as proposed by many authors (Prater, 1933; Gloyd, 1933; Elliot, 1934; Beuchelt, 1936; Curran and Kauffeld, 1937; Volsøe, 1944; Smith, 1951; Wingate, 1956; Klauber, 1956; Barbour, 1962; Lüdicke, 1964; Fitch, 1963, 1965; Price and LaPointe, 1981). If *Coluber constrictor* and *Thamnophis sirtalis* are kept from escaping a predator, they smear the secretion of the glands on the captor by writhing movements of their body (Fitch, 1963, 1965). This behavior is common to many, if not all, species of snakes.

The ability of male desert king snakes, *Lampropeltis getulus splendida,* to follow a trail of female scent gland secretions was investigated in August and December (Price and LaPointe, 1981). The results of the behavioral tests indicated that in both months, males were unable to distinguish the trails of females from those made with the control compound glycerol. However, because these studies were not conducted during the breeding season, conclusions relating the results to normal reproductive behavior are highly speculative and probably do not reflect normal sex behavior. Additional experiments attempted to ascertain whether either sex would differentially tongue-flick male scent gland secretions, female scent gland secretions, or an untreated blank control. Only the number of tongue-flicks by females responding to female secretions varies significantly among stimuli. Paradoxically, the mean tongue-flick rate is higher in response to the glycerol control than to the secretions in all four groups. The suggestion that these results indicate that olfactory rather than vomeronasal mediation of information is occurring seems rather unlikely and is certainly not supported by any data. Price and LaPointe (1981) also point out that females are more responsive to the scent gland secretions than males, relating this to the greater tendency of females to aggregate, but did not test this. Several authors have suggested that aggregating snakes may leave chemical cues on the substrate (see Section II.D).

The muscular sphincter associated with the cloacal scent glands has suggested a defensive or alarm function for the glands (Price and LaPointe, 1981; Kroll and Reno, 1971). The capacity for rapid release is not consistent with the hypothesis that the scent gland secretions are a source of a sexual pheromone, which would presumably necessitate the release of the secretion in small amounts over a long period of time. This contention is supported by the demonstrated failure of secretions to elicit a sexual response in trailing and tongue-flicking experiments.

Another possible use of cloacal scent gland secretion is that of eliciting behavior from conspecifics (Andrén, 1986). During the breeding season, when a male adder, *Vipera berus,* returns from having pursued a rival intruding male, he circles the female with his cloaca open. Such dominant, mate-guarding males continuously press their gaped cloacas on the ground while moving forward; after this, the male returns to courting the female. It has been suggested that this behavior may serve to mark the territory of males (see discussion of territoriality in Section II.D). However, there is currently no evidence for a change in behavior of returning male adders or new intruders crossing this "territorial" border.

A series of experiments on the cloacal secretions of blind snakes demonstrates that many of the glandular secretions discussed in this review may have more than one function. The defensive behavior of the burrowing blind snake, *Leptotyphlops dulcis,* which eats termites and ants, is described by Gehlbach et al. (1968). When attacked by ants, these snakes frequently begin writhing while covering themselves with a clear liquid that originates in the cloacal scent glands (Fig. 4.2). This liquid, when wiped from the body of one blind snake, repelled ants of the species *Labidus coecus, Neivamyrmex nigrescens,* and *Solenopsis geminata.* A follow-up study investigated the effects of cloacal sac secretions as intraspecific sex attractants as well as interspecific ophiophage repellents (Watkins et al., 1969). Cloacal sac material was removed from the body surface after being secreted by the snake or directly from excised glands. The glandular material was then applied to one of two sides in a test arena, and the test animals were scored for how many times a snake entered each side of the arena and how much time was spent on each side. The results indicate that both the whole cloacal secretion (presumably including feces) and the macerated glandular material significantly attract conspecific blind snakes. Blind snake aggregations are known to occur with both mixed and single-sex groups. *Sonora semiannulata, Tantilla gracilis, Virginia striatula, Diadophis punctatus,* and *Lampropeltis triangulum* were also tested, and all were significantly repelled by the cloacal secretions of *L. dulcis.* The *Sonora* and *Tantilla* were also significantly repelled by the glandular material alone, indicating that the repellant properties of the cloacal secretion are contained in the glands themselves. Both *Sonora* and *Tantilla* occur in the same microhabitat as *Leptotyphlops,* and both eat ant broods. This suggests that the repellant effects of the cloacal scent gland secretions may be advantageous in preventing space and food competition between these species, although this seems unlikely given the availability of ant brood. Both *Diadophis* and *Lampropeltis* are ophiophagous. The advantage gained in repelling these potential blind snake predators is obvious. The results represent a clear demonstration of a dual effect for the secretions of the cloacal scent glands in this species.

Both sexes of *Leptotyphlops* and *Typhlops* also have an unpaired median cloacal gland (Robb, 1960; Fox, 1965; Kroll and Reno, 1971), mentioned here only because its secretory activity has been suggested to be correlated with the reproductive season. However, no one has directly tested behavioral responses to the secretion.

The release of musk from the cloacal scent glands of disturbed snakes is usually considered to be a defensive reaction; however, some investigators have attributed an alarm function to the secretions. An adult female chain king snake, *Lampropeltis g. getulus,* was collected as a young hatchling and raised in captivity but never emitted musk despite repeated handling (Brisbin, 1968). When an adult male Florida king snake, *L. g. floridana,* was brought into the lab and measured on a table, it emitted copious amounts of scent gland secretions or musk. The table was subsequently cleaned with water, but

Fig. 4.2 (A) A specimen *Leptotyphlops dulcis.* 135 mm total length, in a writhing coil. (B) A specimen, 183 mm total length, in a stationary position with a glossy appearance caused by feces and presumed anal gland secretions. Note the silvery appearance of A in contrast to B. Photos by F. R. Gehlbach. (From Gehlbach, Watkins, and Reno, © 1968 American Institute of Biological Sciences. All rights reserved.)

the table top still smelled of musk. When the adult female was then placed on the table she became very agitated and emitted musk, presumably for the first time in her life. It has been concluded that the secretions may be acting as an alarm pheromone.

The possibility that cloacal scent gland secretions could act as alarm pheromones has also been investigated (Graves and Duvall, 1984, 1988). Prairie rattlesnakes (*Crotalus v. viridis*) have been exposed to the odors from alarmed conspecifics without cloacal sacs, alarmed conspecifics with sham surgeries, or an empty container control. After experiencing a threatening stimulus, in this case, the breath of the experimenter, the heart rates of experimental animals that had been exposed to the odors of alarmed conspecifics with intact cloacal sacs (shams) increased significantly, whereas the heart rates of those experimental animals that were exposed to alarmed animals with no cloacal sacs did not rise significantly. These results suggest that the cloacal sac secretions of crotalids may serve as an alarm pheromone when experienced in a given behavioral state.

Two other glands that have been described in ophidians warrant a brief discussion. Glandlike structures reported only in the epidermis of the head of *Ramphotyphlops braminus* appear to be analogous to sebaceous glands in mammals (Haas, 1932). A better-documented but still rare occurrence of ophidian glands is the nuchodorsal glands occurring in certain genera of snakes. Nakamura (1935) described an unusual gland in the nuchal or neck region of a Japanese snake, *Rhabdophis tigrina*. He also examined 13 other natricine species but did not find evidence for the gland in any of them. The glands occur in several other natricine species, as well as *Macropisthodon* (Smith, 1938). The glands are not always confined to the neck but in some species extend throughout the whole length of the body. Two types of glandular structures occur, one consisting of sacculated or chains of spherical structures (*Rhabdophis* type), the other a nonsacculated type in which the gland is a single elongated piece of tissue. The gland is found in both sexes as well as in young and adults, although it is more developed in adults. The glands may have a seasonal activity. Smith and Nakamura note that the secretions of the gland are irritating to mucous membranes. Smith claims that the glands cannot be solely for defensive purposes but does not suggest an alternative function for the secretions.

B. Chemical Studies

The chemical composition of the scent glands in any snake were first investigated in the blind snake, *Leptotyphlops dulcis* (Blum et al., 1971). Thin-layer chromatography (TLC) and gas-liquid chromatography (GLC) of samples removed directly from the cloacal scent glands indicated no differences among the sexes in terms of material present. Most of the material identified consisted of free fatty acids, primarily palmitic, oleic, and linoleic acids, accounting for over 90% of the material present. These materials are considered to be the source of the musk odor in the secretion; however, this has not been investigated experimentally. Seventeen amino acids plus glucosamine and galactosamine have also been identified. The repellency and attractive properties of the secretions are possibly due to the fairly volatile fatty acids, although the lack of sexual differences argues against their possible role as sex attractants (Blum et al., 1971).

Fatty acids have also been identified in the cloacal gland secretions of other snakes. By means of defocusing mass spectrometry and the direct analysis of daughter ions (DADI), the major lipid components of the scent gland secretions of the adder, *Vipera berus,* the saw-scaled viper, *Echis carinatus,* and the mamushi, *Agkistrodon halys,* were identified as cholesterol and a series of fatty acids ranging from C_{15} to C_{22} (Razakov and Sadykov, 1986). The vipers possessed only the fatty acids from C_{15} to C_{18}, whereas the crotalid had only C_{20} to C_{22} acids. Taxonomic differences thus do exist in these glandular components as suggested by Oldak (1976).

Gas chromatography–mass spectrometry identifies the cloacal scent gland secretions of the Dumeril's ground boa, *Acrantophis dumerili* (Simpson et al., 1988). The secretions consisted primarily of lactic, palmitic, oleic, and stearic acids, cholesterol, and cholestenone.

Impressed with the marked interspecific differences in cloacal scent gland secretions, Oldak (1976) conducted a TLC study of the secretions of 25 species of snakes. Scent gland secretions were partitioned in diethyl ether and applied in 50-μl aliquots to the TLC plates. The plates were developed in petroleum ether, diethyl ether, and glacial acetic acid (70:30:1) and visualized with mercuric sulphate. Distinct differences appear between the sexes in a species; thus, only sexually mature females were used in this study. In addition, Oldak noted differences related to sexual maturity in the composition of the secretions. The results indicate that there are very clear-cut differences among and within the species. Much of the odor associated with the secretions appears to be associated with the band in the region of a cholesterol ester standard and with the band at the level of the free fatty acid standard. Oldak suggests that identification of the odorous lipids of these various species would complement other taxonomic indicators in elucidating systematic relationships. He argues that it should not be surprising that for the chemosensorily oriented snakes, biochemical cues would be an ideal taxonomic indicator. In an earlier study, similar results were obtained when the proteins extracted from scent gland secretions of 20 species of snakes were separated by gel electrophoresis (Oldak and Hebard, 1966). Again, taxonomic groups show distinct differences; however, no differences were observed between the sexes.

C. Vomeronasal System

Most terrestrial vertebrates have two major chemosensory systems for processing chemical stimuli, the olfactory system, and the vomeronasal system. The essential role of the vomeronasal organ in searching for prey was investigated by Baumann (1927, 1929) and Naulleau (1965, 1967) in European vipers (*Vipera berus* and *V. aspis*) and by Chiszar et al. (see reviews 1980, 1983) in rattlesnakes (*Crotalus*).

The most extensively studied group of snakes in neurophysiological investigations is the garter snakes of the genus *Thamnophis*. They make ideal models for studies of the role of the chemical senses in species-typical behaviors. Investigations into the role of the vomeronasal system in trailing of prey items have been conducted for more than 50 years. Early investigations of the role of the vomeronasal system in trailing prey items suggested that the vomeronasal organ in snakes is activated when odorant cues are delivered by the tongue to the vomeronasal ducts in the roof of the mouth (Broman, 1920; Kahmann, 1932; Abel, 1951). The morphology and histology of the vomeronasal organs in reptiles have been studied extensively (Parsons, 1959a,b, 1967, 1970). Removal of the tongue (Wilde, 1938; Burghardt and Pruitt, 1975), or closure of the vomeronasal ducts by suturing (Kubie and Halpern, 1979), abolishes those behaviors dependent on a functional vomeronasal system such as prey detection. Tongue-flicking is a behavior unique to all snakes and to most lizards (see Gove, 1979, for review of tongue-flicking behavior in squamates). Snakes presented with novel chemical stimuli increase their tongue-flick rates (Gove and Burghardt, 1983). The tongue is capable of delivering chemical cues directly to the vomeronasal organ (Kahmann, 1932; Halpern and Kubie, 1980), although other investigators believe processes in the buccal cavity of snakes and some lizards (varanids) aid in delivery of cues to the vomeronasal ducts (Gillingham and Clark, 1981; Pratt, 1948; Oelofsen and Van den Heever, 1979). This issue does not appear to have been adequately resolved. Electrophysiological evidence suggests that tongue-flicking activates the vomeronasal/accessory olfactory bulb (Meredith and Burghardt, 1978).

Vomeronasal functioning is crucial for male courtship of reproductively active females during the breeding season. A number of studies have clearly demonstrated the essential role played by the vomeronasal organ in the initiation of species-typical courtship behavior. For example, obstruction of the olfactory receptors with cotton plugs coated with Vaseline and inserted into the external nares does not prevent courtship behavior, although the behavior that is exhibited is sporadic (Noble, 1937). If snakes have portions or all of their tongue removed, courtship is completely abolished. This led to the conclusion that male courtship in snakes is controlled primarily by the vomeronasal organ (Noble, 1937).

Because the snakes in Noble's (1937) studies were probably distracted from courting due to the discomfort of the experimental manipulations, Kubie et al. (1978) modified the study, substituting transections of olfactory and vomeronasal nerves for the cotton plugs and tongue removal of the earlier study. The results were unequivocal. Males with a cut vomeronasal nerve do not court, whereas males with a cut olfactory nerve can and do court sexually attractive female garter snakes, as do the control males with no nerve cuts. Courtship behavior in the male cannot be mediated by the gustatory or trigeminal systems because the nerves subserving these systems had not been damaged in the vomeronasal nerve ablation group. In addition, the authors tested and found no evidence supporting Noble's (1937) suggestion that a tactile organ found on the chins of male garter snakes plays a role in courtship.

A functional vomeronasal system is also essential to the male European adder, *Vipera berus*, during the breeding season (see Section II.E and Andrén, 1982). Male adders with anesthetized vomeronasal systems are unable to locate and court females. In addition, previously dominant males with anesthetized vomeronasal systems do not recognize rival males and retreat from conflicts with individuals that have been previously found to be their subordinates.

Clearly the functioning of the vomeronasal system and the complex and wide range of behaviors mediated by this system are of special interest to those interested in reptilian pheromones. An excellent review of this material can be found in Chapter 7.

D. Aggregation

Although snakes have been considered basically nonsocial animals (Prater, 1933) and the least social of all the reptiles (Brattstrom, 1974; Wilson, 1975), this thinking is now being revised (Burghardt, 1980; Carpenter, 1977). Indeed, many studies over the past 50 years have indicated that many snakes exhibit varying degrees of sociality throughout the year. For instance, snakes are known to aggregate in the spring and fall at overwintering dens or hibernacula (Neill, 1948; Woodbury et al., 1951; Carpenter, 1952, 1953; Klauber, 1956; Hirth, 1966; Parker and Brown, 1973; Aleksiuk, 1976; Gregory, 1982; Duvall et al., 1985). Aggregations of 10,000 or more snakes have been reported at individual dens of the Canadian red-sided garter snake, *Thamnophis sirtalis parietalis*, in the Interlake region of Manitoba, Can-

Fig. 4.3 *Thamnophis sirtalis parietalis.* Thousands of males on the floor of a hibernaculum in Narcisse, Manitoba, Canada. (Photo by R. T. Mason.)

ada (Aleksiuk, 1977) (Fig. 4.3). Aggregations of snakes for purposes other than hibernation have also been reported (Noble and Clausen, 1936; Finneran, 1949; Fitch, 1952; Tinkle, 1957; Barbour, 1962; Kropach, 1971; Reichenbach, 1983; Price, 1988). Gravid female snakes have been reported to aggregate, presumably for some benefit in gestating offspring (Blanchard, 1937; Finneran, 1953; Fitch, 1960; Cook, 1964; Fowler, 1966; Gilhen, 1970; Gregory, 1975; Swain and Smith, 1978; Gordon and Cook, 1980; Graves et al., 1986). Two excellent and very thorough reviews have recently appeared (Gillingham, 1987; Gregory et al., 1987).

It is not clear what proximate factors are responsible for aggregations in the field. In the laboratory, temperature, humidity, and stress have been demonstrated to be directly related to the incidence of aggregation (Noble and Clausen, 1936; Dundee and Miller, 1968; Graves et al., 1986). However, whether any or all of these are major factors operating in nature has not been tested directly.

The first experimental investigations into aggregation behavior in snakes were those conducted by Noble and Clausen (1936) using primarily *Storeria dekayi.* Aggregation in this species occurs throughout the year, except that adult females do not aggregate during gestation in the months of June and July. To determine which of the senses is directly involved in the aggregation response, various senses were impaired or eliminated completely for particular experiments. Visual cues were eliminated by covering the eyes with tape or collodion darkened with india ink. Olfaction was eliminated by plugging the nares with cotton plugs coated with Vaseline and then capping the nares with collodion. The vomeronasal system was deprived of input by severing the tongue caudal to the bifurcation. In addition, the Jacobson's organ was destroyed by cauterization. Half the experimental groups were tested in the dark; the other half were tested in the light. During the experiments, aggregation was induced by shaking the cage or tapping the glass sides of the aquarium. The usual reaction of snakes in response to these stimuli was to aggregate. The results indicated that visual cues were important, as the blindfolded animals take longer to aggregate than controls. Animals deprived of olfactory cues never aggregate in the dark but do aggregate in the light. Tongueless animals aggregate, although somewhat more slowly, in the dark. In a blindfolded and nose-stopped group, no aggregations occur under light or dark conditions. Tongueless and nose-stopped groups can aggregate in the light but never in the dark. It was concluded that vision and, less extensively, olfaction, are the chief senses involved in the aggregation response, whereas the vomeronasal system plays little or no part in aggregation behavior.

An additional experiment has examined the species-specific nature of aggregations (Noble and Clausen, 1936). Aggregation tests with *Storeria dekayi* and *Thamnophis sauritus* do not show heterospecific aggregations. Experiments similar to the aforementioned series show no role for the tongue in species identification. Olfaction appears to be the chief sensory mechanism for discriminating among species.

To find the source of the chemical cues that may have been stimulating the olfactory system of the snakes used in the above study, researchers conducted the following manipulations. To eliminate cloacal odors and secretions, the cloacas of the stimulus animals were taped shut and then covered with Vaseline. To eliminate integumental chemical cues, the entire body was smeared with a thin coating of Vaseline. Aggregation experiments similar to the previous ones followed and indicated that the integument and not the cloaca is the source of the chemical cues utilized by snakes in aggregating.

Present-day investigators of aggregation behavior have criticized the pioneering study by Noble and Clausen. For instance, one investigator has pointed out that shaking the cage or tapping the glass are hardly natural stimuli that might elicit aggregation behavior unless earthquakes are presumed to be a common occurrence (Burghardt, 1983). Other criticisms relate to the repeated measures design of the experiments (Heller and Halpern, 1982a). For instance, the snakes were apparently left in the positions in which they were found from one experiment to the next. Thus, one cannot ascertain whether the animals preferred to stay where they were because of the experimental conditions or were simply immobile. An attempt to rectify these problems led to the following experiments (Heller and Halpern, 1982a). Two groups of eight male garter snakes, *Thamnophis s. sirtalis*, were housed in aquaria with newspaper covering the floor and cardboard shelters occupying the four corners (Fig. 4.4). This was done because groups of snakes are found under shelters more often than in the open. In addition, the snakes found under shelters occur in groups more often than would be predicted by chance. Rotation of the test aquaria does not affect aggregation behavior. If the test aquaria were cleaned between trials, snakes would still aggregate more than would be predicted by chance. Furthermore, the chemical cues snakes use in their aggregation preferences were deposited on the filter papers covering the floor of the shelters; the effectiveness of these cues does not fade after periods of 5 to 7 days. Some individuals repeatedly return to the same shelters even if no conspecifics are

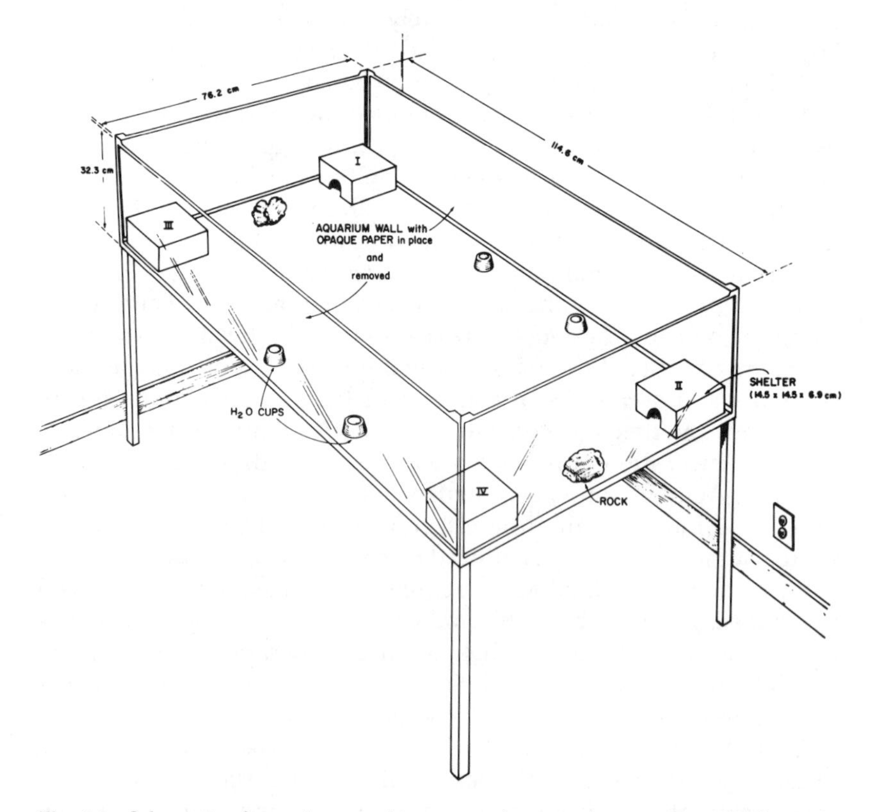

Fig. 4.4 Schematic of aquarium used in aggregation experiments. (From Heller and Halpern, 1982a © The American Psychological Association.)

present; this suggests that these individuals might be able to discriminate their own cues from those deposited by conspecific individuals.

Researchers have also examined the aggregation behavior of ringneck snakes, *Diadophis punctatus arnyi* (Dundee and Miller, 1968). Ringneck snakes released into an arena with equal cover objects do not distribute themselves randomly. Although the tests were not conducted during the breeding season, males were found to aggregate with females, though not exclusively. Males aggregate with other males and females also aggregate with other females. Because the snakes persistently use the same cover objects, it has been hypothesized that the substrate might contain chemical cues that facilitate aggregations, a process referred to as *habitat conditioning*. In one experiment, the substrate from underneath a cover object that had been used by aggregations of snakes over several days was moved to another cover object; some of the snakes then aggregated under the new cover object. It has been speculated that attractant substances, possibly integumentary, cloacal gland, or excretory products, are responsible for this conditioning. It has not been determined whether aggregation is due primarily to visual or to olfactory/vomeronasal-mediated chemical cues.

Ten 17-month-old water snakes, *Nerodia r. rhombifera*, have been examined for their preference for the side of an aquarium that had been soiled by a conspecific versus the side that was clean or soiled by heterospecific *Thamnophis radix* (Porter and Czaplicki, 1974). The results indicate that water snakes tend to avoid an area previously soiled by a conspecific. A concurrent study demonstrates that the plains garter snake, *T. radix*, prefers the side of the aquarium soiled by conspecifics. The opposite response of garter snakes to water snakes may reflect the ecological differences between the two species (Porter and Czaplicki, 1974). For instance, several species of *Thamnophis* are known to aggregate in large numbers (Aleksiuk, 1976, 1977; Gregory, 1975). The study also demonstrates that both species of snakes ignore the chemical cues left by heterospecific individuals. This may reflect the fact that these two species of snakes are not ecological competitors (Porter and Czaplicki, 1974).

The response of neonatal water snakes, *Nerodia s. sipedon*, to conspecific chemical cues has also been assayed as a function of time spent and the rate of tongue-flicking to either side of an aquarium (Scudder et al., 1980). The floor of one side of the aquarium is covered by clean paper towels; the other side is covered by paper towels that have been soiled by two conspecifics for 1 week. The results indicate that neonatal water snakes investigate the soiled side of the aquarium more than the unsoiled side and that tongue-flick rates are higher on the soiled side. In addition, the neonates spend more time aggregating on the soiled side than on the unsoiled side of the aquarium. These results conflict with those of Porter and Czaplicki (1974) recounted above. Neonates may not be important ecological competitors to one another and unlike adults are not competing for potential mates. Thus, in neonates the advantages to aggregation may override the disadvantages that become important in adults.

Allen et al. (1984) argue that the two preceding studies are inconclusive in that species, age, sex, or methodological differences could have been responsible for the behavioral differences. In another study, 24 subjects were divided into groups of 8 individuals, four of each sex, representing members of three sympatric water snakes, *Nerodia r. rhombifera, N. fasciata confluens,* and *N. c. cyclopion.* The animals were 18 months of age and were tested during the breeding season. An aquarium was divided in half with the floor of one-half covered with clean gravel and the other half with gravel that had served as the substrate in the cages of the snakes for 2 weeks. For no species do males and females differ in the number of tongue-flicks emitted to either the stimulus or control gravel. The results indicate that *N. rhombifera* avoids conspecific chemical stimuli as documented by Porter and Czaplicki (1974). In *N. cyclopion* both sexes prefer the conspecific-soiled substrate. Only *N. fasciata* shows a difference between the two sexes; its females prefer the conspecific-soiled substrate, whereas males avoid it. These results also corroborate the findings of Porter and Czaplicki (1974). If the responses of male and female *N. fasciata* are pooled, the species appears to prefer the soiled side of the aquarium. This finding agrees with that of Scudder et al. (1980). Allen et al. conclude that interspecific differences and not methodology or age appear to account for the aforementioned discrepancy between the results of Porter and Czaplicki (1974) and Scudder et al. (1980).

The preference of adult prairie rattlesnakes, *Crotalus v. viridis,* for conspecific-soiled versus unsoiled sides of an aquarium has been tested in individuals collected in the fall and maintained in the laboratory under summer photoregimes (King et al., 1983). The source of the chemical cues was not investigated. Initially (up to 12 hours) both males and females tested in groups prefer the soiled side of the test chamber. However, after 36 hours, almost all the snakes are situated on the unsoiled side of the test chamber. If chemical cues from male and female *Pituophis melanoleucus deserticola* are substituted, male and female rattlesnakes become equally distributed, indicating a neutral response to these chemical cues. Perhaps the snakes avoid conspecific but not interspecific chemical cues and these cues override any tendency to aggregate. King et al. (1983) argue that these findings suggest that chemical cues may be involved in mediating dispersal from hibernacula for feeding purposes.

Ford and Holland (1990) argue convincingly that neither Porter and Czaplicki (1974) nor King et al. (1983) have demonstrated compelling evidence of repulsive pheromones. They cite that neither study specifically evaluates adult males, in which a territorial response would be more likely. The effect on territoriality in snakes of pheromonal cues left on the substrate has been studied using the aggregation behavior of male and female *Thamnophis marcianus,* male *Elaphe guttata,* and male *Agkistrodon contortrix* in large (270 cm) circular tanks. Tests have been conducted both in the breeding and nonbreeding seasons of each of these species. If pheromonal cues deposited on the substrate are involved in mediating territoriality, then the two species displaying male combat (*Elaphe* and *Agkistrodon*) should exhibit male avoidance of conspecific male cues. The garter snakes that lack male combat would then be expected neither to prefer nor to avoid male cues. Unlike expectation, male *Thamnophis marcianus* do not avoid male semiochemicals, either during the breeding or the nonbreeding season. However, neither *Elaphe* nor *Agkistrodon* significantly avoid male semiochemicals, even during the breeding season, at which time such cues might play a role in territorial behavior. Ford and Holland suggest that the results may indicate that snakes are not territorial or at least that no semiochemicals are involved in the spacing of snakes. However, they note that the study does not provide opportunities for males to engage in combat behavior and possibly only loser males avoid conspecific male odors. Such hypotheses certainly need to be tested in the future. Ford and Holland also note the great paucity of studies on spatial relationships, not only in snakes but in reptiles in general. Despite some studies of lizards, this area of research certainly warrants further investigation. The interesting idea that semiochemicals are used to repulse competitors and establish territories needs testing.

The grouping tendencies and species-discriminating abilities of neonatal snakes have been assayed by releasing 20 *Storeria dekayi* and 20 *Thamnophis sirtalis* into an aquarium with a clean gravel substrate and eight cover objects under which the snakes could hide (Burghardt, 1983). Researchers observed the snakes for 11 days and recorded their positions in the cage. The results indicate that both species use cover objects as opposed to remaining in the open. Both species form aggregations primarily with conspecifics. Changing the cover objects has no effect, but rotating the cage does tend to affect the aggregating tendencies of *Storeria,* suggesting that visual cues may be important to this species. Substrate habitat conditioning occurs as indicated by the preference of individuals for certain locations in the cage. Burghardt (1983) argues that the fasting snakes did not produce fecal material as a potential source of chemical cues; however, other chemical cues from the cloaca or integument have not been tested.

Because the vomeronasal system plays a mediating role both in courtship and trailing behavior, a few investigators have reexamined the chemosensory mechanisms involved in the detection of the chemical cues governing aggregation behavior in snakes. Burghardt (1980) has removed the tongues of 12 *Thamnophis sirtalis* shortly after birth to examine the tendency and ability of snakes deprived of their vomeronasal system to aggregate under cover objects. The experimental animals usually are randomly distributed under cover objects. In contrast, most of the controls tend to aggregate under particular ones. Unlike Noble and Clausen (1936), Burghardt (1980) concludes that the vomeronasal system is primarily responsible for aggregation and that visual and olfactory cues play a minor role. Those aggregations that do occur among tongueless snakes are probably assisted by visual or tactile stimulation.

The roles of the visual, olfactory, and vomeronasal systems have been examined concurrently in garter snakes in a fine study by Heller and Halpern (1982b), which challenges Noble and Clausen's (1936) contention that removal of the tongue would eliminate the input to the vomeronasal system. For instance, small amounts of odorants that touch the lips of tongueless snakes can still be detected by the vomeronasal system (Halpern and Kubie, 1980). In addition, the Vaseline-covered cotton plugs used to occlude the nares may cause behavioral deficits for reasons other than sensory deprivation. For example, breathing could have been impaired (Heller and Halpern, 1982b). Two experiments control for these variables. The first experiment identifies the relative roles of the vomeronasal and olfactory systems in enabling snakes to return to previously preferred cover objects. This experiment involves transecting the vomeronasal nerve in one group, severing the olfactory nerve in another group, and giving a sham operation to a third group. The results indicate that snakes with vomeronasal nerve lesions do not return to locations they had preferred previously (Fig. 4.5). Whenever tested with groups of sham animals, they return to previously preferred cover objects. Those animals with olfactory nerve lesions improve in their choices of cover objects through time. The second experiment determines the sensory systems involved in aggregation whenever the substrate has not been conditioned previously. The animals are tested in an aquarium with four cardboard shelters, one in each corner. Groups of subjects are released into the aquarium three times daily and their location noted. The results of the second experiment indicate that blindfolded and anosmic snakes aggregate at levels comparable to those of controls;

however, snakes with vomeronasal ducts sutured closed exhibit significantly depressed aggregation tendencies. This demonstrates that snakes need a functional vomeronasal system in order to return to previously conditioned cover objects and to aggregate there; neither the visual nor olfactory system is essential for this behavior (Heller and Halpern, 1982b).

The sources of these aggregation cues as well as their chemical nature are unclear at this time. Tests in the nonbreeding season show that the cloacal sac substance of blind snakes, *Leptotyphlops dulcis*, attracts conspecifics of both sexes and repels sympatric species and ophiophagous snakes (Watkins et al., 1969). In *Storeria dekayi*, cloacal sac odors are aversive to conspecifics, whereas skin-derived chemical cues serve as a source of attraction. Another alternative chemical source could be fecal material. Snakes will increase their frequency of defecation when placed in a novel environment (Chiszar et al., 1980).

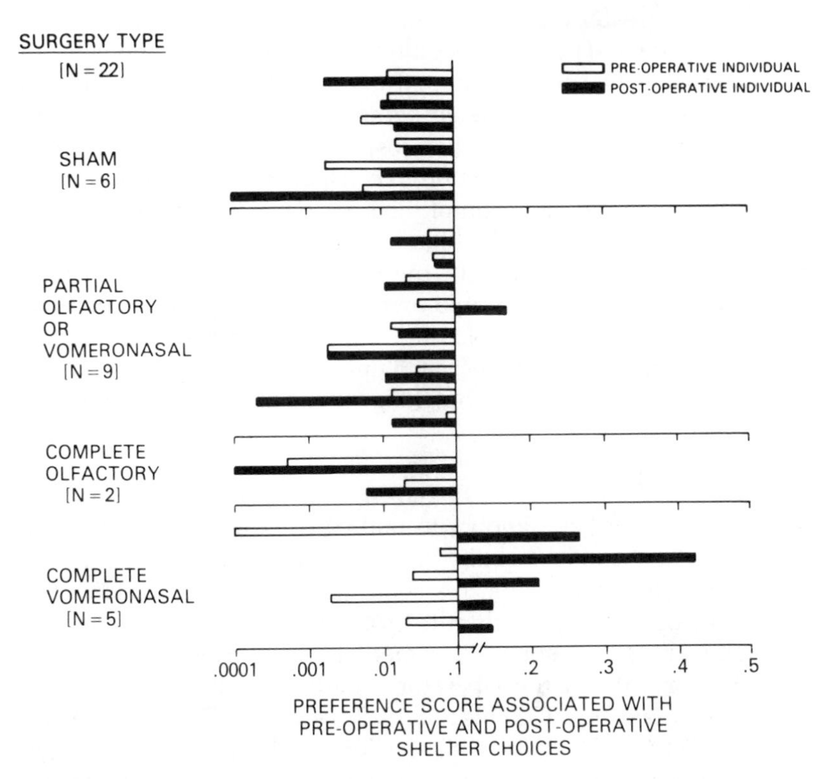

Fig. 4.5 Preference scores (binomial probability values) of 22 animals whose data were individually analyzed for preoperative and postoperative shelter preferences. (From Heller and Halpern, 1982b © The American Psychological Association.)

Once the area is soiled, the frequency of defecation decreases markedly. Another study tends to rule out this hypothesis, at least in neonates (Burghardt, 1983).

Possible sources of semiochemicals that attract snakes to aggregation sites have been investigated by scoring methylene chloride extracts of shed skins of adult *Thamnophis radix,* a congener, *T. s. sirtalis,* and *Crotalus v. viridis* for the ability to induce aggregations of neonatal *T. radix* (Graves and Halpern, 1988). A solution of the snake skin extracts is placed on a filter paper under a shelter on one-half of an aquarium while a solvent control is placed under a shelter on the other half of the aquarium. After 18 hours the number of snakes occupying each side of the arena is counted. The results conclusively demonstrate that neonatal plains garter snakes can both discriminate and significantly prefer to aggregate on the side of the aquarium having conspecific extract. In addition, the mean number of neonates aggregating on the side of the aquarium with conspecific extracts is significantly greater than for the sides having either of the two heterospecific extracts. There is no significant difference in response to the two heterospecific species. Thus, these neonates are attracted to the lipid cues sequestered in the epidermis of adult female conspecifics but not to the skin lipids of heterospecific individuals. It has been suggested that neonate attraction to conspecific odors has the role of chemically mediating location of hibernacula. This is very important for those interested in snake hibernation because neonatal *T. sirtalis parietalis* do not overwinter in the same dens as adults (Gregory, 1982), and one of the great unknowns about this species is the overwintering site of neonates. No one has ever described a large aggregation of neonates in any hibernaculum. Thus, this question remains a very important gap in our knowledge of the life history of this interesting species.

E. Trailing

Snakes have long been known to trail conspecifics. There is a large body of literature on the ability of snakes to trail prey items, and we have several fine reviews of this behavior (Halpern, 1980, 1983; Halpern and Kubie, 1984; Burghardt, 1970, 1980; Ford, 1986). Numerous anecdotal reports and some experimental studies have described trailing behavior both in the laboratory (Baumann, 1929; Noble, 1937; Noble and Clausen, 1936; Pisani, 1967; Devine, 1977b; Gehlbach et al., 1971; Ford, 1978, 1981, 1982; Heller and Halpern, 1981; Burger, 1989; Costanzo, 1989; Ford and Holland, 1990) and field (Bishop, 1927; Truitt, 1927; Fitzsimons, 1962; Davis, 1936; Neill, 1948; Anderson, 1947; Munro, 1948; Finneran, 1949; Carpenter, 1952; Gardner, 1955; Fitch, 1958b, 1960; Aleksiuk and Gregory, 1974; Gregory, 1982;

Duvall et al., 1985; Andrén, 1986; Graves et al., 1986; Gregory et al., 1987; Slip and Shine, 1988). At least 10 species in five families of snakes are known to utilize pheromone trails (Ford and Schofield, 1984). A table of studies documenting trailing behavior in snakes is provided by Ford (1986).

Trailing studies have focused primarily on three behaviors: aggregation (covered in Section II.D), migration to winter hibernacula, and detection and location of conspecifics during the breeding season. Anecdotal reports of trailing by snakes to locate winter hibernacula are common. In some areas of Georgia, "rattlesnake crossings" are observed during the autumn migrations of *Crotalus horridus* to winter hibernacula (Neill, 1948). Rattlesnakes move along these trails, which apparently are used annually. Klauber (1956) is more direct and states (without experimental evidence) that without a doubt rattlesnakes use their acute sense of smell to follow leaders back to overwintering dens and to find mates in the spring. That rattlesnakes locate overwintering hibernacula in Wisconsin by trailing conspecifics has also been proposed by others (Messeling, 1953). Trailing of conspecific chemical cues has been proposed for the homing abilities of *Masticophis t. taeniatus* and *Crotalus viridus lutosus* in Tooele County, Utah (Hirth, 1966). An anecdotal report of the successful return of a blinded *Masticophis taeniatus* to its den, after being displaced 235 m, suggested that chemical cues may be involved (Parker and Brown, 1980). Neonatal rattlesnakes may find hibernacula by trailing adult conspecifics, a hypothesis that has been proposed by several investigators.

Field studies with radiotelemetrically implanted *Crotalus horridus* suggest that neonatal *C. horridus* that maintain close associations with adult conspecifics hibernate in the same burrows (Reinert and Zappalorti, 1988). Whereas trailing behavior enabled the neonates to follow the adults, visual cues do not seem to have been ruled out (Reinert and Zappalorti, 1988). Laboratory experiments investigating the trailing ability of 20 newborn male and female *C. horridus* in autumn, when wild snakes migrate to winter hibernacula, demonstrate that these animals follow scent trails of conspecific newborns and adult males and females in a Y-maze (Brown and MacLean, 1983). This study shows that newborn rattlesnakes follow adult male trails most accurately (95%), adult female trails less accurately (85%), and newborn trails least accurately (72%). These findings have been claimed to support the idea that trailing serves as a mechanism by which newborn snakes locate a hibernaculum. It has also been suggested that spring aggregations of neonatal rattlesnakes at the birthing rookeries may serve to familiarize the young rattlesnakes with the odors of other rattlesnakes and the odors of the den region. Authors suggest that shed skins and the skin lipids released during shedding act as chemical cues important in trailing. Such cues may then serve to guide the rattlesnakes back to the dens in the fall. This idea has also been suggested for juvenile prairie rattlesnakes, *Crotalus viridis* (Graves et al., 1986).

Few experiments have tested trailing in the context of locating overwintering hibernacula. The ability of hatchling *Pituophis m. melanoleucus* to detect and follow the chemical trails of conspecifics has been examined during September and October (Burger, 1989). In this study, 25-35-day-old pine snakes have to choose between conspecific chemical cues and blank arms in a Y-maze. Four regimes have been tested: (1) For controls, no chemical cues are in either arm of the Y-maze; (2) one arm of the maze contains chemical cues from a hatchling conspecific versus no cues; (3) adult conspecific chemical cues are in one arm and no chemical cues are in the other; (4) one arm contains chemical cues from *Lampropeltis getulus,* an ophiophagous predator, and the other arm lacks chemical cues (see Section II.J). Results indicate that the pine snake juveniles do not discriminate in the control experiments, nor do they prefer the hatchling conspecific cues to the blanks. However, they significantly prefer the arm of the maze containing adult conspecific chemical cues. Because pine snakes migrate to hibernacula in the fall, Burger (1989) suggests that juvenile snakes may utilize chemical cues deposited on the ground to detect and trail adult conspecifics back to hibernacula. It also might be beneficial to follow other juvenile chemical trails. Failure to demonstrate this behavior may indicate that it is disadvantageous to follow other naive conspecifics or, perhaps more likely, the chemical cues deposited in these tests are insufficient to elicit detection and choice (Burger, 1989).

Other tests assay the ability of adult garter snakes from a communally denning population in Wisconsin to follow conspecific scent trails under autumn conditions use 28 *Thamnophis sirtalis* captured during October (Costanzo, 1989). The floor of the arena is covered with paper and cat litter; then a large female conspecific is used to create chemical trails, being guided down the arena by means of a bottomless tunnel and through one of a line of 17 nail gates at the end of the arena. The experimental animals try to stay near the arena walls, which confounds the results somewhat. The results indicate that at least some of the animals can follow the trails exactly or almost so. However, most snakes apparently do not trail; they fail to choose the correct gate or refuse to trail. Costanzo (1989) concludes that garter snakes of both sexes can follow the chemical trails of a conspecific. He suggests that these results concur with field observations that *T. sirtalis* use common and well-defined "travel lanes" in their approach to dens (Costanzo, 1986, 1988). Most trail-following studies have been performed under ideal conditions; however, future studies should address the influence on trailing abilities of environmental factors such as temperature, precipitation, humidity, wind velocity, and substrate texture. Costanzo (1989) also provides an interesting discussion of the taxonomic and geographical limits of pheromonal use in trailing to den locations.

The mechanisms involved in the way snakes actually find their way back to dens are still unknown. Some investigators have suggested that visual cues are involved, whereas others favor celestial cues (reviewed by Gregory, 1982; Gregory et al., 1987). Chemical cues left by conspecifics may be used by juveniles (Burger, 1989); however, consideration of chemical trailing in adults always obligates one to explain how such chemical cues could be maintained while being exposed to the environment for periods of up to 1 year (see Graves et al., 1986, for a discussion of this issue). In addition, trailing of conspecifics does not explain how the first animal finds its way back to the den. It has been suggested (Costanzo, 1989) that older, experienced snakes may locate dens by other cues, including solar and celestial information (Landreth, 1973; Newcomer et al., 1974), topographic landmarks (Parker and Brown, 1980), and polarized light (Lawson, 1985). Younger, less experienced snakes then would rely more on chemical cues to locate dens than would adults. Probably several cues are being used concomitantly. These unknowns need to be addressed in order to understand the cues involved in successful trailing to hibernacula.

Most reports of trailing in snakes concern reports of trailing during the breeding season. Male *Vipera aspis* detect and trail females during the breeding season (Baumann, 1929). Places at which the female has opened her cloaca elicit particular attention from the trailing males.

Noble and Clausen (1936) performed one of the earliest laboratory studies on trailing in the snakes *Storeria dekayi*. In the first experiments, conducted in November and December, a trail was generated by either male or female snakes, deriving chemical cues from the cloacal sac secretion or the dorsal integument. The results indicate that the trailing animal does not distinguish sex. That is, males trail both females and males, whereas females trail both males and females. However, males and females do not trail the cloacal sac secretion of conspecifics but tend actively to avoid them.

Storeria dekayi collected in the field in April exhibit active courtship upon return to the laboratory (Noble and Clausen, 1936). The experiments have been repeated using male and female *Storeria* exposed to trails of male and female integumental and cloacal sac cues. During the breeding season, a time of sexual activity, males follow female scent trails much more rapidly than during the nonbreeding season but do not follow trails of other males. Females behave as they do during the autumn trials, showing no preference for the trails of either sex. As in the autumn trials, both males and females actively avoid trails made with cloacal sac secretion of both males and females. Noble and Clausen (1936) conclude that regardless of the time of year, chemical cues present in the skin of these snakes attract conspecifics. However, during the breeding season, females produce these cues or probably another substance, thus attracting males and enabling them to determine the sex of the individual. In addition, contrary to Baumann's (1929) findings, the more obvious cloacal sac secretion apparently plays no role in attraction of conspecifics. Indeed, conspecifics seem to avoid it.

Noble (1937) has tested the trailing abilities of breeding male and female *Thamnophis sirtalis* and *Thamnophis butleri.* Either the dorsal surface of the skin or cloacal secretions are applied onto glass plates. Males consistently prefer the trails of the dorsal integument of females to cloacal secretions. Males do not trail the cloacal or integumental cues of other males. Several male black racers, *Coluber constrictor flaviventris,* have been observed apparently trailing a female during the breeding season in Kansas (Lillywhite, 1985). It has also been speculated that pheromones produced by sexually attractive females allow the well-developed trailing abilities in snakes to structure movements of snakes in conspecific groups (Lillywhite, 1982).

Whenever a male *Vipera berus* encounters the trail of a conspecific during the breeding season, it stops and tongue-flicks with tongue extensions of long duration (Andrén, 1982, 1986). It then follows the trail, tongue-flicking and moving the head from side to side across the trail. If the trail leads to a female, the male begins courtship behavior. If the male encounters another male, it engages in combat behavior. Combat in adders is similar to that observed in crotalines; the males entwine their necks and raise the anterior parts of their bodies off the ground. The ritualized fighting usually ends by one male forcing the body of the other male to the ground. Males mate-guard during the breeding season, and the defending male usually prevails. The loser retreats quickly. During the breeding season, all the males trail attractive females that have shed after emergence from the hibernaculum. To test the role of the vomeronasal organ in trailing abilities of male adders, their vomeronasal organs have been anesthetized by covering the vomeronasal ducts with Xylotape. These males then are unable to trail females, nor can they locate rival males by trailing, although they apparently recognize them by visual cues. Whenever experimental males encounter rival males, the experimental males always retreat. This is in contrast to their aggressive nature both before treatment and after the vomeronasal sense is restored. Thus, male adders tongue-flick chemical cues from the substrate, which informs them that another adder has passed. Chemical cues from the skin of the adder identify the sex and reproductive condition of the individual.

The trailing abilities of several species of snakes exposed to the odors of presumed prey items and to conspecific chemical cues has been investigated on an octagonal course (Gehlbach, Watkins, and Kroll, 1971). For each trial the snake that is to serve as the odor source is allowed to travel five times around the octagonal track. As the deposition of cloacal material is not mentioned, the trail is presumably composed primarily of integumental cues. Snakes, collected between March and June during the presumed breeding season, included male and female *Leptotyphlops dulcis* and *Typhlops pusillus* and three colubrids, *Sonora episcopa, Tantilla gracilis,* and *Virginia striatula.* Ten snakes of each species encountered their own trails as well as trails of conspecific individuals of the same and of opposite sex. The results of these experiments indicate that typhlopid and leptotyphlopid snakes trail further and faster than do colubrids. However, all the snakes tested follow their own trails farther than trails of individuals of the same sex; all follow trails of the opposite sex more than any other trails. Only the males of *Leptotyphlops dulcis* trail significantly better than females in this experiment; however, as males of the other species do not trail better than females it has been suggested that the difference may be due to experimental error. Because this study was conducted during the breeding season, the results may be interpreted as being selectively advantageous for reproduction. The ability of these snakes to trail conspecifics, regardless of sex, may also facilitate aggregational hibernation or estivation. These behaviors may serve to conserve moisture and reduce temperature fluctuations (Dundee and Miller, 1968; Noble and Clausen, 1936). It was suggested that these behaviors may be selectively advantageous as the species used are largely nocturnal and fossorial; however, these hypotheses have not been tested.

Conspecific and congeneric scent trailing in several species of newborn, infant (6 months old), and adult garter snakes (*Thamnophis sirtalis sirtalis, T. s. parietalis, T. couchi,* and *T. elegans*) has also been investigated (Heller and Halpern, 1981). The behavior of subjects is observed in a Y-maze, the floor of which is covered with filter paper and the right arm blocked off. Before each trial, a stimulus animal is allowed free access for 20 hours to all portions of the maze except the blocked right arm. The experimental animals are placed in the start box, and the test ends when the snake reaches the goal box at the end of the right or left arm of the maze. A significantly higher proportion of animals chooses the previously conditioned arm of the maze as opposed to the unconditioned arm. In addition, snakes tend to choose the arm chosen by the previously tested animal. This may be due to the accumulation of chemical cues from a number of the experimental animals as well as the original stimulus animals. A separate experiment, controlling for this possibility, yielded similar results. Snakes tend to follow each other even if the maze is not extensively conditioned. The sexes of the stimulus and experimental animals that followed the trails were not mentioned, making it impossible to assess the involvement of reproductive cues. However, these results concur with the results of previous studies conducted in the nonbreeding season. The trails of congeners, as well as conspecifics, are followed, leading to the hypothesis that chemical cues deposited on the substrate attract snakes to particular locations (i.e., hibernacula or aggregation sites).

The most extensive studies of trailing in snakes have been conducted on several species of *Thamnophis*. A five-armed maze has been used for a trailing study of male *T. proximus* and *T. sirtalis parietalis* during the months of June and October (Ford, 1978). Females were induced to travel up one of the arms and were then removed. Males of both species trail conspecific, but not heterospecific, females. A follow-up study of the same species conducted between April and October used an arena with pegs in the floor traversed by the stimulus females to produce a trail. The ability of male *T. proximus* to trail conspecific females does not change significantly between April and October; however, the ability of male *T. sirtalis* to trail conspecific females declines (Ford, 1981). This result supports the hypothesis that reproductive activity imparts a strong inducement to trail in male garter snakes. The ability during the breeding season in April and May of male *T. butleri* and male *T. sirtalis* to discriminate between the pheromones of conspecific and those of the females of heterospecific congeners has also been tested in a Y-maze (Ford, 1982). Females collected in the field produced the chemical trails followed in each case by 20 males. Male *Thamnophis* discriminate trails of their own species from those of sympatric heterospecific females. However, males of *T. butleri* chose trails of female *T. sirtalis* over blank arms, which suggested the presence of an aggregation drive (Ford, 1982). In contrast, males of *T. butleri* do not discriminate trails of their own females from female trails of the allopatric *T. radix*, suggesting that the trail chemicals produced by females of these two species are the same or very similar. The chemistry of the integumental cues may be subject to selection for reproductive isolation after contact is established between populations previously separated geographically.

Two subspecies, *Thamnophis r. radix* and *T. radix haydeni*, along with *T. sirtalis*, have been used in a series of trailing experiments using a Y-maze (Ford and Schofield, 1984). Reproductively active *T. radix* of both subspecies discriminate the trails of their own females from those of the sympatric and distantly related females of *T. sirtalis*. However, both subspecies of *T. radix* distinguish trails of their own females from those of female *T. marcianus*, which are largely allopatric. Male *T. radix* follow trails of *T. marcianus* when these are paired with a blank arm. Thus, chemical cues of *T. radix* and *T. marcianus* appear to be chemically similar to each other yet different from those of *T. sirtalis*. These results are unexpected because *T. radix* and *T. marcianus* are largely allopatric; thus, there is no reason for a strong selective pressure for species specificity in their trail pheromones. These two species once may have overlapped geographically to a much greater extent than at present (Ford and Schofield, 1984); after separation, males of *T. radix* may have retained the ability to discriminate pheromone trails; however, such hypotheses await further study.

The ability of 25 male *Thamnophis marcianus* to discriminate the pheromone trails of conspecific females during the breeding season from those of the allopatric *T. cyrtopsis* has been tested in a Y-maze (Ford and O'Bleness, 1986). Male *T. marcianus* distinguish pheromone trails of their own females from those of heterospecific females. Male *T. marcianus* follow trails of allopatric females if these are paired with a blank arm. The authors suggest that the geographic ranges of these two species once overlapped more extensively (Ford and O'Bleness, 1986). Conversely, occasional overlapping of ranges in the present may be enough to maintain chemical differences in the trail pheromone. The hypothesis that the pheromone trails are sex specific has also been tested (Ford and O'Bleness, 1986). No males follow male trails, and females do not follow male or female trails in the spring breeding season or in late summer. These results would tend to downplay the idea of Gregory (1974) that chemical cues of gravid females of other snake species are utilized in trailing to aggregation sites. However, the results do not rule out the possibility that chemical cues could be involved in reaching hibernacula during autumn migrations.

A series of ingenious experiments attempts to ascertain how male snakes recognize the direction to go on a trail in order to locate a female that left the trail (Ford and Low, 1983). A test box with pegs set up in the floor (Fig. 4.6) is conditioned by allowing breeding females to crawl through it. The lateral and dorsal surface of the female presumably leaves chemical cues on these pegs as she moves through the test apparatus. During the breeding season, male *Thamnophis radix* follow this pheromone trail in the correct direction after the female has used the pegs to crawl through the apparatus. When the pegs are reversed, the males trail in the opposite direction; however, removal of the pegs no longer allows accurate judgment of directionality. Thus, males determine directionality of pheromone trails from the positions of the chemical cues on surface objects contacted by passing snakes during locomotion.

The evidence that pheromone trails contain information on the reproductive condition of females is ambiguous (Ford and O'Bleness, 1986). It remains unknown whether the decrease in male trailing activity after the breeding season is due to a change in the chemical composition of integumental cues or to a change in the ability of males to perceive the chemical stimuli. It is still unclear whether all trails contain sexual information (Ford and O'Bleness, 1986).

It seems safe to conclude that during the breeding season male snakes preferentially follow female trails. Females in general tend not to trail females or males. Male snakes discriminate conspecific trails from those of closely related species and prefer the former. However, although males recognize heterospecific trails as different from those of their own females, they will still follow the former trails if no other trail is available. This could indicate that males identify the actual species-specificity using the sex attractant pheromone from the dorsal surface of females. The contact pheromone detected by males on the

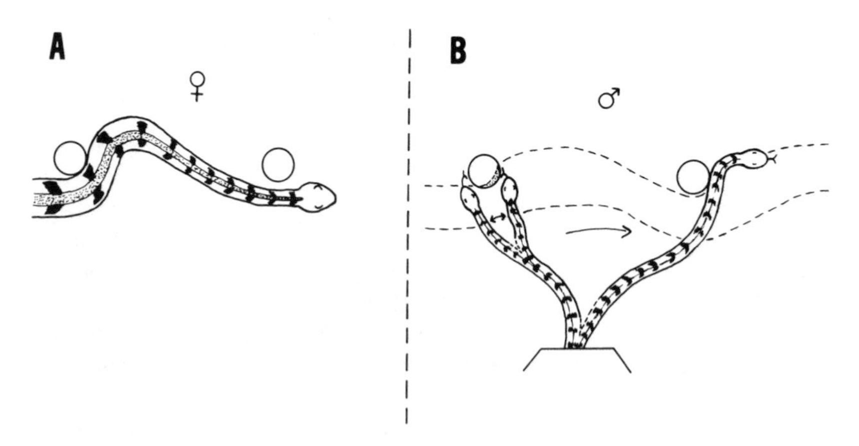

Fig. 4.6 *Thamnophis.* Hypothesized mechanism for male snakes to determine direction from a snake's trail. (A) Female snake laying a trail. Note she pushes against the anterolateral surface of the objects in her environment (pegs). (B) Behavior of a male garter snake when he encounters a female trail. Stipling on the first peg represents the probable deposition of the pheromone on the substrate as a trail. The male tongue-flicks both the anterior and posterior sides of the pegs, and with this sensory assay detects the pheromone on the anterior surface. He then proceeds in this direction, using chemotactic orientation on the trail. (From Ford and Low, 1983.)

skin of females may be the same as the trail pheromone. The ability of males to discriminate species identity would then depend on concentration differences. Thus, a male could detect a trail pheromone and conclude that the individual that left the trail was a female garter snake. However, only by tongue-flicking the dorsum of the female might it gather enough chemical information to discriminate whether the female was a conspecific or heterospecific.

Conversely, the results of these studies may indicate that males are quite effective at discriminating species-specificity and prefer the trails of conspecific females. However, if no other females are available, males will trail and court heterospecific females (Mason, unpublished). If the chances of a male mating in a given breeding season are small, it may be the best strategy to be less choosy of the females with which one chooses to mate. This is certainly the case in many species of anuran amphibians. Whether female snakes would choose to mate with heterospecific males is poorly documented and worthy of investigation.

F. Sexual Behavior

The mating behavior of *Storeria dekayi* and two species of *Thamnophis* was investigated by Noble (1937) 50 years ago. During the breeding season, male *S. dekayi, Thamnophis sirtalis sirtalis,* and *T. butleri,* after encountering an attractive female, start courtship by active investigation of the female with rapid tongue-flicks. The male moves forward along the back of the female, tongue-flicking and pressing his chin onto the dorsum. This behavior continues up the body of the female to the head, reverses abruptly, and is followed by chin-rubbing toward the tail of the female; upon reaching her tail, the male again reverses direction. Eventually, the male aligns his body with that of the female, places his head behind that of the female, and begins searching with his tail for her cloaca. Insertion of one of the hemipenes follows, and mating ensues (Noble, 1937). Equivalent behavior has been described during the courtship activity of many snakes, including boids (Radcliffe and Murphy, 1983), colubrids (Davis, 1936; Noble, 1937; Carpenter, 1977; Bennion and Parker, 1976; Gillingham, 1979; Secor, 1990), natricines (Stein, 1924; Klingelhöffer, 1931; Noble, 1937), viperids (Andrén, 1982, 1986; Schuett and Gillingham, 1988), crotalids (Noble, 1937; Klauber, 1956; Chiszar et al., 1976; Armstrong and Murphy, 1979; Radcliffe and Murphy, 1983), and elapids (Carpenter, 1977).

Early investigators of sexual behavior of *Thamnophis* correctly surmised that males detect a pheromone on the dorsum of the female. Field observations of male garter snakes that vigorously court dead females demonstrate that behavior by the female is unnecessary for elicitation of preliminary male courtship (Truitt, 1927; Aleksiuk and Gregory, 1974; Garstka et al., 1982).

Early investigations into the source of the pheromone that elicits male courtship in snakes focused on the cloacal glands, which emit a powerful odor. Studies on *Vipera aspis* indicate that the locations in which the female had discharged cloacal material are highly attractive and elicit excited searching movements by courting males (Baumann, 1929). However, other workers have disagreed. Cloacal glands occur in both males and females and in young of all ages; the glands secrete actively throughout the year, leading Wall (1921) to argue that they could not serve as the source of attraction, even though many investigators including Darwin had supported the idea that these were the source of attraction. Noble (1937) tested the hypothesis directly by anesthetizing attractive females and placing them on their backs, making sure to exude some of the cloacal sac secretion. The fluid never elicited courtship behavior from courting males. When cloacal sac secretion was rubbed on a glass plate, as was the integument of attractive females, males responded vigorously only to those plates rubbed with the dorsal or lateral skin of the female (Fig. 4.7). Whenever the dorsal skin of males and females is rubbed on the same glass plate, courting males exhibit chin-rubbing only to the skin secretions

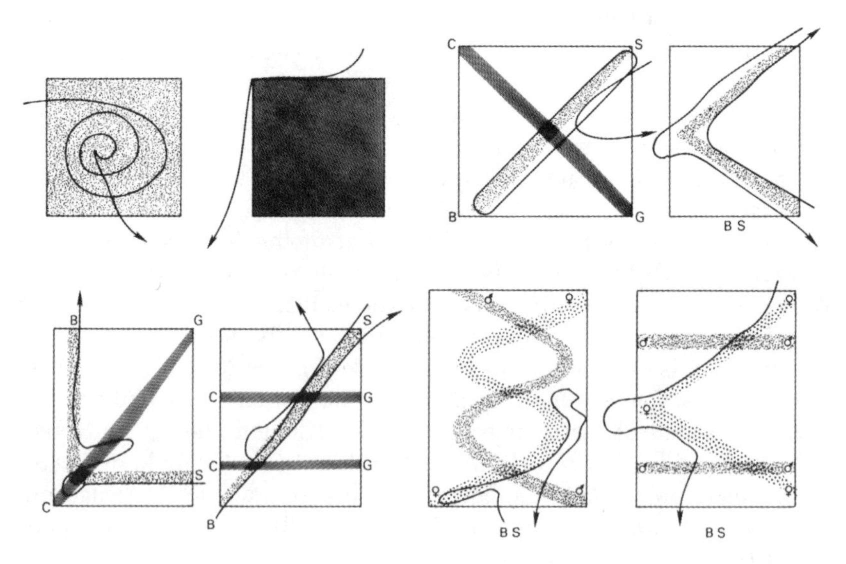

Fig. 4.7 *Thamnophis sirtalis,* Trailing experiments in which various integumental and cloacal secretions of males and females were rubbed on glass plates. Cross-hatched bars represent cloacal gland secretions while stipling represents integumental material. The trails of male garter snakes are marked with arrows. (Redrawn from Noble, 1937. Courtesy, American Museum of Natural History.)

of females and never to those of males. Because blindfolded males chin-rub even the smallest area of the female's back, it is clear that some quality in the skin makes sex discrimination possible (Noble, 1937).

Female common garter snakes, *Thamnophis s. sirtalis,* are attractive in the spring just after emergence from hibernation and before molting. This leads to the hypothesis that some chemical cues attractive to males must be exuded through the skin (Noble, 1937). The nonvolatile nature of the sex attractant pheromone has been confirmed by placing sexually attractive females in a container and covering them with finely perforated paper. Although the males have courted these females minutes earlier and can see them clearly, they do not approach the paper nor attempt to chin-rub it. Furthermore, females in cloth bags placed in the home aquarium of courting males were never investigated. Attempts have also been made to mask the presumed chemical cues originating in the skin of females. Various solutions or solvents (soap and water, 50% or 70% alcohol, ether, quinine sulfate, picric acid, ammonium hydroxide, sodium bicarbonate, oil of wintergreen, Balsam St. Roco) were applied to the backs of attractive females in order to remove the chemical cues affording attractivity; all failed to deter courting males. Only sodium chloride and Vaseline deterred male courtship. The Vaseline undoubtedly blocked the transmission and reception of the skin-derived chemical cues by courting males. The salt was presumably so aversive as to interfere with courtship.

Noble's finding notwithstanding, it is almost axiomatic among zoo workers that snakes breed best after the female has shed her skin (Radcliffe and Murphy, 1983). Indeed, Klauber (1956) describes a male rattlesnake attempting to mate with a freshly shed female skin. The frequency with which courtship and copulation follow skin shedding in snakes strongly suggests that the newly shed snake and skin contain an odor that releases male courtship behavior (Radcliffe and Murphy, 1983).

In this regard, studies of *Vipera berus* are most interesting. Unlike *Thamnophis,* these snakes only start vernal courtship after sexually active males have shed their skins (Nilson, 1980; Andrén, 1982, 1986), although populations in other parts of the range may be more similar to *Thamnophis* in shedding after mating (Saint Girons, 1980). However, as in *Thamnophis,* a sexually active male encountering a newly emerged female immediately begins intense tongue-flicking over the female while courting her by head-nodding (chin-rubbing) up and down her dorsal surface (Fig. 4.8). Behavioral observations (Table 4.1) indicate that a possible pheromone responsible for male courtship is sequestered in or on the dorsal skin of females (Andrén, 1986). Exper-

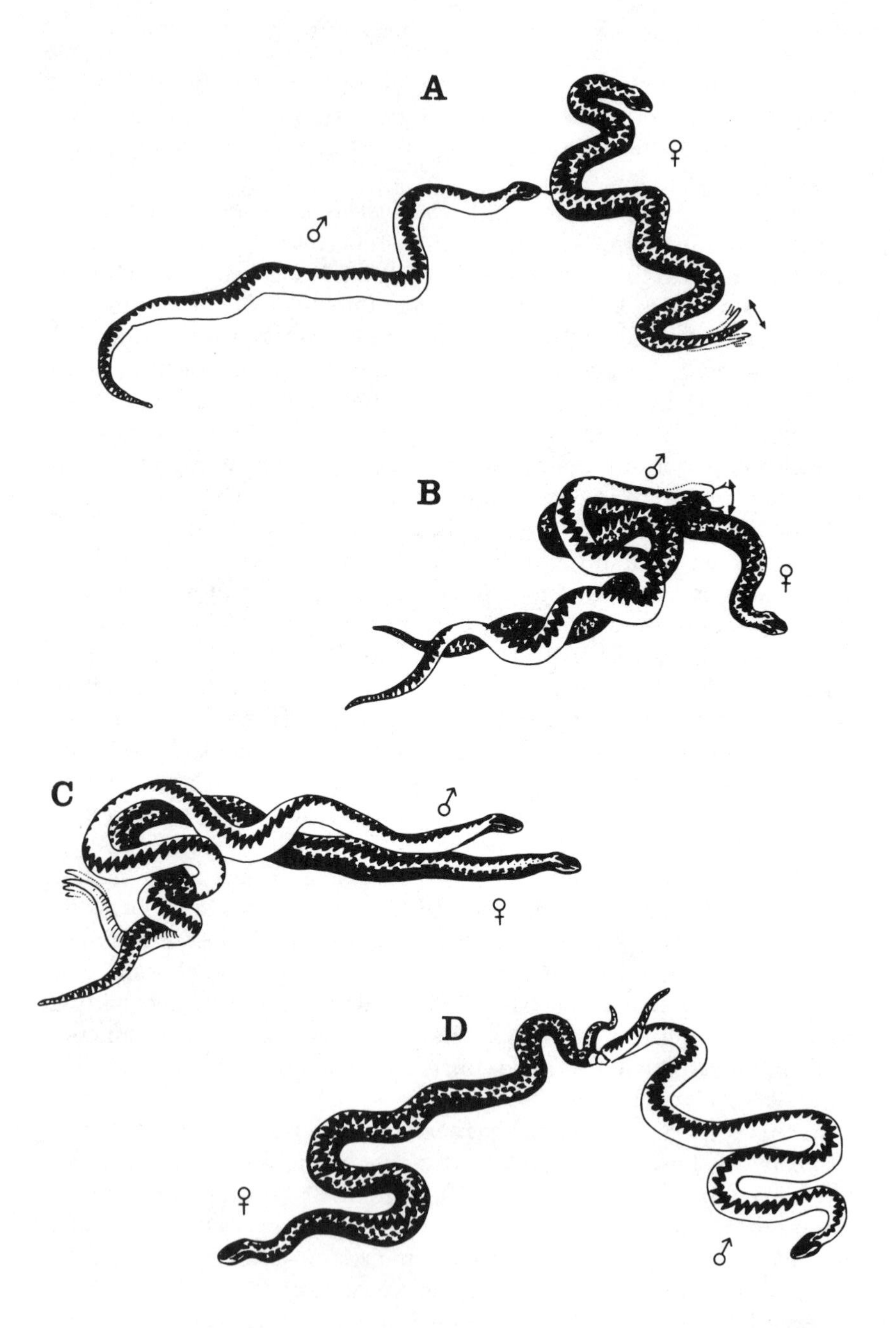

Fig. 4.8 *Vipera berus*. Male-female reproductive behavior patterns. (A) Initial part of courtship, with male tongue-flicking and female tail-vibrating. (B) Intense flicking, with male head-nodding and body contact. (C) Tail-search copulatory-attempt with intense tactile communication. (D) Copulation with female tail-up and male tail-waving. (From Andrén, 1986.)

Table 4.1. Female body parts touched by male tongue-tips during tongue-flicking and intense flicking in courtship interactions in the adder, *Vipera berus* (1708 recordings)

	Male tongue-flicking		Male intense flicking	
	no. of intervals	% of total time	no. of intervals	% of total time
Anterior female body	447	26.0	0	0
Central female body	611	35.8	119	7.0
Posterior female body	87	5.1	21	1.2
Dorsal female body	1,102	64.5	119	7.0
Lateral female body	408	23.9	49	2.9
Ventral female body	7	0.4	0	0

Source: Andrén, 1986.

iments both in the laboratory and the field demonstrate that the vomeronasal system is necessary for inducing courtship behavior in the male adder (see Section II.B). Female adders are courted intensely regardless of whether they have shed. Recently shed males that encounter nonshed males or nonreproductive females generally ignore them (see Mason and Gutzke, 1990, for an analogous situation in lizards).

Another major difference between the behavior of *Thamnophis* and *Vipera* is that male *Vipera* exhibit what some authors have described as territoriality. However, recent debate suggests that there is no convincing evidence for territoriality in any snake (see Ford and Holland, 1990). Duvall et al. (1992) suggest the alternative term, *successive female defensive polygyny,* in which males mate guard females during the breeding season. This term reflects the fact that male *Vipera* guard potential mates but do not seem to defend resources, and thus they cannot be described as being truly territorial.

When a shed male viper encounters another shed male, combat results. The response to the skin-derived chemical cues is very rapid and pronounced. In *Vipera xanthina,* a shed male courted an attractive female in the presence of a nonshed male without any interaction for several days. When the nonshed male abraded a small patch of its old skin, the shed male immediately engaged in combat (G. Nilson as cited in Andrén, 1982). Thus, shedding in male vipers seems to release a chemical cue or pheromone that mediates fighting in males.

Treatment with exogenous estrogen causes intact, reproductively inactive, and ovariectomized female *Thamnophis s. sirtalis* to become attractive to and elicit courtship from sexually active males (Crews, 1976). Thus, ovarian steroids positively affect the production and/or expression of sexual attractivity in female garter snakes. Injection of female *Thamnophis radix* with exogenous estrogen confirms the stimulatory role of the hormone on production of the sex attractant pheromone (Kubie et al., 1978). However, shedding also enhances the sexual attractivity of the females, as it does in *Vipera berus*. Female *T. radix* treated with exogenous estrogen 5 days before shedding are uniformly unattractive. After shedding, these same females are very attractive. In addition, untreated females housed in the same cages as the estrogen-injected females also are more attractive sexually. Presumably, attractant chemical cues from the skin of the treated females are rubbed onto the backs of the sham females, causing them to become more attractive. Even rubbing an unattractive female with the cast skin of an attractive female is sufficient to elicit courtship behavior from breeding males (T. Logan, personal observation, as cited in Radcliffe and Murphy, 1983). Thus, an accumulation of sex attractant between the old and new skin before shedding could provide for the release of a concentrated burst of attractant after shedding (Kubie et al., 1978).

Further confirmation of the skin as the sole source of attraction in garter snakes has been provided by a set of ingenious experiments (Gillingham and Dickinson, 1980). Skin tubes (10 cm in length) from sacrificed animals have been slipped over the bodies of experimental animals (Fig. 4.9). Females with female skin tubes are courted normally by sexually active males. Conversely, males receiving male skin tubes are never courted. However, males that receive female skin tubes are courted as if they were females, which confirms the female skin as the sole source of the attractiveness pheromone. Females receiving male skin tubes cause courting males to abort their courtship

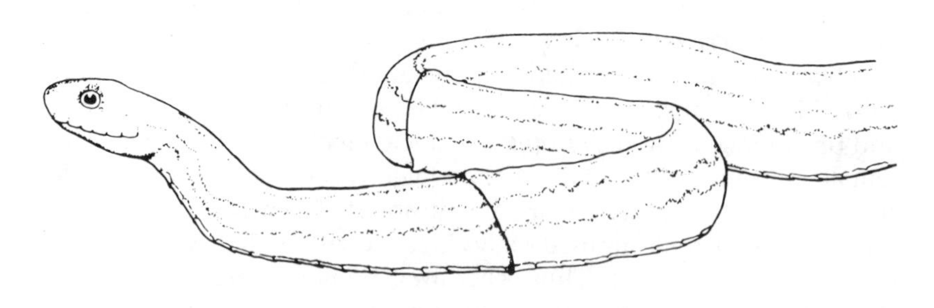

Thamnophis s. sirtalis

(SKIN TUBE IN PLACE)

Fig. 4.9 *Thamnophis s. sirtalis*. Skin tube position on the trunk. Drawn from super-8 motion picture film. (From Gillingham and Dickinson, 1980.)

behavior and to abandon the experimental animal. These results raise another interesting question concerning courtship behavior in *Thamnophis*.

Some authors have suggested that male *Thamnophis* may have a distinct male-identifying pheromone in their skin, thus denoting to other courting males that a male is an inappropriate animal to court (Vagvolgyi and Halpern, 1983). In laboratory studies, some unhibernated male *T. sirtalis parietalis* are attractive to courting males (Vagvolgyi and Halpern, 1983). One explanation for this observation is that the attractive males produce the female sex attractant pheromone (this has also been observed in the field; see Mason and Crews, 1985, and below). However, an alternative hypothesis is that naturally hibernating males produce integumentary chemical cues that are recognized by other courting males and thus avoided. The skin tube experiments (Gillingham and Dickinson, 1980) are consistent with this hypothesis. Males cease courting and leave attractive female garter snakes covered with male skin tubes, even though there is still plenty of female skin exposed to elicit courtship. In addition, mated males that have been in close contact with females for up to several hours are normally not courted. Presumably, these males would be covered with female pheromones. A male identifying pheromone would be beneficial both to the sender and receiver; it would reduce time wasted in courting the wrong individual and unnecessary exposure to predation and limit access to potential mates.

More indirect evidence supports the male-identifying pheromone hypothesis. In field studies in Manitoba, Canada, a small proportion of male red-sided garter snakes consistently elicits courtship from courting males in the same manner as do attractive females (Mason and Crews, 1985, 1986). These "she-males" are anatomically and morphologically male and court and mate with females. Courting males prefer females to she-males in simultaneous choice tests. In competitive mating trails, she-males mate more than twice as often as normal males, indicating that their attractiveness imparts a selective advantage to them. At this time, the only known difference between males and she-males is that she-males have circulating testosterone levels over three times higher than do normal males. Estrogen treatment is known to induce pheromone production in females (Crews, 1976; Kubie et al., 1978), and it has been suggested that high androgen levels in she-males may lead to local elevation of estrogen levels through the conversion of testosterone to estrogen by aromatase enzymes found in the skin of these snakes (Mason and Crews, 1985, 1986). Support for this hypothesis can be found in many hormone-dependent systems. For instance, the induction of vitellogenesis in nonmammalian vertebrates has long been known to be mediated by estrogen. However, several studies have demonstrated that high levels of testosterone can also induce vitellogenesis (Hori et al., 1979; Ho et al., 1981). The mechanisms involved in these processes are still not clearly understood.

The species-specific nature of the female attractiveness pheromone has been addressed several times, although few studies have investigated the problem directly. Noble (1937) describes frequent unsuccessful attempts to induce interspecific courtship in *Thamnophis*. Studies of the sympatric *T. butleri* and *T. sirtalis* never show interspecific courtship which suggests that the sex attractant pheromones are species-specific (Devine, 1977b). No studies report that garter snakes court members of different genera. The tests with the garter snake skin tubes referred to earlier in this section provide further evidence. Without a skin tube, male *Thamnophis* never court female *Nerodia*. Only those *Nerodia* females that have *Thamnophis* skin tubes are courted by male *Thamnophis* (Gillingham and Dickinson, 1980).

However, other reports indicate that female sex attractant pheromones are not species-specific. Several hybrid rattlesnakes have been produced both in the laboratory and in the field (Klauber, 1956). Interspecific courtship has been described in *Thamnophis* (Heller and Halpern, 1981; M. Halpern, personal communication, 1990). In field studies, males of five out of seven species of *Thamnophis* court females of all seven species (R. Mason, unpublished, 1990); however, females of different genera are never courted.

G. Chemistry

Few studies have attempted to isolate, characterize, and identify pheromones in reptiles. The most concerted efforts in this endeavor concern the identification of the female sex attractant pheromone of *Thamnophis sirtalis*. G. K. Noble (1937) was the first to attempt to solve the problem by using solvents to obliterate the attractiveness of the females. He discarded the washes and did not test them for attractivity with courting males. In another study, soapy water used to remove the pheromone from the backs of attractive female *T. sirtalis* and *T. butleri* was rubbed on the backs of females and the reactions of courting males noted (Devine, 1977b). Rubbing the solution of one attractive female on the back of an unattractive conspecific female leads conspecific males to court the previously unattractive female. Rubbing the solution from a heterospecific female on the back of a conspecific female causes males to avoid courtship, leading to the conclusion that soapy water could remove the pheromone and that the pheromone is species-specific (Devine, 1977b). Lipids extracted with trichloroethylene from shed skins of females have been subjected to a number of chemical tests in order to characterize the compound.

Trailing performance in a Y-maze is used as an indicator of the effectiveness of the compounds. Neither acidic nor basic lipids elicit trailing behavior from conspecific males, and only neutral lipids are attractive to males in the Y-maze tests. Extensive TLC investigations and specialized staining leads to the conclusion that the attractive compounds are wax esters (Devine, 1977b). The cues that male snakes use to trail females may be the same ones that elicit courtship behavior when encountered in higher concentrations on the dorsal skin of females; however, this idea has not been tested directly.

Placement of plasma from estrogen-injected pheromone-producing females of *Thamnophis sirtalis parietalis* on the backs of males, which are normally never courted, elicits courtship behavior from courting males (Garstka and Crews, 1981, 1986). Estrogen-injected males are not courted, but their plasma is also attractive to courting males. Chloroform-methanol extractions of female skin are also attractive to courting males, as are the liver and yolk of females. Histochemical studies reveal a lipid-rich layer in the skin of females and estrogen-injected males that stained darkly for lipids. Uninjected males do not exhibit the same staining pattern. In addition, dermal lipid staining in females is concentrated in the hinge region, where the cell layers of the epidermis are thinnest. Thus, skin lipids could be expressed at these points rather than through the scales themselves. These findings lead Garstka and Crews (1981) to conclude that the attractiveness pheromone of the red-sided garter snake was the yolking protein vitellogenin or a lipid-rich subunit of vitellogenin. Subsequent research on the transport, immunoreactivity, and field tests of circulating vitellogenin prevent equating the pheromone with vitellogenin. An alternative approach was therefore undertaken to study the attractiveness pheromone at the skin surface.

In a combination field and laboratory investigation, a series of compounds have been identified that serve as sex pheromones in both male and female *Thamnophis sirtalis parietalis* (Mason et al., 1989). The sex attractiveness pheromones of females have been isolated and identified as a novel series of saturated and monounsaturated methyl ketones. These ketones are relatively nonvolatile and thus are unlike those pheromones commonly found in insects. These compounds have been synthesized and bioassays conducted in the field that verify their identity (Mason et al., 1989, 1990) (Fig. 4.10). These studies represent the first identification and synthesis of a sex pheromone in a reptile. Further studies show that the occurrence of male skin lipids jointly with female lipids causes courting males to cease courtship behavior. Male and female skin lipids differ in that squalene occurs in significantly higher quantities in male than in female skin. Mixtures of squalene with the female sex pheromones significantly diminish

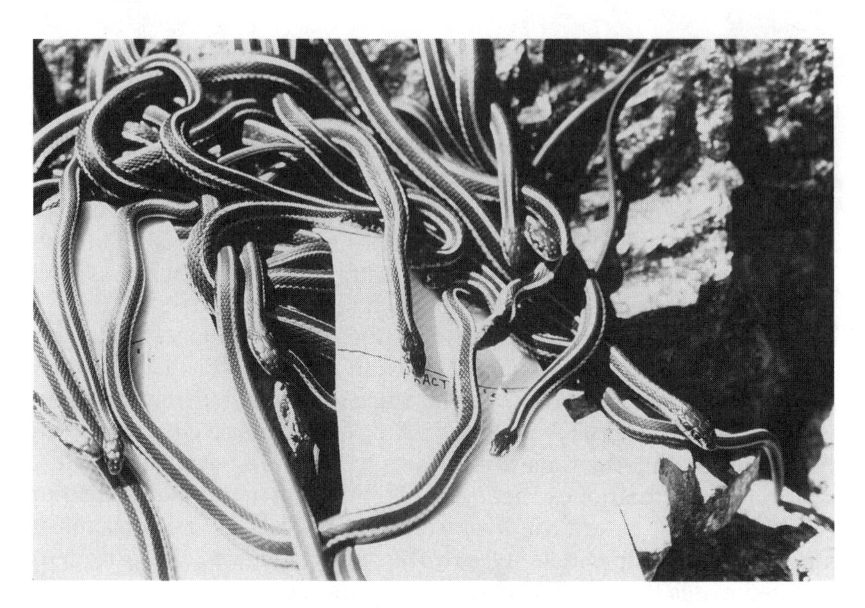

Fig. 4.10 *Thamnophis sirtalis parietalis.* Courting male snakes exhibiting courtship behavior to chemical cues on the paper towel. The number of total males responding with increased tongue-flick rates and chin-rubbing behavior for more than 30 seconds were counted over five minutes and compared to solvent controls. (Photo by R. T. Mason.)

courtship in males. This leads to the conclusion that squalene (and probably other components) are responsible for male sex recognition. Interestingly, the skin lipids of she-males (males that are courted as if they were females (see Section II.F) lack squalene. Thus, the lack of squalene in their skin renders she-males more similar to females than to males.

Using this same rationale, a series of saturated, monounsaturated, and novel diunsaturated ketones have been identified from the Guam brown tree snake, *Boiga irregularis,* that may be the sex attractiveness pheromones of this introduced, vertebrate pest (Murata et al., 1991). Entomologists are investigating possible applications of such studies; for example, in the future, chemical ecologists may be able to use reptilian pheromones to control pest species such as *B. irregularis.*

Nutrition seems to play a powerful role in the attractiveness of female *Thamnophis melanogaster* (Garstka and Crews, 1982). After fattening, females become attractive to males, which begin to court almost immediately. The onset of a breeding bout begins only after the female's nutritional condition becomes optimum (Garstka and Crews, 1982). The effect of nutritional state has also been described by zoo workers for unrelated species of snakes (Radcliffe and Murphy, 1983).

H. Skin Lipids

It is readily apparent from the literature on trailing, aggregation, and courtship in snakes that skin lipids play a central role as pheromones. The ensuing sections show this is also the case for lizards and turtles. Whereas relatively little is known about the chemical nature of the skin lipids of reptiles, some recent studies have investigated the role of skin lipids as semiochemicals.

Unlike the lipids commonly found in internal organs, skin surface lipids from many vertebrate taxa differ dramatically in their composition (Ahern and Downing, 1974). Unusual classes of lipids are common in skin surface extracts. However, no two species have similar compositions of surface lipids (Ahern and Downing, 1974). Nicolaides (1974) states that, indeed, the number of fatty acids that can be synthesized in the skin makes it unlikely that any two individuals would ever make the same substances in the same amounts (Nicolaides, 1974). This may account for the unique "chemical signature" of vertebrates.

Many investigators have identified lipids in the skin secretion of snakes (Ahern and Downing, 1974; Jackson and Sharawy, 1978; Tsumita et al., 1979; Roberts and Lillywhite, 1980; Birkby et al., 1982; Burken et al., 1985a, b; Schell and Weldon, 1985; Mason et al., 1987). Most of these studies focus on the role of skin lipids in retarding transepidermal water loss (Roberts and Lillywhite, 1980; Lillywhite and Maderson, 1982; Baeyens and Rountree, 1983; Burken et al., 1985b).

The nature and quantity of skin lipids have been reported for 11 species of snakes (Roberts and Lillywhite, 1983). Histochemical and ultrastructural investigations indicate that the lipid sheets present in the epidermal layers are sequestered in the mesos region (Fig. 4.11) and originate from mesos granules in immature mesos cells (Landmann, 1979, 1980). The mesos layer provides the principal barrier to water loss (Roberts and Lillywhite, 1983; Landmann, 1979). Polar lipids are the important component in the barrier function of reptilian skin (Baden and Maderson, 1970) and the α-layer is the principal barrier against cutaneous water loss (Maderson et al., 1978). A lacunar cell layer has been described in the skin of the Texas rat snake, *Elaphe obsoleta lindheimeri* (Jackson and Sharawy, 1978). Histochemical and ultrastructural techniques demonstrate that the lacunar cells serve to store lipids, specifically phospholipids and cholesterol.

The skin surface lipids of *Drymarchon corais* have been studied from cast skins extracted in a 2:1 chloroform:methanol solution (Ahern and Downing, 1974). This extraction yields up to 8% lipid from the initial skin material. Thin-layer chromatography (TLC) and gas chromatographic (GC) analysis yields methyl ketones, free primary and secondary alcohols, cholesterol, free fatty acids, and hydrocarbons.

A study of cutaneous water loss in *Elaphe o. obsoleta* removed lipids from sheets of epidermis by extracting them for 24 hours in chloroform:methanol (2:1), ethyl ether and ethanol (8:92), or acetone followed by 24 hours in hexane, or by extraction by all three in succession (Roberts and Lillywhite, 1980). TLC and GC analysis reveals a complex mixture of neutral and polar lipids that resembles those of mammalian epidermis and is similar to the epicuticular lipids of arthropods, except that hydrocarbons are not predominant (Roberts and Lillywhite, 1980). Specifically, the neutral lipids contain sterol esters, triglycerides, free fatty acids, alcohols, and cholesterol. The polar lipids consist of phosphatidylethanolamine, phosphatidylserine, phosphatidylinositol, phosphatidylcholine, and sphingomyelin. It should be noted, however, that those polar compounds frequently identified in studies using TLC methodology have not been corroborated by analytical techniques such as mass spectrometry or nuclear magnetic resonance. Caution is advised in the interpretation of these data.

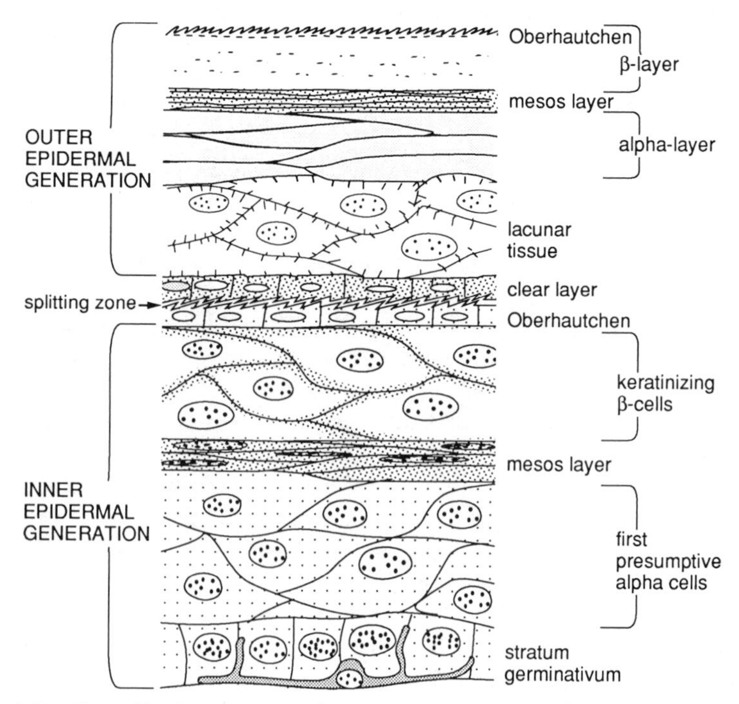

Fig. 4.11 Generalized squamate epidermis several days before shedding. (From Maderson, 1967.)

The skin lipids of *Trimeresurus flavoviridis, Elaphe climacophora,* and *E. quadrivirgata* contain high levels of fatty acids; the content of arachidonic acid is higher than in other vertebrates studied, which include chickens, guinea pigs, rabbits, and mice (Tsumita et al., 1979).

^{13}C-NMR has been utilized to analyze the skin lipids from the cast skins of a boid, four colubrids, and three viperids (Schell and Weldon, 1985). The spectra indicate the presence of free fatty acids, long chain methyl ketones, primary and secondary alcohols, triglycerides, cholesterol, and cholesterol esters.

TLC analyses of lipids in cast skins of 24 species of snakes in four families (extracted with chloroform:methanol [2:1]) show large amounts of cholesterol and free fatty acids; most of the species have apparent triglycerides (Burken et al., 1985b). The polar lipids do not vary much among species. In contrast, the nonpolar lipids vary widely. It may be that the nonpolar lipids have species-specific functions such as pheromonal activity, whereas the polar lipids subserve functions common to all species (Burken et al., 1985b).

The skin lipids of *Thamnophis sirtalis parietalis* have been extracted from the skins of both males and females with hexanes and subjected to analyses by TLC and GC/MS (Mason et al., 1987). The lipids obtained during the breeding season have been compared with those obtained in the nonbreeding season. Clear seasonal and sex differences occur in the skin lipid composition of the integumental lipids. These differences are important in that female garter snakes produce a sex attractant pheromone that can be removed from the skin with hexanes, whereas males normally do not (Mason and Crews, 1985; Mason et al., 1989; see Section II.H).

I. Copulatory Plugs

One very interesting example of chemical communication in reptiles concerns the copulatory plugs of snakes, which have been studied most thoroughly in the genus *Thamnophis.* Field studies of *T. butleri* and *T. sirtalis* show the cloacae of recently mated females to be distended by a gelatinous plug (Devine, 1975) (Fig. 4.12). Dissection indicates that the plug occupies the urodaeum, anterior to the openings of both the ureter and the intestine. The plug is composed of two sections: a firm, translucent plug matrix and an opaque, creamy white, sperm-dense fluid that covers the oviductal orifices and extends into the oviducts (see also Halpert et al., 1982). Histological evidence confirms that the copulatory plug is produced by the sexual segment of male snakes, as hypothesized by Volsøe (1944). The plug is composed primarily of a tryptophan-rich protein and has a large lipid component (Devine, 1975). During mating, male snakes ejaculate sperm stored in the ductus deferens and then inject a fluid containing intact sexual segment secretion granules. These granules subsequently lose their integrity and precipitate to form the plug matrix. The plug remains in place for 2 to 4 days and is then expelled by the female. Presumably, active or passive chemical breakdown along the cloacal walls of the female causes the plug to loosen. Movements by the female expel the plug and possibly force the sperm into the oviducts. Competition among males for female mates may have selected for the evolution of copulatory plugs. In species in which the chances of multiple matings by females are high, males producing copulatory plugs would have a selective advantage. The selective advantage of copulatory plugs would be reduced or eliminated for species in which females are isolated from rival males by territoriality or male combat. Alternatively, if the plugs merely function to retain sperm, they should be equally common in both situations (Devine, 1975). Comparisons necessary to test this hypothesis are lacking in the literature.

The presence of a copulatory plug in the female cloaca may prevent males from courting mated female *Thamnophis radix* (Ross and Crews, 1977). Laboratory studies of mated females retested immediately after

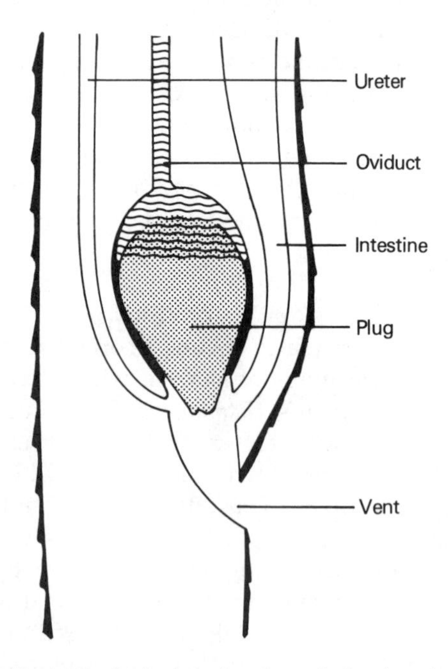

Fig. 4.12 *Thamnophis sirtalis.* Sagittal section through the cloacal region of a female, showing semen (wavy lines) anterior to the plug (stippling) in situ. Diagrammatic reconstruction from gross dissection. (Redrawn from Devine, 1975 © The American Association for the Advancement of Science.)

mating indicates that most sexually active males refuse to court the mated females when exposed to them as early as 9 minutes after mating and up to 48 hours post mating. However, expression of the mating plug from mated females and washing of their cloacae with water restores their attractivity to courting males. However, such females remain unreceptive to courting males for at least 48 hours after mating. Spreading of cloacal material from mated females on the backs of unmated females blocks courtship of most of these females. After the material is washed from their backs, these females again become attractive. Rubbing of cloacal material from unmated females onto the backs of other unmated females causes these experimental animals to be courted. Males exposed to mated females become refractory to courtship for varying times after the exposure. It has been suggested that males protect their reproductive investment by advertising the mated status and nonreceptivity of females to rival males (Ross and Crews, 1977). Also, inhibition of sexual attractivity and receptivity after mating may tend to minimize the exposure of females to predation. Indeed, mortality due to predators is extremely high at the time of spring emergence and mating (Gregory, 1974).

Males can discriminate between mated and unmated females, and males have never been seen to court mated female *Thamnophis sirtalis* (Devine, 1977a). Field observation indicates that many males often court a single female. Whenever one male eventually mates, the other males disperse before the rival male even finishes mating. This occurrence is common and has been described frequently by workers from both the field and the laboratory. There may be a chemical cue responsible for this behavior. However, it is unlikely that the behavior is associated directly with the plug because males never directly investigate the cloacal region of mated females (Devine, 1977a). This behavior may not be the same in all *Thamnophis*.

Captive male *Thamnophis radix* are sexually refractory to courtship for periods of 24 to 72 hours after mating (Ross and Crews, 1978). Further, exposure to a mated female abolishes courtship in unmated males for up to 24 hours. Females that mate with surgically castrated males induce sexual refractoriness in courting males and become unattractive (Ross and Crews, 1978). Females mated with males vasectomized posterior to the kidney become unreceptive to male courtship but do not become unattractive to other courting males. Thus, it appears that the compounds responsible for causing the loss of female attractivity may be produced in the renal sex segment and are deposited by the male in the cloacal plug.

Field studies of *Thamnophis sirtalis parietalis* show that females become unattractive within 12 to 24 hours after mating (Whittier et al.,

1985). Males court mated females less frequently than unmated females and with less intensity. Fifteen percent of the males tested, however, repeatedly courted mated females. This species, in contrast to *T. radix,* shows no residual effects on the courtship behavior of rival males due to exposure to copulating pairs or mated females. Indeed, males will intromit and mate with females with plugged cloacae. The specific copulatory stimuli that bring about changes in the attractivity of females remain unknown.

Male *Vipera berus* do not leave a hard, gelatinous plug in the cloaca of females following the termination of copulation (Nilson and Andrén, 1982). Instead, the uterus of mated females becomes hard posteriorly and may block the entrance of subsequent sperm. However, there is some controversy concerning these proposed actions of the copulatory plug (see Stille et al., 1986, 1987; Andrén and Nilson, 1987). To investigate the possibility of pheromonal effects of the secretion as described in *Thamnophis,* researchers injected the cloaca of six female *V. berus* with a sample of male renal sex secretion; the cloaca of six other females served as vehicle-injected controls. Unlike in *Thamnophis,* tests with breeding males of *Vipera* in a seminatural outdoor terrarium show no significant differences in female sexual attractivity (Nilson and Andrén, 1982). However, female sexual receptivity, defined as acceptance of male courtship, decreases significantly in the experimental animals (Nilson and Andrén, 1982). These results indicate that renal sex secretions may not be producing an inhibitory pheromone such as that found in *Thamnophis.* It is significant, however, that female receptivity changes.

Similar changes in female receptivity also have been documented in *Thamnophis.* In an interesting study on the effects of prostaglandins on the reproductive physiology of female garter snakes, female attractivity and receptivity changes following prostaglandin treatment (Whittier and Crews, 1986). Unmated females treated with prostaglandin $F_{2\alpha}$ become unattractive to courting males and unreceptive within 24 hours of treatment. These effects mimic exactly the changes occurring in females responding to natural matings in the field and the laboratory; this suggests a powerful role for prostaglandins in the loss of attractivity seen in females after mating. Whether prostaglandins are mediating a chemical change in the attractiveness pheromone or causing the production of a courtship-inhibiting pheromone is unknown. Conversely, prostaglandins could be altering the behavior of females, reducing their attractiveness to males. These hypotheses await future experimentation to elucidate the mechanisms involved. For further discussion of the physiological mechanisms of reproduction see Chapter 2.

J. Allomones and Kairomones

Although technically not an area covered by this review, the use of allomones and kairomones by snakes warrants discussion. These semiochemicals differ from true pheromones in that they represent chemical cues which affect the behavior or physiology of heterospecific individuals. One such set of semiochemicals is prey odors and their discrimination by snakes, a topic that has been extensively reviewed for reptiles (Burghardt, 1980; Simon, 1983; Von Achen and Rakestraw, 1984). It seems appropriate here to discuss the recognition and repulsion of potential predators by chemical means.

More than 30 taxa of pit vipers, including members of the genera *Agkistrodon, Bothrops, Crotalus,* and *Sistrurus,* respond to snake-eating, or ophiophagous, snakes by exhibiting a behavior known as body bridging (Weldon, 1982). Rattlesnakes can recognize king snakes as enemies entirely by means of chemical cues (Klauber, 1927). Two species of rattlesnake (*Crotalus viridis oreganus* and *C. cerastes*) respond to the presence of king snakes by posturing their bodies so that a body loop is formed that can then be used to strike the king snake much as "a human elbow could be used to strike a heavy blow" (Cowles, 1938). This posture, in which the head is usually hidden in the coils of the snake, is very different from the response of rattlesnakes to mammalian predators, during which rattlesnakes coil their body and raise the head, ready to strike forward. King snakes seem to be able to identify rattlesnakes by sight or smell. Several other anecdotal reports refer to this behavior in rattlesnakes and other crotalids (Meade, 1940; Neill, 1947; Carpenter and Gillingham, 1975; Weldon and Burghardt, 1979; Marchisin, 1980).

Bogert (1941) was the first experimenter to investigate the cues that rattlesnakes use to identify king snakes. In one experiment, 13 *Crotalus cerastes* and one each of *C. s. scutulatus, C. m. molossus,* and *C. atrox* were placed into a circular enclosure. Subsequently, three subspecies of *Lampropeltis getulus* (*californiae, yumensis, splendida*) were introduced. All of the rattlesnakes except the *C. m. molossus* responded with the defensive body-bridging posture. This behavior was also elicited whenever rattlesnakes were transferred into a gallon jar that had previously held king snakes. It is interesting that rattlesnakes exposed to king snake odors rarely rattle, although they commonly do so in response to mammalian predators. Tests for the sensory mechanisms used to detect king snakes involved blindfolding a group of rattlesnakes and placing them with a king snake. Blindfolded rattlesnakes respond with body bridging as readily as do snakes without blindfolds. These results led to the design of a clever experiment using wooden dowels smeared with one of three chemicals (Bogert,

1941). The first group were smeared with secretions from the cloacal scent glands of king snakes, the second was rubbed on the dorsum of the king snakes, whereas a third group was rubbed on their ventrum. Neither blindfolded nor nonblindfolded rattlesnakes responded to cloacal secretions, whereas all of the snakes responded with the characteristic body-bridging posture to the material from the dorsum of the king snakes. The results with the material from the ventrum were equivocal. Removing the tongues from five individual rattlesnakes also extinguished the capacity to respond to king snakes. Tests for the defensive response also involved many other species of snakes. Positive responses were elicited only from *Masticophis flagellum piceus* and from *Pseustes sulphureus* (Bogert, 1941). Bogert concluded that the bridging posture is mediated by the vomeronasal system via tongue-flicking of chemical cues. Experimental evidence for this claim awaits future investigation.

An olfactometer (a device that presents odors in an airstream to subjects) has been used to investigate the role of olfactory cues in rattlesnake behavior (Cowles and Phelan, 1958). The heart rate of various species of crotalids was measured to assay the response to experimental and control odors (Cowles and Phelan, 1958). One test was of thio-alcohol n-butyl mercaptan, because of the report (Bogert, 1941) that rattlesnakes will exhibit the defensive posture in response to skunk odors. The rattlesnakes did perceive the mercaptan, as indicated by heart rates significantly elevated above control levels. Also, one *Crotalus ruber* responded to mercaptan with the typical body-bridging posture usually seen in response to king snake chemicals. Thus, rattlesnakes can recognize kairomones by olfactory as well as vomeronasally mediated means.

A series of experiments investigating the extent of the body-bridging posture in snake taxa used methanol-soaked cotton balls to extract semiochemicals from the backs of snakes (Weldon and Burghardt, 1979). Twenty-one additional species of New World pit vipers appear to respond to snake skin lipids. Also, several new species of colubrid snakes are shown to elicit the body-bridging behavior. It has been suggested that the phenomenon consequently should be called *ophiophage defensive response* rather than *king snake defense posture*.

The ontogeny of this behavior has also been investigated by testing naive newborn crotaline snakes for their ability to exhibit the defensive posture (Weldon and Burghardt, 1979). A group of four juvenile *Agkistrodon b. bilineatus* do not respond to chemical cues from *Lampropeltis getulus holbrooki*. Young of *Crotalus atrox* and *A. contortrix mokasen* do respond to semiochemicals from the king snakes *L. getulus niger* and *L. g. holbrooki*, respectively. The literature suggests that in rattlesnakes this response tends to attenuate after repeated testing. In addition, there seems to be an age and size effect; older and larger rattlesnakes rarely exhibit this behavior, presumably reflecting the unlikelihood of larger rattlesnakes being effectively attacked and consumed by ophiophagous snakes. The defensive response also can be elicited for species of snake that are not sympatric. Crotaline chemoreceptors may possess low specificity for the chemicals from their various predators; chemicals from several predators may be identical or very similar; or, finally, the different chemicals may elicit the same response. These are important issues for determining the evolutionary history of this response. Understanding of some of the selective pressures involved in this process may lead to a better understanding of snake behaviors in general.

The ability of juvenile garter snakes to discriminate between chemicals from ophiophagous and nonophiophagous snakes has been tested in two experiments (Weldon, 1982). In the first experiment, cotton swabs dipped in methylene chloride were rubbed along the dorsum of a male *Lampropeltis getulus californiae*, a male *Heterodon nasicus*, and a female *Thamnophis radix*. Only the *Lampropeltis* is a known ophiophage. The test subjects were 18 *T. elegans vagrans* and 14 *T. e. terrestris;* all were 7 to 8 months of age. Measurement of tongue-flick rates, which are reliable measures of chemoreception in snakes, indicate that *Thamnophis* tongue-flick at significantly higher rates to the chemical cues from *Lampropeltis* than to those from *Heterodon* or *Thamnophis*. The subjects of a second experiment were 24 *T. s. sirtalis,* all 2 to 3 months old. Testing focused on the ability to respond to chemical cues from a male *Coluber c. constrictor,* a known ophiophage, and a male *Pituophis m. melanoleucus,* a nonophiophage. Again, juvenile garter snakes show a significantly greater response to chemical cues from *Coluber.* This general pattern does show exceptions. *Crotalus viridis* and *Agkistrodon piscivorus* have tongue-flick rates that are significantly lower in response to *Lampropeltis* odors than to those of *Heterodon* (Chiszar et al., 1978). *Agkistrodon piscivorus* tongue-flicks significantly less frequently in response to the odor of *Heterodon* than to a blank, odorless condition. More data are necessary to clarify this situation (Weldon, 1982).

The difference between olfactory and vomeronasal cues remains elusive (See Cooper and Burghardt, 1990, for an interesting discussion of this issue). Tests in an olfactometer using 24 juvenile *Thamnophis sirtalis* in response to airborne chemical cues from *Lampropeltis getulus californiae* and *Elaphe o. obsoleta* show that garter snakes tongue-flick significantly more to the odors of *Lampropeltis* than to *Elaphe* (Weldon, 1982). The olfactometer data may indicate that garter snakes can recognize potential predators by airborne chemical cues alone. However, these results do not indicate which sensory system is mediating the reception and perception of these cues, as a higher tongue-flick rate does not preclude olfaction from playing a role. Thus, rattlesnakes with their mouths taped shut exhibit higher heart rates in response to king snake odors (Cowles and Phelan, 1958). It would thus seem that both the olfactory and vomeronasal system are involved in this interesting behavior. Definitive studies directly addressing this question could easily be designed.

Can snakes, especially juveniles, benefit from being able to detect the chemical trails of potential predators and avoid them? Trailing experiments during September and October with 25- to 35-day-old *Pituophis m. melanoleucus,* which have to choose between conspecific and heterospecific chemical cues in a Y-maze, address this issue (Burger, 1989). The juvenile pine snakes must decide between an arm containing chemical cues from an ophiophagous predator, *Lampropeltis getulus,* and an arm lacking chemical cues. The juvenile snakes do not discriminate if no odor is in either arm. However, the avoidance of the arm of the maze containing the king snake chemical cues is significant. Furthermore, they emit significantly more tongue-flicks in exploring the Y-junction and during passage down the chemical trail of *Lampropeltis* than in the blank arm. They also explore longer and move more slowly in the king snake arm than in the blank arm. This suggests that *Pituophis* can recognize and avoid the chemical cues of *Lampropeltis,* a potential predator. The increased tongue-flick rate in response to *Lampropeltis* odor also corroborates Weldon's (1982) results in other species, although it should be noted that *Heterodon platyrhinos* exhibits neither defensive displays nor changes in tongue-flick rate in response to the odors of *Lampropeltis* (Durham, 1980).

Elaphe is seldom present in stomach content analyses of ophiophagous snakes, suggesting that *Elaphe* may possess mechanisms to reduce predation (Weldon et al., 1990). Nine 5-week-old *Elaphe guttata* show significantly more tongue-flicks in response to methylene chloride–dipped cotton swabs that had been rubbed against the dorsum of *Lampropeltus getulus* than to those rubbed against *Thamnophis radix haydeni.* However, the *Thamnophis* cues do not elicit tongue-flicks differing in number from controls or *Lampropeltis.* Eleven corn snakes presented with a cotton swab containing lipids from a chloroform-extracted shed skin of a *Lampropeltis* or a chloroform-soaked control swab display significantly more tongue-flicks to the *Lampropeltis* lipids. Similar tests with shed skin extracts from *Lampropeltis getulus,* another ophiophage, *Masticophis flagellum,* and a nonophiophagous *E. obsoleta* show significantly more tongue-flicks to snake cues than to solvent control odors. The *E. guttata* apparently cannot distinguish between the chemical cues of *Lampropeltis* and the *E. obsoleta.* If *E. guttata* are subjected to substrate preference tests and have to choose between eight refuges, under two of which is bedding from a stimulus donor snake (*Nerodia erythrogaster, Agkistrodon piscivorus*, or *Lampropeltis getulus*), the corn snakes do not occupy the refuges randomly but reside under the snake-treated tiles at a rate significantly different from that expected by chance (Weldon et al., 1990). This suggests that *E. guttata* may not be able to distinguish between the chemical cues of potential snake predators; however, trail-following experiments are needed to compare cues from ophiophagous versus nonophiophagous snakes. Tests of this sort would measure responses that could be compared more directly to behavior likely to confer selective advantage (Weldon et al., 1990).

A study of predator-prey interactions in snakes contained the first description of a new method by which snake skin chemical cues could be obtained for experimental testing (Weldon and Schell, 1984). Shed skins are collected and extracted in methylene chloride for 2 days in a soxholet extractor. Gram quantities of extracts ranging from 2.4% to 16% of the original dry weight of the shed skins may thus be obtained. Such extracts are then soaked on cotton swabs, and juvenile *Lampropeltis getulus* respond to the prey odors, even to the point of attempting to ingest the cotton swabs (Weldon and Schell, 1984). The technique does not harm the donor animals and yet yields biologically relevant behavioral data.

Cloacal scent gland secretions of *Lampropeltis getulus splendida* also affect mammalian behavior. Western spotted skunk, badger, bobcat, coyote, and gray fox may react if these secretions are applied to food (Price and LaPointe, 1981). Gray fox, bobcat, and skunk become wary in their approach to food, usually refusing to eat treated food, and vigorously wiping their snouts if they do eat treated food. It may be that any distraction of a potential predator would impart a selective advantage to the snake by increasing the likelihood of successful escape.

Other results for cloacal scent gland secretions are less clear (Weldon and Fagre, 1989). Cloacal scent gland material from *Crotalus atrox* and methylene chloride extracts of these secretions have been presented in three field tests (coyotes) and two kennel tests (dogs). The coyotes are not repelled by the secretions but visit the sites and exhibit rubbing and rolling at the scent stations, similar to the well-known canid reaction to putrefying substances. Solvent-extracted material elicits the same behaviors as in controls, suggesting that the extraction procedures may alter or destroy the behaviorally active semiochemicals. Tests with dogs show that solvent-extracted scent gland secretions elicit mouthing behaviors, in which the dogs lick, bite, or eat the test samples. If extracted rattlesnake scent gland secretions and alligator paracloacal gland secretions are offered, the dogs more commonly exhibit urination postures to the secretions than to solvent controls; however, the behavior remains the same between the two secretions. Whether scent gland secretions serve to alert conspecifics (see Section II.A) or interfere with predation by distracting potential predators needs further investigation (Weldon and Fagre, 1989). The possibility that these secretions may have multiple functions has not been adequately investigated.

Rodent responses to snake chemical cues are also unclear. Substrate odors and shed skin extracts of a potential predator, *Elaphe o. obsoleta* and the nonpredatory *Virginia striatula* have been compared (Weldon et al., 1987). Tests involved exposure to papers taken from the floor of the aquaria housing the *Elaphe* and the *Virginia* paired with solvent controls and placed on either side of an aquarium. The rodent behavior shows few significant differences in terms of side preference or activity. However, female laboratory mice left significantly more fecal boli on the *Elaphe* side of the aquarium. This might be a fear-induced behavior; however, why does this occur only in female mice? Shed skin extracts of *Elaphe* cause female mice to reduce intake of food pellets to which the extract (but not the solvent controls) have been applied. Pertinent may be the observation that Siberian chipmunks gnaw off bits of snake skin and apply it to their pelage (Kobayashi and Watanabe, 1986). Does this bizarre behavior interfere with the chemoreceptive abilities of snakes, which are one of their major predators?

Ophidian chemical cues also may act as a signal to *Lacerta vivipara* (Thoen et al., 1986). The lizards have been tested in aquaria that had been previously inhabited by a *Vipera berus* or *Coronella austriaca*, both lizard predators, or *Natrix natrix*, which does not feed on lizards. *Lacerta* respond to the snake odors with increased tongue-flick rates; the highest rates are elicited by the odors of predatory snakes. The odors of predatory snakes generally disrupt the locomotor patterns of the lizards. Lizards tend to freeze once they detect the odors but keep moving after detecting those of nonpredatory snakes.

Thus, the chemicals elaborated by snakes elicit behavioral responses from other snakes, other reptiles such as lizards, and even other vertebrates. The enormous diversity of observed behavioral responses to snake chemical cues demonstrates very clearly that these substances have indirect effects that may play a profound role in the life cycles of not only the snakes themselves but also their potential prey.

III. SAURIA

A. Introduction

Taxonomists have long divided lizards into two divisions, the Autarchoglossa and the Ascalabota (Camp, 1923). Members of the former group possess a rectus superficialis muscle that most ascalabotans lack (Evans, 1961). The autarchoglossans may include the progenitors of the modern snakes. Evans (1961) has stressed a behavioral dichotomy among living lizards that parallels evolutionary trends in locomotory methods. The ascalabotans (which do not tend toward limb loss or body elongation) depend heavily on visual signals. The autarchoglossans (which show evolutionary tendency toward limb loss) depend heavily on chemical signals, which is reflected in their putative derivatives, the snakes (Bellairs and Underwood, 1951; Blair, 1968). Among the ascalabotans, only the nocturnal gekkonids utilize olfaction to any significant degree (Evans, 1959). Diurnal iguanians, including the iguanids and agamids, are noted to be "nearly deprived of osmotic powers" and "show few if any indications of behavioral modifications through the influence of olfaction." Finally, chameleons have been described as being completely anosmic (Haas, 1937).

Within the last 10 years or so, these ideas have been seriously questioned (Estes and Pregill, 1988). The phylogenetic relationship of lizards and snakes is subject to renewed discussion (Rage, 1987; Estes and Pregill, 1988); it has been postulated that snakes may have derived from varanid lizards (Rage, 1984) or gekkonid (Underwood, 1957) lacertilian progenitors. Conversely, lizards and snakes may have independently evolved from a common prelacertilian ancestor (Hoffstetter, 1962, 1968; Rieppel, 1980, 1983; Graves and Halpern, 1989). It is now clear that semiochemical cues are important components in at least courtship and aggression in several lizard families, including iguanids, agamids, teiids, scincids, and lacertids (Carpenter and Ferguson, 1977). Glandular structures are clearly diverse in lizards. Chemical secretions used by lizards in intraspecific communication may derive from the body surface, the cloacal region, specific glands, or other organs such as femoral pores (Simon, 1983). Nevertheless, the chemical identification of and behavioral reactions to specific semiochemical signals are almost completely unknown.

B. Glandular Sources of Semiochemicals

The cloacal glands of lizards are usually numerous and may be isolated or grouped into masses with many ducts. They differ from those of snakes that have only one gland (as in Typhlopidae and Leptotyphlopidae) or two (in Boidae and Colubridae) with a single duct (Saint Girons, 1976).

Only female lizards have urodeal glands. During the breeding season the urogenital fossa or anterodorsal part of the urodaeum becomes greatly hypertrophied (Gabe and Saint Girons, 1965), and in *Lacerta agilis*, the urodeal glands produce a holocrine secretion (Regamey, 1935). Their cyclical development suggests a relation to the reproductive cycle. In *Eumeces laticeps*, the epithelial lining of the urodaeal glands hypertrophies in reproductively active females, whereas that of immature females remains quiescent (Trauth et al., 1987). Lipid components of the urodaeal gland act as sex pheromones in these skinks (Cooper et al., 1986).

Male *Sceloporus g. graciosus* and other sceloporine species have a discolored cloacal area that is moistened during the mating period (Burkholder and Tanner, 1974). Histology shows a gland deep to the integument and the first layer of muscle of the preanal area immediately anterior to the cloacal opening. The secretions appear to be apocrine rather than the more common holocrine type. Because the gland is active only during the breeding season, the secretions may serve to attract females or repulse rival males, but behavioral evidence is lacking.

Many other glands have been described in lizards, including dorsal and ventral glands and hemipenile glands. Although these glands may show seasonal variation in secretion rates, with highest levels evident in the breeding season, no studies have demonstrated direct behavioral effects of these secretions.

These glands are affected by steroid hormones. Pellets of testosterone and estrone implanted into adult male *Sceloporus* produce "cloacal glands" with significantly higher secretory activity than that of controls (Forbes, 1941). Estrone-implanted animals have completely quiescent "cloacal glands," suggesting that certain squamate cloacal glands do respond differentially to seasonal variation in steroid hormone titers.

C. Femoral Pores

Precloacal glands, the femoral pores (or inguinal pores) as first described by Linnaeus (1758), are epidermal structures present in many amphisbaenians and lizards (Whiting, 1969). Precloacal glands and femoral pores are quite similar and probably homologous (Gabe and Saint Girons, 1965). Many, but not all, lacertilian reptiles possess such epidermal structures on the ventral surface of the thigh or the precloacal abdominal area (Fig. 4.13). The glands are formed by an invagination of the stratum germinativum, which forms a follicular unit (Maderson, 1972). The germinal epithelium of the gland is continuous with the generalized body epidermis; however, patterns of cell production and subsequent differentiation are independent of the nonglandular epidermis (Maderson, 1972). The glands produce copious amounts of holocrine secretion as a solid core and may occur in roughly equal numbers in males and females (or be absent in the latter); they are usually larger in males. The amount of secretion produced in different seasons has never been measured quantitatively, but it is generally accepted that femoral glands secrete more actively in the breeding season than during times of the year in which they do not breed and that males produce more secretion than females (Leydig, 1872; Padoa, 1933; Regamey, 1935; Altland, 1941; Smith, 1946; Mahendra, 1953; Cole, 1966a, b; Fergusson et al., 1985).

There have been few studies of the hormonal control of femoral gland secretion, although activity of the femoral glands is correlated positively with testicular activity in male *Sceloporus u. undulatus* (Altland, 1941) and *Lacerta agilis* (Regamey, 1935). Castration of male lizards causes the glands to atrophy (Padoa, 1933; Matthey, 1929; and Regamey, 1932), whereas intramuscular injections of testosterone induce the development of femoral glands in female *Gekko gecko,* which ordinarily lack glands (Chiu et al., 1970). Seasonal activity of the testis, and presumably androgen production, may be correlated to the cyclic development of gekkonine femoral glands (Chiu and Maderson, 1975). A very early study has followed the course of development of male femoral glands in an unspecified species of *Sceloporus* after the implantation of testosterone and estrone pellets (Forbes, 1941). Androgen-treated animals have higher secretion rates than controls, whereas estrone-treated animals show signs of extreme glandular atrophy. Cole (1966b) has suggested that the estrone-treated animals were essentially castrated physiologically; thus, the atrophy of their femoral glands might result from this rather than from estrone. Castration of male *Amphibolurus ornatus* also atrophies

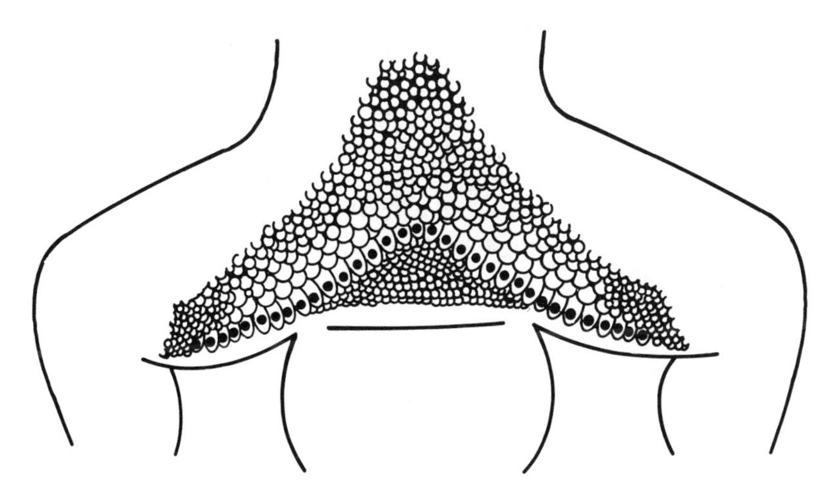

Fig. 4.13 *Gekko vittatus.* Drawing of the posterior abdominal scalation to illustrate the V-line formed by the pre-anal pore bearing scales. (Redrawn from Menchel and Maderson, 1975 © Alan R. Liss, Inc. By permission of Wiley-Liss, a Division of John Wiley and Sons, Inc.)

femoral glands; a single androgen that the authors assumed was testosterone has been isolated from testicular tissue cultures in vitro (Fergusson et al., 1985). Daily injections of testosterone propionate and dihydrotestosterone into castrated males for 1 month stimulate femoral gland secretion over that shown by castrated controls. Intact females, which do not normally have secretory femoral pores, have been stimulated by androgen injections to form femoral gland secretions. These data strongly suggest that femoral glands and their secretions are under direct androgenic control and thus vary seasonally with androgen production.

The secretion of *Amphibolurus,* an agamid, is composed primarily of protein (perhaps keratin) and traces of unidentified volatile substances (Fergusson et al., 1985); however, behavioral effects of the secretions and their chemical components have not been studied. The femoral pore secretion of the iguanid, *Sceloporus,* may be keratin (Hunsaker, personal communication; p. 203 in Cole, 1966b). The material contains little or no lipid but is positive in ninhydrin, Biuret, and Xanthoproteic tests for proteins. Gas chromatographic analyses of extracted femoral pore secretions from both sexes of *Amphibolurus fordi* show no significant differences between the sexes (Cogger, 1978); however, no analyses have been performed during the active part of the breeding season. Femoral gland secretions of *Iguana iguana* contain lipids, which also do not differ between individual adults and juvenile males (Weldon, et al., 1990).

The behavioral responses elicited by femoral gland secretions of *Dipsosaurus dorsalis* have been examined by blindfolding seven males and six females in breeding condition and presenting them with ceramic tiles treated with femoral gland secretions of a donor male, a paraffin wax control, or a blank tile (Alberts, 1989). Lizards tongue-touched the tile bearing femoral gland secretions significantly more than they touched those bearing two controls. Because the secretions are largely nonvolatile, it remains uncertain how the lizards initially locate the secretions. However, the secretions absorb long-wave ultraviolet light selectively between 350 and 400 nm, causing them to fluoresce at wavelengths of around 500 nm. This fluorescence is too faint to be detected by humans under visible light conditions. Lizards were given a choice of a tile coated with femoral gland secretion and an untreated control tile. Whenever the tiles are illuminated by UV light, lizards tongue-touch the secretion-coated one significantly more often than the blank; the difference is less under incandescent light. Thus, these lizards may be utilizing a composite signal; they detect a long-range visual cue that promotes a short-range investigation of femoral gland secretions containing chemical cues (Alberts, 1989). The absence of differences in response to male and female secretions suggests that the secretions may function not for sex recognition but, perhaps, for individual recognition (Alberts, 1989).

Because femoral glands occur in lizards of different families that occupy diverse habitats, the glands provide interesting models for investigating the comparative roles of historical and ecological influences on inter- and intraspecific variation in composition of the secretion. The protein components of femoral gland secretions obtained from 16 species of lizards have been analyzed by gel electrophoresis (Alberts, 1991). Their protein composition indicates that much of the variation is accounted for by phylogeny and much less by the ecological habitat in which individuals are found. Intraspecific variation of the secretion components suggests that chemical differences among individuals could be a basis for conspecific recognition (Cole, 1966b).

The literature contains five hypotheses concerning femoral gland function (Cole, 1966b):

1. The secretion securely fastens males to females during copulation. This is unlikely, and no evidence supports it.
2. The secretion is odorous and males use it for sex recognition and orientation. This idea is testable and is supported by the many accounts of similar pheromones in other reptiles.
3. The secretion is used by males to mark their territory. This idea has been suggested for *Crotaphytus c. collaris* (Hathaway, 1964).
4. The tines of solid secretion act to stimulate females mechanically or quiet her, allowing the male to mount—an idea first postulated for *Ameiva* (Noble and Teale, 1930). Numerous studies of other species fail to bear this out, and the hypothesis is largely disregarded by behaviorists.
5. The glands are vestigial and serve no present function. This seems unlikely and should only be posited if all other hypotheses fail.

D. Chemical Studies

In contrast to the number of studies of skin lipids in snakes, very few studies have examined skin lipids in lizards. The first study utilized TLC and GLC to identify neutral and polar lipids from chloroform:methanol extracts of the skin of *Iguana iguana* (Roberts and Lillywhite, 1980). Such extracts include alcohols, sterols, fatty acids, cholesterol, triglycerides, and wax esters.

The most thorough examination reports on the skin lipids extracted with chloroform:methanol from 23 species (Weldon and Bagnall, 1987). Almost all of the saurian families are included. Analysis by TLC allows tentative identifications of several lipid classes, among these sterols, free fatty acids, phosphatidylcholines, and phosphatidylethanolamines. The skin lipids of some of the lizards include hydrocarbons previously undescribed in reptiles. This is interesting, as hydrocarbons serve as pheromones in many arthropods (Blum, 1981; Hadley, 1985). It remains unclear whether nonpolar lipids in the skin vary less than polar lipids, as has been reported by Burken et al. (1985b).

Chemical analyses, including nuclear magnetic resonance (NMR) and gel electrophoresis, indicate that the femoral gland secretions of *Dipsosaurus dorsalis* are composed of 80% protein and 20% lipid material (Alberts, 1990). The lipid material consists primarily of triglycerides and sterols (most likely cholesterol). The protein constituents of the lizards differ between individuals, across species, and over the course of the year. The protein and/or lipid components of the femoral gland secretions may serve as pheromones mediating individual recognition (Alberts, 1990).

E. Licking Behavior: Integumental and Cloacal Cues

One of the first studies describing the use of chemical communication in lizards investigated the sexual behavior of *Coleonyx variegatus* (Greenberg, 1943). Tongue-testing and tongue-flicking are widespread in the Gekkonidae (Carpenter and Ferguson, 1977). The male begins courtship with the nose low to the ground, body low, and the tail waving. This same type of behavior is also elicited in response to odors from predators (Dial et al., 1989). The male then licks the female repeatedly, takes a neck or shoulder grip, and proceeds to mount her. Males react to other males by biting and fighting. If anesthetized pairs of male and female geckos are placed in the home cage of a test male, the male investigates both the male and female animals by probing with his snout and licking their skin but always returns to court the anesthetized female. Thus males likely identify females through tactile or odor cues in the skin.

Courting males usually grip the tail of the female in their jaws. The tails of four pairs of male and female geckos were broken off and interchanged so that each sex bore the tail of a member of the opposite sex. Whenever courting males grip the "female" tail of a male gecko, they court the male. Female geckos are courted only if the male bites a part of the body and not the "male" tail. Females bitten on their "male" tails are never courted. Presumably, male *Coleonyx* respond to some chemical stimulus emanating from the skin of the female (Greenberg, 1943). Males of *Hemidactylus flaviviridis* also lick females during courtship (Mahendra, 1953).

Male *Eublepharis macularius* frequently lick conspecifics when first encountering them; so do *Coleonyx* (Mason and Gutzke, 1990). If male *Eublepharis* are paired with males, they lick each other, after which the resident male always attacks the other male. In male-female encounters, the response of the male depended upon the state of the female in the shedding cycle. After licking all females introduced into their home cages, males court all but those having cloudy skin just before shedding; these are always attacked. These behaviors may be elicited in response to chemicals sequestered in the skin of the females. During shedding, chemical cues that distinguish females from males may become temporarily unavailable, as the old dry skin of the female provides a physical barrier to the source of chemicals in the new skin. Analysis by GC/MS of hexane-extracted lipids from the skin of male and female specimens shows 40 different compounds, including hydrocarbons, fatty acids, steroids, and methyl ketones. Several steroid analogues of cholesterol are unique to males. These steroids are unusual in that they had previously been thought to be present only in plants. Females have long-chain methyl ketones that are absent in males. Thus, *E. macularius* have clear and recognizable sex differences in the composition of their skin lipids.

In several species of lacertids, males attack females that are painted to look like males. However, if the males lick such females, they then court (Kramer, 1937; Kitzler, 1941). Interestingly, as in *Vipera* (Andrén, 1982), in *Lacerta vivipara,* shedding must occur before reproductive activity (Bauwens et al., 1989). However, the correlation with sexual behavior is with the shedding status of males, not of females. Spring emergence is in April, and before having shed, breeding male *L. vivipara* lack interest in females. After shedding, most males initiate vigorous courtship within 30 seconds of detecting a female. It appears that the shedding status of the female does not influence the onset of courtship (Bauwens et al., 1989). Thus, in at least some lizards the shedding status of males or females or both is important in regulating reproduction.

The mechanism by which males detect females was tested by washing a single female *Lacerta vivipara* with alcohol and coating her with Vaseline (Bauwens et al., 1987). The response did not differ from that of controls. The study found that chemical cues seem to be unimportant in initiating courtship; however, alcohol washing may not remove the nonpolar lipids prevalent in reptiles. In addition, coating an animal with Vaseline not only would impart petroleum odors of its own but would also have a confounding tactile dimension that might be perceived by males as unpleasant. Finally, a report based on a single individual makes it premature to dismiss chemical cues.

Chemical communication has been reported only rarely for either species of *Heloderma,* but sex recognition supposedly involves behavioral or olfactory cues (Bogert and del Campo, 1956). Cloacal rubbing by *Heloderma suspectum cinctum* may leave chemical cues that aid in the establishment of social hierarchies in communal areas (Beck, 1990). Before and during courtship, male *Varanus komodoensis* touch their tongues to the body and head of prospective mates, particularly around the sides of the head, between the eye and nostril, and at the junction of the hind leg with the body (see Fig. 4.14)(Auffenberg, 1978).

Some researchers (Evans, 1961; Ferguson, 1966) claim that iguanid and agamid lizards rely primarily on visual cues as social releasers of behavior; however, others (Duvall, 1979; Simon, 1983) found pheromones to be important in several species. Male *Uta stansburiana* approach estrous females and lick their pelvic area as a prelude to courtship and mating (Ferguson, 1966; Tinkle, 1967). Before copulatory behavior in *Amphibolurus fordi,* males lick the cloacal region of females (Cogger, 1978). It has been speculated that femoral gland or cloacal secretions may facilitate recognition of the sexes. Chemical cues may be important in courtship, intra- and inter-specific interactions, and the marking of sites by *Phrynosoma platyrhinos* and *P. coronatum* (Tollestrup, 1981). Male courtship is characterized by extensive licking of the

Fig. 4.14 *Varanus komodoensis.* Shaded areas are those frequently licked and touched with the snout during courtship. (From Walter Auffenberg, *The Behavioral Ecology of the Komodo Monitor* [University of Florida Press, Gainesville, 1981], p. 177.) Reprinted by permission.

female's vent both before and after mating. Cloacal sniffing appears to be involved in agonistic displays between males (Whitford and Bryant, 1979). Males have been observed licking females during courtship in several other species of iguanids, including *Dipsosaurus dorsalis*, many species of *Uma* (Norris, 1953), and *Crotaphytus collaris* (Fitch, 1956). *Dipsosaurus dorsalis* males lick the ground and occasionally the body of conspecific males during agonistic encounters and are able to distinguish familiar from unfamiliar male conspecifics, possibly by chemical cues obtained during frequent tongue-touching of conspecifics (Glinski and Krekorian, 1985). Tongue extrusions directed at conspecifics, especially after displacement and at burrow entrances, suggest that chemoreception may be involved in species recognition, sex recognition, and possibly individual recognition (Krekorian, 1991). Members of the *Sceloporus torquatus* group frequently "taste" (tongue-touch) foreign objects and other lizards, possibly to identify species through chemoreception (Hunsaker, 1962).

F. Substrate Licking: Cloacal Rubbing and Jaw-Wiping

Tongue-flicking or tongue-touching behavior in captive members of six families of lizards differ significantly (Bissinger and Simon, 1979). The frequency of tongue extrusions of moving lizards was: Cordylidae < Iguanidae < Gerrhosauridae < Scincidae < Helodermatidae < Teiidae. This order may be correlated with the structure of the tongue, which ranges from thick and fleshy to long and slender (snakelike).

Sauromalus o. obesus lick conspecifics during courtship and aggression (Berry, 1974). The feces deposited by male *Sauromalus* in prominent locations may provide olfactory signals that assert their presence (Carpenter, 1975). Both male and female *Sauromalus* rub their jaws frequently on the substrate (Fig. 4.15); conspecifics respond to these sites with behavioral displays. It has been suggested that scent-producing glands may be associated with the jaws or eyes (Berry, 1974); however, no one has described specialized glands in these locations. Jaw-rubbing also occurs *Sceloporus jarrovi* (Simon et al., 1981). Biopsied skin from their jaws discloses no glands. Presumably, the chemical cues originating from epidermal tissues are not specialized into glandular tissues. Several species of chameleon rub a temporal pouch onto the substrate (Ogilvie, 1966). The holocrine gland secretes an odoriferous material. During the breeding season, male chameleons aggregate but maintain territories. The males then leave temporal pouch secretions on branches of trees that are later investigated by the tongue-touching of other males. The chemicals exuded by this jaw-wiping behavior have roles in male territoriality or spacing (Ogilvie, 1966).

Approach of male *Sceloporus cyanogenys* to females during the breeding season often includes dragging the vent area on the substrate (pelvic scrape) (Greenberg, 1977). No such scraping has been observed before male-male encounters, suggesting that males recognize females from a distance and that the pelvic scrape may deposit socially significant chemical cues. Also, males of *Phrynosoma platyrhinos* and *P. coronatum* apparently mark sites by partially extruding the cloaca and rubbing it back and forth on the substrate (Tollestrup, 1981). Other iguanid lizards apparently use chemical communication to gather information about their environment. For instance, hatchling *Iguana iguana* frequently rub their thighs and bellies on the ground immediately after leaving the nest (Burghardt et al., 1977). Also, they frequently tongue-touch the ground and each other. Male

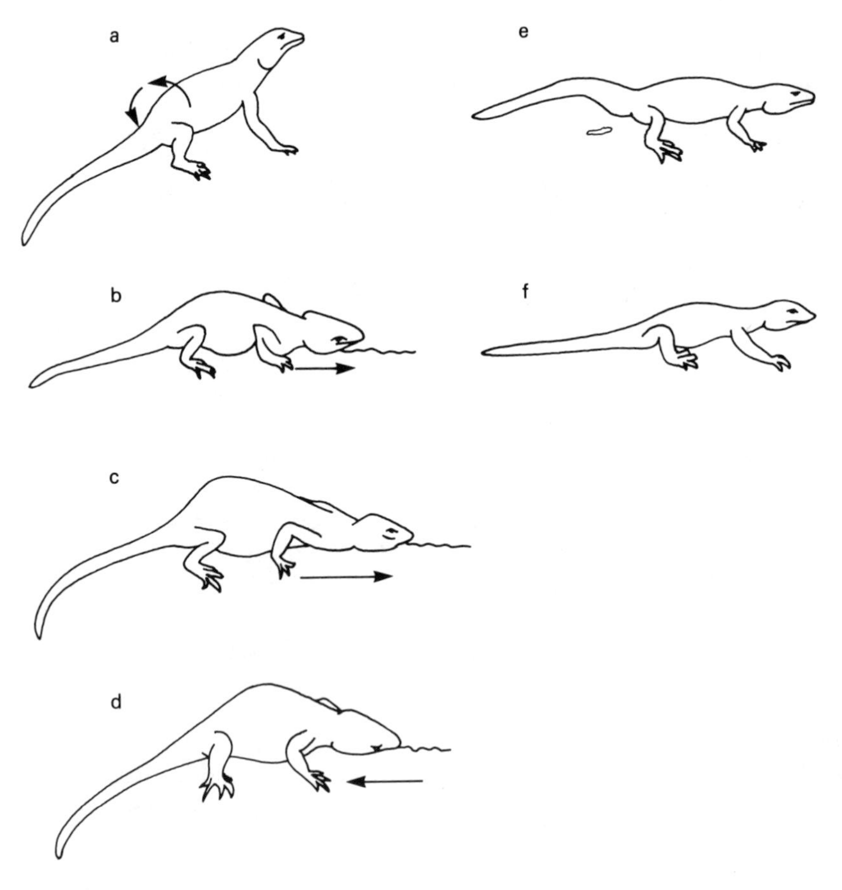

Fig. 4.15 *Sauromalus obesus.* Postures used in possible deposition of chemical signals: (a) hip sway; (b-d) rubbing the jaw on the substrate; (e) defecation; and (f) tail wipe. (Redrawn from Berry, 1974.) Reprinted by permission.

Tropidurus delanonis are stimulated during courtship by chemical cues deriving from the cloaca (Werner, 1978).

During breeding bouts, male *Ameiva chrysolaema* rub their cloacal region from side to side on the ground, repeating this performance frequently. The behavior is followed by attempted copulation within minutes (Noble and Bradley, 1933). Also, males of *A. quadrilineata* rub their cloacal regions on the ground while moving toward females and then mount them during the breeding season (Hirth, 1963). These males are highly aggressive and territorial, chasing other members of either sex; male *Cnemidophorus sexlineatus* rub their cloaca on the ground in a figure-eight pattern before courtship (Noble and Bradley, 1933; Fitch, 1958a; Carpenter, 1962). The behavior is a primary indicator of impending sexual activity and sexual arousal in males and may be used to actively mark their territories. Occasionally males appear to pick up a scent while foraging and immediately begin cloacal rubbing. They also behave this way around the burrow of a female and will chase any males that come near. These behaviors seem to support the idea that the chemical cues denote a breeding territory.

Even the supposedly anosmic chameleonids display substrate licking and cloacal rubbing (Ogilvie, 1966). Substrate licking occurs in the field for males, females, and juveniles of six species of territorial chameleons. This tongue-touching seems to follow cloacal rubbing after defecation.

Between February and December, eight male and female *Sceloporus jarrovi* showed differing frequencies of substrate licking in clean cages and in cages previously occupied by another lizard (DeFazio et al., 1977). Similar results are obtained if a clean rock or a rock taken from the home cage of another lizard is placed in the home cage of a test lizard. Lizards lick the substrate significantly more often if in a cage previously occupied by a conspecific lizard rather than being in a clean cage (Fig. 4.16). Because *S. jarrovi* are known to be highly territorial, the authors suggest that chemical cues left on the substrate may help maintain territories. Whether the cues originate in the cloaca or the femoral pores remains unknown.

In juvenile *Sceloporus jarrovi*, the number of tongue-touches to the substrate does not differ between sexes (Bissinger and Simon, 1981). The juveniles tongue-touch most often if placed into a strange cage that had been previously marked by conspecifics. The juveniles also tongue-touch more in a novel, clean cage than in their own home cage, apparently detecting chemical cues left by other juvenile conspecifics.

Sceloporus jarrovi exhibit most tongue-touching while they are moving (Simon et al., 1981). The tongue-touch rates do not differ significantly between adults and juveniles or males and females. The lizards tongue-touch the substrate significantly more often in the first hour after emergence than later. In addition, the tongue-touch rate increases significantly if the lizards are displaced to a novel environment. They increase the number of chin wipes whenever they move among locations. Frequency of chin wiping does not increase during the breeding season. Histology of the chin region does not identify any glandular sources. Defecation and pelvic rubbing do occur outside of the breeding season; however, they increase in frequency during the breeding season (Gravelle, 1981). In field enclosures, both sexes avoid sections containing chemical cues left by lizards of the same sex; however, males remain in the area of females after these are removed. Both males and females decrease the number of tongue-touches in areas previously inhabited by adults of the same sex, whereas males respond to the home area of the absent female by an

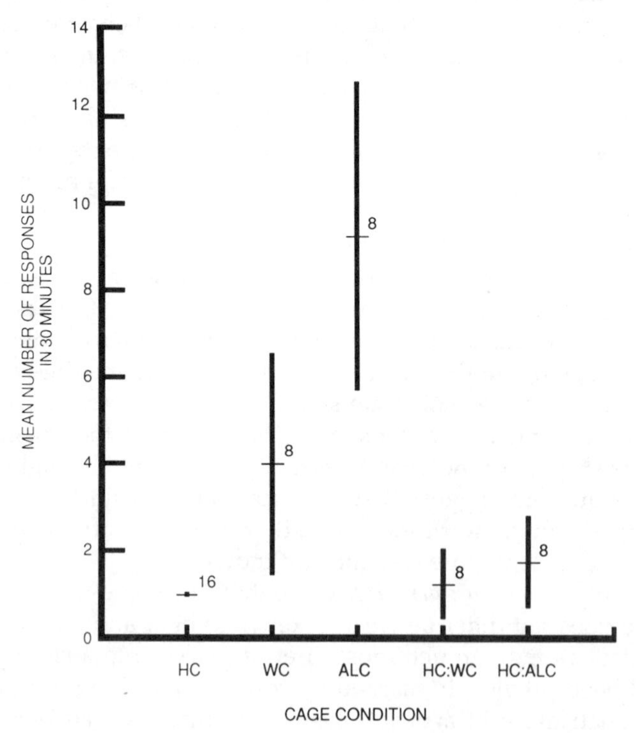

Fig. 4.16 Mean number of substrate licks and standard error of the mean in a 30 min. interval for the following cage conditions: home cage (HC); washed cage (WC); a cage where another lizard had been kept (ALC); home cage after being in a washed cage (HC:WC); and home cage after being in a cage where another lizard had been kept (HC:ALC.) Numbers inside the figure represent sample sizes. (Redrawn from DeFazio, Simon, Middendorf, and Romano, 1977.)

Table 4.2 Fall tongue flick latencies (seconds) of 30 male and female *Sceloporus jarrovi* tested during the fall breeding season: means and 95% confidence limits for treatment effects

Treatment	*S. jarrovi* (male)	Cologne	*S. virgatus* (male)	*S. jarrovi* (female)	Water
Mean	915 ± 158	1,048 ± 123	1,076 ± 112	1,106 ± 109	1,154 ± 42
SNK	________	________	________	________	
		________	________	________	________

Source: Redrawn from Simon and Moakley, 1985.
Note: The results of a Student-Newman-Keuls Test ($p < .05$) are indicated. Means enclosed by the range of a line are not significantly different.

increased number of tongue-touches. They also increase their defecations and pelvic rubs (Gravelle, 1981). Males tongue-touch more than females if they are forced to occupy high-density field areas; however, these rates do not differ from those of controls (Simon et al., 1981), perhaps reflecting territorial behavior in males. On the basis of a separate set of experiments, the authors hypothesized that tongue-touching could serve to detect chemical cues from potential predators. Lizards do not respond with more tongue-touching in the presence of *Lampropeltis pyromelana* or of filter papers on which chemical cues have been deposited by the integument or cloaca of the snake. Lizards do not avoid the snake-exposed paper, indicating that in this species tongue-touching probably does not detect potential predators but conveys information about unfamiliar areas and conspecifics. During the fall breeding season, male and female *S. jarrovi* tongue-touch significantly earlier in response to chemical cues from male *S. jarrovi* than to blank controls (Simon and Moakley, 1985). Males exhibit tongue-touches significantly more than do females and show more tongue extensions (Table 4.2). Because these observations apply during the fall breeding season, males may be more aware of chemical cues as they actively search for breeding females.

Counts of tongue extrusion rates among juvenile and adult male and female *Sceloporus jarrovi* and *Anolis trinitatis* indicate that juveniles have a higher rate of tongue extrusions than adults (Gravelle and Simon, 1980). The *Anolis* were observed in the breeding season but the *Sceloporus* were not, leading to the conclusion that chemical access to the vomeronasal system does not provide behaviorally significant cues involved in reproduction. Gravelle and Simon hypothesize *S. jarrovi* has much larger territories than *A. trinitatis* and possibly cannot visually monitor the territory at all times. If *S. jarrovi* adopted chemical marking of its territory, one would expect greater chemosensory abilities for this species.

Twenty adult male and female (half as behavioral responders and the other half as stimulus source) *Sceloporus occidentalis* have been treated with exogenous testosterone and estradiol to simulate breeding conditions (Duvall, 1979). The stimulus source derives from animals placed into aquaria, each with 10 individuals of a single sex. A brick covered with sandpaper allows the animals to leave femoral gland material, feces, proctodeal exudates, urine and other cloacal material. After 21 days the soiled bricks are placed next to a control, unsoiled brick in the test chamber, and male and female test lizards are exposed to the test chamber for 25 minutes, with the numbers of substrate licks to the bricks observed. Male lizards discriminate the male-soiled bricks, but not the female-soiled bricks, from control bricks. Female test lizards do not show a significant response to either male or female bricks (Fig. 4.17). Immediately after licks of labeled surfaces, but never after licks of unlabeled surfaces, both sexes perform push-ups, the visual display associated with territorial, agonistic, sexual, and defensive behavior in most iguanids. These results indicate a possible coevolution of visual and chemical signaling in the behavioral repertoire and also lend mechanistic support to the earlier-mentioned hypothesis that population density and territorial behavior are affected through semiochemical communication via substrate marking, especially in males.

Sexually active male and female lizards have been presented with four chemical cues on sandpaper, namely male conspecific semiochemicals, female semiochemicals, a cologne pungency control, and a water control (Duvall, 1981). During 15-minute trials male and female lizards show no differences between the sexes in such responses as frequency of rapid nasal inhalation, latency to this behavior, frequency of tongue-flicks, latency to initial tongue-flick, substrate licks, latency to initial substrate lick, and frequency of push-ups. Both sexes respond to male and female markings with significantly more substrate licks than to the cologne or water controls. Unlike the previous investigation, under these conditions males and females do not produce significant differences in tongue-flick rates. Consequently, male and female lizards discriminate conspecific chemical cues. One possible role for the waxy, keratinaceous femoral gland exudate is to serve as a vehicle or substrate for pheromonal cues (Duvall, 1981). Shorter initial tongue-flick latencies to any of the pungent chemical cues support the idea that nasal olfaction detects volatile airborne chemicals and activates the tongue-vomeronasal system for more specific analysis and identification.

An interesting development in the field of lizard chemoreception is the debate over which sensory system actually subserves tongue-flicking and its subsequent delivery of semiochemicals to the oral cavity. An extensive survey of the tongue and oral epithelium of lizards documents taste buds in 37 species (Schwenk, 1985). Taste buds are particularly abundant in the Iguanidae; however, insufficient data exist to distinguish between taste and vomeronasal function as the basis for chemosensory-mediated behavior in lizards. Definitive studies are needed to discriminate between the two sensory modalities; it is prudent to examine the behavioral data bearing this duality in mind.

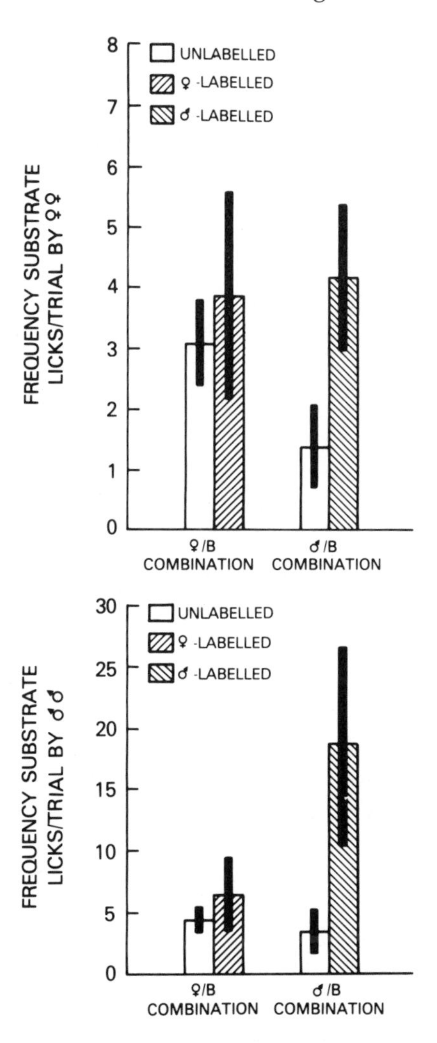

Fig. 4.17 *Sceloporus occidentalis.* The upper portion shows substrate-licking frequencies of adult females; the lower portion depicts those of adult males. Males performed significantly more licks of male-labelled surfaces than of unlabelled ones. Along the abscissa of both plots, designations indicate trials in which lizards were presented with test combinations containing a female-labelled and a "blank" or unlabelled surface, or a male-labelled and blank surface, respectively. (Redrawn from Duvall, 1979.)

Table 4.3. Frequencies of body regions licked by conspecific male and female western fence lizards, *Sceloporus occidentalis biseriatus* ($\bar{X} \pm$ SE)

	Body Regions				
Pair components	Total conspecific licks	Head licks	Flank-back licks	Tail licks	Leg licks
Male-male	7.3 ± 3.1^{a}	3.4 ± 1.5^{a}	3.9 ± 1.5^{a}	0.0^{a}	0.0^{a}
Male-female	0.6 ± 0.2^{b}	0.3 ± 0.2^{b}	0.2 ± 0.1^{b}	0.1 ± 0.1^{a}	0.0^{a}
Female-male	5.3 ± 0.1^{a}	1.8 ± 1.1^{a}	2.9 ± 0.7^{a}	0.2 ± 0.2^{a}	0.4 ± 0.2^{b}
Female-female	2.0 ± 0.6^{b}	0.3 ± 0.1^{b}	0.9 ± 0.3^{b}	0.5 ± 0.3^{a}	0.3 ± 0.2^{b}

Source: Redrawn from Duvall, 1982.
Note: In all groups, $n = 10$. Within a column of data, values with different superscripts are significantly different, based upon Student-Newman-Keuls multiple comparisons with $\alpha_{pc} = 0.05$. See text for complete discussion of statistical significance.

The licking of different areas of the body during social encounters of breeding *Sceloporus occidentalis biseriatus* (Duvall, 1982) suggests which areas of the body may exude or carry pheromones. A study of 28 adult males and 30 adult females during the natural breeding season has noted any lick to any region of the body. Male and female fence lizards lick conspecifics at the same rate, although males are licked significantly more than females. Both sexes lick the head and flank backs of males more than those of females (Table 4.3). The lizards frequently wipe their jaws along the substrate, as do other iguanids. Some iguanids may possess an exocrine gland in the rictus oris, where the upper jaw joins the lower jaw. Secretions from this gland could be wiped on the face and substrate by jaw-wiping (Burghardt, in Duvall, 1982).

Sceloporus occidentalis biseriatus (nine males and eight breeding females) have been tested in a simultaneous two-choice procedure to determine whether conspecific fecal markings possess visual information independent of the pheromonal cues known to be present in feces. Sandpaper surfaces have been treated either with male fecal boli or water as a control and subsequently covered with a plastic coating to eliminate any chemical or olfactory stimuli. Both males and females investigate and hop onto the marked surface rather than the unmarked surface. Although the two groups presumably do not differ chemically, tongue-touches are directed significantly more often to the fecal-marked surface than to the unmarked surface. Few animals display to the plastic-coated surfaces; however, various field observations of lizards show relatively high levels of behavioral displays from lizards directed to uncovered fecal boli emitting chemical cues (Duvall, 1979; Simon, 1983).

In their home cages, individual *Dipsosaurus dorsalis* (10 specimens) were presented with sand taken from the aquaria of stimulus groups that presumably contained chemical cues (Pedersen, 1988): *Dipsosaurus dorsalis, Cnemidophorus tigris,* and kangaroo rats, *Dipodomys deserti.* The *D. dorsalis* were then videotaped for 30 minutes a day for 7 days. Desert iguanas do not increase tongue extrusions to either conspecific or heterospecific lizard cues; however, the iguanas perform significantly greater than random tongue extrusions to the kangaroo rat stimuli. The final result is interesting because it suggests that desert iguanas are able to recognize chemical signals from kangaroo rats; these may help the lizards identify burrows or even food sources. However, the negative responses to conspecifics are difficult to interpret. The animals were not in reproductive condition and were not sexed, although both sex and reproductive condition of individuals should be important in analyzing behaviors associated with chemoreception.

G. Aggregation

Aggregation has been sporadically observed during field investigations for many species of lizards. Fifty-two *Eumeces* were seen hibernating, and the observer suggested that olfactory stimuli guide conspecifics to underground hibernacula (Taylor, 1943). Groups of several dozen skinks have been found hibernating by several authors (Breckenridge, 1943; Hamilton, 1948; Neill, 1948; Fitch, 1954). Interestingly, both Neill (1948) and Scott and Sheldahl (1937) describe large aggregations of skinks that were covered by a spongy, viscous liquid substance that may or may not have been formed by the hibernating lizards. An analysis of such material would be very interesting. Skinks also aggregate while brooding eggs (Cooper et al., 1983). Hatchling *Iguana iguana* have also been described to use olfactory or chemical cues in their social behavior (Burghardt et al., 1977). Upon hatching, iguanas that emerge from the nest in groups display very stereotyped behaviors that include tongue-touches to the substrate and other hatchlings and thigh and belly rubbing on the substrate (Burghardt et al., 1977). A test of grouping tendencies and their possible chemical mediation involved 20 hatchlings from one female and 20 from a second placed together in an enclosure, with 20 more hatchlings from the first female added 3 weeks later (Werner et al., 1987). The lizards form more groups of pairs and triplets of hatchlings from the same mother than do pairs and triplets from different mothers; this suggests that the hatchlings preferentially associate with kin. In more complicated choice experiments, hatchlings exposed to kin or nonkin groups for 24 hours again prefer associating with kin. These choices may be based on the ability of the hatchlings to distinguish kin by body odors sampled by licking or sniffing. This may also be reflected in the preference of hatchlings for cages containing feces from kin versus those with feces from nonkin. The preference seems to be based on olfaction, as only 1 of 32 hatchlings lick the feces, whereas sniffing of feces is common.

Laboratory studies have also documented the tendencies of lizards to aggregate. An example is *Coleonyx variegatus* (Cooper et al., 1985; corroborating Greenberg, 1943; Evans, 1967; and Cooper and Garstka, 1987a). Observations of *Eumeces laticeps* show more aggregations during the winter months than during the early spring, possibly mirroring the behavior observed as animals leave hibernacula in nature. The sexes do aggregate equally, at least during the winter months. These aggregation tendencies are not due to agonistic behavior because no wounds were observed, and agonistic behavior itself only increases in the spring.

H. Skinks

As documented in many studies, scincid lizards utilize chemical cues in many aspects of their behavior, including detection of predators and prey, aggregation, trailing of conspecifics, courtship, and mating. Many species use tongue-flicking and licking (Carpenter and Ferguson, 1977). During the breeding season, male *Eumeces fasciatus* direct the greater part of their activities to a search for females, trailing and finding them by both sight and scent (Fitch, 1954). Upon locating a female, the male first touches her with his tongue, apparently receiving olfactory stimuli that are essential to release breeding. *E. obsoletus* respond to olfactory cues in detecting other males and females (Fitch, 1955; Evans, 1959). Courtship involves contact cues such as nudging, licking, and nipping, which suggests that chemical signals might be perceived. Licking behavior also has been described for *E. multivirgatus* (Everett, 1971). In *E. egregius,* scent is the most important factor, not only in causing sexual excitement in the male but also in enabling him to locate and recognize a female (Mount, 1963). During encounters with females, males rub their pelvic regions, and thus cloaca, against the substrate and in investigating a sexually mature female, the male rushes up quickly and touches his tongue to the nearest part of her anatomy. Perrill (1980) also points out the importance of chemical communication to skink reproduction in his ethological investigation of *E. inexpectatus.* Males tongue-flicked the females in 16 of 17 matings, suggesting the importance of sexual identification and stimulation by means of chemical cues. Perhaps chemical stimuli serve as important isolating mechanisms that prevent hybridization between sibling species.

In *Eumeces fasciatus* and *Scincella lateralis,* responses to airborne conspecific and heterospecific chemical cues have been examined in an olfactometer (Duvall et al., 1980). Individual male and female skinks have been placed in the olfactometer and presented with the odor of a conspecific male, female, or (initially) empty jar. The behaviors measured were approach to the odor, initial movement latency, duration of locomotion, tongue-flick rate, and snout-dip rate (this behavior occurs while the lizards are moving and tongue-flicking). Tongue-flick rate and snout-dip rates can be increased by male and female odors; however, male *Eumeces* tongue-flick and snout-dip more rapidly than do females. Males approach male, but not female, odor streams at rates significantly greater than chance frequency. Females approach female odors but not those of males. Both male and female ground skinks avoid airstreams containing significant odors of males; male approaches to the odors of females are significant. No significant differences occur for movement latency, locomotion duration, or tongue-flick rate, male and female test subjects, or male and female odors.

Use of heterospecific male and female odors in the olfactometer show no significant numbers of approaches to or avoidances by either test species (Duvall, 1982). These results may indicate that the quantity of volatile pheromones exuded by conspecifics is sufficient to be detected in an airstream. Although these studies were conducted in the nonbreeding season, *Eumeces fasciatus* respond as though some aggregation pheromone is present. Why females are attracted to males at this time remains unclear.

In a follow-up study, Cooper and Vitt (1985) have suggested that the air speeds (180 km/hr) used in the previous study are too high, impairing the ecological significance of the results. Greatly reduced air speeds and a higher ambient temperature, using postreproductive *Eumeces fasciatus* and *E. laticeps* treated with exogenous steroid hormones to simulate breeding conditions, produce no significant difference in frequency of approach to sources of air alone or to airborne odors of male or female conspecifics either for both sexes of *E. fasciatus* or for male *E. laticeps*. These observations differ markedly from the previous set (Duvall et al., 1980). Skinks do not seem to be able to detect airborne conspecific odors at air speeds commonly encountered in nature. The discrepancy between the two studies may reflect the use of chemical cues for two different functions. Breeding male skinks appear to locate females by scent-trailing (Goin, 1957) but not to follow male trails. Females have not been reported to follow trails of either sex. Perhaps approaches by both sexes to odors of conspecifics of both sexes reflect the presence of a pheromone that facilitates aggregation (Duvall et al., 1980). This hypothesis is supported by the postreproductive condition of the lizards in the earlier study. *E. laticeps* does aggregate at low temperatures. At higher temperatures, the skinks do not appear to respond to airborne conspecific cues; also breeding males follow female trails laid on the substrate, rather than airborne cues.

The most extensive investigation of pheromonal communication in any reptile is that of skinks, including a systematic analysis of behavioral responses both to conspecific and heterospecific semiochemicals and the endocrine control of synthesis of these pheromones (Cooper and colleagues, below). Hence, the results are summarized in some detail.

Adult female *Eumeces laticeps* flick their tongues repeatedly when investigating conspecifics; they direct these tongue-flicks primarily to the cloaca (Cooper and Vitt, 1984a). Cotton applicators dipped in distilled water and first inserted into the cloaca of adult male and female *E. laticeps* have been presented to postreproductive females implanted with estradiol that induces sexual receptivity; tongue-flick rates then were recorded for 20- and 60-second intervals. Male odor donors were treated with exogenous testosterone and females with estradiol-simulating breeding titers of steroid hormones. Females direct significantly more tongue-flicks to the cloacal odors of conspecifics than to distilled water, but the sexes do not differ otherwise. Males respond with significantly higher tongue-flick rates than do females. Female skinks detect conspecific cloacal cues and probably respond to them. Sex recognition does not involve odor alone. Females do not actively seek mates, and their response to conspecific cues may relate to aggregation tendencies (see Section III.G). As skinks lack femoral glands, semiochemical cues must derive from the cloaca or the integument.

Male *Eumeces laticeps* exhibit significantly more tongue-flicks to conspecific odor cues than to distilled water blanks (Cooper and Vitt, 1984b). Unlike females, males do discriminate cloacal odors, emitting higher numbers of tongue-flicks in response to females than to males. Placement of individual nonresident male or female skinks into the home cage of a male causes the resident male to approach; tongue-flicking starts only at close range, suggesting that discrimination of species and sex requires proximity or actual contact. Analysis of tongue-flick rates of testosterone-treated versus untreated males responding to conspecific cloacal odors indicates no difference between reproductive and postreproductive males; however, reproductively active males much more continuously investigate female cloacal odors. This makes it likely that reproductively active males can locate and discriminate sexually active females during the breeding season. Finally, males discriminate female skin odors from male skin odors (and distilled water controls), but the tongue-flick rates are much lower than those elicited by cloacal stimuli. The difficulty of visual

Table 4.4 Responses of male and female *Eumeces laticeps* to water, cloacal odors of conspecific and heterospecific skinks, and pungency control

		Tongue Flicks		
Responding sex	Stimulus	Mean no.	SD	Range
Male (n = 18)	*laticeps*	26.67[a]	27.67	8–116
	fasciatus	2.44	1.34	1–5
	inexpectatus	4.61	4.15	1–13
	water	2.56	1.76	1–6
	cologne	5.11	4.38	1–16
Female (n = 8)	*laticeps*	10.62	3.38	5–16
	fasciatus	5.38	3.62	1–10
	inexpectatus	5.38	3.96	1–13
	water	3.38	1.30	1–5
	cologne	6.12	4.19	2–1

Source: Redrawn from Cooper and Vitt, 1986a.
[a]Indicates a tongue-flick rate significantly higher than observed in response to all other stimuli.

discrimination of the sexes of *Eumeces* by herpetologists may reflect our use of visual rather than pheromonal signals.

It has been suggested that the sympatric *Eumeces fasciatus* and *E. inexpectatus* use chemical cues to maintain reproductive isolation (Perrill, 1980). The tongue-flick response was noted for postreproductive male and female *E. laticeps* to cloacal semiochemicals of *E. laticeps, E. fasciatus,* and *E. inexpectatus,* conspecific and heterospecific members of the closely related *fasciatus* species group (Cooper and Vitt, 1986a). Cues were offered on cotton swabs containing five chemicals: water control, a cologne pungency control, a conspecific cloacal fluid, and two heterospecific cloacal fluids. Males tongue-flick at significantly higher rates than females. Male *E. laticeps* respond with significantly more tongue-flicks to female conspecifics than either to the controls or the two heterospecifics (Table 4.4). In a like manner, females respond with significantly higher tongue-flick rates to the cloacal odors of conspecific males than to any of the other stimuli. Thus, broad-headed skinks are able to recognize and discriminate heterospecific odors from those of conspecifics. Because tongue-flicking of chemical cues from the cloaca of females is an integral part of courtship, males may detect information about species identity as well as about the sex of the individual encountered. Also, male *E. fasciatus* and *E. inexpectatus* exhibit significantly higher tongue-flick rates to cloacal cues from conspecific females than to heterospecific female cues or to water or cologne controls (Cooper and Vitt, 1986b).

Although snakes often trail conspecifics, this behavior seems rare in lizards; however, lizards can follow conspecific chemical trails on the substrate (Cooper and Vitt, 1986c). The cloaca of male or female *Eumeces laticeps* has been dragged along one arm of a Y-maze, thus contacting the paper covering the floor. Males can follow such trails produced by females. These males frequently tongue-flick the air and tongue-touch the substrate. Upon reaching the choice point, many individuals look about and flick their tongues in both directions before entering one arm or the other. Males do not follow males nor do either estrogen-treated nor untreated females trail either sex. Presumably trailing capacity is mediated by the vomeronasal system; however, the possible role of gustation cannot be ruled out. It is clearly beneficial for males to detect and trail conspecific females, especially during the breeding season. The extreme aggressiveness of males during the breeding season may make it advantageous to avoid trailing males. On the other hand, there are reports of rather large aggregations of skinks, especially during overwintering.

Because male skinks lick the cloacas of adult females during courtship activity and respond with increased tongue-flick rates to cloacal lavages of females, a cloacal pheromone source seems likely. Female *Eumeces laticeps* have a urodaeal gland; Trauth et al. (1987) have described its histology. The gland of adult females enlarges during the breeding season.

Two adult female broad-headed skinks have been treated with estrogen to simulate breeding conditions and to assess the urodaeal gland as a possible pheromone source (Cooper et al., 1986). Their urodaeal glands were removed and macerated in deionized water. Swabs with this material were presented to 24 males, 18 of which had been treated with testosterone to promote courtship behavior. The urodaeal glandular homogenate elicits higher tongue-flick rates by reproductive males than do stimuli prepared by lavage of the cloacal lumen (Fig. 4.18), suggesting that some, if not all, of the pheromonal activity is contained in the urodaeal gland. Whenever postreproductive females are painted with extract of urodaeal glands from sexually attractive females as opposed to unpainted females, males more intensely court the females painted with urodaeal gland semiochemicals. The lipids of the urodaeal pheromone have been extracted and tested by bioassay. Males tongue-flick significantly more to chloroform-methanol extracted lipids of the urodaeal glandular homogenate than to peanut oil controls or to estradiol-17b in peanut oil. Clearly some, if not all, of the pheromonal activity of the urodaeal gland is composed of lipid, but it is unclear whether the pheromone is a single compound or a mixture.

In a further isolation and characterization of urodaeal pheromones, lipid, protein, and carbohydrate fractions have been isolated from the gland of female *Eumeces laticeps,* with the lipid fraction being further partitioned into acidic, basic, and neutral lipid fractions (Cooper and Garstka, 1987b). Sexually active males do not respond to the acidic and basic fractions on cotton applicators with more frequent tongue-flicks. Only the neutral lipids elicit significantly higher rates than controls. The magnitude of tongue-flick rates in response to neutral lipids is similar to the response to unfractionated urodaeal lipids. As neutral lipids are known to be pheromonally active in garter snakes, neutral lipids may serve semiochemical roles here as well.

The ability to distinguish conspecifics from members of other species is basic to maintaining appropriate reproductive, territorial, and agonistic behaviors. Agonistic behavior among male skinks of the *Eumeces fasciatus* group should be directed only at conspecifics and may be elicited by cloacal cues (Cooper and Vitt, 1987a). Encounters among conspecific male *E. laticeps, E. inexpectatus,* and *E. fasciatus* result in species-typical aggressive displays, fights, and establishment of dominance. Encounters among heterospecific males are similar to the conspecific encounters in that males approach and tongue-flick one another. However, in contrast to results in conspecific encounters, very little aggressive behavior occurs. Males emit aggressive head-tilt displays from a distance but cease all aggressive behavior after tongue-flicking heterospecific males. After male *E. inexpectatus* are rubbed on corresponding parts by the head, neck, and trunk skin,

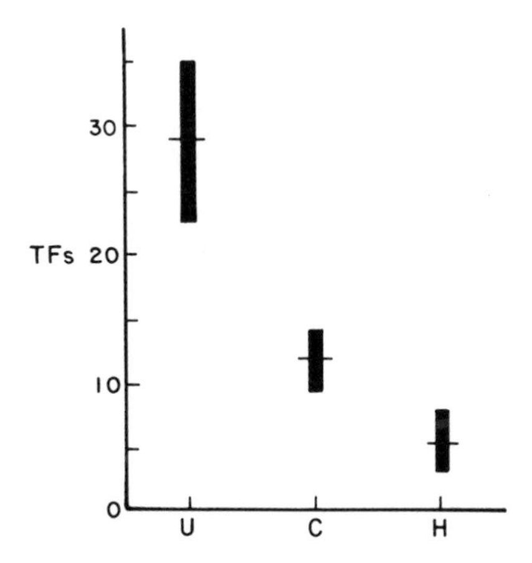

Fig. 4.18 *Eumeces laticeps.* Mean numbers (horizontal lines) of tongue-flicks ± 1 SE (vertical bars) shown for males responding for 60 sec to cotton swabs bearing products of the urodaeal gland (U), female cloacal odors (C), or water (H). All treatment pairs show significantly different tongue-flick rates. (From Cooper, Garstka, and Vitt, 1986.)

including the cloacal margin, of a male *E. fasciatus*, other male *E. inexpectatus* respond with significantly more agonistic behavior toward the male *E. fasciatus* thus treated than to the untreated males. Clearly, male skinks discriminate between conspecific males and those of closely related congeners. Males of all three species tongue-flick each other before terminating contact in heterospecific encounters or initiating agonistic behavior in conspecific encounters. Visual cues are also important in agonistic behavior (Cooper and Vitt, 1987a, 1988); however, visual cues alone may be insufficient for identifying conspecifics. Because such agonistic behavior among conspecifics is restricted to the breeding season, it seems reasonable to conclude that it is reproductive in the broad sense, perhaps involved in territoriality or mate acquisition.

Cloacal odors of reproductive males and males slightly past the reproductive season of *Eumeces laticeps* and *E. fasciatus* have been collected on water-soaked cotton swabs and presented to both conspecific and heterospecific males (Cooper and Vitt, 1987b). Tongue-flick rates of males are significantly higher in response to conspecific cloacal cues than to those of heterospecific males. As urodaeal glands occur only in females, these results indicate that male sex recognition must be based on some other source, perhaps the dorsal gland (Cooper and Vitt, 1987b).

Heterosexual encounters have been observed between conspecific and sympatric heterospecific pairs of *Eumeces inexpectatus, E. fasciatus,* and *E. laticeps,* including those in which heterospecific females had been rubbed with the skin and cloacal odors of conspecific females. Pairs were observed until copulation occurred or the female rejected courtship (Cooper and Vitt, 1987c). All individuals had been brought into breeding condition by treatment with steroid hormones. Males of all three species court and mate with females of their own species. Male *E. fasciatus* and *E. laticeps* never court a heterospecific female. However, male *E. inexpectatus* court heterospecific females, although they prefer conspecific females and court them significantly more frequently than they court those of *E. fasciatus*. Although male *E. inexpectatus* could potentially mate with female *E. fasciatus,* no such matings have been observed. Male *E. inexpectatus* court female *E. laticeps* as often as they court conspecific females. All females reject courtship by heterospecific males.

Male *Eumeces fasciatus* do not court female *E. inexpectatus* that have been coated with skin and cloacal odors of female *E. fasciatus*. In contrast, male *E. inexpectatus* readily court female *E. fasciatus* labeled with female odors of *E. inexpectatus*. In two species, the ethological isolation is maintained at least in part by interspecific differences in pheromones. Failure to court heterospecific females labeled with conspecific female odors suggests that other, probably visual, cues are important in interspecific discrimination in *E. fasciatus*. For example, both male and female *E. laticeps* are much larger than their two congeners.

Males normally respond to individuals with orange heads with aggressive displays, including biting and chasing. After conspecific female skinks had their heads painted orange to mimic the orange head coloration of breeding males, males often bit the painted females, but only before any chemosensory investigation (Cooper and Vitt, 1988). Males that tongue-flick the orange-headed females then cease all aggressive activity and switch to courtship. Apparently, pheromonal identification takes precedence over visual cues. Some females have orange heads during the breeding season (although never as intensely orange as those of males); thus, it is not surprising that chemosensory cues would override the visual identification of an orange-headed animal as a male.

I. Brooding Behavior

Some lizards, notably skinks of the genus *Eumeces* (Noble and Mason, 1933; Cagle, 1940; Fitch, 1954; Evans, 1959; Groves, 1982; Hasegawa, 1985; Vitt and Cooper, 1989), *Mabuya macrorhynca* (Reboucas-Spieker and Vanzolini, 1978), *Elgaria liocephalus* (Greene and Dial, 1966), *Ophisaurus ventralis* (Noble and Mason, 1933), *Anolis carolinensis* (Tokarz and Jones, 1979), and *Calotes versicolor* (Asana, 1941) display behaviors that can loosely be described as brooding their eggs. Relatively little experimental work has addressed the nature of the response or the factors that regulate this behavior (Fitch, 1970; Shine, 1985). In two species of lizards, *Eumeces fasciatus* and *E. laticeps*, females constantly turn the eggs in the nest (Noble and Mason, 1933). Eggs removed from the nests are retrieved by the female and rolled back into the nest by pushing with the head. Females retrieve eggs of both species, despite their different sizes, suggesting that female *Eumeces* will brood any eggs of roughly the same size as her own. However, females of neither species retrieve eggs of *Ophisaurus* or *Sceloporus undulatus* but treat such eggs as they would any foreign object. When presented simultaneously with eggs of *Eumeces* and those of other lizards, female skinks brood only skink eggs, even if these have been moved to another part of the aquarium. If all skink eggs are replaced by the eggs of other lizards, females do not brood any eggs. During initial contact with eggs, the female investigates them with a single flick of the tongue. If the eggs are her own, she returns to brooding them; if they are foreign, she rejects them. The females also ignore paraffin models and shellacked eggs, which suggests that chemical cues are important in the recognition of eggs (Noble and Mason,

1933). Females with their eyes covered with adhesive tape are able to relocate their eggs and begin brooding, even if the eggs have been moved. Hence, visual cues are unnecessary. A brooding female with the tips of her tongue removed did not return to her eggs, although it was claimed that her behavior was otherwise normal. Perhaps the tongue plays an important role in the location, identification, and turning of the eggs and the Jacobson's organ is the chief sensory modality regulating brooding behavior.

Plugging the nostrils of a female *Eumeces laticeps* while she broods her eggs does not prevent her from finding her eggs within 4 hours (Noble and Kumpf, 1936), nor does blindfolding her. Blindfolding and plugging the nostrils still leave the female able to relocate her eggs. With tongue removed, the female is still able to find her eggs and to discriminate them from eggs of *Sceloporus undulatus.* Apparently, *Eumeces* uses the chemical senses for locating and brooding eggs but may use both chemical senses and vision.

Motion pictures of brooding behavior in *Eumeces obsoletus* show females turning and flipping the eggs (Evans, 1959). The female always licks the eggs as she turns them; hence, vomeronasal stimulation may be necessary to elicit and maintain brooding behavior.

Two eggs were removed from the clutches of brooding females of *Eumeces laticeps* and *E. fasciatus* (in the laboratory) and replaced with two eggs from the brood of a conspecific or a heterospecific (Vitt and Cooper, 1989). Females tested accept the eggs of both conspecifics and congeners. Tests involved removing two eggs from the broods and their placement 4.0 cm away from the nest, as well as replacing two eggs by two eggs from a congener or from other genera. Most of the females retrieve their own eggs, whereas some of the females retrieve eggs of congenerics. None of the females retrieve eggs of *Anolis carolinensis* or *Scincella lateralis.* Thus, these results corroborate earlier findings (Noble and Mason, 1933). It does not appear that the females of the two species studied here can distinguish their own eggs from those of conspecifics or congenerics; however, they clearly can distinguish their own eggs from the eggs of other species.

IV. RHYNCHOCEPHALIA

Although there have been several studies of cloacal glands in *Sphenodon* (Gadow, 1887; Gabe and Saint Girons, 1965; Giersberg and Rietschel, 1967), little is known concerning pheromonal communication in this fascinating order.

A study of courtship, mating, and male combat in captive tuatara reported no tongue contacts during any interaction, implying that chemical communication is not critical during such encounters (Gans et al., 1984). Gillingham (personal communication, 1990) reports that

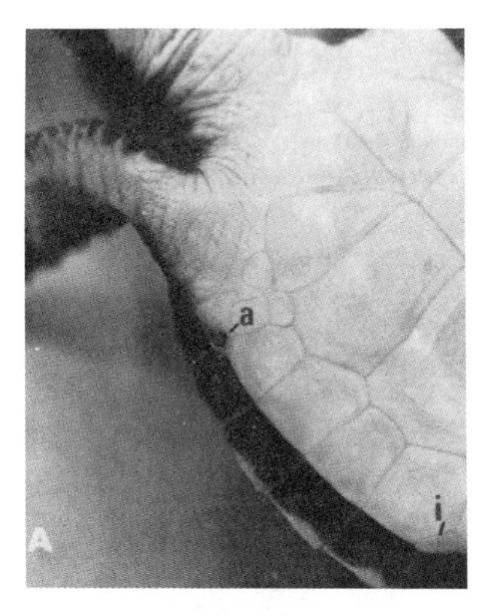

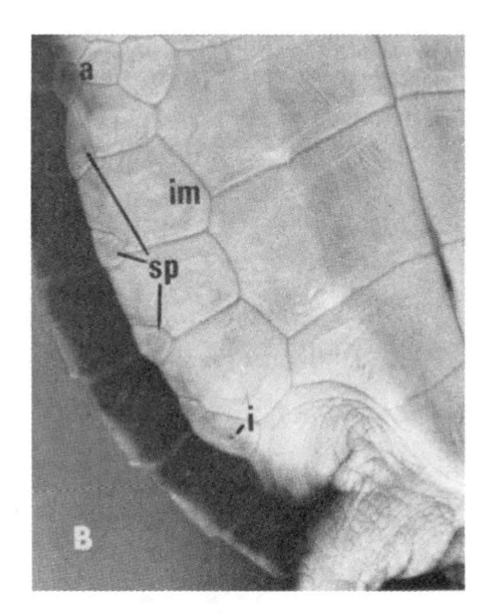

Fig. 4.19 *Chelonia mydas.* Location of pores on plastron. (A) Turtle with pores in axillary (a) and inguinal (i) scutes. (B) Turtle with supernumerary pores (sp) in scutes derived from inframarginal scutes (im). (From Ehrenfeld and Ehrenfeld, 1973.)

he observed few tongue-flicks during mating encounters in the field. In addition, males do not lick females during courtship behavior or during mating.

V. TESTUDINES

A. Introduction

There are many anecdotal accounts of apparent chemoreception of behaviorally significant chemical cues in turtles but few studies of semiochemical communication. Thus, this portion of the review will summarize primarily behavioral studies that suggest the mediation of chemical communication. The role of olfaction and behavior in turtles has been reviewed (Manton, 1979).

B. Glandular Morphology

Turtles have several specialized glands that secrete aromatic, potentially semiochemical, material into the environment. Inguinal and axillary glands, or Rathke's glands—described in Ehrenfeld and Ehrenfeld (1973) and Solomon (1984)—lie medially along the angle formed by the carapace and plastron and secrete their products into the environment via axillary or inguinal pores (Fig. 4.19). Such glands occur in all families of turtles with the exception of the Testudinidae (Loveridge and Williams, 1957). Histological and histochemical studies reveal that the glands of *Chelonia mydas* and *Sternotherus odoratus* are holocrine and consist of a series of secretory lobules embedded in a thick capsule of striated muscle (Ehrenfeld and Ehrenfeld, 1973). The secretion in both species is PAS-positive (PAS-positive substances are protein-carbohydrate complexes), protein-rich, and nonacidic. Analysis of the Rathke's gland secretions of *Caretta caretta* and *Lepidochelys kempii* reveal similar glycoproteins of approximately 55,000 daltons (Radhakrishna et al., 1989). Why these sea turtles produce and secrete this and other high-molecular-weight, water-soluble proteins remains unknown (Radhakrishna et al., 1989). The glands of *S. odoratus*, but not of *C. mydas*, also produce a series of lipids. The absence of lipids from the glands of *C. mydas* deserves further biochemical investigation, as lipid materials would be expected from this type of gland. Perhaps these secretions serve as predator deterrents; the muscular capsule of the gland should allow rapid expulsion of the noxious and malodorous secretions. There has also been speculation on possible pheromonal functions of Rathke's gland components (Legler, 1960).

Another major glandular source of possible semiochemicals are the mental glands (see Fig. 4.20), integumentary organs that open between the mandibular rami and consist of simple or branched invaginations of the epidermis on the skin of the throat (Rathke, 1848; Winokur and Legler, 1975). These holocrine glands occur in 21 of the 69 (of 74 known) genera. They are unique to the three closely related families: Emydidae, Testudinidae, and Platysternidae. The glands of juveniles are quiescent (Winokur and Legler, 1975). Those of *Gopherus* undergo seasonal variation; the size and secretory activity of the glands increase upon administration of gonadal hormones (Weaver, 1970). The glands of *Gopherus* are active between April and November and are sexually dimorphic; those of males are considerably larger than those of females (Rose et al., 1969).

The cloaca is the final source of chemicals with possible behavioral significance. Several anecdotal observations of courtship behavior in turtles suggest that the cloacal discharge of female turtles attracts or stimulates males to court during the breeding season; however, only one kind of glandular tissue has been described from the cloaca of a turtle. This is a small tubulous gland just under the mucous membrane in the wall of the oviductal papilla of a female *Clemmys marmoratus* (Disselhorst, 1904; Whiting, 1969). However, the absence of cloacal glands in turtles may reflect inadequate histological examination of this region (Whiting, 1969).

C. Aggregation

Tortoises have been reported to aggregate in the field; thus, 17 *Gopherus agassizii* hibernated in a single burrow system in Utah during

Fig. 4.20 *Gopherus berlandieri.* View of the head region of a large male showing enlarged chin glands. (From Rose, Drotman, and Weaver, 1969.)

the winter (Woodbury and Hardy, 1940; 1948). Aggregations of *Geochelone denticulata* occur only when the population density is high (Auffenberg, 1969a).

Apparently only one experiment has investigated the role of chemical cues in tortoise aggregating behavior. Eight male and female *Gopherus agassizii* were allowed 6 months to choose sleeping and basking areas (Patterson, 1971). Discharge of urine from male and female tortoises at the two primary sleeping areas does not significantly affect aggregation behavior; however, spreading urine of male *G. berlandieri* over the sleeping site caused four of six tortoises to leave the site and sleep in the open. In contrast, introducing fecal pellets of unfamiliar *G. agassizii* to the sleeping area caused all tortoises to leave; they did not return until the pellets had dried out. The fecal pellets of dominant males also cause subordinate males to leave aggregation sites. In the winter, reduced food consumption dramatically decreases the number of fecal pellets deposited (Patterson, 1971). Furthermore, dominant males are usually the last to enter overwintering hibernacula, other turtles having been previously conditioned to these. The chemical cues may derive from cloacal glands similar to those implicated in courtship behavior of tortoises and aquatic turtles.

D. Aggression and Combat Behavior

Studies of courtship and combat behavior in the genus *Gopherus* provide the best evidence of semiochemical communication in turtles. Auffenberg (1966) was the first to describe courtship behavior in *G. polyphemus*. Male tortoises, on approaching any other tortoise, stop and rapidly bob the head. Slow-motion photography reveals that the swollen mental glands are everted during these displays, suggesting that head bobbing may serve as a means of wafting scents through the air. *Gopherus berlandieri* may utilize secretions from the mental glands to distinguish sex in their own species whenever the glands are actively secreting (Weaver, 1970). Female *G. polyphemus* rub their forearms against their chin glands and extend these forearms toward the male during courtship (Auffenberg, 1969b). The same behavior occurs in males (Weaver, 1970), both sexes possessing an enlarged scale near each elbow. It has been suggested that chin gland secretions are rubbed on the forearms to serve as an olfactory sexual cue or pheromone during courtship and/or combat (Manton, 1979).

The glandular secretions of the chin gland (mental gland) in females of all four species of *Gopherus* (*G. polyphemus*, *G. flavomarginatus*, *G. berlandieri*, and *G. agassizii*) have a protein band not seen in males (Rose et al., 1969). Male secretions, on the other hand, contain esterases that are lacking in secretions of females. Finally, the chin gland secretions of both sexes contain phospholipids, triglycerides, fatty acids, and cholesterol.

The lipid fractions of the mental glands of 30 male *Gopherus berlandieri* have been analyzed with gas chromatography (Rose, 1970). They contain several fatty acids, including caprylic, capric, lauric, myristic, palmitic, palmitoleic, stearic, oleic, and linoleic. The head of a plaster model of a tortoise, placed in an outdoor enclosure, has been painted either with a control solution or with a mixture of the fatty acids. Sixty-six tortoises (37 males and 29 females) were exposed to the model. As head bobbing and ramming are behavioral components of courtship and combat, such activity has been scored as positive. Thirty-nine animals responded positively to the fatty acids, with males responding significantly more than females. All of the males rammed the model, and six overturned it, but none of them attempted to mount the model. Consequently, the fatty acids found in the chin gland secretions of male tortoises appear to elicit combat behavior from conspecific males. No difference in the chin gland secretions of males and females could be determined, as the glands of females are too small to collect secretions (Rose, 1970).

E. Sexual Behavior

Studies of the way *Geochelone carbonaria* and *G. denticulata* recognize conspecifics and determine their sex lead to the conclusion that they determine sex with a simple two-phase behavioral pattern (Auffenberg, 1965). The first phase is entirely visual and involves specific head movements of adult males. The second phase involves the scent of the cloacal region and serves to identify female conspecifics (Auffenberg, 1964; Fig. 4.21). A normal behavioral encounter starts with a male confronting another tortoise with a head challenge. Unless the challenge is returned, the male walks to the posterior part of the second tortoise and smells its cloacal area. If the second individual is a conspecific, sexually mature female, he mounts. Males of either species sometimes mount females of the other species but only in the absence of conspecific females. Smearing of cloacal secretions of sexually active females of either species on the cloacal region of the other species only rarely elicits mounting attempts. The powerful effect of female cloacal pheromones on male sexual behavior in these tortoises is illustrated by two incidental observations. A male *G. carbonaria* mounted a skeletonized shell on which cloacal secretions from a sexually mature female had been rubbed. A male *G. denticulata* mounted and attempted to copulate with a head of lettuce over which a female had clambered with some difficulty just a few moments before (Auffenberg, 1965).

Consequently, terrestrial tortoises may have two primary sources of semiochemicals: the mental or chin glands and the cloaca. The former may elicit combat between males, whereas the latter permits species and sex recognition. However, stronger conclusions require experimental studies.

Aquatic turtles rely primarily on two glandular regions for sources of semiochemicals. Anecdotal accounts suggest that the cloaca discharges semiochemicals that facilitate species and sex recognition. Other sources of semiochemicals are the Rathke's (inframarginal, axillary, or inguinal) glands located along the side of the shell.

Courtship behavior in *Trachemys scripta troostii* has been claimed to depend on sex recognition in response to chemical cues (Cagle, 1950). In *Pseudemys nelsoni,* courtship always begins with the male chasing the female. During these chases the male appears to sniff the cloaca

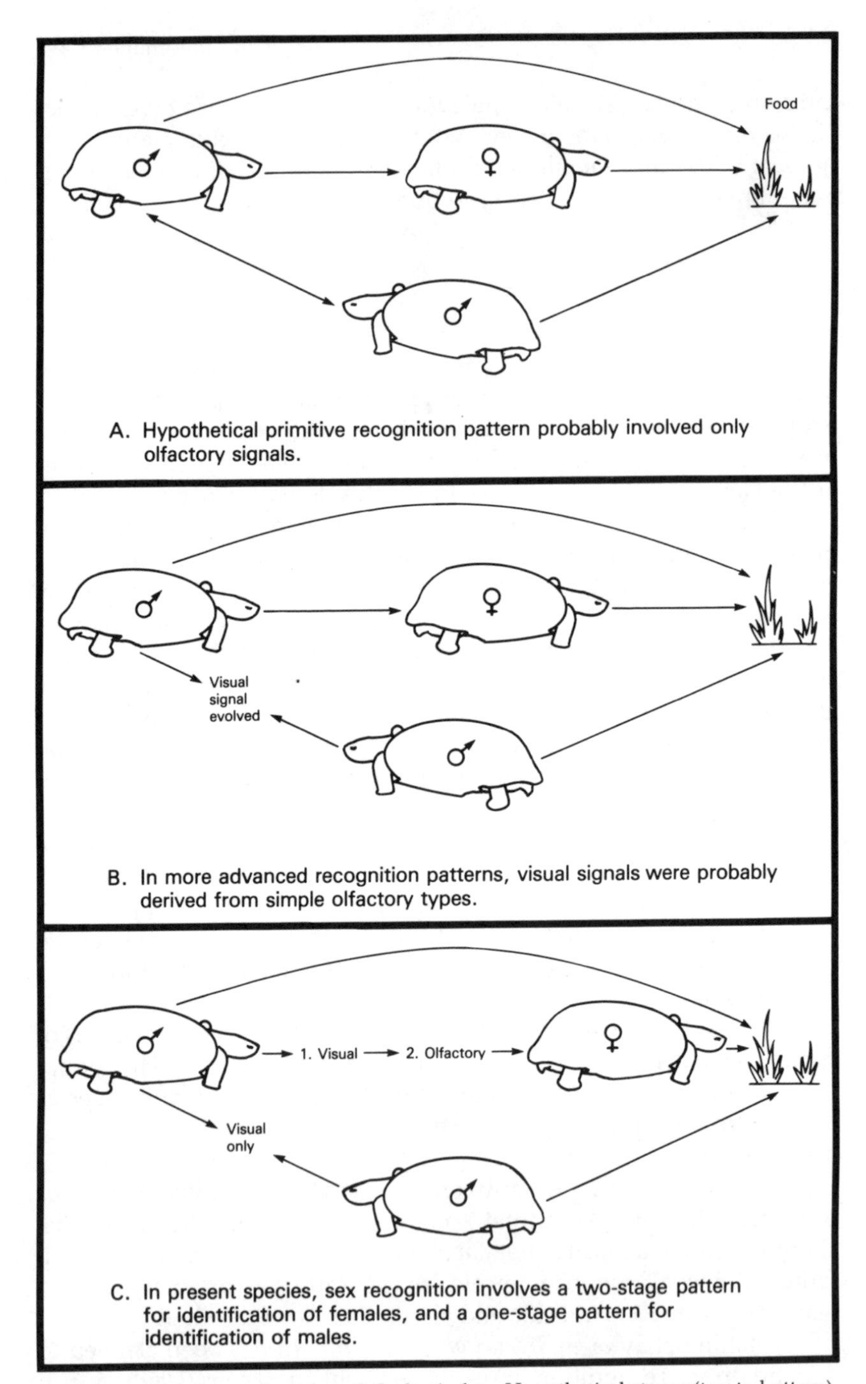

Fig. 4.21 *Geochelone carbonaria* and *G. denticulata.* Hypothetical stages (top to bottom) in the evolution of sex and species discrimination. (From Auffenberg, 1965.)

of the female (Kramer and Fritz, 1989). Similar behavior has been described in *Chrysemys picta* (Ernst, 1971). Laboratory observations of courtship in *Pseudemys concinna suwanniensis* show the male following the female, with snout close to the rear portion of her shell, suggesting the existence of a chemical cue in her cloaca (Jackson and Davis, 1972a). In *Trachemys scripta elegans,* males that sight females in rearward or posterolateral presentations usually move toward their cloacal area, keeping the neck extended and engaging in brief and highly variable periods of cloacal sniffing and trailing (Jackson and Davis, 1972b). Chasing and cloacal sniffing also occurs in *Graptemys* (Ernst, 1974), whereas cloacal sniffing alone is part of the courtship sequence in *Malaclemys terrapin tequesta* (Seigel, 1980).

Very little is known of the behavior of sea turtles. Some evidence suggests that sea turtles use chemoreception in their homing and orientation behavior, but these behaviors have commonly been attributed to chemical cues associated with the natal beach (see Owens et al., 1986; Owens, 1980, for a review of this topic). Extensive behavioral observations of captive *Chelonia mydas* indicate that both male and female adults check the cloacae of breeding females significantly more often than those of nonbreeding females. It has been suggested that females appear to signal reproductive attractiveness by some mechanism, perhaps chemical (Crowell Comuzzie and Owens, 1990). Cloacal checks may be used to ascertain the approximate readiness of the female for mating. A chemical signal or pheromone produced by females may be detected by males from far away and may thus enable the location of females. This could explain the ability of wild male *C. mydas* to find females experimentally tethered at night in murky waters (Crowell Comuzzie and Owens, 1990).

Sternotherus odoratus has glands that induce strikingly different behavioral results, depending on the social context. The courtship behaviors of *Kinosternon subrubrum hippocrepis, K. f. flavescens, S. odoratus, and S. carinatus* are remarkably similar (Mahmoud, 1967). During an investigatory tactile phase the male approaches the female from behind with his neck extending forward (Fig. 4.22). He then touches the cloaca of the female with his nose and sniffs it. If the approached turtle is a male, no courtship occurs. If it is female, the male moves to her side and nudges the area around her bridge with his nose—a behavior perhaps directed at the musk glands located in this area. Whenever males are housed together they occasionally mount each other, but these actions are brief. The sniffing behavior never occurs in such cases. These results strongly suggest that both cloacal and axillary chemical cues may be important in the courtship behavior of *S. odoratus*.

Sternotherus odoratus are called stinkpots because they eject malodorous secretions from their anterior musk glands when disturbed. The secretion is highly pungent and aversive to humans and presumably other mammals. It consists primarily of four ω-phenylalkanoic acids (Eisner et al., 1977). These are phenylacetic, 3-phenylpropionic, 5-phenylpentanoic, and 7-phenylheptanoic acid, the last two being newly described natural products. As the turtles appear to eject the fluid only in response to provocation, a defensive role has always been assumed for these chemicals. Mealworms coated with the ω-phenylalkanoic acids are rejected by swordtails; however, the dose required for rejection is at the limit of the capacity of the gland. Presumably, any piscine predator of *Sternotherus* would be considerably larger than swordtails. In addition, turtles have many other predators, including mammals, that might provide a better test. Perhaps the chemicals act as a kind of aposematic signal that warns predators of the distastefulness of the turtle. The musk of the Australian *Chelodina longicollis* also has been tested for its capacity to deter a wide variety of potential predators, including mammals, birds, and reptiles (Kool, 1981). However, musk applied to food, into the nares, or just present in the vicinity of the predators does not elicit behaviors differing significantly from controls, so that the inguinal musk of the long-

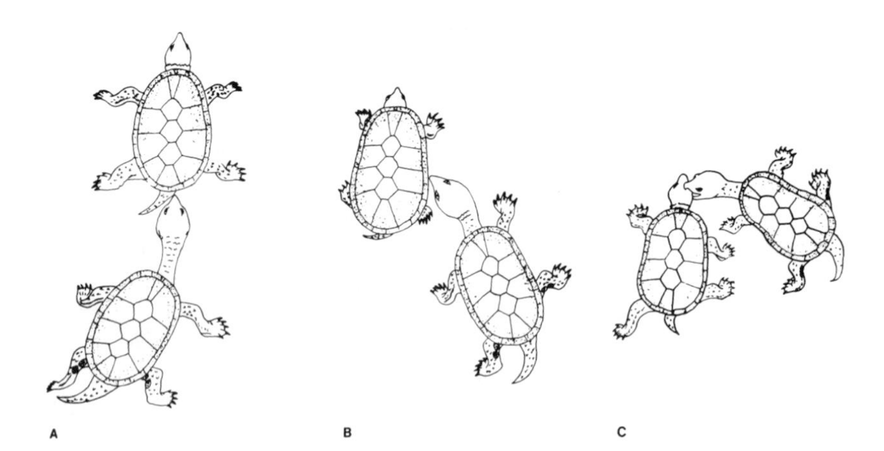

Fig. 4.22 Initial tactile courtship behavior in kinosternid turtles. (A) The approach; (B) nudging at the bridge near the musk gland; (C) biting at the head. (From Mahmoud, 1967.)

necked turtle does not act as a predator deterrent, although it may be a sex attractant (Kool, 1981). Secretions of two adult females test positively for proteins; however, there is no evidence of the presence of carbohydrates or glycerides (Eisner et al., 1978). GC/MS analysis reveals no highly volatile, low-molecular-weight compounds. However, when treated with diazomethane, fatty acids are detected as their corresponding methyl esters, specifically oleic, linoleic, palmitoleic, palmitic, stearic, and citronellic acid. These fatty acids could impart the "stink" for which the tortoise is well known. Ironically, the secretion also contains β-ionone, which is known as lavender oil, the odor of violets.

Secretions from the axillary glands of *Sternotherus* may serve a dual function. They are involved in antipredator mechanisms, whereas during the breeding season, the glandular contents may serve as sex recognition cues by which courting males discriminate females (Eisner et al., 1977). Apparently the musk of a female stinkpot turtle, although repulsive to mammals, is perceived quite differently by courting male *Sternotherus*.

VI. CROCODILIA

A. Glandular Morphology

Crocodilians of both sexes have three types of integumentary glands. The first set is a series of minute glands of unknown function found in two longitudinal rows under the dorsal scales near the midline (Voeltzkow, 1899; Reese, 1921; Wright and Moffat, 1985). Reese (1921) states that the secretions lack odor and suggests that their small size and wide distribution over the dorsal surface might indicate that they are of use in keeping the scales in good condition.

The other two are musk glands that have been investigated sporadically over the last century (Gerhardt, 1933, and references therein). These glands lie in the cloaca and on the underside of the throat (Voeltzkow, 1899; Gadow, 1901; Reese, 1920; 1921; Wright and Moffat, 1985; Weldon and Sampson, 1987; 1988). The cloacal glands are paired, lie within the lips of the cloaca (Gadow, 1901), and develop from the lower layer of the epidermis (Reese, 1921). Their morphology and development were extensively studied in juvenile *Alligator mississippiensis* (Reese, 1921). Other investigators have provided extensive histological studies of posthatchling crocodilians (Disselhörst, 1904; Petit and Geay; 1904; Voeltzkow, 1899). Recent studies using modern histological and histochemical techniques addressed the cloacal glands of adult *A. mississippiensis* (Weldon and Sampson, 1987) and the chin and cloacal glands of juvenile *Crocodylus porosus* (Wright and Moffat, 1985).

In adult *Alligator mississippiensis,* the cloacal or paracloacal glands are large, oval, and encapsulated with a single duct (Weldon and Sampson, 1987) (Fig. 4.23). The capsule is surrounded by a layer of smooth muscle. The glands are holocrine, and the secretion consists of lipids but no mucopolysaccharides. The smooth muscle surrounding the capsule has been observed to evert the gland.

The chin glands of *Crocodylus porosus* open as two narrow fissures or slits toward the posterior end of the underside of the lower jaw (Wright and Moffat, 1985) (Fig. 4.24). They are unique to the order Crocodilia (Weldon and Sampson, 1988). The small, ovoid glands are eversible; the inner portion of the slit and part of the gland then protrudes as a conspicuous black rosette. Surrounding each gland is a region of dense collagenous connective tissue covered by a cup-shaped layer of striated muscle. At the ultrastructural level, chin and cloacal glands are indistinguishable (Wright and Moffat, 1985). Lipids occur in the holocrine secretions of the cloacal glands but not in those of the chin glands (Weldon and Sampson, 1988). The oval chin glands of male *Alligator mississippiensis* are surrounded by striated muscle and lack a distinct lumen or secretory duct (Weldon and Sampson, 1988).

B. Pheromonal Functions

Secretions from the chin and cloacal glands may serve as sex attractant pheromones, territorial markers, interspecific allomones, or defensive secretions. Wright and Moffat (1985) suggest that in these animals the secretions may play a dual role, as do the secretions of *Sternotherus odoratus* (see Section V.E). The secretory products may serve as defensive secretions in juveniles and may be involved in reproductive and territorial behavior of adults. Although neither of these hypotheses has been tested, there are numerous anecdotal reports of behaviors elicited in response to these secretions. For example, addition of the secretion of the chin gland of a juvenile *Caiman crocodylus* into water in which other juveniles have aggregated is reported to cause them to thrash about (Gorzula, 1985). However, repetition of the experiment produced no discernible reaction. It would be interesting to see whether the composition of these secretions changes with age or differs between the sexes.

The secretion of the chin glands has been described as a smeary, pale brownish substance, a concentrated essence of musk (Gadow, 1901). The secretion is most active during the rutting time, at which time the glands are partly everted. Young crocodiles and alligators often evert these glands when handled. The cloacal glands are referred to as strongly scented organs that occur in both sexes and have

Fig. 4.23 *Alligator mississippiensis*. Cloaca of alligator showing region from which paracloacal glands were excised. (Photo by P. J. Weldon.)

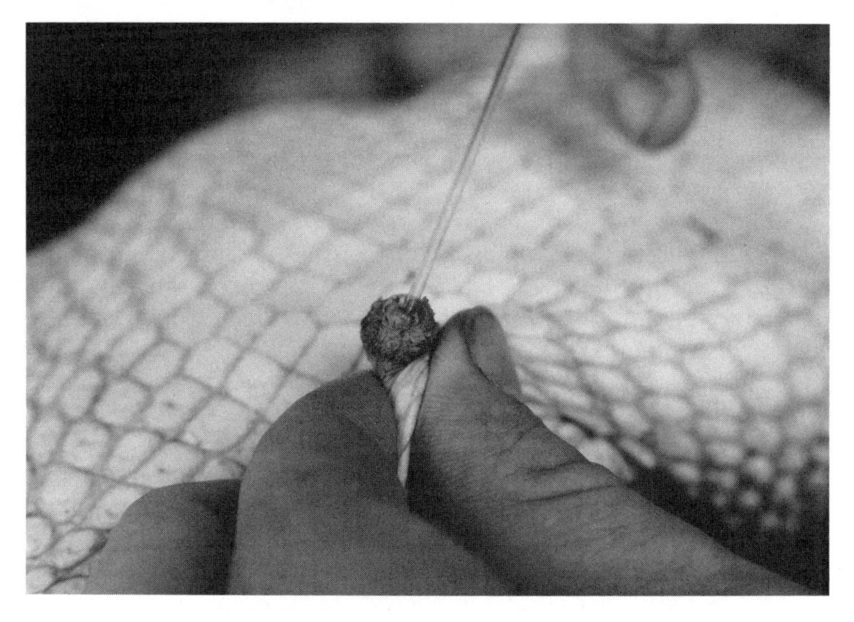

Fig. 4.24 *Alligator mississippiensis*. A gular gland is everted and a capillary tube containing secretions is withdrawn. (Photo by P. J. Weldon.)

an "obviously" hedonic function (Gadow, 1901). It is suggested that the sexes are probably able to follow and find each other thanks to the plume of scented water each individual leaves behind. The glands are supposedly most active during the breeding season, thus implying a reproductive function (Reese, 1920); this assumption is supported further by observations of marking actions (Ditmars, 1910). Male American alligators are stated to attract their mates by roaring and by releasing musky secretions from the glands in the throat and cloaca (Bellairs, 1957). All such reports assume that the secretions are carried on the surface of the water. McIlhenny (1935), who spent his entire life observing the American alligator in his native Louisiana, reports that when the bull alligator roars, it raises its head above the water and throws off a considerable quantity of sweet, pungent-smelling musk. This musk scents not only the air, but the surface of the water as well and lingers in the vicinity for some hours. He suggests that females do not use the chin glands but do produce considerable quantities of musk from their cloacal glands. Male alligators frequently evert the musk glands during the mating season, thus releasing a pungent perfume that is carried by the wind for considerable distances (Evans, 1961). The scent is claimed to attract the female to the male's pool.

Other reports are less clear. Neill (1971) states that it is not known how the sexes of *Alligator mississippiensis* locate each other and suggests that cloacal glands lay a detectable trail. Although it is unknown whether the throat glands play any role in courtship, he claims that the scent of the musk is not intensified during courtship, copulation, bellowing, or fighting. He observes that no detectable scent is given off during bellowing, but LeBuff (1957) suggests just the opposite. Silverstone (1972) never observed fluid being ejected from the throat glands, as reported by McIlhenny (1935). Musk is claimed to be in the air during most bellowing choruses, although eversion of the mandibular glands is rare (Vliet, 1986). However, an oily sheen on the surface of the water often appears near the cloaca (Vliet, 1989). Perhaps alligator musk is mainly transported on the surface of the water and not through the air.

Another report indicates that during the breeding season, the pair of *Alligator mississippiensis* swims about and the male rubs his throat (and presumably the chin glands) across the snout of the female (Evans, 1961; Burrage, 1965). The male has been seen to evert the chin glands and subsequently to rub them over the nares of the female, perhaps introducing a courtship pheromone (Garrick, 1978). Interestingly, most reptiles rely on the vomeronasal organ for the detection of semiochemicals, whereas crocodilians lack this organ.

C. Chemical Studies

Investigation of the chemical nature of crocodilian scent gland secretions began over 50 years ago. A dimethylheptenol was isolated and originally named *yacarol*—the Spanish term for alligator (Fester and Bertuzzi, 1934). Later it was reported that *yacarol* is actually citronellol, whereas cholesterol was verified in the secreted musk (Fester et al., 1937).

Gular gland secretions from seven adult male and one female *Alligator mississippiensis* have been collected outside the breeding season (during September) and extracted in chloroform (Weldon and Sampson, 1988). Both glands of females then contain significantly more secretion than those of the males (Weldon et al., 1990).

Fractionation of lipids on a silica column, to separate polar from nonpolar lipids, discloses far more nonpolar than polar material in both the male and female secretions. Thin-layer chromatography (TLC) of the polar secretions from the gular and paracloacal glands shows bands that comigrate with sterols, free fatty acids, triglycerides, and steryl esters (Weldon and Sampson, 1988; Weldon et al., 1990). ^{13}C-NMR data exhibit absorptions that suggest free and esterified fatty acids. Other peaks suggest the presence of other esters and the glycerol moiety of triglycerides. Cholesterol, free fatty acids, acetate esters (C_{12}–C_{18}), higher-molecular-weight esters and butyl esters, as well as α-tocopherol, have been identified by GC-MS in the paracloacal glands of *A. mississippiensis* (Weldon et al., 1988). A similar set of butanoates, long-chain esters, fatty acids, saturated and monounsaturated alcohols (C_{11}–C_{18}), and citronellol have also been identified by GC-MS in the paracloacal glands of *Paleosuchus palpebrosus* and *P. trigonatus* (Shafagati et al., 1989). The gular gland secretions of the American alligator contain cholesterol, fatty acids, squalene and possibly other hydrocarbons, and α-tocopherol (Weldon et al., 1987). Saturated and unsaturated fatty acids, as well as 10-octadecanolide, occur in glandular secretions from *Crocodylus acutus* and *C. rhombifer* (Polo et al., 1988a, b).

D. Behavioral Studies

Two studies have directly measured the behavioral response to glandular secretions in a crocodilian. The olfactory bulbs of *Caiman crocodylus* respond to chemical stimulation (Huggins et al., 1968). Specifically, the muscles adjacent to the nares show electromyographic responses, and the olfactory bulbs show EEG responses. Only a few sniffs occur in response to artificial stimuli such as oil of cloves or ethyl alcohol, but sniffing behavior lasts for long periods whenever caimans are exposed to musk from the glands of another caiman. In the second study, yearling *A. mississippiensis* respond to airborne odorants in an olfactometer (Johnsen and Wellington, 1982). Gular pumping has been used as a measure of sniffing activity in response to seven stimuli: cloacal gland secretions (from two alligators), mandibular gland (chin gland) secretions (from one alligator), citronellol, androstenol, amyl acetate, and a blank. Alligators respond most strongly to the cloacal gland secretions of an adult male. Responses to the cloacal gland secretions and the chin gland secretions differ significantly from those to the synthetic compounds.

VIII. CONCLUSIONS

This review presents a wide variety of data that share a common feature, that of demonstrating in some way the production or perception of pheromones by reptiles. However, this is the extent of the commonality. Even a brief perusal causes one to note the discrepancies in the relative quantity of information on pheromones for each of the reptilian orders. There has been considerable work on snakes and lizards, some work on turtles and crocodilians, but rhynchocephalians have received little attention and amphisbaenians have not been studied at all.

Even among the snakes and lizards, only few groups have been investigated. For example, we know a great deal about trailing pheromones and courtship pheromones in snakes, but most of the information derives from a few species of *Thamnophis.* What of the many other genera of snakes? It seems likely that many, if not all, of the major groups of snakes respond to integumentary lipid pheromones, for example, in courtship. Demonstration of courtship pheromones in widely divergent groups of snakes would indicate that these pheromones may be an ancient character rather than recently derived. These pheromones are species-specific in most insect species, yet despite some interesting studies on trailing pheromones, this possibility has not been adequately tested in snakes.

For lizards, much information exists on pheromonal communication in one small group of closely related skinks. The unequivocal demonstrations of their behavioral responses to pheromones indicate the feasibility of similar studies in other lizards, especially autarchoglossans. All of the skink studies were conducted in aquaria. The behaviors analyzed, such as tongue-flicking, are robust, easy to quantify, and sometimes repeatable through many trials. This is the case for few other vertebrate models. Brooding behavior is another topic that could yield some interesting data. Several studies suggest that skin lipid as well as cloacal cues may be important in species and sex recognition. As new types of glands elaborating behaviorally active compounds are still being described, exciting avenues remain open for future studies. Finally, the interesting data on the lipid and proteinaceous components of femoral gland secretions hold great promise for clarifying species and individual recognition and territorial marking; they even provide tools for constructing phylogenies.

In terms of phylogenetics, *Sphenodon* is perhaps the most important, yet least studied, taxon in regard to potential pheromones. If *Sphenodon* were to have pheromones, it would certainly argue for an ancient rather than derived history for these compounds and the behaviors elicited by them. These issues remain completely unexplored.

The cloacal and gular glands of crocodilians have been a source of controversy over the years. They produce a lipid and proteinaceous secretion; however, the purpose of these secretions and the behaviors elicited in response to them remain unknown. As crocodilian behavior has been well described, these animals would be good candidates for studies in chemical ecology. Several investigators have identified interesting compounds in crocodilian glands; these compounds have yet to be tested in bioassays.

Turtles also produce and respond to chemical cues. For instance, the mental glands of tortoises elicit social responses. In addition, a source of pheromones in the cloaca of several species of aquatic turtle seem to identify breeding females to males. No one has investigated the behaviors or the source of these putative sex pheromones in the cloaca of any turtle.

The chemical ecology of reptiles, however, shows a clear dichotomy. The many fine studies of reptilian ecology are not paralleled by many chemical studies of the compounds involved. Of course, this trend is prevalent for most, if not all, vertebrates. In striking contrast, studies of pheromonal communication in insects have been successful in isolating and identifying literally thousands of pheromones, as well as in documenting the behaviors they elicit. Indeed, studies involving insect models have proven successful in identifying mechanisms of chemoreception at the molecular level. In contrast, investigators of vertebrate pheromones have had limited success in their investigations of pheromone chemistry. This reflects the association of many mammalian pheromones with urine or feces, thus complicating the isolation procedure. In addition, the behaviors elicited by these pheromones are complex, difficult to reproduce, and frequently nonspecific (such as sniffing).

Studies of reptilian pheromones clearly hold great promise. A number of discrete glands produce semiochemicals that elicit behaviors quantified in laboratory and field studies. Although interpretation of these behaviors is by no means easy, it appears that reptilian behaviors are more stereotyped than their mammalian counterparts. The literature describing the sources of semiochemicals in reptiles also allows a few generalizations. The chemical compounds serving as pheromones seem to be limited to lipids (at least as of this writing)—specifically lipids of high molecular weight and relative nonvolatility. This suggests a dependence on vomeronasal rather than olfactory detection. A concerted effort by zoologists pursuing reptilian natural history and behavior and by natural products chemists holds great promise for understanding the complex aspects of those reptilian behaviors governed by chemical signals.

ACKNOWLEDGMENTS

I would like to thank David Crews for initiating and nurturing my interest in reptilian pheromones. I thank David and Carl Gans for giving me the opportunity to review this subject. Carl Gans and Katherine Vernon have my sincerest gratitude for their great efforts on my behalf in the course of editing this work. Carl's comments and criticisms helped improve this work immensely. I wish to thank several people who have helped me over the years by collaborating on studies, discussing ideas, or reading manuscripts: Allison Alberts, Murray Blum, Jim Bull, Bill Cooper, Dave Duvall, Tom Eisner, Hank Fales, Neil Ford, Bill Gutzke, Mimi Halpern, Tappey Jones, Randy Krohmer, Paul Maderson, Jerry Meinwald, Mary Mendonca, Michael Moore, Yoshiko Murata, Lew Pannell, David Pfennig, Rick Shine, Alan Tousignant, Paul Weldon, and Joan Whittier. Jim and Anna Mason helped proofread the manuscript. Jed Rifkin assisted with computer software. Mitchell Smith, Susan Drew, and Jason Brooks organized the References.

REFERENCES

Abel, E. (1951). Über das Geruchsvermögen der Eidechsen. *Österr. Zool. Zeitschr.* 3, 84–125.

Ahern, D. G., and Downing, D. T. (1974). Skin lipids of the Florida indigo snake. *Lipids* 9, 8–14.

Alberts, A. C. (1989). Ultraviolet visual sensitivity in desert iguanas: implications for pheromone detection. *Anim. Behav.* 38, 129–137.

Alberts, A. C. (1990). Chemical properties and biological functions of femoral gland secretions in the desert iguana, *Dipsosaurus dorsalis*. *J. Chem. Ecol.* 16, 13–25.

Alberts, A. C. (1991). Phylogenetic and adaptive variation in lizard femoral gland secretions. *Copeia* 1991: 69–79.

Aleksiuk, M. (1976). Reptilian hibernation: evidence of adaptive strategies in *Thamnophis sirtalis parietalis*. *Copeia* 1976, 170–178.

Aleksiuk, M. (1977). Cold-induced aggregative behavior in the red-sided garter snake (*Thamnophis sirtalis parietalis*). *Herpetologica* 33, 98–101.
Aleksiuk, M., and Gregory, P. (1974). Regulation of seasonal mating behavior in *Thamnophis sirtalis parietalis*. *Copeia* 1974, 681–688.
Allen, B. A., Burghardt, G. M., and York, D. S. (1984). Species and sex differences in substrate preference and tongue-flick rate in three sympatric species of water snakes, *Nerodia*. *J. Comp. Psychol.* 98, 358–367.
Altland, P. D. (1941). Annual reproductive cycle of the male fence lizard. *J. Elisha Mitchell Sci. Soc.* 57, 73–83.
Anderson, P. (1947). Observation on the denning habits of the prairie rattlesnake, *Crotalus viridis viridis* (Rafinesque). *Chicago Acad. Sci. Nat. Hist. Misc.* 9, 1–2.
Andrén, C. (1982). The role of the vomeronasal organs in the reproductive behavior of the adder, *Vipera berus*. *Copeia* 1982, 148–157.
Andrén, C. (1986). Courtship, mating and agonistic behaviour in a free-living population of adders, *Vipera berus* (L.). *Amphibia-Reptilia* 7, 353–383.
Andrén, C., and Nilson, G. (1987). The copulatory plug of the adder, *Vipera berus:* Does it keep sperm in or out? *Oikos* 49, 230–232.
Armstrong, B. L., and Murphy, J. B. (1979). The natural history of Mexican rattlesnakes. *Spec. Pub., Univ. Kansas Mus. Nat. Hist.*, No. 5, 1–88.
Asana, J. J. (1941). Further observations on the egg-laying habits of the lizard, *Calotes versicolor* (Boulenger). *J. Bombay Nat. Hist. Soc.* 42, 937–940.
Auffenberg, W. (1964). Notes on the courtship of the land tortoise *Geochelone travancorica* (Boulenger). *J. Bombay Nat. Hist. Soc.* 61, 247–253.
Auffenberg, W. (1965). Sex and species discrimination in two sympatric South American tortoises. *Copeia* 1965, 335–342.
Auffenberg, W. (1966). On the courtship of *Gopherus polyphemus*. *Herpetologica* 22, 113–117.
Auffenberg, W. (1969a). Social behavior of *Geochelone denticulata*. *J. Fla. Acad. Sci.* 32, 50–58.
Auffenberg, W. (1969b). *Tortoise Behavior and Survival*. Rand-McNally, Skokie, Ill.
Auffenberg, W. (1978). Social and feeding behavior in *Varanus komodoensis*. In *Behavior and Neurobiology of Lizards* (N. Greenberg and P. D. MacLean, eds.). U.S. Dept. of Health, Education, and Welfare, Rockville, Md., pp. 301–331.
Baden, H. P., and Maderson, P. F. A. (1970). Morphological and biophysical identification of fibrous proteins in the amniote epidermis. *J. Exp. Zool.* 174, 225–232.
Baeyens, D. A., and Rountree, R. L. (1983). A comparative study of evaporative water loss and epidermal permeability in an arboreal snake, *Opheodrys aestivus*, and a semi-aquatic snake, *Nerodia rhombifera*. *Comp. Biochem. Physiol.* 76A, 301–304.
Barbour, R. W. (1962). An aggregation of copperheads *Agkistrodon contortrix*. *Copeia* 1962, 640.
Barbour, T. (1926). *Reptiles and Amphibians: Their Habits and Adaptations*. Houghton Mifflin, Cambridge, Mass.
Baumann, F. (1927). Experimente über den Geruchsinn der Viper. *Rev. Suisse Zool.* 34, 173–184.

Baumann, F. (1929). Experimente über den Geruchsinn und den Beuteerwerb der Viper (*Vipera aspis*). *Z. vergl. Physiol.* 10, 36–119.

Bauwens, D., Nuitjen, K., van Wezel, H., and Verheyen, R. F. (1987). Sex recognition by males of the lizard, *Lacerta vivipara:* an introductory study. *Amphibia-Reptilia* 8, 49–57.

Bauwens, D., Van Damme, R., and Verheyen, R. F. (1989). Synchronization of spring molting with the onset of mating behavior in male lizards, *Lacerta vivipara. J. Herpetol.* 23, 89–91.

Beck, D. D. (1990). Ecology and behavior of the Gila monster in southwestern Utah. *J. Herpetol.* 24, 54–68.

Bellairs, d' A. (1957). *Reptiles: Life History, Evolution, and Structure.* Hutchinson and Co., London.

Bellairs, d' A., and Underwood, G. (1951). The origin of snakes. *Biol. Rev.* 26, 193–337.

Bennion, R. S., and Parker, W. S. (1976). Field observations on courtship and aggressive behavior in desert striped whipsnakes, *Masticophis t. taeniatus. Herpetologica* 32, 30–35.

Berry, K. H. (1974). The ecology and social behavior of the chuckwalla, *Sauromalus o. obesus* Baird. *Univ. Calif. Publ. Zool.* 101, 1–60.

Beuchelt, H. (1936). Bau, Funktion und Entwicklung der Begattungsorgane der männlichen Ringelnatter (*Natrix natrix* L.) und Kreuzotter (*Vipera berus* L.). *Morphol. Jahrb.* 78, 445–516.

Birkby, C. S., Wertz, P. W., and Downing, D. T. (1982). The polar lipids from keratinized tissues of some vertebrates. *Comp. Biochem. Physiol.* 73B, 239–242.

Bishop, S. C. (1927). The amphibians and reptiles of Allegany State Park. *Bull. N.Y. State Mus.*, 3, 1–141.

Bissinger, B. E., and Simon, C. A. (1979). Comparison of tongue extrusions in representatives of six families of lizards. *J. Herpetol.* 13, 133–139.

Bissinger, B. E., and Simon, C. A. (1981). The chemical detection of conspecifics by juvenile Yarrow's spiny lizard *Sceloporus jarrovi. J. Herpetol.* 15, 77–81.

Blair, W. F. (1968). Amphibians and Reptiles. In *Animal Communication: Techniques of Study and Results of Research* (T. A. Sebeok, ed.). Indiana University Press, Bloomington, pp. 289–310.

Blanchard, F. N. (1937). The eggs and natural nests of the Eastern ringnecked snake, *Diadophis punctatus edwardsi. Pap. Mich. Acad. Sci.* 22, 521–532.

Blum, M. S. (1981). *Chemical Defenses of Arthropods.* Academic Press, New York.

Blum, M. S., Byrd, J. B., Travis, J. R., Watkins, J. F., and Gehlbach, F. R. (1971). Chemistry of the cloacal sac secretion of the blind snake *Leptotyphlops dulcis. Comp. Biochem. Physiol.* 38B, 103–107.

Bogert, C. M. (1941). Sensory cues used by rattlesnakes in their recognition of ophidian enemies. *Ann. N.Y. Acad. Sci.* 41, 329–344.

Bogert, C. M., and del Campo, R. M. (1956). The gila monster and its allies. *Bull. Am. Mus. Nat. Hist.* 109, 1–238.

Brattstrom, B. H. (1974). The evolution of reptilian social behavior. *Am. Zool.* 14, 35–49.

Breckenridge, W. J. (1943). The life history of the black-banded skink, *Eumeces septentrionalis* (Baird). *Am. Midl. Nat.* 29, 591–606.

Brisbin, I. L. (1968). Evidence for the use of post anal musk as an alarm device in the king snake, *Lampropeltis getulus*. *Herpetologica* 24, 169–170.

Broman, I. (1920). Das Organon Vomero-nasale Jacobsoni ein Wassergeruchsorgan! *Arb. Anat. Inst., Wiesbaden* (Anat. Hefte, Abt. 1) 58, 134–191.

Brown, W. S., and MacLean, F. M. (1983). Conspecific scent-trailing by newborn timber rattlesnakes, *Crotalus horridus*. *Herpetologica* 39, 430–436.

Burger, J. (1989). Following of conspecific and avoidance of predator chemical cues by pine snakes (*Pituophis melanoleucus*). *J. Chem. Ecol.* 15, 799–806.

Burghardt, G. M. (1970). Chemical perception in reptiles. In *Communication by Chemical Signals* (J. W. Johnston, D. G. Moulton, and M. Turk, eds.). Appleton-Century-Crofts, New York, pp. 241–308.

Burghardt, G. M. (1977). Of iguanas and dinosaurs: social behavior and communication in neonate reptiles. *Am. Zool.* 17, 177–190.

Burghardt, G. M. (1980). Behavioral and stimulus correlates of vomeronasal functioning in reptiles: feeding, grouping, sex, and tongue use. In *Chemical Signals. Vertebrates and Aquatic Invertebrates* (D. Müller-Schwarze and R. M. Silverstein, eds.). Plenum Press, New York, pp. 275–301.

Burghardt, G. M. (1983). Aggregation and species discrimination in newborn snakes. *Z. Tierpsychol.* 61, 89–101.

Burghardt, G. M., Green, H. W., and Rand, A. S. (1977). Social behavior in hatchling green iguanas: life at a reptile rookery. *Science* (N.Y.) 195, 689–691.

Burghardt, G. M., and Pruitt, C. H. (1975). The role of the tongue and senses in feeding of naive and experienced garter snakes. *Physiol. Behav.* 14, 185–194.

Burken, R. R., Wertz, P. W., and Downing, D. T. (1985a). The effect of lipids on transepidermal water permeation in snakes. *Comp. Biochem. Physiol.* 81A, 213–216.

Burken, R. R., Wertz, P. W., and Downing, D. T. (1985b). A survey of polar and nonpolar lipids extracted from snake skin. *Comp. Biochem. Physiol.* 81B, 315–318.

Burkholder, G. L., and Tanner, W. W. (1974). A new gland in *Sceloporus graciosus* males (Sauria: Iguanidae). *Herpetologica* 30, 368–371.

Burrage, B. R. (1965). Copulation in a pair of *Alligator mississippiensis*. *Br. J. Herpetol.* 3, 207–208.

Cagle, F. R. (1940). Eggs and natural nests of *Eumeces fasciatus*. *Am. Midl. Nat.* 23(1), 227–233.

Cagle, F. R. (1950). The life history of the slider turtle, *Pseudemys scripta troosti* (Holbrook). *Ecol. Monogr.* 20:31–54.

Camp, C. L. (1923). Classification of the lizards. *Bull. Am. Mus. Nat. Hist.* 48, 289–481.

Carpenter, C. C. (1952). Comparative ecology of the common garter snake, *Thamnophis sirtalis sirtalis*, the ribbon snake, *Thamnophis sauritus sauritus*, and Butler's garter snake, *Thamnophis butleri*, in mixed populations. *Ecol. Monogr.* 22, 235–258.

Carpenter, C. C. (1953). A study of hibernacula and hibernating associations of snakes and amphibians in Michigan. *Ecology* 34, 74–80.
Carpenter, C. C. (1962). Patterns of behavior in two Oklahoma lizards. *Am. Midl. Nat.* 67, 132–151.
Carpenter, C. C. (1975). Review. *Copeia* 1975, 388–389.
Carpenter, C. C. (1977). Communication and displays of snakes. *Am. Zool.* 17, 217–224.
Carpenter, C. C., and Ferguson, G. W. (1977). Variation and evolution of stereotyped behavior in reptiles. In *Biology of the Reptilia* (C. Gans and D. W. Tinkle, eds.). Academic Press, New York, pp. 335–554.
Carpenter, C. C., and Gillingham, J. C. (1975). Postural responses to kingsnakes by crotaline snakes. *Herpetologica* 31, 293–302.
Chiszar, D., Radcliffe, C. W., Scudder, K. M., and Duvall, D. (1983). Strike-induced chemosensory searching by rattlesnakes: the role of envenomation-related chemical cues in the post-strike environment. In *Chemical Signals in Vertebrates* (D. Müller-Schwarze and R. M. Silverstein, eds.). Plenum Press, New York, pp. 1–24.
Chiszar, D., Scudder, K. M., Knight, L., and Smith, H. M. (1978). Exploratory behavior in prairie rattlesnakes (*Crotalus viridis*) and water moccasins (*Agkistrodon piscivorus*). *Psychol. Rec.* 28, 363–368.
Chiszar, D., Scudder, K. M., Smith, H. M., and Radcliffe, C. W. (1976). Observations of courtship behavior in the Western massasauga (*Sistrurus catenatus tergeminus*). *Herpetologica* 32, 337–338.
Chiszar, D., Wellborn, S., Wand, M. A., Scudder, K. M., and Smith, H. M. (1980). Investigatory behavior in snakes. II. Cage cleaning and the induction of defecation in snakes. *Anim. Learn. Behav.* 8, 505–510.
Chiu, K. W., Lofts, B., and Tsui, H. W. (1970). The effect of testosterone on the sloughing cycle and epidermal glands of the female gecko, *Gekko gecko* L. *Gen. Comp. Endocrinol.* 15, 12–19.
Chiu, K. W., and Maderson, P. F. A. (1975). The microscopic anatomy of epidermal glands in two species of gekkonine lizards, with some observations on testicular activity. *J. Morphol.* 147, 23–40.
Cogger, H. G. (1978). Reproductive cycles, fat body cycles and socio-sexual behaviour in the mallee dragon, *Amphibolurus fordi* (Lacertilia: Agamidae). *Aust. J. Zool.* 26, 653–672.
Cole, C. J. (1966a). Femoral glands of the lizard, *Crotaphytus collaris*. *J. Morphol.* 118, 119–136.
Cole, C. J. (1966b). Femoral glands in lizards: a review. *Herpetologica* 22, 199–206.
Cook, F. R. (1964). Communal egg-laying in the smooth green snake. *Herpetologica* 24, 206.
Cooper, W. E., and Burghardt, G. M. (1990). Vomerolfaction and vomodor. *J. Chem. Ecol.* 16, 103–105.
Cooper, W. E., Caffrey, C., and Vitt, L. J. (1985). Aggregation in the banded gecko, *Coleonyx variegatus*. *Herpetologica* 41, 342–350.
Cooper, W. E., and Garstka, W. R. (1987a). Aggregation in the broad-headed skink (*Eumeces laticeps*). *Copeia* 1987, 807–810.
Cooper, W. E., and Garstka, W. R. (1987b). Lingual responses to chemical

fractions of urodaeal glandular pheromone of the skink, *Eumeces laticeps. J. Exp. Zool.* 242, 249–253.

Cooper, W. E., Garstka, W. R., and Vitt, L. J. (1986). Female sex pheromone in the lizard *Eumeces laticeps. Herpetologica* 42, 361–366.

Cooper, W. E., and Vitt, L. J. (1984a). Detection of conspecific odors by the female broad-headed skink, *Eumeces laticeps. J. Exp. Zool.* 229, 49–54.

Cooper, W. E., and Vitt, L. J. (1984b). Conspecific odor detection by the male broad-headed skink, *Eumeces laticeps:* effects of sex and site of odor source and of male reproductive condition. *J. Exp. Zool.* 230, 199–209.

Cooper, W. E., and Vitt, L. J. (1985). Responses of the skinks *Eumeces fasciatus* and *Eumeces laticeps* to airborne conspecific odors: further appraisal. *J. Herpetol.* 19, 481–486.

Cooper, W. E., and Vitt, L. J. (1986a). Interspecific odour discrimination by a lizard *Eumeces laticeps. Anim. Behav.* 34, 367–376.

Cooper, W. E., and Vitt, L. J. (1986b). Interspecific odour discriminations among syntopic congeners in scincid lizards (genus *Eumeces*). *Behaviour* 97, 1–9.

Cooper, W. E., and Vitt, L. J. (1986c). Tracking of female conspecific odor trails by male broad-headed skinks (*Eumeces laticeps*). *Ethology* 71, 242–248.

Cooper, W. E., and Vitt, L. J. (1987a). Intraspecific and interspecific aggression in lizards of the scincid genus *Eumeces:* chemical detection of conspecific sexual competitors. *Herpetologica* 43, 7–14.

Cooper, W. E., and Vitt, L. J. (1987b). Discrimination of male conspecific from male heterospecific odors by male scincid lizards (*Eumeces laticeps*). *J. Exp. Zool.* 241, 253–256.

Cooper, W. E., and Vitt, L. J. (1987c). Ethological isolation, sexual behavior and pheromones in the *fasciatus* species group of the lizard genus *Eumeces. Ethology* 75, 328–336.

Cooper, W. E., and Vitt, L. J. (1988). Orange head coloration of the male broad-headed skink (*Eumeces laticeps*), a sexually selected social cue. *Copeia* 1988, 1–6.

Cooper, W. E., Vitt, L. J., Vangilder, L. D., and Gibbons, J. W. (1983). Natural nesting sites and brooding behavior of *Eumeces fasciatus. Herp. Rev.* 14(3), 65–66.

Costanzo, J. (1986). Influence of hibernaculum microenvironment on the winter life history of the garter snake (*Thamnophis sirtalis*). *Ohio J. Sci.* 86, 199–204.

Costanzo, J. (1988). "Eco-physiological Adaptations to Overwintering in the Eastern Garter Snake, *Thamnophis sirtalis sirtalis.*" Ph.D. dissertation, Miami University, Oxford, Ohio.

Costanzo, J. (1989). Conspecific scent trailing by garter snakes (*Thamnophis sirtalis*) during autumn: further evidence for use of pheromones in den location. *J. Chem. Ecol.* 15, 2531–2538.

Cowles, R. B. (1938). Unusual defense posture assumed by rattlesnakes. *Copeia* 1938, 13–16.

Cowles, R. B., and Phelan, R. L. (1958). Olfaction in rattlesnakes. *Copeia* 1958, 77–83.

Crews, D. (1976). Hormonal control of male courtship behavior and female attractivity in the garter snake (*Thamnophis sirtalis parietalis*). *Horm. Behav.* 7, 451–460.

Crowell Comuzzie, D. K., and Owens, D. W. (1990). A quantitative analysis of courtship behavior in captive green sea turtles (*Chelonia mydas*). *Herpetologica* 46, 195–202.

Curran, C. H., and Kauffeld, C. (1937). *Snakes and Their Ways*. Harper, New York.

Darwin, C. (1874). *The Descent of Man and Selection in Relation to Sex*. J. Murray, London.

Davis, D. D. (1936). Courtship and mating behavior in snakes. *Field Mus. Nat. Hist. Publ. Zool.*, Ser. 20, 257–290.

DeFazio, A., Simon, C. A., Middendorf, G. A., and Romano, D. (1977). Iguanid substrate licking: a response to novel situations in *Sceloporus jarrovi*. *Copeia* 1977, 706–709.

Devine, M. C. (1975). Copulatory plugs in snakes: enforced chastity. *Science* (N.Y.) 187, 844–845.

Devine, M. C. (1977a). Copulatory plugs, restricted mating opportunities and reproductive competition among male garter snakes. *Nature* (London) 267, 345–346.

Devine, M. C. (1977b). "Chemistry and Source of Sex-attractant Pheromones and their Role in Mate Discrimination by Garter Snakes." Ph.D. dissertation, The University of Michigan, Ann Arbor.

Dial, B. E., Weldon, P. J., and Curtis, B. (1989). Chemosensory identification of snake predators (*Phyllorhynchus decurtatus*) by banded geckos (*Coleonyx variegatus*). *J. Herpetol.* 23, 224–229.

Disselhörst, R. (1904). Aufführapparat und Anhangsdrüsen der männlichen Geschlechtorgane. In *Lehrbuch der vergleichenden mikroskopischen Anatomie der Wirbeltiere*. (A. Opel, ed.). Gustav Fischer, Jena, Vol. 4, pp. 60–89.

Ditmars, R. L. (1910). *Reptiles of the World*. Macmillan, New York.

Dundee, H. A., and Miller, M. C., III. (1968). Aggregative behavior and habitat conditioning by the prairie ringneck snake, *Diadophis punctatus arnyi*. *Tulane Stud. Zool. Bot.* 15, 41–58.

Durham, M. S. (1980). "Analyses of Experimentally Induced Defensive Behavior and Prey-attack Behavior in the Hognose Snake (*Heterodon platyrhinos*)." Ph.D. dissertation, University of Arkansas, Fayetteville.

Duvall, D. (1979). Western fence lizard (*Sceloporus occidentalis*): chemical signals, conspecific discriminations and release of a species-typical visual display. *J. Exp. Zool.* 210, 321–326.

Duvall, D. (1981). Western fence lizard, *Sceloporus occidentalis biseriatus*, chemical signals. 2. A replication with naturally breeding adults and a test of the Cowles and Phelan hypothesis of rattlesnake olfaction. *J. Exp. Zool.* 218, 351–362.

Duvall, D. (1982). Western fence lizard (*Sceloporus occidentalis*) chemical signals. III. An experimental ethogram of conspecific body licking. *J. Exp. Zool.* 221, 23–26.

Duvall, D. (1986). A new question of pheromones: aspects of possible chemical signalling and reception in the mammal-like reptiles. In *The Ecology and Biology of Mammal-like Reptiles* (N. Hotton, III, P. D. MacLean, J. J. Roth, and E. C. Roth, eds.). Smithsonian Institution Press, Washington, D.C., pp. 219–238.

Duvall, D., Arnold, S. J., and Schuett, G. W. (1992). Pitviper mating systems: ecological potential, sexual selection, and microevolution. In *The Biology of the Pitvipers* (J. A. Campbell and E. D. Brodie, Jr., eds.). Selva Press, Tyler, TX In press.

Duvall, D., Graves, B. M., and Carpenter, G. C. (1987). Visual and chemical composite signalling effects of *Sceloporus* lizard fecal boli. *Copeia* 1987, 1028–1031.

Duvall, D., Herskowitz, R., and Trupiano-Duvall, J. (1980). Responses of five-lined skinks (*Eumeces fasciatus*) and ground skinks (*Scincella lateralis*) to conspecific and interspecific chemical cues. *J. Herpetol.* 14, 121–127.

Duvall, D., King, M. B., and Gutzwiller, K. J. (1985). Behavioral ecology and ethology of the prairie rattlesnake. *Nat. Geogr. Res.* 1985, 80–111.

Duvall, D., Müller-Schwarze, D., and Silverstein, R. M. (1986b). Preface. In *Chemical Signals in Vertebrates,* Vol. 4. *Ecology, Evolution, and Comparative Biology* (D. Duvall, D. Müller-Schwarze, and R. M. Silverstein, eds.). Plenum Press, New York, pp. v–vi.

Ehrenfeld, J. G., and Ehrenfeld, D. W. (1973). Externally secreting glands of freshwater and sea turtles. *Copeia* 1973, 305–314.

Eisner, T., Connor, W. E., Hicks, K., Dodge, K. R., Rosenberg, H., Jones, T. H., Cohen, M., and Meinwald, J. (1977). Stink of stinkpot turtle identified: ω-phenylalkanoic acids. *Science* (N.Y.) 196, 1374–1379.

Eisner, T., Jones, T. H., Meinwald, J., and Legler, J. M. (1978). Chemical composition of the odorous secretion of the Australian turtle *Chelodina longicollis. Copeia* 1978, 714–715.

Elliot, R. H. (1934). The adder. *Blackwoods Mag.* 235, 502–516.

Ernst, C. H. (1971). Observations of the painted turtle, *Chrysemys picta. J. Herpetol.* 5, 216–220.

Ernst, C. H. (1974). Observations on the courtship of male *Graptemys pseudogeographica. J. Herpetol.* 8, 377–378.

Estes, R., and Pregill, G. (1988). Phylogenetic relationships of the lizard families: essays commemorating Charles L. Camp (R. Estes and G. Pregill, eds.). Stanford University Press, Stanford, Calif.

Evans, K. J. (1967). Observations on the daily emergence of *Coleonyx variegatus* and *Uta stansburiana. Herpetologica* 23, 217–222.

Evans, L. T. (1959). A motion picture study of maternal behavior of the lizard, *Eumeces obsoletus* (Baird and Girard). *Copeia* 1959, 103–110.

Evans, L. T. (1961). Structure as related to behavior in the organization of population in reptiles. In *Vertebrate Speciation* (W. F. Blair, ed.). University of Texas Press, Austin, pp. 148–178.

Everett, C. T. (1971). Courtship and mating of *Eumeces multivirgatus* (Scincidae). *J. Herpetol.* 5, 189–190.

Ferguson, G. W. (1966). Releasers of courtship and territorial behaviour in the side blotched lizard, *Uta stansburiana. Anim. Behav.* 14, 89–92.

Fergusson, B., Bradshaw, S. D., and Cannon, J. R. (1985). Hormonal control of femoral gland secretion in the lizard, *Amphibolurus ornatus. Gen. Comp. Endocrinol.* 57, 371–376.

Fester, G., and Bertuzzi, F. (1934). Drüsen-Sekret der Alligatoren (Yacarol). *Ber. Deutsch. Chem. Ges.* 67B, 365–370.

Fester, G., Bertuzzi, F., and Pucci, D. (1937). Drüsen-Sekret der Alligatoren (Yacarol) (II. Mitteil). *Ber. Deutsch. Chem. Ges.* 70B, 37–41.

Finneran, L. C. (1949). A sexual aggregation of the garter snake, *Thamnophis butleri*. *Copeia* 1949, 141–144.

Finneran, L. C. (1953). Aggregation behavior of the female copperhead, *Agkistrodon contortrix*, during gestation. *Copeia* 1953, 61–62.

Fitch, H. S. (1952). The University of Kansas Natural History Reservation. *Univ. Kansas Publ., Mus. Nat. Hist.* 4, 1–38.

Fitch, H. S. (1954). Life history and ecology of the five-lined skink, *Eumeces fasciatus*. *Univ. Kansas Publ., Mus. Nat. Hist.* 8, 1–156.

Fitch, H. S. (1955). Habits and adaptations of the great plains skink (*Eumeces obsoletus*). *Ecol. Monogr.* 25, 59–82.

Fitch, H. S. (1956). An ecological study of the collared lizard (*Crotaphytus collaris*). *Univ. Kansas Publ., Mus. Nat. Hist.* 8, 213–274.

Fitch, H. S. (1958a). Natural history of the six-lined racerunner (*Cnemidophorus sexlineatus*). *Univ. Kansas Publ., Mus. Nat. Hist.* 11, 11–62.

Fitch, H. S. (1958b). Home ranges, territories and seasonal movements of vertebrates of the Natural History Reservation. *Univ. Kansas Publ., Mus. Nat. Hist.* 11, 63–326.

Fitch, H. S. (1960). Autecology of the copperhead. *Univ. Kansas Publ., Mus. Nat. Hist.* 13, 85–288.

Fitch, H. S. (1963). Natural history of the racer *Coluber constrictor*. *Univ. Kansas Publ., Mus. Nat. Hist.* 15, 351–468.

Fitch, H. S. (1965). An ecological study of the garter snake, *Thamnophis sirtalis*. *Univ. Kansas Publ., Mus. Nat. Hist.* 15, 493–564.

Fitch, H. S. (1970). Reproductive cycles in lizards and snakes. *Univ. Kansas Mus. Nat. Hist. Misc. Publ.* 52, 1–247.

Fitzsimons, F. W. (1962). *Snakes.* Hutchinson and Co., London.

Forbes, T. R. (1941). Observations on the urogenital anatomy of the adult male lizard, *Sceloporus*, and on the action of implanted pellets of testosterone and of estrone. *J. Morphol.* 68, 31–69.

Ford, N. B. (1978). Evidence for species specificity of pheromone trails in two sympatric garter snakes, *Thamnophis*. *Herpetol. Review.* 9, 10–11.

Ford, N. B. (1981). Seasonality of pheromone trailing behavior in two species of garter snake, *Thamnophis* (Colubridae). *Southwestern Naturalist* 26, 385–388.

Ford, N. B. (1982). Species specificity of sex pheromone trails of sympatric and allopatric garter snakes. *Copeia* 1982, 10–13.

Ford, N. B. (1986). The role of pheromone trails in the sociobiology of snakes. In *Chemical Signals in Vertebrates. 4. Ecology, Evolution, and Comparative Biology* (D. Duvall, D. Müller-Schwarze, and R. M. Silverstein, eds.). Plenum Press, New York, pp. 261–278.

Ford, N. B., and Holland, D. (1990). The role of pheromones in the spacing behaviour of snakes. In *Chemical Signals in Vertebrates.* 5 (D. W. Macdonald, D. Müller-Schwarze, and S. E. Natynczuk, eds.). Oxford University Press, Oxford, pp. 465–472.

Ford, N. B., and Low, J. R., Jr. (1983). Sex pheromone source location by snakes: a mechanism for detection of direction in nonvolatile trails. *J. Chem. Ecol.* 10, 1193–1199.
Ford, N. B., and O'Bleness, M. L. (1986). Species and sexual specificity of pheromone trails of the garter snake, *Thamnophis marcianus. J. Herpetol.* 20, 259–262.
Ford, N. B., and Schofield, C. W. (1984). Species specificity of sex pheromone trails in the plains garter snake, *Thamnophis radix. Herpetologica* 40, 51–55.
Fowler, J. A. (1966). A communal nesting site for the smooth green snake in Michigan. *Herpetologica* 22, 231.
Fox, W. (1965). A comparison of the male urogenital systems of blind snakes, Leptotyphlopidae and Typhlopidae. *Herpetologica* 21, 241–256.
Gabe, M., and Saint Girons, H. (1965). Contribution à la morphologie comparée du cloaque et des glandes épidermoïdes de la région cloacale chez les lépidosauriens. *Mem. Mus. Natl. Hist., Nat. Paris, Ser. A. Zool.* 33, 149–292.
Gadow, H. (1887). Remarks on the cloaca and on the copulatory organs of the Amniota. *Phil. Trans. Roy. Soc. London* B178, 5–37.
Gadow, H. (1901). Amphibia and Reptiles. In *The Cambridge Natural History.* Vol. VIII (S. F. Harmer and A. E. Shipley, eds.). Repr. Wheldon and Wesley, Codicote, England.
Gans, C., Gillingham, J. C., and Clark, D. L. (1984). Courtship, mating, and male combat in Tuatara, *Sphenodon punctatus. J. Herpetol.* 18, 194–197.
Gardner, J. B. (1955). A ball of garter snakes. *Copeia* 1955, 310.
Garrick, L. D. (1978). Love among the alligators. *Animal Kingdom* 78, 2–8.
Garstka, W., Camazine, B., and Crews, D. (1982). Interactions of behavior and physiology during the annual reproductive cycle of the red-sided garter snake (*Thamnophis sirtalis parietalis*). *Herpetologica* 38, 104–123.
Garstka, W., and Crews, D. (1981). Female sex pheromone in the skin and circulation of a garter snake. *Science* (N.Y.) 214, 681–683.
Garstka, W., and Crews, D. (1982). Female control of male reproduction in a Mexican snake. *Science* (N.Y.) 217, 1159–1160.
Garstka, W., and Crews, D. (1986). Pheromones and reproduction in garter snakes. In *Chemical Signals in Vertebrates. IV. Ecology, Evolution, and Comparative Biology* (D. Duvall, D. Müller-Schwarze, and R. M. Silverstein, eds.). Plenum Press, New York, pp. 243–260.
Gehlbach, F. R., Watkins, J. F., and Kroll, J. C. (1971). Pheromone trail-following studies of typhlopid, leptotyphlopid, and colubrid snakes. *Behaviour* 40, 282–294.
Gehlbach, F. R., Watkins, J. F., and Reno, H. W. (1968). Blind snake defensive behavior elicited by ant attacks. *BioScience* 18, 784–785.
Gerhardt, U. (1933). Kloake und Begattungsorgane. In *Handbuch der vergleichenden Anatomie der Wirbeltiere* (L. Bolk, E. Göppert, E. Kallius, and W. Lobosch, eds.). Urban and Schwarzenberg, Berlin, 6, 267–350.
Giersberg, H., and Rietschel, P. (1967). *Vergleichende Anatomie der Wirbeltiere.* Gustav Fischer, Jena.
Gilhen, J. (1970). An unusual Nova Scotia population of the northern ringneck snake, *Diadophis punctatus edwardsi* (Merrum). *Nova Scotia Museum Occas. Pap.*, No. 9, Science Series No. 6.

Gillingham, J. C. (1979). Reproductive behavior of the rat snakes of Eastern North America, genus *Elaphe*. *Copeia* 1979, 319–331.
Gillingham, J. C. (1987). Social behavior. In *Snakes: Ecology and Evolutionary Biology* (R. A. Seigel, J. T. Collins, and S. S. Novak, eds.). Macmillan, New York, pp. 184–209.
Gillingham, J. C., and Clark, D. L. (1981). Snake tongue-flicking: transfer mechanics to Jacobson's organ. *Can. J. Zool.* 59, 1651–1657.
Gillingham, J. C., and Dickinson, J. A. (1980). Postural orientation during courtship in the eastern garter snake, *Thamnophis sirtalis sirtalis*. *Behav. Neural Biol.* 28, 211–217.
Glinski, T. H., and Krekorian, C. O. (1985). Individual recognition in free-living adult male desert iguanas, *Dipsosaurus dorsalis*. *J. Herpetol.* 19, 541–544.
Gloyd, H. K. (1933). Studies on the breeding habits and young of the copperhead *Agkistrodon mokasen* (Beauvois). *Pap. Mich. Acad. Sci., Arts Let.* 19, 587–604.
Goin, O. B. (1957). An observation of mating in the broad-headed skink *Eumeces laticeps*. *Herpetologica* 13, 155–156.
Gordon, D. M., and Cook, F. R. (1980). An aggregation of gravid snakes in the Quebec Laurentians. *Can. Field-Nat.* 94, 456–457.
Gorzula, S. (1985). Are caimans always in distress? *Biotropica* 17, 343–344.
Gove, D. (1979). A comparative study of snake and lizard tongue-flicking, with an evolutionary hypothesis. *Z. Tierpsychol.* 51, 58–76.
Gove, D., and Burghardt, G. M. (1983). Context-correlated parameters of snake and lizard tongue-flicking. *Anim. Behav.* 31, 718–723.
Gravelle, K. (1981). "Chemical Communication in the Iguanid Lizard *Sceloporus jarrovi*." Ph.D. dissertation, Hunter College, City University of New York.
Gravelle, K., and Simon, C. A. (1980). Field observations on the use of the tongue-Jacobson's organ system in two iguanid lizards, *Sceloporus jarrovi* and *Anolis trinitatis*. *Copeia* 1980, 359–362.
Graves, B. M., and Duvall, D. (1984). An alarm pheromone from the cloacal sacs of prairie rattlesnakes. *Am. Zool.* 24, 17A.
Graves, B. M., and Duvall, D. (1988). Evidence of an alarm pheromone from the cloacal sacs of prairie rattlesnakes. *Southwestern Nat.* 33, 339–345.
Graves, B. M., Duvall, D., King, M. B., Lindstedt, S. L., and Gern, W. A. (1986). Initial den location by neonatal prairie rattlesnakes: functions, causes, and natural history in chemical ecology. In *Chemical Signals in Vertebrates. IV. Ecology, Evolution, and Comparative Biology* (D. Duvall, D. Müller-Schwarze, and R. M. Silverstein, eds.). Plenum Press, New York, pp. 285–304.
Graves, B. M., and Halpern, M. (1988). Neonate plains garter snakes (*Thamnophis radix*) are attracted to conspecific skin extracts. *J. Comp. Psychol.* 102, 251–253.
Graves, B. M., and Halpern, M. (1989). Chemical access to the vomeronasal organs of the lizard *Chalcides ocellatus*. *J. Exp. Zool.* 249, 150–157.
Greenberg, B. (1943). Social behavior of the western banded gecko, *Coleonyx variegatus baird*. *Physiol. Zool.* 16, 110–122.

Greenberg, N. (1977). An ethogram of the blue spiny lizard *Sceloporus cyanogenys* (Reptilia, Lacertilia, Iguanidae). *J. Herpetol.* 11, 177–195.
Greene, H. W., and Dial, B. E. (1966). Brooding behavior by female Texas alligator lizards. *Herpetologica* 22, 303.
Gregory, P. T. (1974). Patterns of spring emergence of the red-sided garter snake, *Thamnophis sirtalis parietalis* in the Interlake region of Manitoba. *Can. J. Zool.* 52, 1063–1069.
Gregory, P. T. (1975). Aggregation of gravid snakes in Manitoba, Canada. *Copeia* 1975, 185–186.
Gregory, P. T. (1982). Reptilian hibernation. In *Biology of the Reptilia,* Vol. 13 (C. Gans and F. H. Pough, eds.). Academic Press, New York, pp. 53–154.
Gregory, P. T., Macartney, J. M., and Larsen, K. W. (1987). Spatial patterns and movements. In *Snakes: Ecology and Evolutionary Biology* (R. A. Seigel, J. T. Collins, and S. S. Novak, eds.). Macmillan, New York, pp. 366–395.
Groves, J. D. (1982). Egg-eating behavior of brooding five-lined skinks, *Eumeces fasciatus. Copeia* 1982, 969–971.
Guibé, J. (1970). La peau et les productions cutanées. In *Traité de Zoologie,* Vol. 14. *Reptiles* (P. -P. Grassé, ed.). Masson, Paris, pp. 6–32.
Haas, G. (1932). Über drüsenähnliche Gebilde der Epidermis am Kopfe von *Typhlops braminus. Z. Zellforsch. Mikr. Anat. Berlin* 16B, 745–752.
Haas, G. (1937). The structure of the nasal cavity of the chameleon. *J. Morphol.* 61, 433–451.
Hadley, N. F. (1985). *The Adaptive Role of Lipids in Biological Systems.* Wiley-Interscience, New York.
Halpern, M. (1980). Chemical ecology of terrestrial vertebrates. In *Animals and Environmental Fitness* (R. Giles, ed.). Pergamon Press, Oxford, pp. 263–282.
Halpern, M. (1983). Nasal chemical senses in snakes. In *Advances in Vertebrate Neuroethology* (J. P. Ewert, R. R. Capranica, and D. J. Ingle, eds.). Plenum Press, New York, pp. 141–176.
Halpern, M., and Kubie, J. L. (1980). Chemical access to the vomeronasal organ of the garter snakes. *Physiol. Behav.* 24, 367–371.
Halpern, M., and Kubie, J. L. (1984). The role of the ophidian vomeronasal system in species-typical behavior. *Trends Neurosci.* 1984 (December), 472–477.
Halpert, A. P., Garstka, W. R., and Crews, D. (1982). Sperm transport and storage and its relation to the annual sexual cycle of the female red-sided garter snake, *Thamnophis sirtalis parietalis. J. Morphol.* 174, 149–159.
Hamilton, W. J., Jr., (1948). Hibernation site of the lizards *Eumeces* and *Anolis* in Louisiana. *Copeia* 1948, 211.
Hasegawa, M. (1985). Effect of brooding on egg mortality in the lizard *Eumeces okadae* on Miyake-Jima, Izu Islands. *Copeia* 1985, 497–500.
Hathaway, L. M. (1964). Suggested function of femoral glands in *Crotaphytus collaris. Bull. Ecol. Soc. Am.* 45, 117.
Heller, S., and Halpern, M. (1981). Laboratory observations on conspecific and congeneric scent trailing in garter snakes (*Thamnophis*). *Behav. Neural Biol.* 33, 372–377.
Heller, S. B., and Halpern, M. (1982a). Laboratory observations of aggrega-

tive behavior of garter snakes, *Thamnophis sirtalis. J. Comp. Physiol. Psychol.* 96, 967–983.

Heller, S. B., and Halpern, M. (1982b). Laboratory observations of aggregative behavior of garter snakes, *Thamnophis sirtalis:* roles of the visual, olfactory, and vomeronasal senses. *J. Comp. Physiol. Psychol.* 96, 984–989.

Hirth, H. F. (1963). The ecology of two lizards on a tropical beach. *Ecol. Monogr.* 33, 83–112.

Hirth, H. F. (1966). The ability of two species of snakes to return to a hibernaculum after displacement. *Southwestern Nat.* 11, 49–53.

Ho, S-M, Danko, D., and Callard, I. P. (1981). Effect of exogenous estradiol 17-β on plasma vitellogenin levels in male and female *Chrysemys* and its modulation by testosterone and progesterone. *Gen. Comp. Endocrinol.* 43, 413–421.

Hoffstetter, R. (1962). Revue des recentes acquisitions concernant l'histoire et la systematique des squamates. *Coll. Intern. CNRS* 104, 243–278.

Hoffstetter, R. (1968). Review of "A Contribution to the Classification of Snakes" by G. Underwood. *Copeia* 1968, 201–213.

Hori, S. H., Kodama, T., and Tanahashi, T. (1979). Induction of vitellogenesis synthesis in goldfish by massive doses of androgens. *Gen. Comp. Endocrinol.* 37, 306–320.

Huggins, S. E., Parsons, L. C., and Peña, R. V. (1968). Further study of the spontaneous electrical activity of the brain of *Caiman sclerops:* olfactory lobes. *Physiol. Zool.* 41, 371–383.

Hunsaker, D. (1962). Ethological isolating mechanisms in the *Sceloporus torquatus* group of lizards. *Evolution* 16, 62–74.

Jackson, C. G., and Davis, J. D. (1972a). Courtship display behavior of *Chrysemys concinna suwanniensis. Copeia* 1972, 385–387.

Jackson, C. G., and Davis, J. D. (1972b). A quantitative study of the courtship display of the red-eared turtle, *Chrysemys scripta elegans* (Wied). *Herpetologica* 28, 58–64.

Jackson, M. K., and Sharawy, M. (1978). Lipids and cholesterol clefts in the lacunar cells of snake skin. *Anat. Rec.* 190, 41–46.

Johnsen, P. B., and Wellington, J. L. (1982). Detection of glandular secretions by yearling alligators. *Copeia* 1982, 705–708.

Kahmann, H. (1932). Sinnesphysiologische Studien an Reptilien, I. Experimentelle Untersuchungen über das Jakobson'sche Organ der Eidechsen and Schlangen. *Zool. Jahrb., Physiol.* 51, 173–238.

Karlson, P., and Luscher, M. (1959). Pheromones: a new term for a class of biologically active substances. *Nature* (London) 183, 55–56.

King, M., McCarron, D., Duvall, D., Baxter, G., and Gern, W. (1983). Group avoidance of conspecific but not interspecific chemical cues by prairie rattlesnakes (*Crotalus viridis*). *J. Herpetol.* 17, 196–198.

Kitzler, G. (1941). Die Paarungsbiologie einiger Eidechsen. *Z. Tierpsychol.* 4, 353–402.

Klauber, L. M. (1927). Some observations on rattlesnakes of the extreme Southwest. *Bull. Antivenin Inst. Am.* 1, 7–21.

Klauber, L. M. (1956). *Rattlesnakes: Their Habits, Life Histories, and Influence on Mankind.* University of California Press, Berkeley.

Klingelhöffer, W. (1931). *Terrarienkunde.* Julius E. G. Wegner, Stuttgart.
Kluge, A. G. (1967). Higher taxonomic categories of gekkonid lizards and their evolution. *Bull. Am. Mus. Nat. Hist.* 135, 1–60.
Kobayashi, T., and Watanabe, M. (1986). An analysis of snake-scent application behavior in Siberian chipmunks (*Eutamias sibiricus asiaticus*). *Ethology* 72, 40–52.
Kool, K. (1981). Is the musk of the long-necked turtle, *Chelodina longicollis,* a deterrent to predators? *Aust. J. Herpetol.* 1, 45–53.
Kramer, G. (1937). Beobachtungen über Paarungsbiologie und soziales Verhalten von Mauereidechsen. *Z. Morphol. Ecol. Tiere* 32, 752–783.
Kramer, M., and Fritz, U. (1989). Courtship of the turtle, *Pseudemys nelsoni. J. Herpetol.* 23, 84–86.
Krekorian, C. O. (1991). Field and laboratory observations on chemoreceptive functions in the desert iguana, *Dipsosaurus dorsalis. J. Herpetol.* In press.
Kroll, J. C., and Reno, H. W. (1971). A re-examination of the cloacal sacs and gland of the blind snake, *Leptotyphlops dulcis* (Reptilia: Leptotyphlopidae). *J. Morphol.* 133, 273–280.
Kropach, C. (1971). Sea snake (*Pelamis platurus*) aggregations on slicks in Panama. *Herpetologica* 27, 131–135.
Kubie, J. L., and Halpern, M. (1979). The chemical senses involved in garter snake prey trailing. *J. Comp. Physiol. Psychol.* 93, 648–667.
Kubie, J. L., Vagvolgyi, A., and Halpern, M. (1978). The roles of the vomeronasal and olfactory systems in the courtship behavior of male snakes. *J. Comp. Physiol. Psychol.* 92, 627–641.
Landmann, L. (1979). Keratin formation and barrier mechanisms in the epidermis of *Natrix natrix* (Reptilia: Serpentes): an ultrastructural study. *J. Morphol.* 162, 93–126.
Landmann, L. (1980). Zonulae occludentes in the epidermis of the snake *Natrix natrix* L. *Experientia* 36, 110–112.
Landreth, H. (1973). Orientation and behavior of the rattlesnakes, *Crotalus atrox. Copeia* 1973, 26–31.
Lawson, P. (1985). "Preliminary Investigations into the Roles of Visual and Pheromonal Stimuli on Aspects of Behavior of the Western Plains Garter Snake, *Thamnophis radix haydeni.*" M.S. thesis, University of Regina, Saskatchewan, Canada.
LeBuff, C. R., Jr. (1957). Observations on captive and wild North American crocodiles. *Herpetologica* 13, 25–28.
Legler, J. M. (1960). Natural history of the ornate box turtle, *Terrapene ornata* Aggassiz. *Univ. Kansas Publ., Mus. Nat. Hist.* 11, 327–669.
Leydig, F. (1872). *Die in Deutschland lebenden Arten der Saurier.* H. Laupp, Tübingen, Germany.
Lillywhite, H. B. (1982). Tracking as an aid in ecological studies of snakes. In *Herpetological Communities: a Symposium for the Society for the Study of Amphibians and Reptiles and the Herpetologists League, August 1977* (N. J. Scott, Jr., ed.), U.S. Fish and Wildlife Service, Wildlife Res. Rep. 13, 181–191.
Lillywhite, H. B. (1985). Trailing movements and sexual behavior in *Coluber constrictor. J. Herpetol.* 19, 306–308.

Lillywhite, H. B., and Maderson, P. F. A. (1982). Skin structure and permeability. In *Biology of the Reptilia,* Vol. 12 (C. Gans and F. H. Pough, eds.). Academic Press, New York, pp. 397–442.

Linnaeus, C. (1758). *Systema Naturae.* Laurentius Salvius, Stockholm.

Loveridge, A., and Williams, E. E. (1957). Revision of the African tortoises and turtles of the suborder Cryptodira. *Bull. Mus. Comp. Zool. Harvard* 115, 161–557.

Lüdicke, M. (1964). Ordnung der Klasse Reptilia, Serpentes. In *Handbuch der Zoologie* (Helmcke, Lengerken, Starck, eds.). Walter de Gruyter and Co., Berlin, pp. 129–298.

Maderson, P. F. A. (1967). The histology of the escutcheon scales of *Gonatodes* (Gekkonidae) with a comment on the squamate sloughing cycle. *Copeia* 1967, 743–752.

Maderson, P. F. A. (1972). The structure and evolution of holocrine epidermal glands in sphaerodactyline and eublepharine gekkonid lizards. *Copeia* 1972, 559–571.

Maderson, P. F. A. (1986). The tetrapod epidermis: a system protoadapted as a semiochemical source. In *Chemical Signals in Vertebrates,* Vol. 4. *Ecology, Evolution, and Comparative Biology* (D. Duvall, D. Müller-Schwarze, and R. M. Silverstein, eds.). Plenum Press, New York, pp. 13–26.

Maderson, P. F. A., Zucker, A. H., and Roth, S. I. (1978). Epidermal regeneration and percutaneous water loss following cellophane stripping of reptile epidermis. *J. Exp. Zool.* 204, 11–32.

Madison, D. M. (1977). Chemical communication in amphibians and reptiles. In *Chemical Signals in Vertebrates* (D. Müller-Schwarze and M. M. Mozell, eds.). Plenum Press, New York, pp. 135–168.

Mahendra, B. C. (1953). Contributions to the bionomics, anatomy, reproduction, and development of the Indian house gecko, *Hemidactylus flaviviridis. Proc. Indian Acad. Sci.* 38B, 215–230.

Mahmoud, I. Y. (1967). Courtship behavior and sexual maturity in four species of kinosternid turtles. *Copeia* 1967, 314–319.

Manton, M. L. (1979). Olfaction and behavior. In *Turtles: Perspectives and Research* (M. Harless and H. Marlock, eds.). John Wiley and Sons, New York, pp. 289–304.

Marchisin, A. (1980). "Predator-prey Interactions: Snake Eating Snakes and Pit Vipers." Ph.D. dissertation, Rutgers University, New Brunswick, N.J.

Mason, R. T., Chinn, J. W., Jr., and Crews, D. (1987). Sex and seasonal differences in the skin lipids of garter snakes. *Comp. Biochem. Physiol.* 87B, 999–1003.

Mason, R. T., and Crews, D. (1985). Female mimicry in garter snakes. *Nature* (London) 316, 59–60.

Mason, R. T., and Crews, D. (1986). Pheromone mimicry in garter snakes. In *Chemical Signals in Vertebrates,* Vol. 4. *Ecology, Evolution, and Comparative Biology* (D. Duvall, D. Müller-Schwarze, and R. M. Silverstein, eds.). Plenum Press, New York, pp. 279–283.

Mason, R. T., Fales, H. M., Jones, T. H., Pannell, L. K., Chinn, J. W., Jr., and Crews, D. (1989). Sex pheromones in garter snakes. *Science* (N.Y.) 245, 290–293.

Mason, R. T., and Gutzke, W. H. N. (1990). Sex recognition in the Leopard gecko, *Eublepharis macularius* (Sauria: Gekkonidae): possible mediation by skin-derived semiochemicals. *J. Chem. Ecol.* 16, 27–36.

Mason, R. T., Jones, T. H., Fales, H. M., Pannell, L. K., and Crews, D. (1990). Characterization, synthesis, and behavioral response to sex pheromones in garter snakes. *J. Chem. Ecol.* 16, 2353–2369.

Matthey, R. (1929). Caractères sexuels secondaires du lézard mâle. *Bull. Soc. Vaud. Sci. Nat.* 57, 71–81.

McIlhenny, E. A. (1935). *The Alligator's Life History.* Christopher Publishing House, Boston.

Meade, G. P. (1940). Observations on Louisiana captive snakes. *Copeia* 1940, 165–168.

Menchel, S., and Maderson, P. F. A. (1975). The post-natal development of holocrine epidermal specializations in Gekkonid lizards. *J. Morphol.* 147, 1–8.

Meredith, M., and Burghardt, G. M. (1978). Electrophysiological studies of the tongue and accessory olfactory bulb in garter snakes. *Physiol. Behav.* 21, 1001–1008.

Messeling, E. (1953). Rattlesnakes in southwestern Wisconsin. *Wisc. Cons. Bull.* 18, 21–23.

Mount, R. H. (1963). The natural history of the red-tailed skink, *Eumeces egregius* (Baird). *Am. Midl. Nat.* 70, 356–385.

Munro, D. F. (1948). Mating behavior and seasonal cloacal discharge of a female *Thamnophis sirtalis parietalis. Herpetologica* 4, 185–188.

Murata, Y., Jones, T. H., Pannell, L. K., Yeh, H., Fales, H. M., and Mason, R. T. (1991). New ketodienes from the integumental lipids of the Guam Brown tree snake, *Boiga irregularis. J. Nat. Products.* 54, 233–240.

Nakamura, K. (1935). On a new integumental poison gland found in the nuchal region of a snake: *Natrix tigrina. Mem. Coll. Sci. Kyoto Imp. Univ.* B10, 229–246.

Naulleau, G. (1965). La biologie et le comportement prédateur de *Vipera aspis* au laboratoire et dans la nature. *Bull. Biol. Fr. Belg.* 99, 395–524.

Naulleau, G. (1967). Le comportement de prédation chez *Vipera aspis. Rev. Comp. Anim.* 2, 41–96.

Neill, W. T. (1947). Size and habits of the cottonmouth moccasin. *Herpetologica* 3, 203–205.

Neill, W. T. (1948). Hibernation of amphibians and reptiles in Richmond County, Georgia. *Herpetologica* 4, 107–114.

Neill, W. T. (1971). *The Last of the Ruling Reptiles—Alligators, Crocodiles, and Their Kin.* Columbia University Press, New York.

Newcomer, R., Taylor, D., and Guttman, S. (1974). Celestial orientation in two species of water snakes (*Natrix sipedon* and *Regina septemvittata*). *Herpetologica* 30, 194–200.

Nicolaides, N. (1974). Skin lipids: their biochemical uniqueness. *Science* (N.Y.) 186, 19–26.

Nilson, G. (1980). Male reproductive cycle of the European adder, *Vipera berus,* and its relation to annual activity periods. *Copeia* 1980, 729–737.

Nilson, G., and Andrén, C. (1982). Function of renal sex secretion and male hierarchy in the adder, *Vipera berus,* during reproduction. *Horm. Behav.* 16, 404–413.

Noble, G. K. (1937). The sense organs involved in the courtship of *Storeria, Thamnophis,* and other snakes. *Bull. Am. Mus. Nat. Hist.* 73, 673–725.

Noble, G. K., and Bradley, H. T. (1933). The mating behavior of lizards: its bearing on the theory of sexual selection. *Ann. NY Acad. Sci.* 35, 25–100.

Noble, G. K., and Clausen, H. J. (1936). The aggregation behavior of *Storeria dekayi* and other snakes, with especial reference to the sense organs involved. *Ecol. Monogr.* 6, 271–316.

Noble, G. K., and Kumpf, K. F. (1936). The function of Jacobson's organ in lizards. *J. Genet. Psychol.* 48, 371–382.

Noble, G. K., and Mason, E. R. (1933). Experiments on the brooding habits of the lizards *Eumeces* and *Ophisaurus. Am. Mus. Novitates* 619, 1–29.

Noble, G. K., and Teale, H. K. (1930). The courtship of some iguanid and teid lizards. *Copeia* 1930, 54–56.

Norris, K. S. (1953). The ecology of the desert iguana, *Dipsosaurus dorsalis. Ecology* 34, 265–287.

Oelofsen, B. W., and Van den Heever, J. H. (1979). Role of the tongue during olfaction in varanids and snakes. *S. African J. Sci.* 75, 365–366.

Ogilvie, P. W. (1966). "An Anatomical and Behavioral Investigation of a Previously Undescribed Pouch Found in Certain Species of the Genus *Chameleo.*" Ph.D. dissertation, University of Oklahoma, Norman.

Oldak, P. D. (1976). Comparison of the scent gland secretion lipids of twenty five snakes: implications for biochemical systematics. *Copeia* 1976, 320–326.

Oldak, P. D., and Hebard, W. B. (1966). Electrophoretic studies of the scent gland secretions of snakes. *Georgetown Med. Bull.* 20, 103.

Owens, D. W. (1980). Introduction to the symposium: behavioral and reproductive biology of sea turtles. *Am. Zool.* 20, 485–486.

Owens, D. W., Comuzzie, D. C., and Grassman, M. (1986). Chemoreception in the homing and orientation behavior of amphibians and reptiles, with special reference to sea turtles. In *Chemical Signals in Vertebrates,* Vol. 4. *Ecology, Evolution, and Comparative Biology* (D. Duvall, D. Müller-Schwarze, and R. M. Silverstein, eds.). Plenum Press, New York, pp. 341–355.

Padoa, E. (1933). Ricerche sperimentali sui pori femorali e sull'epididimo della lucertola (*Lacerta muralis* Laur.) considerati come caratteri sessuali secondari. *Arch. Italiano Anat. Embriol.* 31, 205–252.

Parcher, S. R. (1974). Observations on the natural histories of six Malagasy Chamaeleontidae. *Z. Tierpsychol.* 34, 500–523.

Parker, W. S., and Brown, W. S. (1973). Species composition and population changes in two complexes of snake hibernacula in Northern Utah. *Herpetologica* 29, 319–326.

Parker, W. S., and Brown, W. S. (1980). Comparative ecology of two colubrid snakes, *Masticophis t. taeniatus* and *Pituophis melanoleucus deserticola,* in Northern Utah. *Milwaukee Pub. Mus. Publ. Biol. Geol.,* No. 7, 1–104.

Parsons, T. S. (1959a). Nasal anatomy and the phylogeny of reptiles. *Evolution* 13, 175–187.

Parsons, T. S. (1959b). Studies on the comparative embryology of the reptilian nose. *Bull. Mus. Comp. Zool.* 120, 101–277.

Parsons, T. S. (1967). Evolution of the nasal structure in the lower tetrapods. *Am. Zool.* 7, 397–413.
Parsons, T. S. (1970). The nose and Jacobson's organ. In *Biology of the Reptilia*, Vol. 2 (C. Gans and T. S. Parsons, eds.). Academic Press, New York, pp. 99–191.
Patterson, R. (1971). Aggregation and dispersal behavior in captive *Gopherus agassizi*. *J. Herpetol.* 5, 214–216.
Pedersen, J. M. (1988). Laboratory observations on the function of tongue-extrusion in the desert iguana (*Dipsosaurus dorsalis*). *J. Comp. Psychol.* 102, 193–196.
Perrill, S. A. (1980). Social communication in *Eumeces inexpectatus*. *J. Herpetol.* 14, 129–135.
Petit, A., and Geay, F. (1904). Sur la glande cloacale du caiman (*Jacaretinga sclerops* Schneid.). *C. R. Soc. Biol. Paris* 56, 1087–1089.
Pisani, G. R. (1967). Notes on the courtship and mating behavior of *Thamnophis brachystoma* (Cape). *Herpetologica* 23, 112–115.
Polo, C. N., Rosado, A., Magraner, J., and Rodriguez, G. (1988a). Estudio quimico del almizcle del genero *Crocodylus* que habita en Cuba. I-A. Algunos acidos grasos saturados presentes. *Rev. Cuban Farm.* 22, 61–68.
Polo, C. N., Rosado, A., Magraner, J., and Rodriguez, G. (1988b). Estudio quimico del almizcle del genero *Crocodylus* que habita en Cuba. I-B. Algunos acidos grasos insaturados presentes. *Rev. Cuban Farm.* 22, 69–75.
Porter, R. H., and Czaplicki, J. A. (1974). Responses of water snakes (*Natrix r. rhombifera*) and garter snakes (*Thamnophis sirtalis*) to chemical cues. *Anim. Learn. Behav.* 2, 129–132.
Portier, M. (1894). Sur les sacs annaux des ophidiens. *C. R. Acad. hebd. Sci. Paris* 118, 662–663.
Prater, S. H. (1933). The social life of snakes. *J. Bombay Nat. Hist. Soc.* 36, 469–476.
Pratt, C. W. Mc. E. (1948). The morphology of the ethmoidal region of *Sphenodon* and other lizards. *Proc. Zool. Soc. Lond.* 118, 171–201.
Price, A. H. (1988). Observations on maternal behavior and neonate aggregation in the Western diamondback rattlesnake, *Crotalus atrox* (Crotalidae). *Southwestern Nat.* 33, 370–373.
Price, A. H., and LaPointe, J. L. (1981). Structure-functional aspects of the scent gland in *Lampropeltis getulus splendida*. *Copeia* 1981, 138–146.
Quay, W. B. (1972). Integument and the environment: glandular composition, function, and evolution. *Am. Zool.* 12, 95–108.
Radcliffe, C. W., and Murphy, J. B. (1983). Precopulatory and related behaviors in captive crotalids and other reptiles: suggestions for future investigation. *Intl. Zoo Yearbook* 1983, 163–166.
Radhakrishna, G., Chin, C. C. Q., Wold, F., and Weldon, P. J. (1989). Glycoproteins in Rathke's gland secretions of loggerhead (*Caretta caretta*) and Kemp's Ridley (*Lepidochelys kempi*) sea turtles. *Comp. Biochem. Physiol.* 94B, 375–378.
Rage, J. (1984). Serpentes. Handbuch der Paleoherpetologie/Encyclopedia of Paleoherpetology, part II. Gustav Fischer, Stuttgart.
Rage, J. (1987). Fossil history. In *Snakes: Ecology and Evolutionary Biology* (R. A.

Seigel, J. T. Collins, and S. S. Novak, eds.). Macmillan, New York, pp. 51–76.
Rathke, H. (1848). *Ueber die Entwicklung der Schildkröten.* F. Vieweg und Sohn, Braunschweig, Germany.
Razakov, R. R., and Sadykov, A. S. (1986). An investigation of complex mixtures of natural substances by the defocusing and DADI methods. VI. Components in the secretion from preanal glands of some venomous snakes. *Khim. Prir. Soedin* 4, 421–423.
Reboucas-Spieker, R., and Vanzolini, P. E. (1978). Parturition in *Mabuya macrorhynca* Hoge, 1946 (Sauria, Scincidae), with a note on the distribution of maternal behavior in lizards. *Papeis Avulsos* (Sao Paulo) 32, 95–99.
Reese, A. M. (1920). The integumental glands of *Alligator mississippiensis. Anat. Rec.* 20, 203.
Reese, A. M. (1921). The structure and development of the integumental glands of the Crocodilia. *J. Morphol.* 35, 581–611.
Regamey, J. (1932). Caractères sexuels secondaires du *Lacerta agilis* Linné. *Bull. Soc. Vaud. Sci. Nat.* 57, 589–591.
Regamey, J. (1935). Les caractères sexuels du lézard (*Lacerta agilis* L.). *Rev. Suisse Zool.* 42, 87–168.
Reichenbach, N. G. (1983). An aggregation of female garter snakes under corrugated metal sheets. *J. Herpetol.* 17, 412–413.
Reinert, H. K., and Zappalorti, R. T. (1988). Field observations of the association of adult and neonatal timber rattlesnakes, *Crotalus horridus,* with possible evidence for conspecific trailing. *Copeia* 1988, 1057–1059.
Rieppel, O. (1980). The sound-transmitting apparatus in primitive snakes and its phylogenetic significance. *Zoomorphology* 96, 45–62.
Rieppel, O. (1983). A comparison of the skull of *Lanthanotus borneensis* (Reptilia: Varanoidea) with the skull of primitive snakes. *Z. Zool. Syst. Evolutionsforsch* 21, 142–153.
Robb, J. (1960). The internal anatomy of *Typhlops schneider* (Reptilia). *Aust. J. Zool.* 8, 181–216.
Roberts, J. B., and Lillywhite, H. B. (1980). Lipid barrier to water exchange in reptile epidermis. *Science* (N.Y.) 1980, 1077–1079.
Roberts, J. B., and Lillywhite, H. B. (1983). Lipids and the permeability of epidermis from snakes. *J. Exp. Zool.* 228, 1–9.
Rose, F. L. (1970). Tortoise chin gland fatty acid composition: behavioral significance. *Comp. Biochem. Physiol.* 32, 577–580.
Rose, F. L., Drotman, R., and Weaver, G. W. (1969). Electrophoresis of chin gland extracts of *Gopherus* (Tortoises). *Comp. Biochem. Physiol.* 29, 847–851.
Ross, P., and Crews, D. (1977). Influence of the seminal plug on mating behavior in the garter snake. *Nature* (London) 267, 344–345.
Ross, P., and Crews, D. (1978). Stimuli influencing mating behavior in the garter snake, *Thamnophis radix. Behav. Ecol. Sociobiol.* 4, 132–142.
Roux, J. (1930). Note sur un reptile scincidé des iles Salomon, présentant des pores pédiaux. *Verh. naturforsch. Ges. Basel* 41, 129–135.
Saint Girons, H. (1976). Comparative histology of the endocrine glands, nasal cavities and digestive tract in anguimorph lizards. In *Morphology and Biology*

of Reptiles (A. Bellairs and C. B. Cox, eds.). Academic Press, London, pp. 205–216.

Saint Girons, H. (1980). Le cycle de mues chez les Vipères Européennes. *Bull. Soc. Zool. Fr.* 105, 551–559.

Schell, F. M., and Weldon, P. J. (1985). ^{13}C-NMR analysis of snake skin lipids. *Agric. Biol. Chem.* 49, 3597–3600.

Schuett, G. W., and Gillingham, J. C. (1988). Courtship and mating of the copperhead, *Agkistrodon contortrix. Copeia* 1988, 374–381.

Schwenk, K. (1985). Occurrence, distribution, and functional significance of taste buds in lizards. *Copeia* 1985, 91–101.

Scott, T. G., and Sheldahl, R. B. (1937). Black-banded skink in Iowa. *Copeia* 1937, 192.

Scudder, K. M., Stewart, N. J., and Smith, H. M. (1980). Response of neonate water snakes (*Nerodia sipedon sipedon*) to conspecific chemical cues. *J. Herpetol.* 14, 196–198.

Secor, S. M. (1990). Reproductive and combat behavior of the Mexican kingsnake, *Lampropeltis mexicana. J. Herpetol.* 24, 217–221.

Seigel, R. A. (1980). Courtship and mating behavior of the diamondback terrapin *Malaclemys terrapin tequesta. J. Herpetol.* 14, 420–421.

Seigel, R. A., Collins, J. T., and Novak, S. S. (eds.) (1987). *Snakes: Ecology and Evolutionary Biology.* Macmillan, New York.

Shafagati, A., Weldon, P. J., and Wheeler, J. W. (1989). Lipids in the paracloacal gland secretions of dwarf (*Paleosuchus palpebrosus*) and smooth-fronted (*P. trigonatus*) caimans. *Biochem. Syst. Ecol.* 17, 431–435.

Shine, R. (1985). The evolution of viviparity in reptiles: an ecological analysis. In *Biology of the Reptilia,* Vol. 15, Development B (C. Gans and F. Billet, eds.). John Wiley and Sons, New York, pp. 605–694.

Silverstone, P. A. (1972). Final report of a study of the behavior of the American alligator at the Fort Pierce, FL Bureau of the Smithsonian Institute. Unpublished report, Smithsonian Institute.

Simon, C. A. (1983). A review of lizard chemoreception. In *Lizard Ecology: Studies of a Model Organism* (R. B. Huey, E. R. Pianka, and T. W. Schoener, eds.). Harvard University Press, Cambridge, Mass., pp. 119–133.

Simon, C. A., Gravelle, K., Bissinger, B. E., Eiss, I., and Ruibal, R. (1981). The role of chemoreception in the iguanid lizard *Sceloporus jarrovi. Anim. Behav.* 29, 46–54.

Simon, C. A., and Moakley, G. P. (1985). Chemoreception in *Sceloporus jarrovi:* does olfaction activate the vomeronasal system? *Copeia* 1985, 239–242.

Simpson, J. T., Weldon, P. J., and Sharp, T. R. (1988). Identification of major lipids from the scent gland secretions of Dumeril's ground boa (*Acrantophis dumerili,* Jan) by gas chromatography-mass spectrometry. *Z. Naturforschung* 43C, 914–917.

Slip, D. J., and Shine, R. (1988). The reproductive biology and mating system of Diamond pythons, *Morelia spilota* (Serpentes, Boidae). *Herpetologica* 44, 396–404.

Smith, H. M. (1946). Hybridization between two species of garter snakes. *Univ. Kansas Pub., Mus. Nat. Hist.* 1, 97–100.

Smith, M. A. (1938). The nucho-dorsal glands of snakes. *Proc. Zool. Soc. Lond.* 108, 575–583.
Smith, M. (1951). *The British Amphibians and Reptiles.* Collins, London.
Solomon, S. E. (1984). The characterisation and distribution of cells lining the axillary gland of the adult green turtle (*Chelonia mydas* L.). *J. Anat.* (London) 138, 267–279.
Spieker-Reboucas, R., and Vanzolini, P. E. (1978). Parturition in *Mabuya macrorhynca* Hoge, 1946 (Sauria, Scincidae), with a note on the distribution of maternal behavior in lizards. *Papéis Avulsos Zool.* (Sao Paulo) 32, 95–99.
Stein, J. (1924). Von der Ringelnatter. *Bl. Aquar. Terr.* 35, 303–306.
Stille, B., Madsen, T., and Niklasson, M. (1986). Multiple paternity in the adder, *Vipera berus. Oikos* 47, 173–175.
Stille, B., Niklasson, M., and Madsen, T. (1987). Within season multiple paternity in the adder, *Vipera berus:* a reply. *Oikos* 49, 232–233.
Swain, R. A., and Smith, H. M. (1978). Communal nesting in *Coluber constrictor* in Colorado (Reptilia: Serpentes). *Herpetologica* 34, 175–177.
Taylor, E. H. (1943). Mexican lizards of the genus *Eumeces* with comments on the recent literature of the genus. *Univ. Kans. Sci. Bull.* 29, 269–300.
Thoen, C., Bauwens, D., and Verheyen, R. F. (1986). Chemoreceptive and behavioral responses of the common lizard *Lacerta vivipara* to snake chemical deposits. *Anim. Behav.* 34, 1805–1813.
Tinkle, D. W. (1957). Ecology, maturation, and reproduction of *Thamnophis sauritus proximus. Ecology* 38, 69–77.
Tinkle, D. W. (1967). The life and demography of the side-blotched lizard, *Uta stansburiana. Misc. Publ. Mus. Zool. Univ. Mich.* 132, 1–182.
Tokarz, R. R., and Jones, R. E. (1979). A study of egg-related maternal behavior in *Anolis carolinensis* (Reptilia, Lacertilia, Iguanidae). *J. Herpetol.* 13, 283–288.
Tollestrup, K. (1981). The social behavior and displays of two species of horned lizards, *Phrynosoma platyrhinos* and *Phrynosoma coronatum. Herpetologica* 37, 130–141.
Trauth, S. E., Cooper, W. E., Vitt, L. J., and Perrill, S. A. (1987). Cloacal anatomy of the broad-headed skink, *Eumeces laticeps,* with a description of a female pheromonal gland. *Herpetologica* 43, 458–466.
Truitt, R. V. (1927). Notes on the mating of snakes. *Copeia* 1927, 21–24.
Tsumita, T., Niwa, T., Shimoyama, Y., and Tomita, H. (1979). Fatty acid composition of skin tissues of snakes. *The Snake* 11, 19–21.
Tyson, E. (1683). The anatomy of a rattlesnake. *Phil. Trans. Roy. Soc. London* 13, 25–54.
Underwood, G. (1957). On lizards of the family Pygopodidae: a contribution to the morphology and phylogeny of the Squamata. *J. Morphol.* 100, 207–268.
Vagvolgyi, A., and Halpern, M. (1983). Courtship behavior in garter snakes: effects of artificial hibernation. *Can. J. Zool.* 61, 1171–1174.
Vitt, L. J., and Cooper, W. E. (1989). Maternal care in skinks (*Eumeces*). *J. Herpetol.* 23, 29–34.
Vliet, K. A. (1986). Social behavior of the American alligator. In *Crocodiles,* IUCN Publ., News Service, Caracas, Venezuela.

Vliet, K. A. (1989). Social displays of the American alligator (*Alligator mississippiensis*). *Am. Zool.* 29, 1019–1031.

Voeltzkow, A. (1899). Beiträge zur Entwicklungsgeschichte der Reptilien. I. Biologie und Entwicklung der äusseren Körperform von *Crocodylus madagascariensis* Grand. *Abh. senckb. naturforsch. Ges.* 26, 1–150.

Volsøe, H. (1944). Structure and seasonal variation of the male reproductive organs of *Vipera berus* (L.). *Spolia zool. Mus. hauniensis.* 5, 1–157.

Von Achen, P. H., and Rakestraw, J. L. (1984). The role of chemoreception in the prey selection of neonate reptiles. In *Vertebrate Ecology and Systematics: A Tribute to Henry S. Fitch* (R. A. Seigel, L. E. Hunt, J. L. Knight, L. Malaret, and N. L. Zuschlag, eds.). Museum of Natural History, University of Kansas, Lawrence, Kans., pp. 163–172.

Wall, F. (1921). *Ophidia Taprobanica or the Snakes of Ceylon.* H. R. Cottle, Colombo.

Watkins, J. F., Gehlbach, F. R., and Kroll, J. C. (1969). Attractant-repellant secretions of blind snakes (*Leptotyphlops dulcis*) and their army ant prey (*Neivamyrmex nigrescens*). *Ecology* 50, 1098–1102.

Weaver, W. G. (1970). Courtship and combat behavior in *Gopherus berlandieri. Bull. Fla. State Mus. Biol. Sci.* 15, 1–43.

Weldon, P. J. (1982). Responses to ophiophagous snakes by snakes of the genus *Thamnophis*. *Copeia* 1982, 788–794.

Weldon, P. J., and Bagnall, D. (1987). A survey of polar and nonpolar skin lipids from lizards by thin-layer chromatography. *Comp. Biochem. Physiol.* 87B, 345–349.

Weldon, P. J., and Burghardt, G. M. (1979). The ophiophage defensive response in Crotaline snakes: extension to new taxa. *J. Chem. Ecol.* 5, 141–151.

Weldon, P. J., Divita, F. M., and Middendorf, G. A., III. (1987). Responses to snake odors by laboratory mice. *Behav. Proc.* 14, 137–146.

Weldon, P. J., Dunn, B. S., Jr., McDaniel, C. A., and Werner, D. I. (1990). Lipids in the femoral gland secretions of the green iguana (*Iguana iguana*). *Comp. Biochem. Physiol.* 95B, 541–543.

Weldon, P. J., and Fagre, D. B. (1989). Responses by canids to scent gland secretions of the western diamondback rattlesnake (*Crotalus atrox*). *J. Chem. Ecol.* 15, 1589–1604.

Weldon, P. J., Ford, N. B., and Perry-Richardson, J. J. (1990). Responses by corn snakes (*Elaphe guttata*) to chemicals from heterospecific snakes. *J. Chem. Ecol.* 16, 37–44.

Weldon, P. J., and Sampson, H. W. (1987). Paracloacal glands of *Alligator mississippiensis:* A histological and histochemical study. *J. Zool.* (London) 212, 109–115.

Weldon, P. J., and Sampson, H. W. (1988). The gular glands of *Alligator mississippiensis:* histology and preliminary analysis of lipoidal secretions. *Copeia* 1988, 80–86.

Weldon, P. J., and Schell, F. M. (1984). Responses by king snakes (*Lampropeltis getulus*) to chemicals from colubrid and crotaline snakes. *J. Chem. Ecol.* 10, 1509–1520.

Weldon, P. J., Scott, T. P., and Tanner, M. J. (1990). Analysis of gular and paracloacal gland secretions of the American alligator (*Alligator mississip-*

piensis) by thin-layer chromatography: gland, sex, and individual differences in lipid components. *J. Chem. Ecol.* 16, 3–12.

Weldon, P. J., Shafagati, A., and Wheeler, J. W. (1987). Lipids in the gular gland secretion of the American alligator (*Alligator mississippiensis*). *Z. Naturforsch.* 42C, 1345–1346.

Weldon, P. J., Shafagati, A., and Wheeler, J. W. (1988). Lipids from the paracloacal glands of the American alligator (*Alligator mississippiensis*). *Lipids* 23, 727–729.

Werner, D. I. (1978). On the biology of *Tropidurus delanonis*, Baur (Iguanidae). *Z. Tierpsych.* 47, 337–395.

Werner, D. I., Baker, E. M., Gonzalez, E. del C., and Sosa, I. R. (1987). Kinship recognition and grouping in hatchling green iguanas. *Behav. Ecol. Sociobiol.* 21, 83–89.

Whitford, W. B., and Bryant, M. A. (1979). Behavior of a predator and its prey: the horned lizard (*Phrynosoma cornutum*) and harvester ants (*Pogonomyrmex* spp.). *Ecology* 60, 686–694.

Whiting, A. M. (1969). "Squamate Cloacal Glands: Morphology, Histology, and Histochemistry." Ph.D. dissertation, The Pennsylvania State University, College Park.

Whittier, J. M., and Crews, D. (1986). Effects of prostaglandin $F_{2\alpha}$ on sexual behavior and ovarian function in female garter snakes (*Thamnophis sirtalis parietalis*). *Endocrinology* 119, 787–792.

Whittier, J. M., Mason, R. T., and Crews, D. (1985). Mating in the red-sided garter snake, *Thamnophis sirtalis parietalis:* Differential effects on male and female sexual behavior. *Behav. Ecol. Sociobiol.* 16, 257–261.

Wickler, W. (1978). A special constraint on the evolution of composite signals. *Z. Tierpsychol.* 48, 345–348.

Wilde, W. S. (1938). The role of Jacobson's organ in the feeding reaction of the common garter snake, *Thamnophis sirtalis sirtalis* (Linn.). *J. Exp. Zool.* 77, 445–465.

Wilson, E. O. (1975). *Sociobiology: The New Synthesis.* Harvard University Press, Cambridge, Mass.

Wingate, L. R. (1956). Mambas strong smell. *African Wildlife* 10, 256–257.

Winokur, R. M., and Legler, J. M. (1975). Chelonian mental glands. *J. Morphol.* 147, 275–292.

Woodbury, A. M., and Hardy, R. (1940). The dens and behavior of the desert tortoise. *Science* (N.Y.) 92, 529.

Woodbury, A. M., and Hardy, R. (1948). Studies of the desert tortoise, *Gopherus agassizi*. *Ecol. Monogr.* 18, 145–200.

Woodbury, A. M., Vetas, B., Julian, G., Glissmeyer, H. R., Hayrand, F. L., Call, A., Smart, E. W., and Sanders, R. T. (1951). Symposium: a snake den in Tooele County, Utah. *Herpetologica* 7, 1–52.

Wright, D. E., and Moffat, L. A. (1985). Morphology and ultrastructure of the chin and cloacal glands of juvenile *Crocodylus porosus* (Reptilia, Crocodilia). In *Biology of Australasian Frogs and Reptiles* (G. Grigg, R. Shine, and H. Ehmann, eds.). Royal Zoological Society of New South Wales, Surrey Beatty and Sons Pty. Ltd., Chipping Norton, pp. 411–422.

5

Endogenous Rhythms

Herbert Underwood

CONTENTS

I. INTRODUCTION

A. Endogenous Rhythms

It is now widely recognized that all organisms, whether unicellular or vertebrate, can exhibit daily rhythms in a veritable host of behavioral, physiological, and biochemical parameters. The origin of this rhythmicity is obscure; however, it has been speculated that the selective pressure generating this rhythmicity is the daily shower of radiation from the sun (Pittendrigh, 1965a; Paietta, 1982). In primordial organisms the daily pulse of radiation, particularly in the ultraviolet, could have impaired some biochemical reactions; consequently, such processes were confined to the dark phase of the light-dark cycle (Pittendrigh, 1965a). The increasing level of free oxygen in the Precambrian may also have led to damaging photooxidative processes that varied with the daily flux of visible and near-visible light (Paietta, 1982). Consequently, light-impaired cellular processes were confined to times of the day that would minimize the photochemical hazard.

Temporal organization of biological processes confers a selective advantage; that is, particular events occur at the "proper" time of day. Significantly, even if the organism is maintained under constant conditions, most daily rhythms will persist with periods approximating the period of the earth's rotation (i.e., 24 hours). Endogenous rhythms therefore must be overt expressions of an internal "biological clock." The term *circadian* (circa, about; *dies,* a day) has been coined to describe such rhythms. Under constant conditions circadian rhythms are said to be "freerunning"; that is, they show persistent cycles with periods approximately, but rarely exactly, 24 hours long. Circadian rhythms can be synchronized, or "entrained," by certain environmental stimuli, such as regular changes of light or temperature. Light or temperature cycles entrain the circadian clock because it has a phase-dependent sensitivity to these stimuli. Light or temperature either phase-advances or phase-delays the motion of the endogenous clock during each cycle so that the period (τ) of the clock is matched to the period (T) of the light or temperature cycle (Pittendrigh, 1965b). Entraining agents such as light or temperature therefore control not only the *period* of the rhythm but also its *phase.* A glossary providing definitions of terms and symbols used in the discussion of biological rhythm is provided in Table 5.1.

In cases in which more than one circadian rhythm has been examined in an individual organism, the different rhythms typically show the same freerunning period and bear fixed phase relationships with each other. However, under certain conditions an organism might express two different circadian periodicities. For example, following pinealectomy, the circadian activity rhythm of the lizard *Sceloporus olivaceus* can "split" into two circadian components, each of which freeruns with a different periodicity (Underwood, 1977). Furthermore, the pineal organ of the lizard *Anolis carolinensis* exhibits a circadian rhythm of melatonin secretion in vitro (Fig. 5.1) (Menaker and Wisner, 1983). These observations show that a multicellular organism, such as a lizard, represents a "multioscillator"; that is, individuals possess more than a single circadian clock. Because all of an

Table 5.1. Glossary of terms and symbols used to describe biological rhythms

circadian rhythm: an endogenous rhythm with a period that is close to the period of the solar day (24 hours)
circadian system: the sum of all the circadian oscillators, including the pacemakers (driving oscillators) and overt circadian rhythms of an organism
entrainment: the process by which an endogenous rhythm becomes coupled to, and assumes the period of, an environmental cycle
freerunning period: the period of an endogenous rhythm that is not entrained by an external cycle
limits of entrainment: the range of frequencies within which a self-sustained oscillation can be entrained
period: the length of time between recurring phase reference points of a rhythm, i.e., the time required for a rhythm to complete one full cycle
phase shift: a single displacement of an oscillation along its time axis
phase-response curve: a plot of the magnitude and direction (+/−) of the phase shift caused by a perturbation applied at different phases of a freerunning rhythm
photoperiod: day length; the length (hours) of the light phase in a light-dark (LD) cycle
resonance: designates photoperiod experiments in which the period of a light cycle is 24, 36, 48, 60, or 72 hours

Symbols

LD X:Y light-dark cycle with X hours of light (L) and Y hours of dark (D) per cycle
DD continuous dark
LL continuous light
τ period of a freerunning circadian rhythm
T period of the entraining stimulus; also used to designate photoperiod experiments in which the period of a light cycle is varied within the limits of entrainment

organism's circadian rhythms are normally phase-locked one to another, some mechanism must ensure a fairly rigid organization among different rhythms. Conceptually, such organization could occur if all the circadian clocks of an organism were mutually coupled to one another or, alternatively, if a central circadian pacemaker existed that imposed phase and frequency upon subordinate oscillators. These two alternatives are not mutually exclusive, and circadian organization may involve elements of both.

The understanding of endogenous rhythms in vertebrates has utilized a variety of approaches. For example, the mechanisms generating the circadian system have been probed by analyzing the response of rhythms to perturbations by various entraining stimuli, most notably light and temperature. These experiments form the basis for a number of formal models of circadian systems. More recently, the physiology of circadian systems has been studied by various techniques, including ablation of neural sites and exogenous administration of drugs and putative hormones. Within recent years the idea that discrete areas may serve a "pacemaker" function within the multioscillator circadian system of vertebrates has gained wide acceptance. In vertebrates at least two sites have been shown to be of major importance in organizing circadian systems: the pineal organ and the suprachiasmatic nuclei of the hypothalamus.

The aim of this review is to synthesize our current knowledge about the role of endogenous rhythms in reptiles; it focuses mainly on the role of circadian rhythms, but endogenous annual (circannual) rhythms are discussed as well. Much of the literature on reptiles is devoted to describing their life histories, including their daily activity and thermoregulatory behaviors. This review attempts to glean from this literature those aspects of behavior that are likely to be influenced by the circadian system, to discuss the significance of the circadian control of these behaviors, and to discuss the way in which a formal

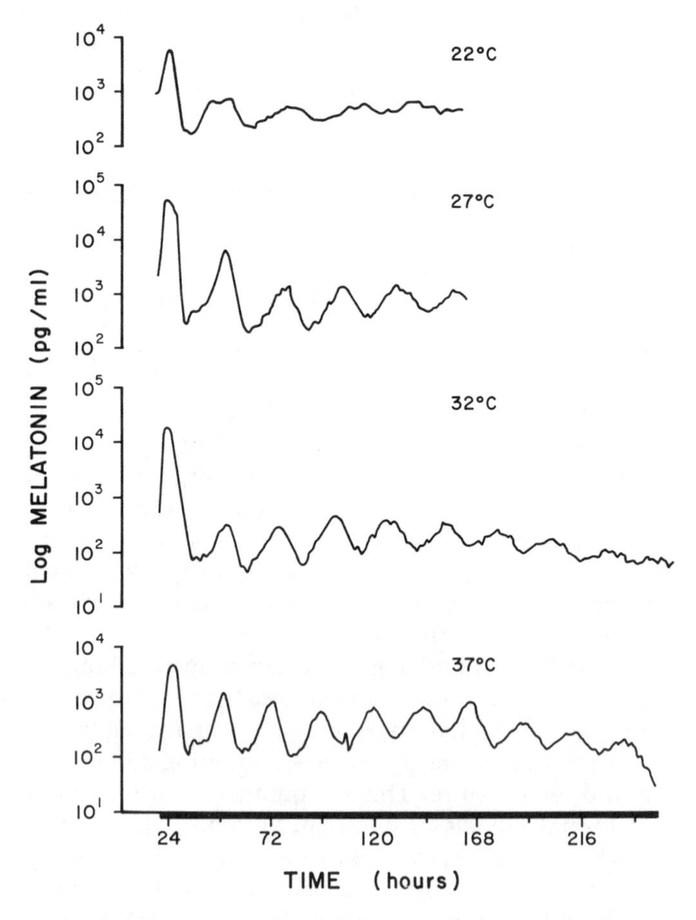

Fig. 5.1. *Anolis carolinensis.* Circadian rhythms of melatonin secretion in four individual pineal glands held in continuous darkness under four different constant temperatures. (After Menaker and Wisner, 1983.)

"model" of circadian organization proposed by Pittendrigh and his co-workers (Pittendrigh, 1974; Pittendrigh and Daan, 1976) to describe the circadian system of higher vertebrates may be applicable to reptiles as well. Recent studies of the physiology of the circadian system of reptiles are also discussed, focusing mainly on the role of the pineal organ and its hormones. These studies are used to support a model for the physiology of the circadian system.

B. Pineal Systems

Discussion of the role of the pineal organ in the circadian system of reptiles requires a brief description of the anatomy, physiology, and biochemistry of the pineal organ. Beyond the earlier account in this series (Quay, 1979), a number of reviews can be consulted for additional information (Hamasaki and Eder, 1977; Collin and Oksche, 1981; Collin et al., 1986).

The pineal organ originates as an evagination of the roof of the diencephalon and remains attached to the brain via a stalk. As far as is known pineal organs are present in every species of reptile except for members of the subclass Archosauria (Roth et al., 1980). In some species of lizards, the pineal "system" has a second component. This component, termed the *parietal eye,* also originates from the roof of the diencephalon. During development, the parietal eye migrates to a position beneath a foramen in the roof of the skull. The parietal eye has a well-defined retina, lens, and cornea and is strikingly eyelike in morphology (Quay, 1979; Hamasaki and Eder, 1977). Typically there is a parietal eye nerve that innervates several areas of the brain, including the habenular area, the dorsolateral nucleus of the thalamus, pretectal areas, the hypothalamus (periventricular nucleus, preoptic area), and possibly the pineal as well (Engbretson et al., 1981; Korf and Wagner, 1981; Isabekova et al., 1987).

All reptilian pineals contain photosensory cells, supporting cells, and neurons, but the ultrastructure and number of these cells varies among species (Quay, 1979; Hamasaki and Eder, 1977; Collin and Oksche, 1981; Collin et al., 1986). Typically, the outer segments of the photosensory cells are poorly organized. They are concentric or whorllike, lacking the regular stacked lamellar disks characteristic of the outer segments of rods or cones. They suggest "degenerate" photosensory cells and have been termed *secretory rudimentary photoreceptors* (Collin, 1979). The snake pineal is an exception. It lacks photoreceptivelike cells in the adult, and the principal cell type is the pinealocyte or pineal parenchymal cell. Pinealocytes, however, are considered to belong to the same cell line as the photosensory cells (Collin and Oksche, 1981). Neurophysiological investigations confirm the morphological evidence of photoreception by both the parietal eye and the pineal organ (Hamasaki and Eder, 1977; Hamasaki and Dodt, 1969; Hamasaki, 1969). The pineal of lizards tends to show achromatic responses because visible light of all wavelengths tends to inhibit ongoing electrical activity; however, the parietal eye can show chromatic responses to visible light. The parietal eye shows both graded (slow) potentials as well as action potentials. The graded potentials presumably reflect the well-organized structure of the photosensory and neural elements in the parietal eye.

The innervation of the reptilian pineal organ appears to be mainly pinealofugal (afferent). These afferent nerves, presumably carrying photic information from the rudimentary photoreceptors, project to pretectal and tegmental areas of the brain; the exact terminations of these nerves is unknown (Quay, 1979; Hamasaki and Eder, 1977; Ueck, 1979). Pinealopetal (efferent) nerves have been seen in the pineal organ of reptiles; both the origin and function of this innervation remains unknown (Collin et al., 1986; Ueck, 1979).

II. PINEAL BIOCHEMICAL RHYTHMS

A. General

The rudimentary photosensory cells of the pineal organs of lizards and turtles also give evidence of considerable secretory activity, as they often possess a well-developed Golgi complex, numerous free ribosomes, and rough and smooth endoplasmic reticulum (Quay, 1979; Collin, 1979; Collin and Oksche, 1981). Among pineal compounds, the indoleamines, which are derived from the amino acid tryptophan, have attracted the most interest. The indoleamine melatonin (*N*-acetyl 5-methoxytryptamine) appears to play a role in a variety of systems, including those involved in reproductive and circadian rhythms; it has been studied extensively. The synthesis of melatonin involves the 5-hydroxylation of tryptophan followed by its decarboxylation by aromatic amino acid decarboxylase to produce serotonin. Subsequently, serotonin is acetylated by the enzyme *N*-acetyltransferase (NAT) to produce *N*-acetylserotonin. It is then O-methylated by the enzyme hydroxyindole-O-methyltransferase (HIOMT) to produce melatonin (Quay, 1974). Immunohistochemical methods indicate that indoleamine synthesis within the reptilian pineal occurs in the rudimentary photoreceptors (Collin, 1979). This suggests a direct photic influence on the synthesis and secretion of indoleamines and perhaps of other compounds as well.

The last step in the synthesis of methoxyindoles, such as melatonin, is catalyzed by HIOMT. Previously, HIOMT was thought to occur only in pineal tissue; consequently, the O-methylated indoleamines such as melatonin have received the greatest attention as potentially unique pineal hormones. Although a few other areas in some vertebrates have now been shown to possess HIOMT activity, among them being the eyes or Harderian glands (Ralph, 1980; Pang and Allen, 1986), in most cases the pineal organ is the main, if not exclusive, source of blood-borne melatonin (Ralph, 1980). Because melatonin is not stored within the pineal but is rapidly secreted into the blood (Rollag et al., 1980), blood levels of melatonin faithfully reflect pineal melatonin synthesis, at least in those species in which extrapineal sites of melatonin do not secrete significant amounts of melatonin into the blood.

Although the details of indoleamine metabolism have been studied most extensively in mammals, the metabolism of these compounds in other vertebrates probably involves similar pathways. One of the most striking aspects of indoleamine synthesis is the presence of daily rhythms in many of the substrate concentrations and enzyme activities that comprise the biosynthetic pathways.

B. Testudines and Crocodilia

Table 5.2 lists those species of the Testudines and Crocodilia that have been examined for the presence of daily rhythms in pineal or blood indoleamine or catecholamine content. One example of day-night variations in pineal levels of both serotonin and melatonin occurs in the tortoise *Testudo hermanni* (Vivien-Roels, 1983, 1985; Vivien-Roels and Arendt, 1979, 1981, 1983; Vivien-Roels et al., 1971, 1979). Under natural conditions melatonin levels peak at night, whereas serotonin levels peak during the day. However, during winter hibernation, the levels of both pineal melatonin and of serotonin are very low and nonrhythmic (Vivien-Roels and Arendt, 1981; Vivien-Roels et al., 1979). Laboratory studies in which either the photoperiod or the level of constant temperature to which *T. hermanni* were exposed was varied showed: (1) Tortoises exposed to cold (0 to 10°C) temperatures for 3 weeks in the winter exhibited a serotonin rhythm if maintained at LD 16:8, but not if maintained at LD 8:16, whereas no melatonin rhythm occurred under either photoperiod, and (2) a melatonin rhythm could be observed in tortoises maintained at LD 8:16 in the winter if the temperature was kept at 30°C (Vivien-Roels et al., 1979; Vivien-Roels and Arendt, 1981, 1983). These studies have been interpreted as showing that photoperiod is more important for expressing the rhythm in pineal serotonin, whereas temperature is more important for expressing the rhythm in melatonin production. Such conclusions must remain tentative, however, until the potential roles of light and temperature in the control of pineal indoleamine metabolism are studied in detail. Some of the studies described above utilized a single daytime and a single nighttime measurement of pineal melatonin levels. Such "two-sample" experiments would not permit discrimination

Table 5.2. Indoleamine content, catecholamine content, or HIOMT activity in several tissues of reptiles exposed to daily (natural or artificial) light-dark (LD) cycles and/or temperature (T) cycles, or to continuous dark (DD) or continuous light (LL)

Species	Compound	Rhythm	Tissue	Condition	Reference
Testudo hermanni	Serotonin	+	Pineal	LD, LD + T	Vivien-Roels, 1983, 1985;
	Melatonin	+	Pineal	LD, LD + T	Vivien-Roels & Arendt, 1979, 1981, 1983;
	Melatonin	+	Blood	LD	Vivien-Roels et al., 1971, 1979
Lissemys punctata	Serotonin	+	Pineal	LD	Mahapatra et al., 1986, 1988
	Norepinephrine	+	Pineal	LD	
	Epinephrine	+	Pineal	LD	
Chelonia mydas	Melatonin	+	Blood	LD	Owens et al., 1980
	Melatonin	+	Cerebro-spinal fluid	LD + T	
Terrapene carolina	Melatonin	+	Pineal	LD	Vivien-Roels et al., 1988;
	Melatonin	+	Blood	LD	Skene et al., 1989
	5-Methoxy-tryptophol	+	Pineal	LD	
Lampropholis guichenoti	HIOMT	+	Pineal	LD	Joss, 1978
	HIOMT	+	Eye	LD	
Podarcis muralis	Serotonin	+	Pineal	LD + T	Petit & Vivien-Roels, 1977
	Serotonin	−	Pineal	LL	
	Serotonin	+	Pineal	DD	
Trachydosaurus rugosus	Melatonin	+	Pineal	LD, LD + T	Firth and Kennaway, 1980, 1987, 1989;
	Melatonin	+	Parietal eye	LD, LD + T	Firth et al., 1979, 1989a; Kennaway et al.,
	Melatonin	+	Blood	LD, LD + T	1977
	Melatonin	−	Blood	DD, LL	

continued.

Table 5.2 *(continued)*

Anolis carolinensis	Melatonin	+	Pineal (in vitro)	LD	Menaker & Wisner, 1983
Anolis carolinensis	Melatonin	+	Pineal	LD,LD+T,DD	Underwood, 1985a; Underwood & Calaban, 1987a,b; Underwood & Hyde, 1989a,b
	Melatonin	−	Eye	LD, LD+T	
Sphenodon punctatus	Melatonin	+	Blood	LD + T	Firth et al., 1989a
Alligator mississippiensis	Melatonin	−	Blood	LD + T	Roth et al., 1980

Note: Plus (+) or minus (−) indicates whether or not the compound exhibited a daily rhythm.

between an effect of light or temperature on the *phase* of the rhythm as opposed to an effect on the *amplitude* of the rhythm.

In the box turtle, *Terrapene carolina triunguis,* both light and temperature can affect pineal and blood melatonin levels; the duration of the nighttime rise in melatonin is longer under LD 8:16 than under LD 18:6, and the amplitude of the melatonin rhythm is influenced by ambient temperatures (Vivien-Roels et al., 1988). No melatonin rhythm is observed in the pineal or blood of box turtles held under 24-hour LD cycles at 5°C, a small increase in nighttime pineal and blood melatonin levels is seen at 15°C, and a large amplitude rhythm is observed at 20°C and 27°C (Vivien-Roels et al., 1988). A rhythm in pineal 5-methoxytryptophol levels is seen in *T. carolina triunguis,* which also seems to be influenced by photoperiod; the level of pineal 5-methoxytryptophol is significantly higher during the day than during the night in turtles exposed to LD 18:6 (constant 27°C), but no rhythm is shown in LD 8:16 (Skene et al., 1989).

Under natural conditions, the blood and cerebrospinal fluid of the turtle *Chelonia mydas* shows a rhythm of melatonin content (Owens et al., 1980). Daily variations in serotonin, norepinephrine, and epinephrine levels have also been observed in the pineal-paraphyseal complex of the turtle *Lissemys punctata punctata* held under natural lighting conditions in March; serotonin levels peak during the day and norepinephrine and epinephrine levels peak at night (Mahapatra et al., 1986). Similar patterns in the levels of serotonin, norepinephrine, and epinephrine were seen in juvenile *L. punctata punctata* after 3 days' exposure to LD 12:12 in the laboratory in March (Mahapatra et al., 1988). The alligator (*Alligator mississippiensis*) lacks a pineal organ and shows only arrhythmic, low levels of melatonin in the blood (Roth et al., 1980).

C. Sauria and Rhynchocephalia

Indoleamine levels in the pineals of several species of lizards exhibit daily variations (see Table 5.2). Spectrofluorimetric measurements of the pineal serotonin content of *Podarcis muralis* show that (1) a daily rhythm of serotonin was present under natural conditions with maximal serotonin levels occurring at the end of the lighting period, although during winter months the serotonin levels were low and nonrhythmic, and (2) the pineal serotonin rhythm persists for 3 weeks under constant temperature and constant darkness (Petit and Vivien-Roels, 1977). Plasma melatonin in the scincid lizard *Trachydosaurus rugosus* shows a significant daily variation with maximal melatonin levels occurring during the dark phase of the photoperiod (Kennaway et al., 1977; Firth and Kennaway, 1980, 1987; Firth et al., 1979); the rhythm has a low amplitude in the spring and a higher amplitude in the summer and autumn (Firth et al., 1979; Kennaway et al., 1977). The plasma melatonin rhythm is reported to be absent in *T. rugosus* held in constant darkness (DD) for 9 days in the spring (Firth et al., 1979). However, the plasma was sampled only four times (every 6 hours) over a 24-hour period, and a peak may have been missed. In addition, the 9 days' exposure to DD may have allowed the melatonin rhythm in the individual freerunning lizards to become out of phase with each other, thereby obscuring the population rhythm.

When sampled under natural conditions in the autumn, the blood melatonin rhythm in the tuatara *Sphenodon punctatus* had a much lower amplitude than did the lizard *Trachydosaurus rugosus* (Firth et al., 1989a). Although both species can attain similar plasma melatonin levels in the laboratory, the threshold at which a significant daily rhythm occurs is less than 17°C in *S. punctatus* compared with 25°C in *T. rugosus* (Firth et al., 1989a). Firth et al. (1989a) suggest that this difference reflects the different thermoregulatory adaptations of these two species; under natural conditions in the autumn *S. punctatus* exhibited a mean body temperature of about 12°C when active, whereas *T. rugosus* exhibited a temperature of around 23°C.

The daily rhythms in melatonin in pineal, parietal eye, and plasma of *Trachydosaurus rugosus* show a higher amplitude after exposure to a 24-hour LD cycle and a temperature cycle (warm photophase/cool scotophase) than after exposure to a 24-hour LD cycle and a constant temperature (Firth and Kennaway, 1980, 1987). Exposure to constant cold (15°C) temperatures abolishes melatonin rhythms in pineals, parietal eyes, or plasma of *T. rugosus* held on LD 12:12 (Firth and Kennaway, 1987). Removal of the parietal eye has no effect on plasma melatonin rhythms in *T. rugosus* exposed to LD cycles and constant temperatures, but the level of plasma melatonin shows no rhythm in parietalectomized lizards held in LD and cyclic temperatures (Firth and Kennaway, 1980). The mechanism by which parietalectomy abolishes the plasma melatonin rhythm is unknown. As the parietal eye of *T. rugosus* appears to contain as much melatonin as does the pineal (Firth and Kennaway, 1987), parietalectomy may have removed a significant melatonin-secreting source. Also, the parietal eye may have a neural input to the pineal that influences the ability of this organ to secrete melatonin. However, the observed effect of parietalectomy on the plasma melatonin should be viewed with caution because the sampling procedure employed (samples every 6 hours) may have missed a shift in the phase of the rhythm induced by parietalectomy.

A parietal eye–pineal interaction is supported by the observation that in the lizard *Sceloporus occidentalis,* parietalectomy tends to reverse the inhibition of pineal HIOMT activity induced by constant light exposure (Bethea and Walker, 1978). However, little is known about the relationship between HIOMT activity and melatonin production in reptiles. In homeotherms it is generally accepted that the rhythm in melatonin synthesis is due to a rhythm in the activity of the NAT enzyme because HIOMT activity remains relatively constant over 24 hours. In at least one lizard, *Lampropholis guichenoti,* however, HIOMT activity shows daily bimodal fluctuations in both the eyes and the pineal (Joss, 1978). In this species HIOMT activity peaks both 2 hours after the onset of dark and 1 hour before the onset of light. Pineal HIOMT activity has been reported in several kinds of reptiles, including turtles, lizards, and snakes, but the possibility of daily rhythmicity in HIOMT activity has not been considered (Quay, 1965; Quay et al., 1971).

Maintenance of the pineal gland of the lizard *Anolis carolinensis* in a superfused organ culture discloses a daily rhythm of melatonin secretion that can persist for up to 10 cycles in DD (see Fig. 5.1) (Menaker and Wisner, 1983). This rhythm is entrainable by 24-hour LD cycles in vitro and shows temperature compensation (Q_{10}~1.14). Accordingly, the pineal gland of *A. carolinensis* possesses a fully competent circadian clock that is capable of driving a rhythm in melatonin synthesis.

A number of studies have examined the roles played by light and temperature in the control of the in vivo pineal melatonin rhythm (PMR) of *Anolis carolinensis* (Underwood, 1985a, 1986a, 1990; Underwood and Calaban, 1987a, b; Underwood and Hyde, 1989a,b). These studies demonstrate that both daily light cycles and daily temperature cycles are effective entraining stimuli for the PMR. Furthermore, the length of the photoperiod or the length of the thermoperiod (i.e., the duration of the warm or cool phase of the 24-hour temperature cycle) affects several features of the PMR, including the amplitude and duration of the PMR and the phase-relationship between the melatonin rhythm and the light or temperature cycles.

The light intensity experienced during the day can affect the phase and amplitude of the PMR in *A. carolinensis* (Underwood and Hyde, 1989b). The phase of the PMR of *A. carolinensis* also shows systematic changes as the length of the photoperiod changes; for example, melatonin levels peak before dusk on LD 18:6 (constant 32°C), around middark on LD 12:12, and late in the dark on LD 6:18 (Underwood and Hyde, 1989a). The duration of the elevated pineal melatonin levels, and the total amount of melatonin produced, are less during short photoperiods (LD 6:18 and LD 8:16) than during longer photoperiods. The influence of photoperiod duration on pineal melatonin levels in *A. carolinensis,* therefore, is in marked contrast to that observed in birds and mammals. In higher vertebrates melatonin levels cannot rise during the day, and the duration of melatonin production is typically longer during short photoperiods than during longer photoperiods. Another interesting difference has also been observed between the response of the pineal of *A. carolinensis* to light as compared to higher vertebrates; a light pulse placed in the middle of the night (on LD 12:12) does not suppress pineal melatonin levels in *A. carolinensis* (Underwood, 1986a), whereas night interruptions in birds or mammals elicit a rapid fall in pineal melatonin levels (e.g., Hamm et al., 1983; Brainard et al., 1983). A similar observation has been made in the box turtle, *Terrapene carolina triunguis;* a light pulse placed in the dark of an LD 16:8 cycle had no effect on pineal or blood melatonin levels (Vivien-Roels et al., 1988). However, in the sea turtle, *Chelonia mydas,* a 2-hour nocturnal exposure to light significantly depresses melatonin levels in the blood (Owens et al., 1980). The reason for the difference in light responsiveness between *A. carolinensis* and *T. carolina triunguis* and homeotherms is presently unknown.

Temperature cycles as low as 2°C in amplitude will entrain the PMR in *Anolis carolinensis* and, whenever this species is exposed to 24-hour temperature cycles, the melatonin levels peak during the cool phase of the cycle. As melatonin levels peak at night in every vertebrate species examined to date, the peak of melatonin levels during the cool phase of a daily temperature cycle is, perhaps, not unexpected in a poikilotherm such as *A. carolinensis.* In the field, of course, nocturnal temperatures would be cooler than those experienced during the day.

Information on the relative strengths of temperature and light as entraining stimuli were obtained by "conflicting" the phase of the light and temperature cycles and examining the effects on the PMR in *Anolis carolinensis* (Underwood and Calaban, 1987a). As expected, in a "normal" type phase-relationship (warm days and cool nights) the PMR peaks during the (cool) night (Fig. 5.2). However, in an "abnormal" type phase-relationship (cool days and warm nights) the phase of the melatonin peak is determined by the relative strength of the two entraining stimuli (i.e., the amplitude of the temperature cycle) (Fig. 5.2). Altering the phase-relationship between a light cycle (LD

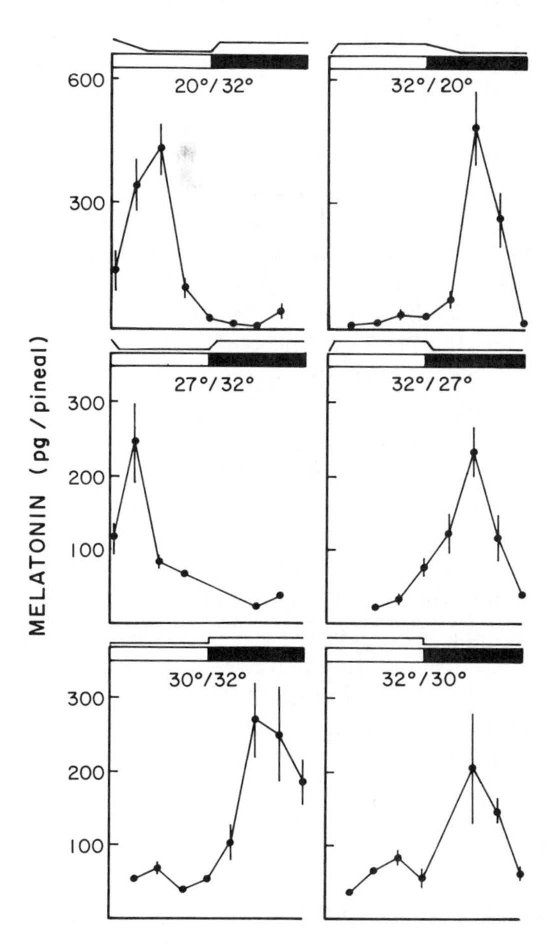

Fig. 5.2. *Anolis carolinensis*. Effects of light and temperature cycles on the pineal melatonin rhythm. The anoles were exposed to 24-hour temperature cycles of various amplitudes in which the cool phase of the cycle occurred either during the day (left panels) or night (right panels). (From Underwood and Calaban, 1987a.)

12:12) and a 24-hour temperature cycle (33°C/15°C) can also affect the phase (and amplitude) of the plasma melatonin rhythm in the lizard, *Trachydosaurus rugosus* (Firth and Kennaway, 1989).

The way in which light or temperature is transduced by the pineal may be explicable in terms of the multioscillator nature of the pineal itself (Underwood, 1987). The pineal organ of birds is clearly a multioscillator; small fragments of chicken pineals exhibit circadian rhythms in melatonin secretion in vitro (Takahashi and Menaker, 1984). Although the theory is unproven, the lizard pineal may also be comprised of multiple circadian oscillators (individual cells?) that are normally coupled; the mutual phase-relationships among these oscillators will determine the duration and amplitude of their output (i.e., melatonin levels). Because light and temperature are effective entraining stimuli for circadian clocks, changes in the photoperiod or thermoperiod would be reflected in altered phase-relationships among the coupled oscillators; these, in turn, would generate changes in the phase, duration, and amplitude of the melatonin rhythm.

In *Anolis carolinensis*, removal of the parietal eye or the lateral eyes fails to affect the PMR of lizards entrained either to LD 12:12 (constant 32°C) or to LD 12:12 and a temperature cycle (32°C day, 20°C night) (Underwood and Calaban, 1987b). The observation that parietalectomy did not affect the PMR in anoles is in contrast to the observation that parietalectomy does affect the plasma melatonin rhythm in *Trachydosaurus rugosus:* in the latter species parietalectomy abolishes the plasma rhythm after exposure both to an LD cycle and a temperature cycle (Firth and Kennaway, 1980). In addition, in blinded *A. carolinensis* the PMR reentrains to a 10-hour shift in the phase of an entraining LD 12:12 light cycle as rapidly as in sighted anoles. Furthermore, blocking light penetration to the pineals of anoles but leaving the eyes exposed completely inhibits the ability of the PMR to reentrain to a 10-hour shift in the phase of an LD 12:12 light cycle (Underwood and Calaban, 1987b). These studies therefore support the idea that light controls the PMR directly and inputs from the eyes or parietal eye are not involved. Direct photic entrainment of the PMR is also supported by the in vitro entrainment of the PMR by LD cycles (Menaker and Wisner, 1983), as well as by ultrastructural and neurophysiological evidence of photoreception by lizard pineals (e.g., Hamasaki and Eder, 1977; Quay, 1979).

III. RETINAL PHOTORECEPTION, EXTRARETINAL PHOTORECEPTION, AND OCULAR RHYTHMICITY

The ease of measurement has led biologists to use the circadian rhythm of locomotor activity as an overt measure of the state of an animal's circadian clock. Observations of the effects of 24-hour LD cycles on the circadian activity rhythm of blinded lizards documents that extraretinal photoreceptors are fully capable of mediating entrainment of the circadian system in nine species of lizards representing four lizard families (Iguanidae: *Anolis carolinensis, Sceloporus olivaceus, S. magister, S. occidentalis, S. clarki;* Xantusiidae: *Xantusia vigilis;* Gekkonidae: *Hemidactylus turcicus, Coleonyx variegatus;* Lacertidae: *Podarcis sicula*) (Underwood, 1973, 1985b; Underwood and Menaker, 1970, 1976). Furthermore, removal of the pineal organ of *C. variegatus*, or of both the pineal organ and the parietal eye of *S. olivaceus* and *S. magister,* does not prevent entrainment to LD cycles of blinded individuals (Underwood, 1973). The extraretinal receptors must be quite sensitive because blind lizards can be entrained to an LD cycle as dim as 1 lux (Underwood, 1973). The extraretinal receptors are located in the brain, although their number and location are unknown (Underwood, 1973; Underwood and Menaker, 1976). However, several reasons make the hypothalamus a likely site for the location of at least some of these extraretinal photoreceptors. First, direct illumination of the medial-ventral hypothalamus of birds can elicit photoperiodic stimulation of the reproductive system (Yokoyama et al., 1978). Second, a neural circadian pacemaker may reside in the hypothalamus (see Section IV), and a close anatomical connection between a clock and its photoreceptors may be advantageous. Third, the diencephalon is known to give rise to the only other identifiable photoreceptors in vertebrates (lateral eyes, pineal system).

Examination of the ability of light to penetrate to the base of the brain of a variety of vertebrates, including the lizard *Podarcis muralis*, shows that light of longer wavelengths (700 to 750 nm) penetrates much more effectively than light of shorter wavelengths (Hartwig and vanVeen, 1979). Although visible light of longer wavelengths can readily reach extraretinal receptors, nothing is known about the actual photopigments involved in these photoreceptors of reptiles.

Encephalic photoreceptors mediating entrainment have also been documented in juvenile alligators (*Alligator mississippiensis*). Capping the eyes does not prevent entrainment of the activity rhythm by light; however, entrainment is abolished in alligators having both eyes and skulls capped (Kavaliers and Ralph, 1981). This is interesting, as crocodilians are among the few vertebrates that lack a pineal organ (Roth et al., 1980).

The lateral eyes are also involved in mediating entrainment to LD cycles. For example, *Sceloporus olivaceus* entrained to a very dim green light (0.05 lux) will freerun if blinded (Underwood, 1973). However, the eyes may have a more complex role in the circadian system of lizards than simple provision of additional photic inputs to the system. At least potentially, light can entrain circadian rhythms via two mechanisms: entrainment may be effected either by the transition between the light and dark phases of the LD cycle (termed the *nonparametric effect of light*) or by the duration of the light (termed the *parametric effect*). For example, parametric entrainment could occur if the light phase of an LD cycle parametrically increases the velocity of the circadian clock and the dark phase decreases its velocity so that an exact 24-hour periodicity is obtained. Empirical observations also have shown a consistent relationship between the parametric effects of light and the period of freerunning rhythms. In diurnal animals, an increase in the intensity of constant light (LL) shortens the freerunning period (i.e., increases the velocity of the clock) and, furthermore, the period of the rhythm in LL is shorter than the period expressed in DD (Aschoff, 1979). In nocturnal animals opposite relationships hold between the free-running period and light. These generalizations have been termed *Aschoff's Rule.*

Blind *Sceloporus olivaceus* and *S. occidentalis* can sense the parametric effects of light because the freerunning period lengthens after the lizards are switched from LL to DD (Underwood and Menaker, 1976); that is, these two diurnal species obey Aschoff's Rule even after blinding. However, although sighted *Podarcis sicula* can obey Aschoff's Rule, blind *P. sicula* do not because the freerunning period expressed in LL is the same as that expressed in DD. Blind *P. sicula,* however, are still entrainable by LD cycles (Underwood and Menaker, 1976). Accordingly, *P. sicula,* but not *S. olivaceus* or *S. occidentalis,* show extraretinal entrainment that involves a nonparametric mechanism. In *P. sicula* the eyes must be solely responsible for mediating the parametric effects of light.

Another observation documenting the complex role of the eye in circadian rhythms is that blinding *Sceloporus occidentalis* by optic nerve section actually renders the circadian system more sensitive to the phase-shifting effects of light (Underwood, 1985b). Since light pulses administered to lizards freerunning in DD causes predictable effects on the subsequent freerunning rhythm; that is, depending on the temporal position of the light pulse relative to the activity-rest cycle, it may cause the rhythm to phase-advance, to phase-delay, or to lack effect (Underwood, 1983a, 1985b). A graph of the relationship between the phase at which the light pulse is administered versus the direction and magnitude of the phase shift it induced produces a phase-response curve (PRC). Phase-response curves have had great utility in describing how light entrains circadian rhythms (Pittendrigh, 1965b). Remarkably, the amplitude of the PRC is severalfold larger in blind *S. occidentalis* than in sighted lizards (Underwood, 1985b). Consistent with this result is the observation that blind *S. oc-*

cidentalis reentrain more rapidly than sighted lizards to a shift in the phase of an LD cycle. Also the limits of entrainment of blind lizards are larger than those exhibited by sighted lizards (Underwood, 1985b).

Several studies have documented daily changes in physiology or morphology within the eye itself. For example, the shedding of disks from cone outer segments is rhythmic in the eye of *Sceloporus occidentalis* (Young, 1977; Bernstein et al., 1984). In lizards held on LD 12:12 the burst of shedding occurs 2 hours after light offset. The rhythm in disk shedding will also persist for at least two cycles in constant darkness (Bernstein et al., 1984). A freerunning rhythm in the b-wave component of the electroretinogram (ERG) has also been demonstrated in the lizard *Anolis carolinensis;* this rhythm can persist for up to 8 days in DD, with the largest response occurring at projected midday (Fowlkes et al., 1984, 1987). In about 30% of the eyes, the a-wave of the ERG exhibits a circadian rhythm in DD (Fowlkes et al., 1987). In a nocturnal lizard (*Gekko gecko*) the b-wave of the ERG shows a rhythm in DD with a peak amplitude near projected sunset but no detectable rhythm in the a-wave (Fowlkes et al., 1987). Turtles, *Pseudemys* sp., show diurnal morphological changes in synaptic ribbons of the rod cells if they are maintained on LD 12:12 (Abe and Yamamoto, 1984). At middark, synaptic ribbons lie close to, and perpendicular to, the presynaptic membrane and show single stick-shaped profiles. From middark to midlight the stick-shaped ribbons grow into large multilayered ribbon complexes, which then degenerate between midday and the onset of dark.

The demonstration of diurnal changes in ocular physiology and anatomy, combined with the demonstration that at least some of these events will persist in constant conditions, shows that these rhythms are driven by a circadian clock. However, the location (intra- versus extraocular) of this clock has not been established. Clearly, the clock could be located within the eye itself, or it could be located extraocularly, and drive ocular rhythms via neural or hormonal inputs. In this context, it is important to note that an intraocular clock has been identified in the amphibian, *Xenopus laevis,* which can drive rhythms in NAT activity and disk shedding (Besharse and Iuvone, 1983; Flannery and Fisher, 1984). The existence of an intraocular clock in lizards would explain some of the "complex" roles of the eyes within circadian organization. For example, the increased sensitivity of blind *Sceloporus occidentalis* to light (Underwood, 1985b) might represent an effect of removing one oscillatory component from a normally coupled multioscillator system. The responsiveness to light of the remaining components might then increase.

It seems likely that extraretinal photoreception existed before the evolution of lateral eyes. More primitive eyeless organisms undoubtedly used the daily light cycle as an entraining stimulus. Once the eyes had evolved, they were "wired" into the multioscillator system. The persistence of extraretinal receptors and their probably universal presence in reptiles implies that they likely have some selective advantage. For example, eyes and extraretinal receptors may extract different kinds of information from the photic environment (i.e., parametric versus nonparametric). Clearly more work is needed: (1) to examine other orders of reptiles for the presence of extraretinal photoreception (after all, to date all studies have focused exclusively on lizards); (2) to gain insight into the selective advantages of using both retinal and extraretinal photoreceptors in controlling circadian organization; (3) to localize the extraretinal receptors and study their responses to light and their connections to the rest of the brain; and (4) to ascertain whether the eyes can act as circadian pacemakers and, if so, how they contribute toward the total circadian organization of animals.

IV. PHYSIOLOGICAL BASIS OF ENDOGENOUS RHYTHMICITY: A MODEL

A number of studies support a major role for the pineal organ within the circadian system of lizards (Underwood, 1990). Pinealectomy of lizards (*Sceloporus olivaceus, S. occidentalis, Anolis carolinensis*) that are freerunning in constant conditions elicits either a "splitting" of the activity rhythm into two circadian components, a change in the period of the freerunning rhythm, or arrhythmicity (continuous activity with no discernible activity-rest rhythm) (Underwood, 1977, 1981a, 1983b). These effects of pinealectomy are not species-specific; that is, a single species such as *S. olivaceus* may exhibit all three kinds of effects of pinealectomy (Underwood, 1977). The pineal of *A. carolinensis* must contain a circadian clock, because organ-cultured pineals show a circadian rhythm of melatonin secretion (see Fig. 5.1) (Menaker and Wisner, 1983). Clearly, lizards have a multioscillator circadian system; one circadian clock (or clocks) lies in the pineal, and two (or more) are located elsewhere. The diverse effects of pinealectomy are understandable in terms of the multioscillator structure of the lizard's circadian system. For example, if it is assumed that the pineal acts as a pacemaker within the multioscillator system of lizards, its removal could uncouple the remaining oscillators. This might lead to splitting or arrhythmicity. If the residual coupling is sufficient to prevent disassociation, the period of the coupled system will likely change. The observation that pinealectomy produces a major distortion of the phase response curve to light pulses is also consistent with a role for the pineal as a pacemaker within a multioscillator circadian system (Underwood, 1983a).

The pineal probably communicates with the rest of the circadian system in lizards via the secretion of melatonin. Several kinds of evidence support a role for melatonin within the lizard circadian system. First, exogenous melatonin administered via subcutaneous Silastic capsules causes the freerunning period to lengthen, or produce arrhythmicity, in *Sceloporus olivaceus* and *S. occidentalis* (Underwood, 1979, 1981a). Second, daily injections of melatonin can entrain the activity rhythm of *S. occidentalis* (Underwood and Harless, 1985). Third, a phase-response curve can be generated in response to single injections of melatonin into freerunning *S. occidentalis* (Underwood, 1986b). These data suggest that the daily melatonin pulse may provide the mechanism whereby the pacemaker in the pineal entrains other circadian oscillators.

The suprachiasmatic nuclei (SCN) of the hypothalamus may contain at least some of the extrapineal oscillators comprising the circadian system of lizards. The SCN have been shown to be a major (or perhaps exclusive) circadian pacemaker in mammals (Rusak and Zucker, 1979), and the SCN may play a pacemaking role in birds as well (Takahashi and Menaker, 1979). The mammalian and avian SCN receive direct photic input via a retinohypothalamic (RH) pathway. The possibility that the SCN are also an important part of the circadian system of lizards is suggested by studies showing that lizards and snakes also have a RH pathway (Repérant et al., 1978; Butler, 1974; Janik, 1987; Korf and Wagner, 1981; Schroeder, 1981); furthermore, lesions of an area presumed to be homologous to the mammalian SCN induce arrhythmicity in the activity rhythm of the lizard *Dipsosaurus dorsalis* (Janik, 1987). Interestingly, removal of the pineal organ or the lateral eyes in *D. dorsalis* fails to affect the freerunning activity rhythm (Janik, 1987).

Taken together the data suggest a multioscillator model for the circadian system of lizards (Fig. 5.3). The pineal is itself autonomously rhythmic and may comprise a population of coupled oscillators. The pineal transduces light and temperature stimuli into a hormonal signal (melatonin) so that the phase, amplitude, and duration of the melatonin pulse faithfully reflect the ambient conditions of light and temperature. The daily melatonin pulse entrains circadian clocks located elsewhere (possibly in the SCN), so that these oscillators and the overt rhythms they control assume appropriate phase-relationships. Light can entrain extrapineal (SCN?) oscillators via extraretinal receptors and via the eyes, independently of the pineal. The possibility that the eyes serve as clocks can be hypothesized, but this is unproven. The model shown in Fig. 5.3 is consistent with observations made on several species of iguanids (*Sceloporus occidentalis, S. olivaceus, Anolis carolinensis*). However, the recent finding that SCN lesions abolish circadian rhythmicity in the iguanid *Dipsosaurus dorsalis*, but that blinding or pinealectomy appear to have little or no effect (Janik, 1987), suggests that different species, even of the same family, may differently emphasize the several elements of the circadian system.

Current knowledge about the anatomy and physiology of the circadian system of reptiles is still sketchy. In fact, no information is available on the physiology of the circadian system of any nonsaurian reptile. Some of the major unanswered questions include: Do circadian clocks lie in the suprachiasmatic nuclei? Do additional areas, either within the central nervous system or in other organs (such as the eyes), have pacemaker roles? How are these areas coupled to one another and to the external environment? What factors are responsible for the observation that the role of the pineal may be more important in some species than in others?

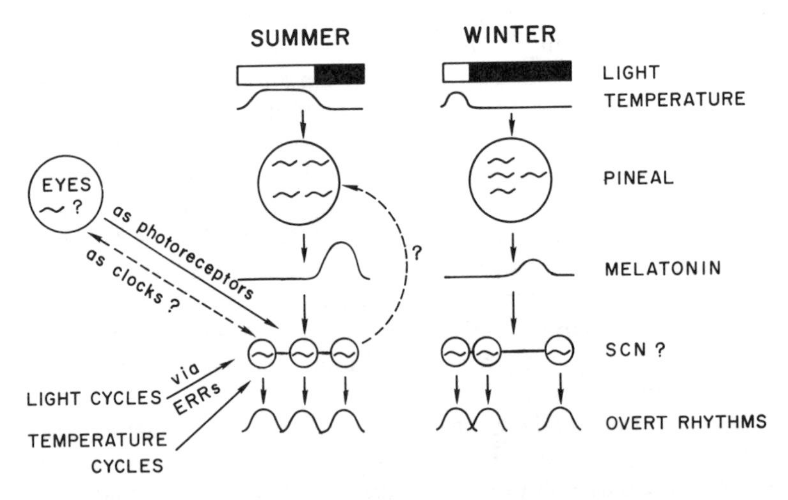

Fig. 5.3. A model for the circadian system of lizards. The left panel depicts the response of the system to a "summer type" light/temperature condition, and the right panel depicts the response to "winter type" conditions. These stimuli are transduced by the (multioscillator) pineal into a melatonin pulse, which in turn controls both the period and the phase of circadian oscillators located elsewhere (i.e., in SCN?). Other components of the system (shown in the left panel only) include (1) the eyes, which act as photoreceptors and (possibly) as biological clocks, (2) "direct" entrainment of extrapineal oscillators by temperature and light (via extraretinal photoreceptors), and (3) possible feedback loops between central oscillators and pineal oscillators. This system allows coordination between internal circadian rhythms and between internal rhythms and events in the external world.

V. ACTIVITY AND THERMOREGULATORY RHYTHMS IN THE FIELD AND LABORATORY

A. Models and Formal Properties

Although a variety of formal models has been proposed to describe circadian systems in vertebrates, one of the most fruitful models was developed by Pittendrigh and co-workers to describe the circadian system of rodents (Pittendrigh, 1974; Pittendrigh and Daan, 1976). According to this model the rhythm of locomotor activity is controlled by two circadian oscillators (or sets of oscillators) that are normally coupled in a stable phase relationship. The onset of activity and the end of activity are each under the control of one of these oscillators. The freerunning period of the coupled system is determined by the spontaneous frequencies of the individual oscillators (i.e., the frequencies they would show in the absence of coupling) and by the strength of coupling between the two oscillators. The two-oscillator model implies that the phase-relationship between the two coupled oscillators driving the overt rhythm of activity changes as the length of the photoperiod of an entraining LD cycle changes. Therefore, a change in the length of the photoperiod affects the length of the daily activity period.

The two-oscillator model has utility in explaining several features of the activity rhythms of lizards. It requires the assumption that the pineal organ may, at least in some species, also play a role in maintaining appropriate phasing of the two oscillators (or sets of oscillators). For example, the splitting and period changes observed following pinealectomy in the lizard *Sceloporus olivaceus* (Underwood, 1977) are compatible with the behavior of two oscillators. These either become uncoupled or remain coupled but exhibit their own (albeit coupled) endogenous freerunning periodicity in the absence of their pacemaker. Other aspects of the reptilian activity patterns are also interpretable in terms of the two-oscillator model. Many species of reptiles (e.g., Testudines: *Gopherus berlandieri,* Rose and Judd, 1975; *Emydoidea blandingii,* Graham, 1979; Sauria: *Aporosaura anchietae,* Holm, 1973; *Uta stansburiana,* Irwin, 1965) show a bimodal activity pattern both in the laboratory under constant conditions and in the field under natural conditions. Each of the two components of the locomotor pattern may represent the overt expression of one of the two oscillators.

In poikilotherms, both light and temperature cycles have been shown to be effective entraining agents for the circadian system. In the laboratory, either stimulus can entrain the activity rhythm of reptiles (e.g., Mangelsdorff and Hauty, 1972; Hoffmann, 1968; Underwood, 1973; Graham and Hutchison, 1979a). Obviously, in the field an animal will be exposed to both stimuli simultaneously; however, the quality (e.g., wavelength) and intensity of these two stimuli will vary throughout the day. Caution must be observed in correlating any overt rhythm, such as activity, with the state of an animal's biological clock. Exogenous stimuli such as temperature or light can exert direct effects on activity either by stimulating activity or by suppressing it. This direct effect of a stimulus on the overt expression of a rhythm is typically referred to as *masking*.

B. Activity Rhythm

1. TESTUDINES One of the major selection pressures affecting the behavior of any animal is the daily periodicity of its environment. Any species adapts, in a specific manner, to daily and seasonal changes of light and temperature. One of the most commonly studied aspects of this adaptation is the daily pattern of activity. For turtles, there are relatively few studies of activity patterns in the laboratory or the field; far fewer studies than have been carried out, for instance, in lizards.

Most species of turtles are diurnally active, but some, such as *Chelydra serpentina osceola* and *Terrapene ornata*, are nocturnal (Carr, 1952). Although the persistence of a rhythm under constant conditions is the generally accepted criterion for its identification as being truly circadian, few studies have examined the activity rhythms of turtles under constant conditions. Persistent rhythms under constant light and temperature conditions have been observed in *Gopherus polyphemus* (Gourley, 1972), *Trachemys scripta* (Brett, 1971), and *Geochelone sulcata* (Cloudsley-Thompson, 1970). However, most workers in the field of circadian biology would accept the premise that activity rhythms are commonly driven by a circadian clock, even if persistence of the rhythm in constant conditions has not been demonstrated for a particular species.

Most field studies of reptiles, including turtles, have focused on the active (locomotor) phase of their daily activity pattern. Less information is available on the temporal distribution of other components of the active phase, such as emergence from dens or basking. The daily pattern of locomotor activity is often characterized as unimodal or bimodal, depending on whether activity appears to be relatively continuous throughout the day or whether there are morning and afternoon peaks of activity separated by periods of inactivity. In south Texas, *Gopherus berlandieri* shows a bimodal pattern; however, the phase relationship between the morning and afternoon peaks differs depending on season (Rose and Judd, 1975). The midday inactive period is presumed to be the result of the high temperatures that occur at that time of day. However, *G. berlandieri* expresses a bimodal pattern even when environmental temperature conditions are compatible with activity throughout the day (Rose and Judd, 1975). In the field, bimodal patterns of activity have also been observed in *Chelydra serpentina* (Obbard and Brooks, 1981) and western members of the genus *Gopherus* (e.g., McGinnis and Voigt, 1971). *G. polyphemus* in Florida shows a unimodal pattern throughout the year, with peaks occurring during the hottest part of the day (Douglass and Layne, 1978). Laboratory studies of some species of turtles have shown that the expression of bimodal versus unimodal activity pattern is temperature dependent. Under LD 8:16 and LD 16:8, *Chrysemys picta, Clemmys guttata,* and *Sternotherus odoratus* show a unimodal rhythm of locomotor activity at 15°C and a bimodal rhythm at 25°C (Graham and Hutchison, 1979a); *Emydoidea blandingii* exposed to LD 14:10 shows a bimodal pattern of activity at 25°C, with but a single peak at midday at 15°C (Graham, 1979).

The onset of the active phase of the activity-rest cycle after emergence from nocturnal retreats is usually preceded by basking (e.g., *Clemmys guttata,* Ernst, 1982; *Chrysemys picta,* Ernst, 1972; *Gopherus polyphemus,* Douglass and Layne, 1978). In *C. picta* the average duration of basking is inversely proportional to temperature (Ernst, 1972).

The phase of the activity rhythm (activity onsets) shows systematic seasonal variations. In Pennsylvania, for example, *Clemmys guttata* shifted the peak of daily activity from the afternoon in March to the morning in summer; this presumably avoids increasing environmental temperatures (Ernst, 1982). Other examples of seasonal changes in activity onsets and offsets include *Gopherus berlandieri* in Texas (Rose and Judd, 1975), *Chelydra serpentina* in Ontario (Obbard and Brooks, 1981), *Chrysemys picta* in Pennsylvania (Ernst, 1972), and *Gopherus polyphemus* in Florida (Douglass and Layne, 1978). In the laboratory, activity onset occurred before light onset in *Chrysemys picta* and *Clemmys guttata* that had been entrained to LD 8:16 (constant temperature); it occurred after light onset if this species had been entrained to LD 16:8 (Graham and Hutchison, 1979a). In more temperate areas, turtles hibernate in the colder months (e.g., *Trionyx muticus,* Plummer, 1977; *Chelydra serpentina,* Obbard and Brooks, 1981; *Clemmys guttata,* Ernst, 1982; *Chrysemys picta,* Ernst, 1972).

The length of the photoperiod (or thermoperiod) strongly influences the phase that an overt circadian rhythm assumes with respect to an entraining agent such as light or temperature (Pittendrigh, 1965b; Pittendrigh and Daan, 1976). For example, activity onsets in diurnal animals often precede light onsets whenever the animals are entrained to shorter photoperiods, whereas onsets of daily activity may occur after light onsets in animals that are entrained to long photoperiods. The seasonal changes in the onset of activity noted above then are markedly affected by the way seasonal changes in photoperiod (and thermoperiod) entrain the circadian system of the turtle.

Presumably, other environmental factors, such as high environmental temperatures or predators, will also modify the phase relationships among various circadian rhythms and the cyclic environment. The way selection pressures modify phase relationships between the circadian system and its entraining stimuli is poorly understood.

2. SAURIA AND SERPENTES

Various aspects of the circadian properties of the activity rhythms of lizards have been studied in the laboratory. Some of the conclusions include:

1. The activity rhythm is innate. *Podarcis sicula* held in either LL or DD since conception exhibit circadian rhythms of locomotor activity (Hoffmann, 1955).

2. *P. sicula* obey Aschoff's Rule for diurnal animals; in LL of 60 lux the freerunning period shortens (to 23.67 hours) from an average of 24.67 hours in LL of 5.5 lux (Hoffmann, 1960). Interestingly, two other diurnal lizards (*Sceloporus occidentalis* and *Heloderma suspectum*) obey Aschoff's Rule for nocturnal animals; in LL their freerunning periods are longer than in DD (Underwood, 1985b; Lowe et al., 1967).

3. Sinusoidal 24-hour temperature cycles as low as 0.9°C in amplitude can entrain the activity rhythm of *P. sicula* held in LL (Hoffmann, 1968).

4. Twenty-four hour LD cycles are also highly effective entraining agents for the activity rhythms of lizards held under conditions of constant temperature (e.g., Underwood, 1973).

5. Phase-response curves exist both to light and to injections of melatonin (Underwood, 1983a, 1985b, 1986b).

6. The phase of a rhythm entrained to a sinusoidal 24-hour temperature cycle (6°C in amplitude) under LL depends on the period of the rhythm expressed under constant conditions (Hoffmann, 1963). This relationship can be explained on the basis of the entrainment model proposed by Pittendrigh (1965b).

7. The circadian activity rhythm shows temperature compensation. Whereas most metabolic processes show a Q_{10} in the range of 2 to 4, the circadian clock is temperature compensated and typically exhibits a Q_{10} in the range of 0.9 to 1.4. *Podarcis sicula,* for example, exhibit freerunning periods of 25.2, 24.34, and 24.19 hours under constant temperatures of 16°C, 25°C, and 35°C, respectively (Hoffmann, 1957). The period of the activity rhythms of *Lacerta agilis* and *P. muralis* remains around 24 hours in lizards held in LL or DD between 15°C and 25°C (Marx and Kayser, 1949). Temperature compensation has clear adaptive value because, in order to measure the passage of time accurately, the circadian clock must run with a constant frequency regardless of the level of ambient temperature. Perhaps the clearest ex-

ample of a requirement for temperature compensation is seen in the role of the circadian clock in time-compensated celestial orientation (see Section VI.A). An animal could not accurately compensate for the apparent movement of celestial cues by referring to its circadian clock as long as the frequency of the clock is temperature dependent.

The simultaneous presentation of light and temperature cycles has an interesting effect on the activity rhythm of reptiles (Graham and Hutchison, 1978; Evans, 1966; Lee, 1974). Whenever presented with an LD cycle of a period (T) of 23 hours and a temperature cycle with T = 25 hours (*Coleonyx variegatus*), or whenever presented with an LD cycle with T = 25 hours and a temperature cycle with T = 23 hours (*Xantusia vigilis*), the activity patterns reflect an interaction between light and temperature (Evans, 1966; Lee, 1974). Of particular interest is the behavior exhibited whenever the phases of the light and temperature cycles are out of phase by 180° to 360° (i.e., whenever the animals are exposed to cool days and warm nights). The activity rhythm "jumps" abruptly across this zone and becomes phase-locked to the LD cycle. A similar "phase jumping" has been observed in both intact and pinealectomized *Sceloporus occidentalis* (Underwood, unpublished) and in a turtle *Chrysemys picta* (Graham and Hutchison, 1978).

These effects of light and temperature can be explained on the basis of a two-oscillator model developed by Pittendrigh (1960) and Pittendrigh and Bruce (1959) for *Drosophila*. According to this model, one oscillator (the A oscillator) is entrainable by light and the other (the B oscillator) by temperature. Normally the A oscillator is the pacemaker and drives B. However, the B oscillator can be directly entrained by temperature, and the model supposes some feedback of B on A. The behavior of the B oscillator is reflected in the overt rhythm of activity. Shifts of the phase of the daily temperature cycle relative to the light cycle (i.e., to the right) cause the activity rhythm to shift, as this reflects the motion of the B oscillator. The activity rhythm follows the temperature cycle until the onset of activity reaches about 12 to 15 hours after dawn. Then there is a discrete "phase jump" to the next dawn. This phase jump implies a zone of forbidden coupling between A and B because a small change in the phase of the two entraining stimuli results in a large change in the phase of the activity rhythm. This is consistent with a principle of physics; that is, the phase relations of any coupled oscillator system involves 180° of forbidden phase relations. If the driven system (B) assumes coupling in the forbidden zone, it would be transferring energy into, not drawing it from, the driver (Pittendrigh, 1960). How this kind of two-oscillator model can be related to the model described in Fig. 5.3 is uncertain. However, the putative A and B oscillators must lie at an extrapineal site, as pinealectomized *Sceloporus occidentalis* are entrainable by either light or temperature cycles and exhibit the phase-jumping phenomenon (Underwood, unpublished). One could speculate that the extrapineal oscillators shown in Fig. 5.3 are themselves comprised of a mutually coupled set of both light-sensitive (A) and temperature-sensitive (B) oscillators.

In the field, lizards exhibit a variety of phase relationships between the activity rhythm and the cyclic environment. The phase adopted by a particular species presumably reflects response to a variety of selection pressures on the circadian system, within constraints imposed by the morphology and physiology of the species. Most lizards are day active but some, notably members of the Gekkonidae, are night active. Seasonal changes occur in the phase relationships between activity and the cyclic environment, but lizards show no large changes, such as a switch from completely diurnal to completely nocturnal. In general, the type of activity seen in the field can also be observed in the laboratory. Day-active species are also diurnal when exposed to 24-hour LD cycles in the laboratory (e.g., *Sceloporus olivaceus, S. occidentalis, S. clarki, S. magister, Anolis carolinensis,* Underwood, 1973; *Gallotia galloti galloti,* Molina-Borja et al., 1986; García-Díaz et al., 1989; *Uta stansburiana,* Evans, 1966; *Dipsosaurus dorsalis,* Gelderloos, 1976; *Cnemidophorus sexlineatus,* Barden, 1942; *Acanthodactylus schmidti,* Constantinou and Cloudsley-Thompson, 1985; *Phrynosoma coronatum* and *P. cornutum,* Heath, 1962; *Aporosaura anchietae,* Holm, 1973; *Podarcis sicula,* Underwood and Menaker, 1976; *Lacerta agilis* and *P. muralis,* Marx and Kayser, 1949; *Mabuya quinquetaeniatus,* Cloudsley-Thompson, 1965), whereas nocturnal species are mainly night active in the laboratory (e.g., *Coleonyx variegatus,* Evans, 1966; Underwood, 1973; *Gehyra variegata,* Bustard, 1967; *Diplodactylus vittatus,* Bustard, 1968; *Tarentola annularis,* Cloudsley-Thompson, 1965). Laboratory exposure of diurnal and nocturnal lizards to temperature cycles under conditions of constant light results in the activity of the day-active species (e.g., *Anolis carolinensis,* Underwood, 1985a; *Acanthodactylus schmidti,* Constantinou and Cloudsley-Thompson, 1985; *Podarcis sicula,* Hoffmann, 1963, 1968; *Mabuya quinquetaeniatus,* Cloudsley-Thompson, 1965) during the warm phase of the temperature cycle and activity of the nocturnal species during the cool phase of the temperature cycle (e.g., *Ptyodactylus hasselquisti,* Frankenberg, 1979; *Tarentola annularis,* Cloudsley-Thompson, 1965). Studies on species that in the field are neither strictly diurnal nor nocturnal show consistency between the phases of the activity patterns seen under artificial cycles of light and temperature in the laboratory and those noted in the field (Frankenberg, 1979). Therefore, a species in the laboratory will assume phase relationships with light or temperature cycles that are approximations of the phases observed in the field.

However, one should be cautious in extrapolating activity patterns recorded in the laboratory to patterns noted in the field. Numerous studies of the formal properties of circadian rhythms have shown that important parameters of activity rhythms, such as the initiation of activity, are a function of several properties of entraining light or temperature cycles. Among these properties are the *length* of the photoperiod or thermoperiod, the *quality* of the stimulus (i.e., spectral characteristics), the *intensity* of the stimulus (i.e., light intensity, amplitude of the temperature cycle), and the *phase relationship* between the light and temperature cycle (Bruce, 1960; Saunders, 1977; Aschoff, 1981; Pittendrigh, 1960, 1965b, 1974; Pittendrigh and Daan, 1976). In the laboratory it is difficult to generate light and temperature cycles that precisely mimic those found in the field. In addition, there is a behavioral component, as reptiles in the field can often affect their exposure to light or temperature conditions by behavioral means.

The activity patterns exhibited by lizards in the field undoubtedly reflect both endogenous and exogenous factors. The circadian clock has an important role in initiating onsets of activity, and perhaps its cessation, in many species. In diurnal species, the cyclic activity observed in the field is often preceded by a morning emergence from a burrow or den, followed by a period of basking, after which the lizard becomes active. The basking behavior allows the animal to reach an optimal (generally species-specific) temperature for activity. Once this "mean preferred body temperature" is attained, it is maintained during the day by behavioral means (Huey, 1982). It is important that, at least in some species, the initial emergence from the nocturnal retreat appears to be clock controlled and is temperature compensated. *Phrynosoma coronatum* and *P. cornutum* have been exposed to an infrared light source for 8 hours per day plus an additional 4 hours of light per day (LD 12:12) and held at either 18°C or 27°C at night (Heath, 1962). The lizards then spend the night buried in the sand and commonly emerge in the dark before light onset. Significantly, times of emergence are comparable at 18°C and 27°C; this supports the idea that emergence was timed by a temperature-compensated circadian clock. Morning emergence at times of low temperatures occurs in a number of species, suggesting that clock-timed emergence may be widespread (Stebbins and Barwick, 1968; Heath, 1965; Huey, 1974; Case, 1976). This mechanism may have an adaptive value in allowing lizards to initiate activity earlier and thereby maximize foraging time. If emergence is delayed until burrow temperatures have risen to levels appropriate for activity, valuable foraging time may be lost.

Activity offsets may be clock timed in some species. Lizards may retire to burrows, although light and temperature conditions are still permissive for activity (Stebbins and Barwick, 1968; Spellerberg, 1974; Stebbins, 1963). Lizards of the genus *Phrynosoma* may respond to cool temperatures by burrowing in the evening, but not at midday (Heath, 1965); this suggests a diurnal change in their behavioral response to cool temperatures. This behavior may allow lizards to seek shelter before locomotor activity is impaired by rapidly falling temperatures, leaving them vulnerable to predation.

Some diurnal species emerge and retire at specific environmental temperatures; their times of emergence and retirement vary from day to day depending on thermal or other environmental conditions (Heatwole, 1976; Bradshaw and Main, 1968; King, 1980). In this situation, the role of the circadian clock in initiating activity is not obvious. However, we lack information about the behavior of these lizards in their nocturnal retreats before emergence. It is possible that the clock initiates arousal from sleep and perhaps some initial burrow activity, "priming" the individual for emergence whenever temperature conditions are appropriate (e.g., *Uta stansburiana*, Evans, 1967). This possibility is also supported by a study on *Dipsosaurus dorsalis* in which it was observed that the lizard initiates activity before emergence by moving to shallow areas of its burrow, in which it raises its temperature through conduction from warm substrates near the surface (McGinnis and Dickson, 1967). When the temperature of the body reaches the requisite level, the lizard emerges. Therefore, initiation of activity appears to be temperature independent, whereas emergence is temperature dependent. Another example of the role of the circadian clock in initiating activity in a burrow is seen in the snake *Chionactis occipitalis* (Norris and Kavanau, 1966). Whenever they are allowed to burrow in sand in the laboratory under constant illumination, they show a daily rhythm of activity. However, actual emergence onto the surface of the sand depends on the surface temperature; this nocturnal snake delays emergence until sand temperatures drop to a habitable level.

Emergence and activity onset are usually simultaneous in nocturnal geckos (Evans, 1967; Cooper et al., 1985; Bustard, 1967, 1968; Marcellini, 1971). Nocturnal lizards lack a primary heat source during their active period that severely limits the degree to which they can practice behavioral thermoregulation. Outside of the tropics, geckos are faced with a falling temperature during their period of activity; therefore, they cease nocturnal activity and retreat to shelter whenever temperatures drop below critical levels. The temperature requirements for nocturnal geckos are illustrated by *Diplodactylus vittatus,* a small nocturnal gecko with an extensive distribution in southern Australia (Bustard, 1968). This species becomes active at sunset if the temperature lies within a prescribed range. Temperatures in excess of 25 to 26°C delay initiation of evening activity, and the minimum foraging temperatures for a southern and northern population, respectively, are 13°C and 17°C. If the environmental temperature remains above these minimums, the animals remain active until dawn; however, throughout much of the active season the temperatures curtail activity in late evening and early morning. Many geckos show a peak of activity shortly after dusk and a gradual decline until dawn (Evans, 1966; Bustard, 1967; King, 1958; Park, 1938; Pianka and Pianka, 1976; Pianka and Huey, 1978; Huey, 1979; Huey et al., 1977).

In addition to the onset, and possibly the offset of activity, at least one other characteristic of the activity pattern of lizards may be affected by the circadian system. In diurnal lizards under natural conditions, the activity pattern often shows a bimodal pattern peaking during the morning and afternoon; the lizards will be less active, or inactive, during midday (e.g., *Callisaurus draconoides*, Packard and Packard, 1972; *Phrynosoma platyrhinos*, Pianka and Parker, 1975; *Uta stansburiana*, Irwin, 1965; Tinkle, 1967; *Holbrookia propinqua*, Judd, 1975; *Uma notata*, Mayhew, 1964). Often, however, lizards that are bimodally active in hot weather will be unimodally active at cooler times of year (e.g., Pianka and Parker, 1975; Pianka, 1973; Tinkle, 1967; Huey et al., 1977; Mayhew, 1964; Kay, 1972; Judd, 1975). As suggested for turtles, ceasing activity at midday is assumed to help lizards avoid heat stress during the hottest times of day and may also help them avoid the possibly harmful effects of solar radiation (Porter, 1967).

The possibility that the cessation of activity at midday is more than a merely behavioral response to high temperatures and rather involves an endogenous component is suggested by the persistence of a bimodal activity pattern under laboratory conditions. Such bimodal patterns can be observed under LD cycles (lizards: *Dipsosaurus dorsalis*, Gelderloos, 1976; *Xantusia henshawi*, Lee, 1974; snakes: *Thamnophis radix*, Heckrotte, 1962, 1975), as well as under constant conditions (e.g., *Aporosaura anchietae*, Holm, 1973; *X. henshawi*, Lee, 1974). Interestingly, the pattern can be switched between unimodal and bimodal patterns, respectively, by lowering or raising the ambient temperature (Holm, 1973; Heckrotte, 1962, 1975). The duration of daily activity also appears to be greater in animals displaying bimodal patterns than in those displaying unimodal patterns.

The effects of environmental factors on the bimodal activity pattern of snakes may be explicable in terms of Pittendrigh's two-oscillator model (see Section V.A). Each activity peak may be the overt expression of one of these two oscillators. Because the phase relationship between these two putative oscillators changes as a function as the length of the photoperiod, the decreasing photoperiods occurring in the fall and winter may drive the switch from a bimodal to a unimodal pattern during cooler seasons in the field. As a drop in the ambient temperature also switches activity from a bimodal to a unimodal pattern, and decreases the duration of daily activity, temperature as well as photoperiod may be able to affect the phase relationship between the two oscillators controlling activity. Insofar as the period of circadian oscillators is temperature compensated, it is likely that temperature does not exert an effect on the innate frequencies of the oscillators but rather on some other property, such as the coupling strength between the oscillators.

There have been numerous field observations on various aspects of snake behavior, but few laboratory studies of the circadian properties of snake activity rhythms (Griffiths, 1984; Heckrotte, 1962, 1975). Although the data are sparse, they suggest that the phase relationships assumable by the activity rhythms with respect to the day of some species of snakes are far more labile than those observed in lizards or turtles. In the laboratory, *Thamnophis radix* are diurnal at 15°C and 20°C, but they become mainly nocturnal when held at 31°C (Heckrotte, 1975). Field observations indicate that *Crotalus mitchelli*, *C. atrox*, and *Agkistrodon contortrix* are mainly diurnal in the spring and fall and nocturnal during the summer (Sanders and Jacob, 1981; Moore, 1978; Landreth, 1973). Three species of water snakes (*Nerodia rhombifera*, *N. cyclopion*, *N. fasciata*) also shift to nocturnal activity in the summer (Mushinsky et al., 1980). These changes in activity patterns would allow the snakes to avoid high daytime temperatures and imply that (1) snakes are morphologically and physiologically able to adapt both to a diurnal and a nocturnal environment and (2) snakes can shift from feeding on diurnal prey to prey that are active at night. The ability to switch between diurnal and nocturnal entrainment raises some interesting questions about the circadian systems of these snakes but, unfortunately, little is known about their circadian organization.

3. Summary

Analyses of the formal properties of circadian rhythms in the laboratory have been of great utility in understanding some of the observed properties of activity patterns of reptiles in the field. For example, the phase of circadian rhythms is a function of the length of the photoperiod or thermoperiod. Undoubtedly, the seasonal changes noted in the onset of activity in the field are, at least partly, a function of the effects of seasonal changes in photoperiod and thermoperiod on the circadian system. The temperature compensation of circadian rhythms may explain the temperature-"independent" emergence of reptiles from nocturnal retreats. The two-oscillator model may help explain some aspects of activity patterns in the field; one such aspect is bimodality, which may be viewed as an adaptation of reptiles to their thermal environment. Such observations also suggest that the circadian system may have utility not only in timing activity onsets but also in adapting the organism to other aspects of the environment. However, laboratory studies on reptiles other than Sauria are few, and even the Sauria have been poorly studied in comparison to birds and mammals. Much of the behavior noted in the field reflects the interplay between endogenous (i.e., clock-mediated) and exogenous factors. Only careful laboratory studies complemented by field studies with the same species can lead to an understanding of the roles that endogenous and exogenous factors play in controlling activity patterns in the field.

C. Thermoregulatory Rhythms

1. Circadian Aspects of Thermoregulatory Behavior The following discussion of thermoregulatory behavior in reptiles will focus on those aspects of thermoregulation that may be regulated, at least in part, by the circadian system. Comprehensive reviews deal with other aspects of thermoregulation in reptiles (e.g., Huey, 1982; Avery, 1982; Heatwole, 1976; Bartholomew, 1982; Firth and Turner, 1982; Heatwole, 1976; Cloudsley-Thompson, 1971).

Ectotherms depend on the physical environment to provide the heat necessary for metabolism because their metabolic heat production generally does not significantly affect their body temperature. The primary means of thermoregulation in reptiles is behavioral, although some large reptiles are capable of some physiological modification of body temperatures. Under natural conditions, reptiles are presented with a variety of options, in terms of time and space, that let them achieve and maintain body temperatures compatible with their metabolism and let them keep body temperatures below lethal limits (Sturbaum, 1982).

Few studies have examined the rhythmic aspects of thermoregulation in turtles. However, one aspect of their thermoregulatory behavior that has been shown to exhibit a diurnal change is the *critical thermal maximum* (CTMax). This term is defined as "the thermal point at which locomotory activity becomes disorganized and the animal loses its ability to escape from conditions that will promptly lead to its death" (Cowles and Bogert, 1944). The CTMax is a function of acclimation and acclimatization and also exhibits a distinct daily rhythm. *Chrysemys picta* exposed to LD 16:8 or LD 8:16 and either constant 10°C or 20°C showed greater thermal tolerance in the day than in the night.

If the exposure was to LD 16:8, the peak tolerance occurred around 10 hours after the onset of the light; on LD 8:16 the peak occurred about 6 hours after the onset of the light (Kosh and Hutchison, 1968). Therefore, the phase of the CTMax is a function of the length of the photoperiod, which is a characteristic of a circadian rhythm. A daily rhythm in CTMax has also been recorded in *Chelodina longicollis* acclimated to LD 16:8 and 20°C (Webb and Witten, 1973). However, the circadian nature of the CTMax in turtles has not been definitively established because the persistence of a rhythm in the CTMax has not been examined under constant conditions.

The possibility of daily rhythms in thermal preferences has also been examined in turtles (Jarling et al., 1984; Graham and Hutchison, 1979b; Crawshaw et al., 1980). Whenever fasted *Chrysemys picta, Clemmys guttata,* or *Sternotherus odoratus* are maintained on thermal gradients in the laboratory and exposed to LD 16:8 or LD 8:16, they show no significant variation in body temperatures (Graham and Hutchison, 1979b). Although it has been reported that *Trachemys scripta* does not show a rhythm of thermal preference (Crawshaw et al., 1980), other studies have demonstrated a rhythm of temperature preference in this species (Jarling et al., 1984, 1989). The body temperatures of *T. scripta* held on a thermal gradient correlate significantly with time of day; turtles select warmer temperatures later in the day (Jarling et al., 1984, 1989). Furthermore, a daily (circadian) rhythm of thermal preference continues in *T. scripta* exposed to LL (Jarling et al., 1989).

Laboratory studies on lizards show that some species will voluntarily and actively select warm temperatures during the photophase and cool temperatures during the scotophase of 24-hour LD cycles (Regal, 1967; Grenot and Loirat, 1973; Hutchison and Kosh, 1974; Sievert and Hutchison, 1989a, b; Rismiller, 1987; Rismiller and Heldmaier, 1982, 1988; Cowgell and Underwood, 1979; Engbretson and Hutchison, 1976; Kluger et al., 1973; Spellerberg, 1974; Spellerberg and Smith, 1975; Roth and Ralph, 1976; Hammel et al., 1967; Myhre and Hammel, 1969; Firth et al., 1989b). This response may predispose lizards to seek shelter at the end of the day before ambient temperatures drop below critical levels; it may also play a role in initiating heat-seeking behavior in the morning. The fact that in *Sceloporus occidentalis* the rhythm of behavioral thermoregulation persists in DD shows that the rhythm is circadian rather than being exogenously induced (Cowgell and Underwood, 1979).

An extensive series of studies on freely moving *Lacerta viridis* on thermal gradients has revealed a number of facts about the daily rhythm of thermoregulation (Rismiller, 1987; Rismiller and Heldmaier, 1982, 1985, 1988). Whenever lizards are held under LD cycles that approximate natural photoperiodic conditions, they show entrained rhythms of body temperature selection; higher body temperatures are selected during the photophase than during the scotophase. Body temperatures are maintained within narrow ranges both during the photophase and the scotophase. Significantly, this pattern shows seasonal variations in the amplitude of the rhythm of body temperature, in the duration of the scotophase and photophase body temperatures, and in the phase relationship between the rhythm and the LD cycle (Fig. 5.4) (Rismiller, 1987; Rismiller and Heldmaier, 1982, 1988). Lower body temperatures are selected in the fall, and higher temperatures are selected in the spring, which may facilitate acclimatization to conditions of winter and summer temperature, respectively. Interestingly, laboratory simulations of fall photoperiods lead to an overall decline in body temperatures until rhythmicity disappears; this supports the importance of photoperiod in initiating winter dormancy (brumation). Subsequent exposure to long photoperiods (LD 16:8) reinitiates the rhythm in body temperature selection. These results suggest that photoperiod is an important stimulus for seasonal cueing of temperature selection (Rismiller and Heldmaier, 1988). Metabolic rates of *L. viridis* also show a daily rhythm with higher rates in the photophase than in the scotophase, and a pronounced annual cycle is evident for both day and nighttime metabolic rates (Rismiller, 1987; Rismiller and Heldmaier, 1985). Thermal conductance in *L. viridis* also exhibits both daily and seasonal rhythmicity (Rismiller, 1987; Rismiller and Heldmaier, 1985).

There have been few additional studies on the effects of photoperiod or thermal acclimation on preferred body temperatures in lizards. Monthly measurements of the body temperatures of field-caught *Sceloporus orcutti* show that the body temperatures were lowest from January through April; however, the mean body temperature is not related to the ambient temperature conditions prevalent at the time of capture (Mayhew and Weintraub, 1971). Lizards have been shown to exhibit significant seasonal variations in the body temperatures at which they are active in the field (e.g., McGinnis, 1966; DeWitt, 1967). When collected in May or July, *S. undulatus* acclimated to LD 12:12 (but not to LD 6:18) select higher body temperatures than do field-caught controls (Ballinger et al., 1969). No seasonal shift in self-selected body temperatures can be detected in field-caught *Sauromalus obesus* that are held on thermal gradients and LD 12:12 in the laboratory before testing (Case, 1976).

A circadian rhythm in the critical minimum temperature (CTMin) has been described in *Podarcis sicula* (Spellerberg and Hoffmann, 1972). The CTMin of *P. sicula* is higher at the end of the active phase of a freerunning rhythm of locomotor activity than at other phases of the activity rhythm. *Anolis carolinensis* acclimated to LD 12:12 and

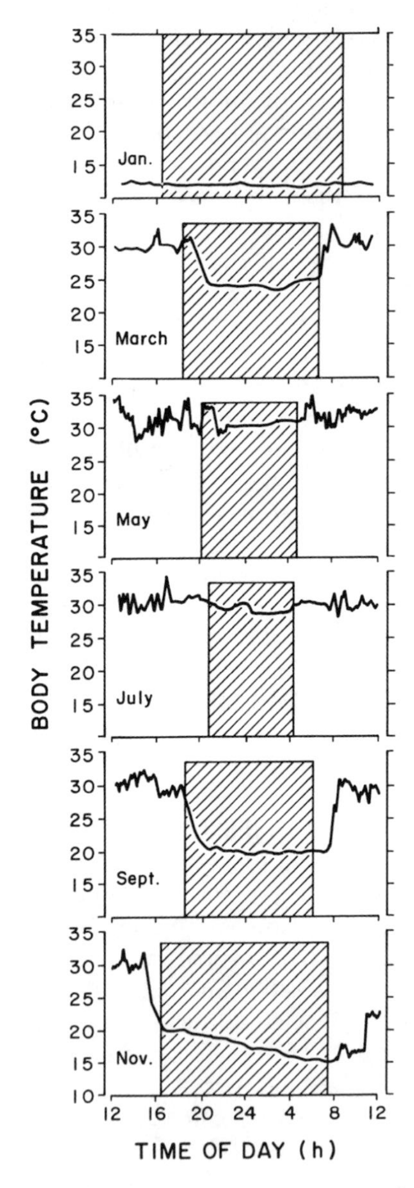

Fig. 5.4. *Lacerta viridis.* Daily patterns of body temperature selection of lizards held on thermal gradients and exposed to natural photoperiod conditions (fluorescent illumination). The hatched area represents the dark phase of the LD cycle. (After Rismiller and Heldmaier, 1988.)

either a constant temperature (25°C) or a temperature cycle (25°C/15°C) shows significant daily cycles in the critical thermal maximum (CTMax); the highest CTMax occurs at the end of the photophase in the constant temperature condition or in the scotophase in the cyclic temperature groups (Kosh and Hutchison, 1972). Parietalectomized anoles exposed to LD 12:12 (cyclic temperatures) also showed a daily rhythm in CTMax, but the parietalectomized anoles had a significantly lower CTMax than had controls at four of the six phases tested (Kosh and Hutchison, 1972). *Sceloporus occidentalis* maintained in a thermoperiod approximating environmental conditions at the time of capture have a higher CTMax after 10 to 25 days in LD 16:8 than lizards exposed to LD 8:16 (Lashbrook and Livezey, 1970). Survival times have been determined in *A. carolinensis* held at 42.5°C after thermal acclimation under various photoperiods (Licht, 1968). Lizards maintained on LD 14:10 tend to be more heat resistant than those on short day lengths (LD 6:18) or DD; however, the only significant difference is between the LD 14:10 and LD 6:18 lizards maintained at 20°C. No differences in survival times are noted between anoles acclimated to a constant 32°C and those acclimated to a daily temperature cycle (32°C day, 20°C night) (Licht, 1968). No significant differences in CTMax are noted in *Phrynosoma cornutum* caught in September and acclimated in LL to 5°C, to 27°C, and to 40°C (Ballinger and Schrank, 1970).

In summary, good evidence indicates that the CTMax of lizards often shows significant daily changes. It is uncertain whether these changes in thermal tolerance represent an adaptation to the thermal environment of the animals or whether they are a "byproduct" of circadian fluctuations in other aspects of their physiology. Significantly, some of the studies described above did not examine the CTMax for diurnal rhythmicity, so the changes, or lack of changes, noted in the CTMax after acclimation to various photoperiods or temperature cycles may be misleading. Because there is strong evidence that CTMax shows daily variations, studies of this aspect of thermal biology must control for rhythmicity. In addition, discrepancies among different studies of thermoregulatory physiology of reptiles could also result from (1) inattention to the seasonal aspects of thermoregulation, including the use of seasonally irrelevant photoperiods or thermoperiods in the laboratory, or (2) the use of gradients with inappropriate temperatures, including the use of photothermal gradients that shut down at night and prevent the voluntary selection of scotophase temperatures.

Enough evidence has accumulated to show that rhythms in the body temperatures of lizards noted both in the laboratory and the field do not merely reflect a passive response to exogenous temperature cycles but reflect an active selection of body temperatures. Furthermore, this process of selection exhibits circadian properties. The concept of *mean preferred body temperature* must be expanded to recognize the fact that there are two "preferred" states (a photophase and a scotophase state) and that prehistory can modify both of them. Clearly, the physiology of reptiles is very different at different times of day and at different times of year.

Few studies have dealt with the circadian properties of thermoregulation in snakes. Daily rhythms in behaviorally selected body temperatures have been noted in *Vipera aspis* and *Nerodia erythrogaster* under LD conditions (Naulleau, 1976, 1979; Gehrmann, 1971). Under continuous light, *V. aspis* may have exhibited a freerunning rhythm in temperature selection, but the amplitude of the rhythm was small (Naulleau, 1976, 1979).

2. ROLE OF THE PARIETAL EYE AND PINEAL SYSTEM

A number of studies have examined the role of the pineal system in thermoregulation in lizards. Most of these studies have focused on the role of the parietal eye, undoubtedly due to its more accessible location. Both field and laboratory studies have supported a role for the pineal system in thermoregulation, but the exact nature of its involvement has not been established.

Probably the beginning of serious interest in the thermoregulatory function of the parietal eye was sparked by a series of field and laboratory studies that observed the behavior of lizards after shielding or ablation of the parietal eye (Stebbins, 1963, 1970; Stebbins and Cohen, 1973; Stebbins and Eakin, 1958; Stebbins and Wilhoft, 1966; Eakin, 1973). For example, shielding or removing the parietal eye in *Sceloporus occidentalis* and *S. undulatus* is followed by an earlier onset of surface activity in the morning and a later offset in the afternoon than controls (Stebbins and Eakin, 1958). Increased surface activity also occurs in several species of iguanid lizards, whenever held on photothermal gradients in the laboratory (Stebbins and Eakin, 1958). Shielding the parietal eye of *Callisaurus draconoides* held in an outdoor enclosure causes increased surface exposure of males, but not females (Packard and Packard, 1972). However, in *Sceloporus virgatus* held in a photothermal gradient and exposed to daily LD cycles, shielding or removing the parietal eye does not modify their frequency of exposure on the surface; however, lower body temperatures are selected before the animals bury themselves in the sand (Stebbins, 1963). Shielding the parietal eye of *Phrynosoma douglassi* significantly elevated the upper voluntary temperatures associated with shuttling behavior (Phillips and Harlow, 1981). This response to shielding may be mediated hormonally because there is a 4- to 6-day lag between shielding and significant elevations in selected temperatures (Phillips and Harlow, 1981). Removing the parietal eye generally accelerates the seasonal activation of the thyroid (Stebbins and Eakin, 1958; Stebbins and Wilhoft, 1966; Stebbins and Cohen, 1973) and possibly the gonads (Stebbins, 1970; Stebbins and Eakin, 1958; Eakin, 1973).

In summary, these studies suggest that the parietal eye may act as a "radiation dosimeter" and that shielding or ablating this structure causes an increased exposure of lizards to photothermal stimuli (i.e., the sun). This increased exposure to photothermal stimuli accelerates metabolism, thereby stimulating endocrine (i.e., thyroidal, gonadal) function. Under field conditions the effects of manipulations of the parietal eye are more difficult to interpret because of the numerous variables to which the animals are exposed.

Recent laboratory investigations into parietal eye function have examined the effects of removing or shielding the parietal eye on the daily rhythm of body temperature (Roth and Ralph, 1976; Engbretson and Hutchison, 1976; Hutchison and Kosh, 1974). Following parietalectomy, *Anolis carolinensis* exposed to a photothermal gradient for 12 hours per day selects significantly higher temperatures during the photophase than controls (Roth and Ralph, 1976). When recorded over 24-hour periods, *A. carolinensis* shows a rhythm in body temperature selection on a thermal gradient; higher body temperatures are selected (on LD 12:12) during the photophase. Parietalectomized anoles also exhibit a daily rhythm, but they select significantly higher temperatures than do controls during both the photophase and scotophase (Hutchison and Kosh, 1974). A similar result occurs in parietalectomized *Sceloporus magister* held on thermal gradients and exposed to 24-hour LD cycles (Engbretson and Hutchison, 1976). Shielding the parietal eye of *Lacerta viridis* appears to have little or no effect on body temperatures during the photophase; however, in the laboratory, shielding reduces the scotophase body temperatures in lizards held on thigmothermal gradients and exposed to "natural" LD cycles (Rismiller, 1987). The nature of the lighting regime used when determining the role of the parietal eye in temperature selection can affect the outcome (Sievert and Hutchison, 1989b). For example, shielding the parietal eye does not affect the pattern of body temperature selection in *Crotaphytus collaris* held on a thigmothermal gradient if the gradient is exposed to uniform illumination; however, shielding the parietal eye elevates preferred body temperature and abolishes the daily cycle of temperature selection if the lizards are exposed to a point source of light over the hot end of the gradient (Sievert and Hutchison, 1989b). The lizards therefore must be selecting for light as well as heat.

The possibility that the role of the parietal eye in thermoregulation does not necessarily involve light perception per se is supported by the observation that parietalectomized *Anolis carolinensis* tend to congregate around a heat source more than sham-parietalectomized controls (Roth and Ralph, 1977); also parietalectomized anoles select warmer temperatures than do controls even during the scotophase of LD cycles (Hutchison and Kosh, 1974; Engbretson and Hutchison, 1976).

In an attempt to assess the role of the parietal eye in reproduction, parietalectomized and sham-parietalectomized *Anolis carolinensis* have been held either on thermal gradients and 24-hour LD cycles or on photic gradients and a constant temperature (Underwood, 1981b, 1985c). Parietalectomized anoles that are allowed to thermoregulate show an accelerated gonadal response to LD cycles, but exposure to a photic gradient (constant temperature) does not change their gonadal response from that of controls. Therefore, parietalectomized lizards will only show an increased gonadal response if they are allowed to thermoregulate.

Interestingly, a latitudinal trend has been noted in the presence of the parietal eye (Gundy et al., 1975). Lizards found in lower latitudes (e.g., Gekkonidae, Teiidae) generally lack a parietal eye, whereas those inhabiting mid and high latitudes (e.g., Agamidae, Iguanidae) usually possess one. Those members of the Scincidae and Lacertidae lacking a parietal eye are confined to within 10° of the equator, whereas species possessing a parietal eye exhibit much larger ranges. This has led to the suggestion that the parietal eye may be required to meet the increased thermoregulatory challenges occurring at higher latitudes.

In summary, there is strong support for the role of the parietal eye in the thermoregulation of lizards; however, the precise nature of this role is controversial. The obvious photosensory capacity of the parietal eye (Quay, 1979; Hamasaki and Eder, 1977) suggests that its role in thermoregulation involves the perception of visible light. However, laboratory studies documenting that parietalectomized lizards respond to thermal stimuli even in the absence of light suggest that the parietal eye may have inputs into thermoregulatory centers that do not depend on photoreception.

Pineal involvement in the thermoregulation of lizards has received much less attention, but the data suggest that pinealectomy lowers behaviorally selected body temperatures. Pinealectomized collared lizards, *Crotaphytus collaris,* held on thermal gradients select lower temperatures than do controls during both the scotophase and photophase of an imposed LD 12:12 cycle (Firth et al., 1980, 1989b). Pinealectomy, or a combination of pinealectomy and parietalectomy, significantly lowers body temperatures of collared lizards during the day when tested in outdoor runways (Firth et al., 1988). Both in the field and in seminatural outdoor environments, pinealectomized *Sceloporus occidentalis* and *C. collaris* also tend to select lower body temperature than do sham-operated ones (Stebbins, 1960; Ralph et al., 1979).

During the first 2 days following melatonin injection, *Crotaphytus collaris* select higher body temperatures during the photophase than do controls; however, continued injections cause the lizards to select lower scotophase temperatures (Cothran and Hutchison, 1979). Chronic melatonin implants in *Lacerta viridis* can increase or decrease both day- and nighttime-selected body temperatures, depending on the phase of the annual cycle (Rismiller, 1987). Melatonin implants also induce *L. viridis* held on LD 14:10 to begin exhibiting self-selected body temperature rhythms similar to those exhibited by lizards held on LD 8:16 (Rismiller and Heldmaier, 1987). Melatonin-injected turtles, *Terrapene carolina*, select lower body temperatures (Erskine and Hutchison, 1981). The melatonin studies conducted to date therefore support the idea that the pineal may be involved in the transduction of lighting information into a melatonin pulse and that this then acts on thermoregulatory centers to generate a body temperature rhythm. However, the data are as yet too sparse to validate this hypothesis.

Any analysis of the role of the pineal system (pineal and parietal eye) in thermoregulation should include the following considerations:

1. Shielding and ablating the parietal eye are often considered to be equivalent operations. However, input of the parietal eye to brain centers is probably neural (Hamasaki and Eder, 1977; Engbretson et al., 1981; Korf and Wagner, 1981); consequently neural "messages" being relayed after shielding may well be different from those following parietalectomy.

2. Shielding the parietal eye also decreases the input of light to the pineal and to extraretinal receptors located in the brain (Underwood, 1973; Underwood and Menaker, 1976).

3. A parietal eye–pineal interaction has been suggested (Engbretson and Lent, 1976; Bethea and Walker, 1978), so parietal eye shielding or ablation may affect pineal function.

4. Pinealectomy will also sever the parietal eye nerve, which complicates interpretation of the effects of pinealectomy.

5. Behavioral temperature selection in lizards is a circadian rhythm; thus, any manipulation having the potential to disrupt circadian organization, such as pinealectomy or SCN lesions, also has the potential to affect thermoregulation.

6. Although no one has demonstrated that the parietal eye serves as a photoreceptor having an input to the circadian system (Under-

wood and Menaker, 1970; Underwood, 1973, 1977), such a role cannot be excluded. If such a role exists, it must be considered in interpreting the effect of parietalectomy on thermoregulation.

7. The role of the parietal eye may differ among species; thus the behaviorally selected body temperature commonly increases after parietalectomy of iguanids but decreases in a lacertid (e.g., Roth and Ralph, 1976; Engbretson and Hutchison, 1976; Rismiller, 1987). However, these differences may reflect experimental procedures; the lacertids were held under natural photoperiodic conditions and the iguanids were not.

VI. OTHER CLOCK-MEDIATED PROCESSES

A. Orientation

Orientation is defined as a selective process in which environmental cues elicit a response sequence that results in a pattern of locomotion or change in the direction of the body axis of an animal (Adler, 1970). One type of orientation, compass orientation, is manifested by an individual showing a preferred direction of travel; after displacement the animal returns to the original direction. Compass orientation requires that an animal possess three sources of information: a directional component, such as a shoreline or compass bearing; a useful celestial cue, such as the sun; and a circadian clock adjusted to local time in order to compensate for the apparent movement of the celestial body due to the rotation of the earth. Such orientational abilities probably arose because the home ranges of many vertebrates were too large to permit sensory contact with all areas, or contact was limited by obstacles that might obscure a particular area or resource (DeRosa and Taylor, 1982). Compass orientation thus becomes an advantageous mechanism that increases the ability to exploit resources or to seek home or shelter.

Recently, compass orientation has been demonstrated in turtles, lizards, snakes, and alligators. The possibility that turtles can utilize sun compass orientation was suggested by studies in which turtles continued to show a directional preference even after displacement from home ranges (Gould, 1957, 1959; Gibbons and Smith, 1968). Under overcast skies, however, homing ability disappears. To compensate for the apparent movement across the horizon (an average of 15°/hr) of the sun (or some other celestial cue), the animal needs a circadian clock entrained to local time. Accordingly, a shift in the phase of the circadian clock should cause a corresponding shift in the direction of orientation. This prediction has been fully confirmed in turtles because exposure to an artificial LD cycle shifts the phase of the circadian clock, thus shifting the direction of orientation (Gourley, 1974; DeRosa and Taylor, 1978, 1982). Gopher tortoises, *Gopherus polyphe-*

mus, exhibit individual directional preferences when displaced and released into either an open field or a test arena that excludes all directional landmarks (Gourley, 1974). Tortoises subjected to a 6-hour phase advance by exposing them to an artificial LD cycle in the laboratory for 7 days show a 90° shift in their preferred headings.

In two studies turtles have been trained to a specific direction of orientation in arenas (DeRosa and Taylor, 1978, 1982). *Chrysemys picta* trained for 10 days on an E-W axis, after being placed in a circular arena filled with water and provided with a gravel shoreline, exhibit a unipolar response; whenever subsequently tested in an outdoor aquatic arena, they moved in the direction of the original shoreline (DeRosa and Taylor, 1978). Significantly, *C. picta* show a 90° shift in orientation in a clockwise direction after entrainment to an LD 12:12 light cycle delayed 6 hours from natural sunrise. A second study utilized a similar training protocol on three species of turtles (*Trionyx spiniferus, C. picta,* and *Terrapene carolina*) (DeRosa and Taylor, 1982). All three species could be trained to orient in a specific direction; this orientational ability is evident only whenever the sun is visible. After entrainment to an LD 12:12 light cycle delayed 6 hours from the time of natural sunrise, the turtles show a 90° clockwise shift in the direction of orientation.

Time-compensated celestial orientation in lizards has been shown by studies on *Lacerta viridis* (Fischer, 1960, 1961; Fischer and Birukow, 1960) and *Uma notata* (Adler and Phillips, 1985). In one study it was possible to train *L. viridis* to seek a certain direction with respect to an artificial celestial cue (a lamp) at several different times of day (Fischer and Birukow, 1960). A second study showed that a 6-hour shift in the phase of the circadian clock of *L. viridis* produces a 90° shift in orientation (Fischer, 1960). *Uma notata* also can be trained to seek a certain direction whenever exposed to celestial cues during the day; their orientation shifts 90° following reentrainment to LD cycles shifted 6 hours with respect to sunrise (Adler and Phillips, 1985). Interestingly, *U. notata* can also be trained to orient to plane-polarized light and exhibit time-compensated orientation with respect to this plane (Adler and Phillips, 1985). As patterns of polarized light vary systematically according to the position of the sun, animals that can perceive these patterns can utilize them as a reference point for compass orientation. Exactly which cue—the position of the sun or the polarized patterns produced by the sun—is used by *U. notata* is unknown, but it is clear that polarized light cues alone are sufficient (Adler and Phillips, 1985). The development of sun compass orientation in *U. notata* probably reflects their desert environment; it has few reliable landmarks, and the landscape is constantly altered by wind action.

The use of a sun compass has also been demonstrated in snakes

(Landreth, 1973; Newcomer et al., 1974) and juvenile alligators (Murphy, 1981). Water snakes (*Nerodia sipedon* and *Regina septemvittata*) can be trained to orient to the sun, and this orientation can be altered by 90° after reentrainment to LD cycles shifted by 6 hours (Newcomer et al., 1974). Juvenile alligators can utilize both solar and stellar compass mechanisms for orientation, but only the solar compass appears to be time compensated (Murphy, 1981).

Although time-compensated celestial orientation can be demonstrated in some turtles, lizards, snakes, and alligators, it is unlikely that all species of these groups have developed this capability. Only animals living in relatively featureless environments, having large home ranges, or migrating long distances might be expected to utilize time-compensated orientation. Furthermore, species that possess time-compensated compass orientation may also use other modalities for orientational purposes, including landmark recognition and olfactory cues. For example, it has been suggested that olfactory cues play a large role in snake orientation (Newcomer et al., 1974).

B. Photoperiodism

Many vertebrates use the annual change in the length of the day to time such important events as fattening, molting, migration, and reproduction. Photoperiod is often used for timing, presumably because it represents a very consistent and noise-free stimulus. The adaptive significance of such "photoperiodic" responses is obvious; animals can anticipate and prepare for adverse conditions and they can confine reproduction to the time of year most conducive to the survival of the organism and its offspring. However, as is not surprising, temperature often also is important in the control of annual gonadal cycles in poikilotherms (Heatwole, 1976).

The circadian system is an important component of the photoperiodic response of vertebrates, including reptiles. An endogenous circadian rhythm of responsiveness to light is the basis for photoperiodic time measurement. As first formulated (Bünning, 1936), this rhythm was hypothesized to execute a complete cycle approximately every 24 hours; during part of the cycle, light is photoperiodically inductive (i.e., the organism is photosensitive), but during the remainder of the cycle, the organism is insensitive to light. It has been emphasized that light must have a dual role in this system. First, the LD cycle acts as an entraining agent for all the circadian rhythms of the organism, including the circadian rhythm of photoperiodic photosensitivity (CRPP). Second, light is photoperiodically inductive only if the CRPP of the organism is entrained in such a way that illumination coincides with its photosensitive phase (Pittendrigh and Minis, 1964; Pittendrigh, 1972). The term *external coincidence* has been coined to describe this model because it demands the coincidence of an external stimulus (light) with the photosensitive phase of an internal circadian rhythm (Pittendrigh, 1972). An *internal coincidence* model has also been postulated. This states that the phase relationships between two (or more) internal circadian oscillators determine the occurrence of induction (Pittendrigh, 1972). In this model, light controls the phase relationships between these oscillators, and the occurrence of inductive or noninductive phase relationships depends upon the length of the photoperiod.

Two classic lighting protocols, termed *T-experiments* and *resonance experiments,* have supported a role for the circadian system in *Anolis carolinensis* (Underwood and Hyde, 1990). A detailed discussion of the theory behind these lighting protocols will not be attempted here (for additional information consult Pittendrigh and Minis, 1964; Elliott, 1981; Saunders, 1977). Briefly, however, resonance experiments utilize a subcritical photoperiod coupled to varying durations of dark yielding LD cycles with periods of 24, 36, 48, and 60 hours (i.e., LD 8:16, LD 8:28, LD 8:40, LD 8:52). A "subcritical" photoperiod is one that is shorter than the photoperiod that will stimulate the reproductive system (i.e., be inductive) whenever presented in an LD cycle with a period length of 24 hours. The critical photoperiod in *A. carolinensis* is about 13.5 hours (LD 13.5:10.5). If a CRPP is involved in photoperiodic time measurement, LD cycles with periods of 24 and 48 hours would be predicted to be noninductive. By contrast, the 36- and 60-hour cycles should stimulate the reproductive system because the 8-hour light pulse will coincide every other cycle with the photosensitive phase of the CRPP. The T-cycle protocol involves entraining animals to LD cycles with periods (T) within their range of entrainment (i.e., T = 18 hours to T = 28 hours) using a subcritical photoperiod. If the circadian system is involved in the measurement of photoperiodic time, some of these T-cycles would be predicted to entrain the circadian system in such a way that light falls during the photosensitive phase of the CRPP.

Importantly, resonance and T-cycle lighting protocols are only effective in *Anolis carolinensis* if the light phase of the LD cycle is 10 to 11 hours long (Underwood and Hyde, 1990). Previous studies using resonance and T-cycle photoperiods 6 to 8 hours in duration did not show circadian involvement in *A. carolinensis* (Underwood, 1978; Underwood and Hall, 1982). Subsequently, however, it has been shown that the circadian system of *A. carolinensis* is "disorganized" when it is entrained to LD cycles with photoperiods of short (6 to 8 hour) duration (Underwood, 1983b). The circadian system in *A. carolinensis* acts as a loosely coupled multioscillator system, and photoperiods of relatively long duration are required to keep the system organized (Underwood, 1983b). The effectiveness of resonance and T-cycle lighting protocols using photoperiods lasting 10 to 11 hours (Underwood and Hyde, 1990) and the ineffectiveness of these protocols whenever photoperiods lasting 6 to 8 hours are used (Underwood, 1978; Underwood and Hall, 1982) therefore may reflect the ability (or inability) of these light cycles to keep the circadian system organized. Only an organized system can accurately measure photoperiodic time.

Night-break lighting protocols have also been used to demonstrate a photosensitive phase of a CRPP in *Anolis carolinensis*. Daily light pulses placed during certain times of the dark of an LD 10:14 light cycle can stimulate testicular growth (Ferrell, 1984; Underwood and Hyde, 1990). A previous night-break study in *A. carolinensis* (Licht, 1971) fails to demonstrate a photoinducible phase, but in this study night breaks were introduced to animals entrained to LD 6:18. As proposed for the negative resonance and T-cycle experiments reported above, the 6 hours of light used in this study may have been insufficient to keep the circadian system organized.

The interaction of light and temperature cycles in the control of reproductive physiology of *Anolis carolinensis* has also been examined (Noeske and Meier, 1977, 1983); the results have been interpreted to support the existence of a circadian rhythm in thermosensitivity. According to this hypothesis, the occurrence of warm temperatures during the thermoinducible phase of an entrained (by light) rhythm of thermosensitivity leads to a physiological event such as fattening. However, these kinds of experiments are difficult to interpret, as temperature cycles not only entrain the circadian system, but temperature also profoundly affects general metabolism, which indirectly affects the expression of a fattening or gonadal response.

Little is known about the physiological substrate of photoperiodic time measurement in reptiles. Studies on photoperiodic mammals have demonstrated that the pineal organ transduces the photoperiodic stimulus into a daily melatonin pulse, the duration (and perhaps phase) of which controls the reproductive response. In lizards, only a few studies have examined a reproductive role for the pineal or melatonin (Levey, 1973; Haldar and Thapliyal, 1977; Thapliyal and Haldar, 1979; Misra and Thapliyal, 1979; Underwood, 1981c, 1985d). These studies support such a role for the pineal, insofar as pinealectomy tends to stimulate the gonadal response in two species of lizard, *Anolis carolinensis* and *Calotes versicolor,* and administration of melatonin tends to inhibit gonadal development.

Photoperiodic photoreception in the lizard *Anolis carolinensis* may be mediated entirely by extraretinal receptors located in the brain. The eyes may not be involved, as (1) the reproductive response of blinded anoles to stimulatory photoperiods is comparable to that observed in sighted anoles (Underwood, 1975), and (2) blocking light penetration to the brains of anoles with the eyes exposed blocks their ability to respond to stimulatory photoperiods (Underwood, 1980). Tail regeneration in the gekkonid lizard, *Hemidactylus flaviviridis*, is also under the control of photoperiod, and blinded lizards can respond to photoperiod as well as sighted controls, suggesting that this photoperiodic response may also be mediated exclusively by extraretinal photoreceptors (Ndukuba and Ramachandran, 1988). Whether the extraretinal photoreceptors mediating photoperiodic responses are identical to those mediating entrainment of the circadian system of lizards is not known.

C. Other Circadian Rhythms

Although many studies of reptiles, particularly laboratory studies, have focused on the easily measured circadian rhythm of activity, other circadian rhythms have received little attention. The studies that do exist, however, serve to underscore the importance of the circadian system in the biology of reptiles.

Day-night differences in skin color due to the dispersion or concentration of the melanin granules within dermal melanophores has long been observed in lizards held under daily LD conditions (e.g., Redfield, 1918). The control of dermal melanophores is species specific and complex; it involves neural control, hormonal control, or both. Also, the number of environmental factors that can influence skin color are numerous and include illumination, background, temperature, social cues, and stress (Binkley et al., 1987). A role for the circadian system is suggested by the fact that the skin color may show day-night differences; two studies in *Anolis carolinensis* report a persistent daily rhythm in color change for anoles held in DD (Rahn and Rosendale, 1941; Binkley et al., 1987).

Daily rhythms have also been observed in gastrointestinal contraction (for *Ctenosaura pectinata* and *Varanus flavescens*, Mackay, 1968); in plasma or adrenal corticosterone levels (for *Dipsosaurus dorsalis*, *Lacerta vivipara*, *Lissemys punctata*, *Anolis carolinensis* and *Alligator mississippiensis*, Chan and Callard, 1972; Dauphin-Villemant and Xavier, 1987; Mahapatra et al., 1987; Summers and Norman, 1988; Lance and Lauren, 1984); in adrenal norepinephrine and epinephrine levels (for *Lissemys punctata*, Mahapatra et al., 1987); in white and red blood cell numbers and blood glucose levels (for *Calotes versicolor*, Haldar and Thapliyal, 1981); and in blood and intracellular pH (for *Dipsosaurus dorsalis*, Bickler, 1986). Diurnal rhythms also occur in breathing frequency or oxygen consumption of turtles (Glass et al., 1979); snakes (Hicks and Riedesel, 1983; Gratz and Hutchison, 1977); and lizards

(Cragg, 1978; Songdahl and Hutchison, 1972; Feder and Feder, 1981; Jameson et al., 1977; Wood et al., 1978; Brownlie and Loveridge, 1983; Mautz, 1979; Rismiller and Heldmaier, 1985). In most of the studies cited above, the rhythms were measured under daily LD cycles or under daily photothermal cycles, but in a few cases, the rhythms were shown to persist under constant conditions (i.e., breathing frequency, Hicks and Riedesel, 1983; oxygen consumption, Mautz and Case, 1974; Cragg, 1978; Jameson et al., 1977).

It is likely that most aspects of the physiology and behavior of reptiles show daily rhythmicity. Most studies of reptilian biology have paid scant attention to the possibility that the parameter being measured may exhibit a significant daily variation. It is important that, at a minimum, measurements be made at the same time of day and in animals held under conditions of well-defined light and temperature in order to compensate for possible circadian variations in the parameter under study. Prior to measurement the animals should be held under daily light and/or temperature cycles long enough to allow steady-state entrainment to those conditions. Alternatively, if animals are being taken directly from the field, the time of the measurement relative to the solar day should be noted, as should the time of year.

One potential methodological problem with the measurement of circadian rhythms has cropped up numerous times. This involves the constant conditions under which free-running animals are held. Animals are often held under continuous illumination; however, this should be avoided, as constant light often dampens circadian rhythms. Hence, the absence of a rhythm in constant light does not support the concept that the rhythm under study is not circadian. Another potential methodological problem involves sampling populations comprised of individuals maintained under constant conditions. Under such constant conditions the freerunning period expressed by any individual animal will likely vary somewhat from those of its neighbors. As the freerunning periods of the individuals begin drifting out of phase, the mean rhythm shown by the population will begin to dampen out. In general, the absence of a population rhythm in animals held under constant conditions for more than a few days should not be interpreted to imply that the rhythm under study is not circadian.

D. Circannual Rhythms

The photoperiodic control of physiological processes represents the use of a proximate factor (photoperiod) to anticipate and prepare for future conditions in the environment of the animal. Somewhat surprisingly, however, some species display seasonal changes in physiology in the absence of variations in external stimuli (i.e., of light or temperature). These endogenous rhythms exhibit a periodicity around a year long and have been termed *circannual* (Gwinner, 1986). Under natural conditions the period of these rhythms matches that of the natural year, indicating that they may be entrained by some aspect of the environment such as light or temperature.

The use of the term *constant* conditions with respect to circannual rhythmicity can be controversial; that is, some consider that an unvarying 24-hour LD cycle (or temperature cycle) fulfills the criterion of constant conditions. After all, these cycles contain no information about the duration of the year. However, other students consider only DD or LL (constant temperature) conditions appropriate for the demonstration of circannual rhythms. If one accepts the premise that an unvarying LD or temperature cycle represents constant conditions, it is important to realize that the LD or temperature cycle may itself affect the circannual rhythm, for example, by affecting the period of the rhythm (Gwinner, 1986). Furthermore, one hypothesis for circannual rhythmicity postulates the occurrence of a circadian rhythm of photosensitivity that can spontaneously change its phase relationship to the (unvarying) LD cycle. Illumination during a photosensitive phase may initiate a circannual event such as gonadal development. This hypothesis therefore incorporates the external coincidence model for photoperiodic time measurement (see Section VI.B). For poikilotherms, such a model could also be expanded to include the existence of a circadian rhythm of sensitivity to temperature and the coincidence of this rhythm with an external temperature cycle.

As discussed previously (Section VI.B) good evidence indicates that the circadian system is important in the measurement of photoperiodic time. Insofar as the process of entrainment of circannual rhythms with the natural year may involve measuring the annual change in photoperiod, the circadian system may somehow also be involved in either generating or entraining circannual rhythms. However, direct experimental evidence of a substantial involvement of the circadian clock in circannual rhythmicity has proven elusive (Gwinner, 1986). However, circannual rhythms can exhibit a number of properties analogous to those shown by circadian rhythms; among these are entrainment, the dependency of the phase of the rhythm on the period of the entraining agent, and a limited range of entrainment.

Some species of vertebrates, including reptiles, show a postbreeding "refractoriness" to photic (or thermal) stimulation (Licht, 1984). The reproductive system collapses under conditions of light or temperature that are normally stimulatory to it. Refractoriness prevents the animal from producing young at a time of year that may not be conducive to their survival. A period of short days or cold temperatures is often required to break refractoriness and allow the animal to regain sensitivity to stimulatory conditions of light or temperature. In animals that hibernate during the winter, dark (and cold) conditions break refractoriness. In some species, the gonads may recrudesce spontaneously, even in the absence of "stimulatory" photic or thermal conditions (e.g., *Podarcis sicula,* Fischer, 1970). The breaking of refractoriness after the passage of a certain number of days under constant conditions, whether in a hibernaculum or in the laboratory, suggests an endogenous ability to measure time. This ability is often assumed to represent a component of a circannual rhythm. Among reptiles, such postbreeding refractoriness to photothermal stimulation has been reported for various temperate zone species (e.g., *Uta stansburiana,* Duvall et al., 1982; *Cnemidophorus uniparens,* Cuellar and Cuellar, 1977a; *Podarcis sicula,* Fischer, 1970; *Lacerta vivipara,* Gavaud, 1983; *Thamnophis sirtalis parietalis,* Whittier et al., 1987; Nelson et al., 1987; Crews et al., 1988). In *T. sirtalis parietalis,* for example, a prolonged exposure to cold temperature, followed by exposure to warm ambient temperatures, is required to initiate courtship and mating (Nelson et al., 1987; Crews et al., 1988). Furthermore, removal of the pineal prior to exposure to cold temperatures blocks the subsequent vernal courtship behavior in *T. sirtalis parietalis* (Nelson et al., 1987). Whether or not the "breaking" of the photo- or thermorefractoriness in these species could occur spontaneously is, of course, an important question. In most of these studies, however, stimulatory photothermal conditions were imposed following exposure to short days or cold temperatures, so the ability to break refractoriness by referring to an internal circannual clock was not tested. The generation of a circannual cycle of reproduction in any of these species would also require that endogenous mechanisms both generate and terminate a refractory period at yearly intervals. Accordingly, demonstration of circannual rhythmicity in reptiles requires holding animals under constant conditions for periods of a year or more. Few such studies have been conducted on reptiles.

A circannual rhythm of reproduction has been observed in *Cnemidophorus uniparens* and a circannual rhythm of activity has been noted in *Sceloporus virgatus* (Stebbins, 1963; Cuellar, 1981; Cuellar and Cuellar, 1977b). In *S. virgatus* held under an LD 9:15 photothermal cycle for 21 months, activity, as measured by the frequency with which each lizard is seen on the surface, shows a circannual rhythm with a period of about 10 months (Stebbins, 1963). The parthenogenetic lizard *C. uniparens* shows a reproductive refractory period from early July to mid-December (Cuellar, 1978). Whenever *C. uniparens* are held under either LD 10:14 or LD 14:10 photothermal cycles for about 20 months they show two, or in several cases three, cycles of ovulation

(Cuellar and Cuellar, 1977b). In another study *C. uniparens* held under an LD 12:12 photothermal cycle for over 3 years show reproductive cycles for the duration of their lives (Cuellar, 1981). Typically, however, the duration of the first reproductive cycle in *C. uniparens,* held under unvarying photothermal cycles, is significantly longer than subsequent cycles (Cuellar, 1981; Cuellar and Cuellar, 1977b). The shorter cycles exhibit periods in the range of 115 to 165 days. Whether or not the shorter-duration cycles could be considered to be within the range of a circannual rhythm is problematical. However, in some homeotherms it has also been observed that an animal may go through one or two typical circannual cycles followed by irregular oscillations with much shorter periods (Gwinner, 1986).

In *Cnemidophorus uniparens* held in DD (constant 35°C) for 7 months starting in the fall, large oocytes, or ovulation, subsequently appear between December and March; this is comparable to what is observed in the reproductive cycles of controls held on an LD 14:10 photothermal regime (Cuellar and Cuellar, 1977a). Both pinealectomized and sham-operated *C. uniparens,* maintained for about a year beginning during the latter part of their reproductive refractory period on an LD 10:14 photothermal schedule, become reproductive starting in mid-November; this suggests that the expression of the circannual rhythm of reproduction in this species does not require the pineal organ (Cuellar, 1978). However, there is no information on the requirement of the pineal for the timing of subsequent cycles of reproduction. That is, an involvement of the pineal in transducing entrainment information from the environment might be demonstrated only after observation of several circannual cycles (e.g., see Phillips and Harlow, 1982).

The timing of laying of first clutches by individual *Cnemidophorus uniparens* depends on the length of the exposure of the lizards to "winterlike" conditions (LD 10:14, 15°C/10°C daily temperature cycle) in the laboratory (Moore et al., 1984). That is, clutches are laid 2 or 3 weeks after return to summerlike conditions in lizards exposed to either 4 weeks, or to 17 weeks, of winterlike conditions. Although this result is not incompatible with entrainment of an endogenous rhythm of reproduction, Moore and his co-workers favor the idea that the winterlike conditions exogenously drive the reproductive response, because reproduction commences at a fixed interval after termination of the winter conditions (Moore et al., 1984). In addition, these workers suggest that alternative explanations to the circannual hypothesis might be entertained to explain the cycling of reproduction in *C. uniparens* exposed to constant conditions. For example, they suggest that the laying of clutches exhausts stored fat reserves; consequently, lizards may pause to replenish stored reserves before initiating the next cycle of reproduction. The cycles of reproduction generated in this fashion, therefore, may not be related to reproductive cycles shown under natural conditions.

Seasonal changes have been documented in a variety of physiological parameters of turtles, including plasma concentrations of sodium, glucose, uric acid, sulfate, magnesium, phosphorus, amino acid nitrogen, lipoid-phosphorus, and testosterone, as well as in heart rate and thyroid activity (Kim et al., 1987; Claire, 1956; Hutton, 1960; Hutton and Goodnight, 1959; Licht et al., 1985). The possible occurrence of a circannual component in these seasonal changes cannot be determined at present; with a single exception (Hutton, 1960), the studies were conducted on animals maintained under seasonally changing conditions of light and temperature.

In summary, few studies have examined circannual rhythmicity of reptiles. Even in birds and mammals, which have been more extensively studied, the mechanisms generating circannual rhythms and the ways these rhythms are coupled to the external environment are poorly understood. However, reptiles may provide excellent models for understanding the evolution of circannual rhythmicity. Many reptiles of the mid and high latitudes overwinter in hibernacula that provide few, if any, light or temperature cues. Yet the animals must be able to measure the passage of time, as they anticipate, and prepare for, emergence and postemergence activities such as breeding. It seems likely that selection in reptiles may have favored the development of endogenous circannual rhythmicity and that homeotherms may have subsequently used this mechanism to generate circannual rhythmicity. Present-day reptiles therefore may offer valuable insights into the evolution of circannual rhythmicity.

VII. CONCLUSION: AREAS FOR FUTURE INVESTIGATION

In general, the behavior and biology of reptiles have been more poorly studied than the behavior and biology of homeotherms. The body of knowledge currently available on the rhythmic aspects of reptilian biology is much less than that available for birds or mammals. However, the reptiles represent a pivotal class in the evolution of the vertebrates. They show the origins of many of the anatomical, behavioral, and physiological characteristics of birds and mammals. A good understanding of the way present-day reptiles have adapted to their temporal environment is crucial to understanding circadian organization in higher vertebrates as well.

A couple of examples should suffice to show the value of reptiles as model systems for probing the circadian organization of vertebrates. First, the circadian "system" of birds and mammals appears to differ substantially. The avian system seems to be organized similarly to that of lizards, with the pineal, the SCN, and the eyes being major components. In mammals, however, the SCN has a role, but the pineal and eyes seem not to play a part. Second, the circadian system appears to be important in the measurement of photoperiodic time in reptiles, birds, and mammals. However, whereas the pineal of mammals (and possibly reptiles) transduces the photoperiodic signal into a hormonal signal in the form of the melatonin rhythm, the pineal and melatonin have a far more equivocal role in birds. It seems likely that a complete understanding of these differences between the avian and mammalian systems will be greatly facilitated by studying the reptiles, the ancestors of which gave rise to the homeotherms.

The body of knowledge currently available on endogenous rhythms in reptiles yields some strong guidelines to areas of future investigation:

1. Clearly, we need more studies on the physiology of the circadian pacemakers themselves, particularly on their locations and the manner in which these multiple clocks communicate with each other. Of particular value would be the inclusion of snakes and turtles among the species studied.

2. It would be valuable to investigate the roles of photoreceptors, both retinal and extraretinal, within the circadian system. The possibility that the eyes themselves are the loci of circadian pacemakers should be tested.

3. Further investigation is needed of the ways in which external stimuli, notably light and temperature, control the phase and period of circadian pacemakers. Laboratory studies complemented with field studies should focus on the physical qualities of light and temperature stimuli that are experienced by reptiles occupying different niches. Such studies should yield insight into the ways that reptiles have adapted their circadian systems to different photic and thermal environments. Also, the possible influence on circadian rhythms of other stimuli, such as social cues or predation, needs to be explored.

4. Field studies should be designed to discriminate between clock-timed behaviors (such as initiation of burrow activity) and behaviors more directly reflecting environmental stimuli (such as actual emergence from nocturnal retreats). Such studies should include the recording of behavioral (i.e., activity, thermoregulatory) rhythms via telemetry or actographs in the field. They should attempt to bridge the gap between laboratory studies and field observations.

5. The circadian system of snakes deserves particular attention for at least two reasons. First, field investigations have shown that some snakes show a remarkable ability to switch from diurnal to nocturnal activity at different seasons; this suggests that they have some interesting entrainment properties. Second, the snake pineal contains

cells that are more similar in appearance to the pineal cells of birds and mammals than to the pineal cells of other groups of reptiles. It would be interesting to determine if this morphological difference is associated with any differences in the role of the pineal within the circadian system.

6. We need additional studies on the role of the circadian system in photoperiodism and of the way in which the photoperiodic signal is transduced. Currently, *Anolis carolinensis* is the only species studied in detail in this regard.

7. The roles of seasonal adjustments in physiology (i.e., preparation for hibernation) that compensate for changing climatic conditions may be as important in reptiles as they are in birds or mammals. It is becoming increasingly apparent that seasonal adaptations in reptiles do not solely reflect a direct response to environmental conditions but involve more complex mechanisms. Circannual rhythms, for example, allow animals to anticipate changes in their physical environment. The nature of the system eliciting these adjustments has been poorly explored. It remains to be determined whether the circadian system is involved in generating these seasonal changes or in entraining them to the external world.

REFERENCES

Abe, H., and Yamamoto, T. Y. (1984). Diurnal changes in synaptic ribbons of rod cells of the turtle. *J. Ultrastr. Res.* 86, 246–251.

Adler, H. E. (1970). Ontogeny and phylogeny of orientation. In *Development and Evolution of Behavior* (L. R. Aronson, E. Tobach, D. S. Lehrman, and J. S. Rosenblatt, eds.). Freeman and Company, San Francisco, pp. 303–336.

Adler, K., and Phillips, J. B. (1985). Orientation in a desert lizard (*Uma notata*): time-compensated compass movement and polarotaxis. *J. Comp. Physiol.* A156, 547–552.

Aschoff, J. (1979). Circadian rhythms: influence of internal and external factors on the period measured in constant conditions. *Z. Tierpsychol.* 49, 225–249.

Aschoff, J. (1981). Freerunning and entrained circadian rhythms. In *Handbook of Behavioral Neurobiology,* Vol. 4, *Biological Rhythms* (J. Aschoff, ed.). Plenum Press, New York, pp. 81–93.

Avery, R. A. (1982). Field studies of body temperatures and thermoregulation. In *Biology of the Reptilia,* Vol. 12 (C. Gans and F. H. Pough, eds.). Academic Press, London and New York, pp. 93–166.

Ballinger, R. E., Hawker, J., and Sexton, O. J. (1969). The effect of photoperiod acclimation on thermoregulation of the lizard, *Sceloporus undulatus. J. Exp. Zoöl.* 171, 43–47.

Ballinger, R. E., and Schrank, G. D. (1970). Acclimation rate and variability of the critical thermal maximum in the lizard *Phrynosoma cornutum. Physiol. Zool.* 43, 19–22.

Barden, A. (1942). Activity of the lizard *Cnemidophorus sexlineatus. Ecology* 23, 336–344.

Bartholomew, G. A. (1982). Physiological control of body temperature. In *Biology of the Reptilia,* Vol. 12 (C. Gans and F. H. Pough, eds.). Academic Press, London and New York, pp. 167–211.

Bernstein, S. A., Breding, D. J., and Fisher, S. K. (1984). The influence of light on cone disk shedding in the lizard, *Sceloporus occidentalis. J. Cell Biol.* 99, 379–389.

Besharse, J. C., and Iuvone, P. M. (1983). Circadian clock in *Xenopus* eye controlling retinal serotonin N-acetyltransferase. *Nature* (London) 305, 133–135.

Bethea, C. L., and Walker, R. F. (1978). Parietal eye-pineal gland interactions in the lizard *Sceloporus occidentalis* (Reptilia, Lacertilia, Iguanidae). *J. Herpetol.* 12, 83–87.

Bickler, P. E. (1986). Day-night variations in blood and intracellular pH in a lizard, *Dipsosaurus dorsalis. J. Comp. Physiol.* B156, 853–857.

Binkley, S., Reilly, K. B., Hermida, V., and Mosher, K. (1987). Circadian rhythm of color change in *Anolis carolinensis:* reconsideration of regulation, especially the role of melatonin in dark-time pallor. *Pineal Res. Rev.* 5, 133–151.

Bradshaw, S. D., and Main, A. R. (1968). Behavioural attitudes and regulation of temperature in *Amphibolurus* lizards. *J. Zool.* (London) 154, 193–221.

Brainard, G. C., Richardson, B. A., King, T. S., Matthews, S. A., and Reiter, R. J. (1983). The suppression of pineal melatonin content and N-acetyltransferase activity by different light irradiances in the Syrian hamster: a dose-response relationship. *Endocrinology* 113, 293–296.

Brett, W. J. (1971). Persistent rhythms of locomotor activity in the turtle *Pseudemys scripta. Comp. Biochem. Physiol.* 40A, 925–934.

Brownlie, S., and Loveridge, J. P. (1983). The oxygen consumption of limbed and limbless African skinks (Sauria: Scincidae): circadian rhythms and effect of temperature. *Comp. Biochem. Physiol.* 74A, 643–647.

Bruce, V. G. (1960). Environmental entrainment of circadian rhythms. *Cold Spring Harbor Symp. Quant. Biol.* 25, 29–48.

Bünning, E. (1936). Die endogene Tagesrhythmik als Grundlage der photoperiodischen Reaktion. *Ber. Dtsch. Bot. Ges.* 54, 590–607.

Bustard, H. R. (1967). Activity cycle and thermoregulation in the Australian gecko *Gehyra variegata. Copeia* 1967, 753–758.

Bustard, H. R. (1968). Temperature dependent activity in the Australian gecko *Diplodactylus vittatus. Copeia* 1968, 606–612.

Butler, A. B. (1974). Retinal projections in the night lizard, *Xantusia vigilis* Baird. *Brain Res.* 80, 116–121.

Carr, A. (1952). *Handbook of Turtles.* Cornell University Press, Ithaca, N.Y.
Case, T. J. (1976). Seasonal aspects of thermoregulatory behavior in the chuckwalla, *Sauromalus obesus* (Reptilia, Lacertilia, Iguanidae). *J. Herpetol.* 10, 85–95.
Chan, S. W. C., and Callard, I. P. (1972). Circadian rhythm in the secretion of corticosterone by the desert iguana, *Dipsosaurus dorsalis. Gen. Comp. Endocrinol.* 18, 565–568.
Claire, M. (1956). The uptake of I^{131} by the thyroid gland of turtles after treatment with thiourea. *Biol. Bull.* 111, 190–203.
Cloudsley-Thompson, J. L. (1965). Rhythmic activity, temperature-tolerance, water-relations and mechanism of heat death in a tropical skink and gecko. *J. Zool.* (London) 146, 55–69.
Cloudsley-Thompson, J. L. (1970). On the biology of the desert tortoise *Testudo sulcata* in Sudan *J. Zool.* (London) 160, 17–33.
Cloudsley-Thompson, J. L. (1971). *The Temperature and Water Relations of Reptiles.* Merrow Publishing Co., Watford Herts, England.
Collin, J. P. (1979). Recent advances in pineal cytochemistry: evidence of the production of indoleamines and proteinaceous substances by rudimentary photoreceptor cells and pinealocytes of amniota. In *The Pineal Gland of Vertebrates Including Man* (J. A. Kappers and P. Pévet, eds.). Elsevier/North Holland Biomedical Press, Amsterdam, pp. 271–296.
Collin, J. P., and Oksche, A. (1981). Structural and functional relationships in the nonmammalian pineal gland. In *The Pineal Gland: Anatomy and Biochemistry* (R. J. Reiter, ed.). CRC Press, Boca Raton, Fl., pp. 27–67.
Collin, J. P., Falcón, J., Voisin, P., and Brisson, P. (1986). The pineal organ: ontogenetic differentiation of photoreceptor cells and pinealocytes. In *The Pineal Gland During Development from Fetus to Adult* (D. Gupta and R. J. Reiter, eds.). Croom Helm, London, pp. 14–30.
Constantinou, C., and Cloudsley-Thompson, J. L. (1985). The circadian rhythm of locomotory activity in the desert lizard *Acanthodactylus schmidti. J. Interdiscipl. Cycle Res.* 16, 107–111.
Cooper, W. E., Jr., Caffrey, C., and Vitt, L. J. (1985). Diel activity patterns in the banded gecko, *Coleonyx variegatus. J. Herpetol.* 19, 308–311.
Cothran, M. L., and Hutchison, V. H. (1979). Effect of melatonin on thermal selection by *Crotaphytus collaris* (Squamata: Iguanidae). *Comp. Biochem. Physiol.* A63, 461–466.
Cowgell, J., and Underwood, H. (1979). Behavioral thermoregulation in lizards: a circadian rhythm. *J. Exp. Zool.* 210, 189–194.
Cowles, R. B., and Bogert, C. M. (1944). A preliminary study of the thermal requirements of desert reptiles. *Bull. Am. Mus. Nat. Hist.* 83, 265–296.
Cragg, P. A. (1978). Oxygen consumption in the lizard genus *Lacerta* in relation to diel variation, maximum activity and body weight. *J. Exp. Biol.* 77, 33–56.
Crawshaw, L. I., Johnston, M. H., and Lemons, D. E. (1980). Acclimation, temperature selection, and heat exchange in the turtle *Chrysemys scripta. Am. J. Physiol.* 238, R443–R446.
Crews, D., Hingorani, V., and Nelson, R. J. (1988). Role of the pineal gland in the control of annual reproductive behavioral and physiological cycles in

the red-sided garter snake (*Thamnophis sirtalis parietalis*). *J. Biol. Rhythms* 3: 293–302.

Cuellar, H. S. (1978). Continuance of circannual reproductive refractoriness in pinealectomized parthenogenetic whiptails (*Cnemidophorus uniparens*). *J. Exp. Zool.* 206, 207–214.

Cuellar, H. S., and Cuellar, O. (1977a). Refractoriness in female lizard reproduction: a probable circannual clock. *Science* (N.Y.) 197, 495–497.

Cuellar, H. S., and Cuellar, O. (1977b). Evidence for endogenous rhythmicity in the reproductive cycle of the parthenogenetic lizard *Cnemidophorus uniparens*. *Copeia* 1977, 554–557.

Cuellar, O. (1981). Long-term analysis of reproductive periodicity in the lizard *Cnemidophorus uniparens*. *Am. Mid. Naturalist* 105, 93–101.

Dauphin-Villemant, C., and Xavier, F. (1987). Nychthemeral variations of plasma corticosteroids in captive female *Lacerta vivipara* Jacquin: influence of stress and reproductive state. *Gen. Comp. Endocrinol.* 67, 292–302.

DeRosa, C. T., and Taylor, D. H. (1978). Sun-compass orientation in the painted turtle, *Chrysemys picta* (Reptilia, Testudines, Testudinidae). *J. Herpetol.* 12, 25–28.

DeRosa, C. T., and Taylor, D. H. (1982). A comparison of compass orientation mechanisms in three turtles (*Trionyx spinifer, Chrysemys picta* and *Terrapene carolina*). *Copeia* 1982, 394–399.

DeWitt, C. B. (1967). Precision of thermoregulation and its relation to environmental factors in the desert iguana, *Dipsosaurus dorsalis*. *Physiol. Zoöl.* 40, 49–66.

Douglass, J. F., and Layne, J. N. (1978). Activity and thermoregulation of the gopher tortoise (*Gopherus polyphemus*) in southern Florida. *Herpetologica* 34, 359–374.

Duvall, D., Guillette, L. J., Jr., and Jones, R. E. (1982). Environmental control of reptilian reproductive cycles. In *Biology of the Reptilia*, Vol. 13 (C. Gans and F. H. Pough, eds.). Academic Press, London and New York, pp. 201–231.

Eakin, R. M. (1973). *The Third Eye*. University of California Press, Berkeley.

Elliott, J. A. (1981). Circadian rhythms, entrainment and photoperiodism in the Syrian hamster. In *Biological Clocks in Seasonal Reproductive Cycles* (B. K. Follett and D. E. Follett, eds.). Scientechnica, Bristol, England, pp. 202–217.

Engbretson, G. A., and Hutchison, V. H. (1976). Parietalectomy and thermal selection in the lizard *Sceloporus magister*. *J. Exp. Zool.* 198, 29–38.

Engbretson, G. A., and Lent, C. M. (1976). Parietal eye of the lizard: neuronal photoresponses and feedback from the pineal gland. *Proc. Natl. Acad. Sci. USA* 73, 654–657.

Engbretson, G. A., Reiner, A., and Brecha, N. (1981). Habenular asymmetry and the central connections of the parietal eye of the lizard. *J. Comp. Neurol.* 198, 155–165.

Ernst, C. H. (1972). Temperature-activity relationship in the painted turtle, *Chrysemys picta*. *Copeia* 1972, 217–222.

Ernst, C. H. (1982). Environmental temperatures and activities in wild spotted turtles, *Clemmys guttata*. *J. Herpetol.* 16, 112–120.

Erskine, D. J., and Hutchison, V. H. (1981). Melatonin and behavioral thermoregulation in the turtle *Terrapene carolina triungius. Physiol. Behav.* 26, 991–995.

Evans, K. J. (1966). Responses of the locomotor activity rhythms of lizards to simultaneous light and temperature cycles. *Comp. Biochem. Physiol.* 19, 91–103.

Evans, K. J. (1967). Observations on the daily emergence of *Coleonyx variegatus* and *Uta stansburiana. Herpetologica* 23, 217–222.

Feder, M. E., and Feder, J. H. (1981). Diel variation of oxygen consumption in three species of Philippine gekkonid lizards. *Copeia* 1981, 204–209.

Ferrell, B. (1984). Photoperiodism in the lizard, *Anolis carolinensis. J. Exp. Zool.* 232, 19–27.

Firth, B. T., and Kennaway, D. J. (1980). Plasma melatonin levels in the scincid lizard *Trachydosaurus rugosus:* the effects of parietal eye and lateral eye impairment. *J. Exp. Biol.* 85, 311–321.

Firth, B. T., and Kennaway, D. J. (1987). Melatonin content of the pineal, parietal eye and blood plasma of the lizard, *Trachydosaurus rugosus:* effect of constant and fluctuating temperature. *Brain Res.* 404, 313–318.

Firth, B. T., and Kennaway, D. J. (1989). Thermoperiod and photoperiod interact to affect the phase of the plasma melatonin rhythm in the lizard, *Tiliqua rugosa. Neurosci. Lett.* 96, 125–130.

Firth, B. T., and Turner, J. S. (1982). Sensory, neural, and hormonal aspects of thermoregulation. In *Biology of the Reptilia,* Vol. 12 (C. Gans and F. H. Pough, eds.). Academic Press, London and New York, 213–274.

Firth, B. T., Kennaway, D. J., and Rozenbilds, M. A. M. (1979). Plasma melatonin in the scincid lizard, *Trachydosaurus rugosus:* diel rhythm, seasonality, and the effect of constant light and constant darkness. *Gen. Comp. Endocrinol.* 37, 493–500.

Firth, B. T., Mauldin, R. E., and Ralph, C. L. (1988). The role of the pineal complex in behavioral thermoregulation in the collared lizard *Crotaphytus collaris* under seminatural conditions. *Physiol. Zoöl.* 61, 176–185.

Firth, B. T., Ralph, C. L., and Boardman, T. J. (1980). Independent effects of the pineal and a bacterial pyrogen in behavioural thermoregulation in lizards. *Nature* (London) 285, 399–400.

Firth, B. T., Thompson, M. B., Kennaway, D. J., and Belan, I. (1989a). Thermal sensitivity of reptilian melatonin rhythms: "cold" tuatara vs. "warm" skink. *Am. J. Physiol.* 256, R1160–R1163.

Firth, B. T., Turner, J. S., and Ralph, C. L. (1989b). Thermoregulatory behaviour in two species of iguanid lizards (*Crotaphytus collaris* and *Sauromalus obesus*): diel variation and the effect of pinealectomy. *J. Comp. Physiol.* B159, 13–20.

Fischer, K. (1960). Experimentelle Beeinflussung der inneren Uhr bei der Sonnenkompassorientierung und der Laufaktivität von *Lacerta viridis* (Laur.). *Naturwissenschaften* 47, 287–288.

Fischer, K. (1961). Untersuchungen zur Sonnen-Kompassorientierung und Laufaktivität von Smaragdeidechsen (*Lacerta viridis* Laur.). *Z. Tierpsychol.* 18, 450–470.

Fischer, K. (1970). Untersuchungen zur Jahresperiodik der Fortpflanzung bei männlichen Ruineneidechsen (*Lacerta sicula campestris* Betta). III. Spontanes Einsetzen und Ausklingen der Gonadenaktivität; ein Beitrag zur Frage der circannualen Periodik. *Z. vergl. Physiol.* 66, 273–293.

Fischer, K., and Birukow, G. (1960). Dressur von Smaragdeidechsen auf Kompassrichtungen. *Naturwissenschaften* 47, 93–94.

Flannery, J. G., and Fisher, S. K. (1984). Circadian disc shedding in *Xenopus* retina *in vitro*. *Invest. Ophthalmol. Vis. Sci.* 25, 229–232.

Fowlkes, D. H., Karwoski, C. L., and Proenza, L. M. (1984). Endogenous circadian rhythm in electroretinogram of free-moving lizards. *Invest. Ophthalmol. Visual Sci.* 25, 121–124.

Fowlkes, D. H., Karwoski, C. L., and Proenza, L. M. (1987). Circadian modulation of the electroretinogram (a- and b-waves) in the diurnal lizard *Anolis carolinensis*. *J. Interdiscipl. Cycle Res.* 18, 147–168.

Frankenberg, E. (1979). Influence of light and temperature on daily activity patterns of three Israeli forms of *Ptyodactylus* (Reptilia: Gekkonidae). *J. Zool.* (London) 189, 21–30.

García-Díaz, C., Molina-Borja, M., and González-González, J. (1989). Circadian rhythm and ultradian oscillations in the motor activity of the lacertid lizard *Gallotia galloti* in continuous light. *J. Interdiscipl. Cycle Res.* 20, 97–105.

Gavaud, J. (1983). Obligatory hibernation for completion of vitellogenesis in the lizard *Lacerta vivipara* J. *J. Exp. Zool.* 397–405.

Gehrmann, W. H. (1971). Influence of constant illumination on thermal preference in the immature water snake, *Natrix erythrogaster transversa*. *Physiol. Zoöl.* 44, 84–89.

Gelderloos, O. G. (1976). Circadian activity patterns in the desert iguana, *Dipsosaurus dorsalis*. *Zoology* 49, 100–108.

Gibbons, J. W., and Smith, M. H. (1968). Evidence for orientation by turtles. *Herpetologica* 24, 331–333.

Glass, M. L., Hicks, J. W., and Riedesel, M. L. (1979). Respiratory responses to long-term temperature exposure in the box turtle, *Terrapene ornata*. *J. Comp. Physiol.* 131, 353–359.

Gould, E. (1957). Orientation in box turtles, *Terrapene c. carolina*. *Biol. Bull.* 112, 336–348.

Gould, E. (1959). Studies on the orientation of turtles. *Copeia* 1959, 174–176.

Gourley, E. V. (1972). Circadian activity rhythm of the gopher tortoise (*Gopherus polyphemus*). *Anim. Behav.* 20, 13–20.

Gourley, E. V. (1974). Orientation of the gopher tortoise, *Gopherus polyphemus*. *Anim. Behav.* 22, 158–169.

Graham, T. E. (1979). Locomotor activity in the Blanding's turtle, *Emydoidea blandingii* (Reptilia, Testudines, Emydidae): the phasing effect of temperature. *J. Herpetol.* 13, 363–365.

Graham, T. E., and Hutchison, V. H. (1978). Locomotor activity in *Chrysemys picta:* response to asynchronous cycles of temperature and photoperiod. *Copeia* 1978, 364–367.

Graham, T. E., and Hutchison, V. H. (1979a). Turtle diel activity: response to different regimes of temperature and photoperiod. *Comp. Biochem. Physiol.* 63A, 299–305.

Graham, T. E., and Hutchison, V. H. (1979b). Effect of temperature and photoperiod acclimatization on thermal preferences of selected freshwater turtles. *Copeia* 1979, 165–169.

Gratz, R. K., and Hutchison, V. H. (1977). Energetics for activity in the diamondback water snake, *Natrix rhombifera*. *Physiol. Zoöl.* 50, 99–114.

Grenot, C., and Loirat, F. (1973). L'activité et le comportement thermorégulateur du lézard saharien *Uromastix acanthinurus* Bell. *Terre et Vie* 27, 435–455.

Griffiths, R. A. (1984). The influence of light and temperature on diel activity rhythms in the sand boa, *Eryx conicus*. *J. Herpetol.* 18, 374–380.

Gundy, G. C., Ralph, C. L., and Wurst, G. Z. (1975). Parietal eyes in lizards: zoogeographical correlates. *Science* (N.Y.) 190, 671–673.

Gwinner, E. (1986). *Circannual Rhythms*. Springer-Verlag, New York.

Haldar, C., and Thapliyal, J. P. (1977). Effect of pinealectomy on the annual testicular cycle of *Calotes versicolor*. *Gen. Comp. Endocrinol.* 32, 395–399.

Haldar, C., and Thapliyal, J. P. (1981). Chronohaematological changes in the lizard *Calotes versicolor* after pinealectomy. *Ann. d'Endocrinol.* (Paris) 42, 35–41.

Hamasaki, D. I. (1969). Spectral sensitivity of the parietal eye of the green iguana. *Vision Res.* 9, 515–523.

Hamasaki, D. I., and Dodt, E. (1969). Light sensitivity of the lizard's epiphysis cerebri. *Pflügers Arch. ges. Physiol.* 313, 19–29.

Hamasaki, D. I., and Eder, D. J. (1977). Adaptive radiation of the pineal system. In *Handbook of Sensory Physiology* (F. Crescitelli, ed.). Springer-Verlag, New York, 7(5), 497–548.

Hamm, H., Takahashi, J. S., and Menaker, M. (1983). Light-induced decrease of serotonin N-acetyltransferase activity and melatonin in the chicken pineal gland and retina. *Brain Res.* 266, 287–293.

Hammel, H., Caldwell, F. T., and Abrams, R. M. (1967). Regulation of body temperature in the blue-tongued lizard. *Science* (N.Y.) 156, 1260–1262.

Hartwig, H. G., and vanVeen, T. (1979). Spectral characteristics of visible radiation penetrating into the brain and stimulating extraretinal photoreceptors. *J. Comp. Physiol.* 130, 277–282.

Heath, J. E. (1962). Temperature-independent morning emergence in lizards of the genus *Phrynosoma*. *Science* (N.Y.) 138, 891–892.

Heath, J. E. (1965). Temperature regulation and diurnal activity in horned lizards. *Univ. Calif. Publ. Zool.* 64, 97–136.

Heatwole, H. (1976). *Reptile Ecology.* University of Queensland Press, St. Lucia, Queensland, Australia.

Heckrotte, C. (1962). The effect of the environmental factors in the locomotory activity of the plains garter snake (*Thamnophis radix radix*). *Anim. Behav.* 10, 193–207.

Heckrotte, C. (1975). Temperature and light effects on the circadian rhythm and locomotory activity of the plains garter snake (*Thamnophis radix haydendi*). *J. Interdiscipl. Cycle Res.* 6, 279–290.

Hicks, J. W., and Riedesel, M. L. (1983). Diurnal ventilatory patterns in the garter snake, *Thamnophis elegans. J. Comp. Physiol.* 149, 503–510.

Hoffmann, K. (1955). Aktivitätsregistrierungen bei frisch geschlüpften Eidechsen. *Z. Vergl. Physiol.* 37, 253–262.

Hoffmann, K. (1957). Über den Einfluss der Temperatur auf die Tagesperiodik bei einem Poikilothermen. *Naturwissenschaften* 44, 358.

Hoffmann, K. (1960). Versuche zur Analyse der Tagesperiodik. I. Der Einfluss der Lichtintensität. *Z. Vergl. Physiol.* 43, 544–566.

Hoffmann, K. (1963). Zur Beziehung zwischen Phasenlage und Spontanfrequenz bei der endogenen Tagesperiodik. *Z. Naturforsch.* 18b, 154–157.

Hoffmann, K. (1968). Synchronisation der circadianen Aktivitätsperiodik von Eidechsen durch Temperaturcyclen verschiedener Amplitude. *Z. Vergl. Physiol.* 58, 225–228.

Holm, E. (1973). The influence of constant temperatures upon the circadian rhythm of the Namib desert dune lizard *Aporosaura anchietae* Bocage. *Madoqua,* Series II, 2, 33–41.

Huey, R. B. (1974). Winter thermal ecology of the iguanid lizard *Tropidurus peruvianus. Copeia* 1974, 149–154.

Huey, R. B. (1979). Parapatry and niche complementarity of Peruvian desert geckos (*Phyllodactylus*): the ambiguous role of competition. *Oecologia* 38, 249–259.

Huey, R. B. (1982). Temperature, physiology, and the ecology of reptiles. In *Biology of the Reptilia,* Vol. 12 (C. Gans and F. H. Pough, eds.). Academic Press, London and New York, pp. 25–91.

Huey, R. B., Pianka, E. R., and Hoffman, J. A. (1977). Seasonal variation in thermoregulatory behavior and body temperature of diurnal Kalahari lizards. *Ecology* 58, 1066–1075.

Hutchison, V. H., and Kosh, R. J. (1974). Thermoregulatory function of the parietal eye in the lizard *Anolis carolinensis. Oecologia* 16, 173–177.

Hutton, K. E. (1960). Seasonal physiological changes in the red-eared turtle, *Pseudemys scripta elegans. Copeia* 1960, 360–362.

Hutton, K. E., and Goodnight, C. J. (1959). Variations in the blood chemistry of turtles under active and hibernating conditions. *Physiol. Zoöl.* 30, 198–207.

Irwin, L. N. (1965). Diel activity and social interaction of the lizard *Uta stansburiana stejnegeri. Copeia* 1965, 99–101.

Isabekova, S. B., Tlepbergenova, L. N., and Veselkin, N. P. (1987). Central connections of parietal eye of the lizard. *J. Evol. Biochem. Physiol.* 23, 258–263.

Jameson, E. W., Jr., Heusner, A. A., and Arbogast, R. (1977). Oxygen consumption of *Sceloporus occidentalis* from three different elevations. *Comp. Biochem. Physiol.* 56A, 73–79.

Janik, D. S. (1987). "Circadian Organization in the Desert Iguana." Ph.D. dissertation, University of Oregon, Eugene.

Jarling, C., Scarperi, M., and Bleichert, A. (1984). Thermoregulatory behavior of the turtle, *Pseudemys scripta elegans,* in a thermal gradient. *Comp. Biochem. Physiol.* 77A, 675–678.

Jarling, C., Scarperi, M., and Bleichert, A. (1989). Circadian rhythm in the temperature preference of the turtle, *Chrysemys* (=*Pseudemys*) *scripta elegans*, in a thermal gradient. *J. Therm. Biol.* 14, 173–178.

Joss, J. M. P. (1978). A rhythm in hydroxyindole-O-methyltransferase (HIOMT) activity in the scincid lizard, *Lampropholas guichenoti*. *Gen. Comp. Endocrinol.* 36, 521–525.

Judd, F. W. (1975). Activity and thermal ecology of the keeled earless lizard, *Holbrookia propinqua*. *Herpetologica* 31, 137–150.

Kavaliers, M., and Ralph, C. L. (1981). Encephalic photoreceptor involvement in the entrainment and control of circadian activity of young American alligators. *Physiol. Behav.* 26, 413–418.

Kay, F. R. (1972). Activity patterns of *Callisaurus draconoides* at Saratoga Springs, Death Valley, California. *Herpetologica* 28, 65–69.

Kennaway, D. J., Frith, R. G., Phillipou, G., Matthews, C. D., and Seamark, R. F. (1977). A specific radioimmunoassay for melatonin in biological tissues and fluids and its validation by gas chromatography–mass spectrometry. *Endocrinology* 101, 119–127.

Kim, S. H., Cho, K. W., and Koh, G. Y. (1987). Circannual changes in renin concentration, plasma electrolytes, and osmolality in the freshwater turtle. *Gen. Comp. Endocrinol.* 67, 383–389.

King, D. (1980). The thermal biology of free-living sand goannas (*Varanus gouldii*) in southern Australia. *Copeia* 1980, 755–767.

King, W. (1958). Observations on the ecology of a new population of the Mediterranean gecko in Florida. *J. Fla. Acad. Sci.* 21, 317–318.

Kluger, M. J., Tarr, R. S., and Heath, J. E. (1973). Posterior hypothalamic lesions and disturbances in behavioral thermoregulation. *Physiol. Zoöl.* 46, 79–84.

Korf, H.-W., and Wagner, U. (1981). Nervous connections of the parietal eye in adult *Lacerta s. sicula* Rafinesque as demonstrated by anterograde and retrograde transport of horseradish peroxidase. *Cell Tissue Res.* 219, 567–583.

Kosh, R. J., and Hutchison, V. H. (1968). Daily rhythmicity of temperature tolerance in eastern painted turtles, *Chrysemys picta*. *Copeia* 1968, 244–246.

Kosh, R. J., and Hutchison, V. H. (1972). Thermal tolerances of parietalectomized *Anolis carolinensis* acclimatized at different temperatures and photoperiods. *Herpetologica* 28, 183–191.

Lance, V. A., and Lauren, D. (1984). Circadian variation in plasma corticosterone in the American alligator, *Alligator mississippiensis*, and the effects of ACTH injections. *Gen. Comp. Endocrinol.* 54, 1–7.

Landreth, H. F. (1973). Orientation and behavior of the rattlesnake, *Crotalus atrox*. *Copeia* 1973, 26–31.

Lashbrook, M. K., and Livezey, R. L. (1970). Effects of photoperiod on heat tolerance in *Sceloporus occidentalis*. *Physiol. Zoöl.* 43, 38–46.

Lee, J. C. (1974). The diel activity cycle of the lizard, *Xantusia henshawi*. *Copeia* 1974, 934–940.

Levey, I. L. (1973). Effects of pinealectomy and melatonin injections at different seasons on ovarian activity in the lizard *Anolis carolinensis*. *J. Exp. Zool.* 185, 169–174.

Licht, P. (1968). Responses of the thermal preferendum and heat resistance to thermal acclimation under different photoperiods in the lizard *Anolis carolinensis*. *Am. Midl. Naturalist* 79, 149–158.

Licht, P. (1971). Response of the male reproductive system to interrupted-night photoperiods in the lizard *Anolis carolinensis*. *Z. Vergl. Physiol.* 73, 274–284.

Licht, P. (1984). Reptiles. In *Marshall's Physiology of Reproduction*, Vol. 1 (G. E. Lamming, ed.). Churchill Livingstone, New York, pp. 206–282.

Licht, P., Wood, J. F., and Wood, F. E. (1985). Annual and diurnal cycles in plasma testosterone and thyroxine in the male green sea turtle, *Chelonia mydas*. *Gen. Comp. Endocrinol.* 57, 335–344.

Lowe, C. H., Hinds, D. S., Lardner, P. J., and Justice, K. E. (1967). Natural free-running period in vertebrate animal populations. *Science* (N.Y.) 156, 531–534.

Mackay, R. S. (1968). Observations on peristaltic activity versus temperature and circadian rhythms in undisturbed *Varanus flavescens* and *Ctenosaura pectinata*. *Copeia* 1968, 252–259.

Mahapatra, M. S., Mahata, S. K., and Maiti, B. R. (1986). Circadian rhythms in serotonin, norepinephrine, and epinephrine contents of the pineal-paraphyseal complex of the soft-shelled turtle (*Lissemys punctata punctata*). *Gen. Comp. Endocrinol.* 64, 246–249.

Mahapatra, M. S., Mahata, S. K., and Maiti, B. R. (1987). Influence of age on diurnal rhythms of adrenal norepinephrine, epinephrine, and corticosterone levels in soft-shelled turtles (*Lissemys punctata punctata*). *Gen. Comp. Endocrinol.* 67, 279–281.

Mahapatra, M. S., Mahata, S. K., and Maiti, B. R. (1988). Circadian rhythms and influence of light on serotonin, norepinephrine, and epinephrine contents in the pineal-paraphyseal complex of soft-shelled turtles (*Lissemys punctata punctata*). *Gen. Comp. Endocrinol.* 71, 183–188.

Mangelsdorff, A. D., and Hauty, G. T. (1972). Effects of temperature and illumination on the locomotor activity and internal temperature of the turtle (*Terrapene carolina carolina*). *J. Interdiscipl. Cycle Res.* 3, 167–177.

Marcellini, D. L. (1971). Activity patterns of the gecko *Hemidactylus frenatus*. *Copeia* 1971, 631–635.

Marx, C., and Kayser, C. (1949). Le rythme nycthéméral de l'activité chez le lézard (*Lacerta agilis, Lacerta muralis*). *C.r. Soc. Biol. Paris* 143, 1375–1377.

Mautz, W. J. (1979). The metabolism of reclusive lizards, the Xantusiidae. *Copeia* 1979, 577–584.

Mautz, W. J., and Case, T. J. (1974). A diurnal activity cycle in the granite night lizard, *Xantusia henshawi*. *Copeia* 1974, 243–251.

Mayhew, W. W. (1964). Photoperiodic responses in three species of the lizard genus *Uma*. *Herpetologica* 20, 95–113.

Mayhew, W. W., and Weintraub, J. D. (1971). Possible acclimatization in the lizard *Sceloporus orcutti*. *J. Physiologie* (Paris) 63, 336–340.

McGinnis, S. M. (1966). *Sceloporus occidentalis:* preferred body temperature of the western fence lizard. *Science* (N.Y.) 152, 1090–1091.

McGinnis, S. M., and Dickson, L. L. (1967). Thermoregulation in the desert iguana *Dipsosaurus dorsalis*. *Science* (N.Y.) 156, 1757–1758.

McGinnis, S. M., and Voigt, W. G. (1971). Thermoregulation in the desert tortoise, *Gopherus agassizii. Comp. Biochem. Physiol.* 40A, 119–126.

Menaker, M., and Wisner, S. (1983). Temperature-compensated circadian clock in the pineal of *Anolis. Proc. Natl. Acad. Sci. USA* 80, 6119–6121.

Misra, C., and Thapliyal, J. P. (1979). Time of administration of indolamines and testicular response of Indian garden lizard *Calotes versicolor. Ind. J. Exp. Biol.* 17, 1383–1385.

Molina-Borja, M., González, J. G., Soutullo, T. G., and Diaz, C. G. (1986). 24-H entrainment and ultradian fluctuations in the activity of the lizard *Gallotia galloti galloti* (Sauria-Lacertidae). *J. Interdiscipl. Cycle Res.* 17, 295–305.

Moore, M. C., Whittier, J. M., and Crews, D. (1984). Environmental control of seasonal reproduction in a parthenogenetic lizard *Cnemidophorus uniparens. Physiol. Zoöl.* 57, 544–549.

Moore, R. G. (1978). Seasonal and daily activity patterns and thermoregulation in the southwestern speckled rattlesnake (*Crotalus mitchelli pyrrhus*) and the Colorado desert sidewinder (*Crotalus cerastes laterorepens*). *Copeia* 1978, 439–442.

Murphy, P. A. (1981). Celestial compass orientation in juvenile American alligators (*Alligator mississippiensis*). *Copeia* 1981, 638–645.

Mushinsky, H. R., Hebrard, J. J., and Walley, M. G. (1980). The role of temperature on the behavioral and ecological associations of sympatric water snakes. *Copeia* 1980, 744–754.

Myhre, K., and Hammel, H. T. (1969). Behavioral regulation of internal temperature in the lizard *Tiligua scincoides. Am. J. Physiol.* 217, 1490–1495.

Naulleau, G. (1976). La thermorégulation chez vipère aspic (*Vipera aspis*) étudiée par biotélémétrie dans différentes conditions artificielles expérimentales. *Bull. Soc. Zool. France* 101, 726–728.

Naulleau, G. (1979). Etude biotélémétrique de la thermorégulation chez *Vipera aspis* (L.) elevée on conditions artificielles. *J. Herpetol.* 13, 203–208.

Ndukuba, P. I., and Ramachandran, A. V. (1988). Extraretinal photoreception in lacertilian tail regeneration: the lateral eyes are not involved in photoperiodic photoreception in the gekkonid lizard *Hemidactylus flaviviridis. J. Exp. Zool.* 248, 73–80.

Nelson, R. J., Mason, R. T., Krohmer, R. W., and Crews, D. (1987). Pinealectomy blocks normal courtship behavior in red-sided garter snakes. *Physiol. Behav.* 39, 231–233.

Newcomer, R. T., Taylor, D. H., and Guttman, S. I. (1974). Celestial orientation in two species of water snake (*Natrix sipedon* and *Regina septemvittata*). *Herpetologica* 30, 194–200.

Noeske, T. A., and Meier, A. H. (1977). Photoperiodic and thermoperiodic interaction affecting fat stores and reproductive indices in the male green anole, *Anolis carolinensis. J. Exp. Zool.* 202, 97–102.

Noeske, T. A., and Meier, A. H. (1983). Thermoperiodic and photoperiodic influences on daily and seasonal changes in the physiology of the male green anole, *Anolis carolinensis. J. Exp. Zool.* 226, 177–184.

Norris, K. S., and Kavanau, J. L. (1966). The burrowing of the western shovel-nosed snake, *Chionactis occipitalis* Hallowell, and the undersand environment. *Copeia* 1966, 650–664.

Obbard, M. E., and Brooks, R. J. (1981). A radio-telemetry and mark-recapture study of activity in the common snapping turtle *Chelydra serpentina*. *Copeia* 1981, 630–637.

Owens, D. W., Gern, W. A., and Ralph, C. L. (1980). Melatonin in the blood and cerebrospinal fluid of the green sea turtle (*Chelonia mydas*). *Gen. Comp. Endocrinol.* 40, 180–187.

Packard, G. C., and Packard, M. J. (1972). Photic exposure of the lizard *Callisaurus draconides* following shielding of the parietal eye. *Copeia* 1972, 695–701.

Paietta, J. (1982). Photooxidation and the evolution of circadian rhythmicity. *J. Theor. Biol.* 97, 77–82.

Pang, S. F., and Allen, A. E. (1986). Extra-pineal melatonin in the retina: Its regulation and physiological function. In *Pineal Research Reviews,* Vol. 4 (R. J. Reiter, ed.). Allen R. Liss, New York, pp. 55–95.

Park, O. (1938). Studies on nocturnal ecology. Preliminary observations on rain forest animals. *Ecology* 19, 208–223.

Petit, A., and Vivien-Roels, B. (1977). Les rythmes circadien et circannuel du taux de 5-hydroxytryptamine (sérotonine) épiphysaire chez *Lacerta muralis,* Laurenti (Reptiles, Lacertiliens). *Arch. Biol.* (Bruxelles) 88, 217–234.

Phillips, J. A., and Harlow, H. J. (1981). Elevation of upper voluntary temperatures after shielding the parietal eye of horned lizards (*Phrynosoma douglassi*). *Herpetologica* 37, 199–205.

Phillips, J. A., and Harlow, H. J. (1982). Long-term effects of pinealectomy on the annual cycle of golden-mantled ground squirrels, *Spermophilus lateralis*. *J. Comp. Physiol.* 146, 501–505.

Pianka, E. R. (1973). The structure of lizard communities. *Ann. Rev. Ecol. Syst.* 4, 53–74.

Pianka, E. R., and Huey, R. B. (1978). Comparative ecology, resource utilization and niche segregation among gekkonid lizards in the southern Kalahari. *Copeia* 1978, 692–702.

Pianka, E. R., and Parker, W. S. (1975). Ecology of horned lizards: a review with special reference to *Phrynosoma platyrhinos*. *Copeia* 1975, 141–162.

Pianka, E. R., and Pianka, H. D. (1976). Comparative ecology of twelve species of nocturnal lizard (Gekkonidae) in the western Australian desert. *Copeia* 1976, 125–142.

Pittendrigh, C. S. (1960). Circadian rhythms and the circadian organization of living systems. *Cold Spring Harbor Symp. Quant. Biol.* 25, 159–184.

Pittendrigh, C. S. (1965a). Biological clocks: the functions, ancient and modern, of circadian oscillations. In *Science in the Sixties* (D. L. Arm, ed.). University of New Mexico Press, Albuquerque, pp. 96–111.

Pittendrigh, C. S. (1965b). On the mechanism of entrainment of a circadian rhythm by light cycles. In *Circadian Clocks* (J. Aschoff, ed.). Elsevier/North Holland, Amsterdam, pp. 277–297.

Pittendrigh, C. S. (1972). Circadian surfaces and the diversity of possible roles of circadian organization in photoperiodic induction. *Proc. Natl. Acad. Sci. USA* 69, 2734–2737.

Pittendrigh, C. S. (1974). Circadian oscillations in cells and the circadian organization of multicellular systems. In *The Neurosciences: Third Study Pro-*

gram (F. O. Schmitt and F. G. Worden, eds.). MIT Press, Cambridge, Mass., pp. 437–458.

Pittendrigh, C. S., and Bruce, V. G. (1959). Daily rhythms as coupled oscillator systems and their relation to thermoperiodism and photoperiodism. In *Photoperiodism and Related Phenomena in Plants and Animals* (R. B. Withrow, ed.). A.A.A.S., Washington, D.C., pp. 475–505.

Pittendrigh, C. S., and Daan, S. (1976). A functional analysis of circadian pacemakers in nocturnal rodents. V. Pacemaker structure: a clock for all seasons. *J. Comp. Physiol.* 106, 333–355.

Pittendrigh, C. S., and Minis, D. H. (1964). The entrainment of circadian oscillations by light and their role as photoperiodic clocks. *Am. Nat.* 98, 261–294.

Plummer, M. V. (1977). Activity, habitat and population structure in the turtle, *Trionyx muticus*. *Copeia* 1977, 431–440.

Porter, W. P. (1967). Solar radiation through the living body walls of vertebrates with emphasis on desert reptiles. *Ecol. Monogr.* 37, 273–296.

Quay, W. B. (1965). Retinal and pineal hydroxyindole-O-methyltransferase activity in vertebrates. *Life Sci.* 4, 983–991.

Quay, W. B. (1974). *Pineal Chemistry in Cellular and Physiological Mechanisms.* Charles C. Thomas, Publisher, Springfield, Ill.

Quay, W. B. (1979). The parietal eye-pineal complex. In *Biology of the Reptilia,* Vol. 9 (C. Gans, R. G. Northcutt, and P. Ulinski, eds.). Academic Press, London and New York, 245–406.

Quay, W. B., Stebbins, R. C., Kelley, T. D., and Cohen, N. W. (1971). Effects of environmental and physiological factors on pineal acetylserotonin methyltransferase activity in the lizard *Sceloporus occidentalis*. *Physiol. Zoöl.* 44, 241–248.

Rahn, H., and Rosendale, F. (1941). Diurnal rhythm of melanophore hormone secretion in the *Anolis* pituitary. *Proc. Soc. Exp. Biol. Med.* 48, 100–102.

Ralph, C. L. (1980). Melatonin production by extra-pineal tissues. In *Melatonin: Current Status and Perspectives* (N. Birau and W. Schloot, eds.). Pergamon, Oxford, pp. 35–46.

Ralph, C. L., Firth, B. T., and Turner, J. S. (1979). The role of the pineal body in ectotherm thermoregulation. *Am. Zool.* 19, 273–293.

Redfield, A. C. (1918). The physiology of the melanophores of the horned toad *Phrynosoma*. *J. Exp. Zool.* 26, 275–323.

Regal, P. J. (1967). Voluntary hypothermia in reptiles. *Science* (N.Y.) 155, 1551–1553.

Repérant, J., Rio, J. P., Miceli, D., and Lemire, M. (1978). A radioautographic study of retinal projections in type I and type II lizards. *Brain Res.* 142, 401–411.

Rismiller, P. D. (1987). "Thermal Biology of *Lacerta viridis:* Seasonal Aspects." Ph.D. dissertation, Philipps-Universität, Marburg, Germany.

Rismiller, P. D., and Heldmaier, G. (1982). The effect of photoperiod on temperature selection in the European green lizard, *Lacerta viridis*. *Oecologia* 53, 222–226.

Rismiller, P. D., and Heldmaier, G. (1985). Thermal behavior as a function of the time of day: heat exchange rates and oxygen consumption in the lacertid lizard *Lacerta viridis*. *Physiol. Zoöl.* 58, 71–79.

Rismiller, P. D., and Heldmaier, G. (1987). Melatonin and photoperiod affect body temperature selection in the lizard *Lacerta viridis*. *J. Therm. Biol.* 12, 131–134.

Rismiller, P. D., and Heldmaier, G. (1988). How photoperiod influences body temperature selection in *Lacerta viridis*. *Oecologia* 75, 125–131.

Rollag, M. D., Panke, E. S., Trakulrungsi, W. K., Trakulrungsi, C., and Reiter, R. J. (1980). Quantification of daily melatonin synthesis in the hamster pineal gland. *Endocrinology* 106, 231–236.

Rose, F. L., and Judd, F. W. (1975). Activity and home range size of the Texas tortoise, *Gopherus berlandieri*, in south Texas. *Herpetologica* 31, 448–456.

Roth, J. J., and Ralph, C. L. (1976). Body temperature of the lizard (*Anolis carolinensis*): effect of parietalectomy. *J. Exp. Zool.* 198, 17–28.

Roth, J. J., and Ralph, C. L. (1977). Thermal and photic preferences in intact and parietalectomized *Anolis carolinensis*. *Behav. Biol.* 19, 341–348.

Roth, J. J., Gern, W. A., Roth, E. C., Ralph, C. L., and Jacobson, E. (1980). Nonpineal melatonin in the alligator (*Alligator mississippiensis*). *Science* (N.Y.) 210, 548–550.

Rusak, B., and Zucker, I. (1979). Neural regulation of circadian rhythms. *Physiol. Rev.* 59, 449–526.

Sanders, J. S., and Jacob, J. S. (1981). Thermal ecology of the copperhead (*Agkistrodon contortrix*). *Herpetologica* 37, 264–270.

Saunders, D. S. (1977). *An Introduction to Biological Rhythms*. John Wiley and Sons, New York.

Schroeder, D. M. (1981). Retinal afferents and efferents of an infrared sensitive snake, *Crotalus viridis*. *J. Morphol.* 170, 29–42.

Sievert, L. M., and Hutchison, V. H. (1989a). Influences of season, time of day, light and sex on the thermoregulatory behaviour of *Crotaphytus collaris*. *J. Therm. Biol.* 14, 159–165.

Sievert, L. M., and Hutchison, V. H. (1989b). The parietal eye and thermoregulatory behavior of *Crotaphytus collaris* (Squamata: Iguanidae). *Comp. Biochem. Physiol.* 94A, 339–343.

Skene, D. J., Vivien-Roels, B., and Pévet, P. (1989). Pineal 5-methoxytryptophol rhythms in the box turtle: effect of photoperiod and environmental temperature. *Neurosci. Lett.* 98, 69–73.

Songdahl, J. H., and Hutchison, V. H. (1972). The effect of photoperiod, parietalectomy and eye enucleation on oxygen consumption in the blue granite lizard, *Sceloporus cyanogenys*. *Herpetologica* 28, 148–156.

Spellerberg, I. (1974). Influence of photoperiod and light intensity on lizard voluntary temperatures. *Br. J. Herpetol.* 5, 412–420.

Spellerberg, I. F., and Hoffmann, K. (1972). Circadian rhythm in lizard critical minimum temperature. *Naturwissenschaften* 59, 517–518.

Spellerberg, I., and Smith, N. D. (1975). Inter- and intra-individual variation in lizard voluntary temperatures. *Br. J. Herpetol.* 5, 496–504.

Stebbins, R. C. (1960). Effects of pinealectomy in the western fence lizard *Sceloporus occidentalis*. *Copeia* 1960, 276–283.

Stebbins, R. C. (1963). Activity changes in the striped plateau lizard with evidence on influence of the parietal eye. *Copeia* 1963, 681–691.

Stebbins, R. C. (1970). The effect of parietalectomy on testicular activity and exposure to light in the desert night lizard (*Xantusia vigilis*). *Copeia* 1970, 261–270.

Stebbins, R. C., and Barwick, R. E. (1968). Radiotelemetric study of thermoregulation in a lace monitor. *Copeia* 1968, 541–547.

Stebbins, R. C., and Cohen, N. W. (1973). The effect of parietalectomy on the thyroid and gonads in free-living western fence lizards (*Sceloporus occidentalis*). *Copeia* 1973, 662–668.

Stebbins, R. C., and Eakin, R. M. (1958). The role of the "third eye" in reptilian behavior. *Am. Mus. Novitates* (1870), 1–40.

Stebbins, R. C., and Wilhoft, D. C. (1966). Influence of the parietal eye on activity in lizards. In *The Galápagos, Proceedings of the Symposium of the Galápagos International Scientific Project* (R. I. Bowman, ed.). University of California Press, Berkeley, pp. 258–268.

Sturbaum, B. A. (1982). Temperature regulation in turtles. *Comp. Biochem. Physiol.* 72A, 615–620.

Summers, C. H., and Norman, M. F. (1988). Chronic low humidity-stress in the lizard *Anolis carolinensis:* changes in diurnal corticosterone rhythms. *J. Exp. Zool.* 247, 271–278.

Takahashi, J. S., and Menaker, M. (1979). Physiology of avian circadian pacemakers. *Fed. Proc.* 38, 2583–2588.

Takahashi, J. S., and Menaker, M. (1984). Multiple redundant circadian oscillators within the isolated avian pineal gland. *J. Comp. Physiol.* A154, 435–440.

Thapliyal, J. P., and Haldar, C. (1979). Effect of pinealectomy on the photoperiodic gonadal response of the Indian garden lizard, *Calotes versicolor. Gen. Comp. Endocrinol.* 39, 79–86.

Tinkle, D. S. (1967). The Life and Demography of the Side-blotched Lizard, *Uta stansburiana. Misc. Publ. Mus. Zool.*, University of Michigan, 132, 1–182.

Ueck, M. (1979). Innervation of the vertebrate pineal. *Prog. Brain Res.* 52, 45–87.

Underwood, H. (1973). Retinal and extraretinal photoreceptors mediate entrainment of the circadian locomotor rhythm in lizards. *J. Comp. Physiol.* 83, 187–222.

Underwood, H. (1975). Extraretinal light receptors can mediate photoperiodic photoreception in the male lizard *Anolis carolinensis. J. Comp. Physiol.* 99, 71–78.

Underwood, H. (1977). Circadian organization in lizards: The role of the pineal organ. *Science* (N.Y.) 195, 587–589.

Underwood, H. (1978). Photoperiodic time measurement in the male lizard *Anolis carolinensis. J. Comp. Physiol.* 125, 143–150.

Underwood, H. (1979). Melatonin affects circadian rhythmicity in lizards. *J. Comp. Physiol.* 130, 317–323.

Underwood, H. (1980). Photoperiodic photoreception in the male lizard *Anolis carolinensis:* the eyes are not involved. *Comp. Biochem. Physiol.* 67A, 191–194.

Underwood, H. (1981a). Circadian organization in the lizard *Sceloporus occidentalis:* the effects of pinealectomy, blinding, and melatonin. *J. Comp. Physiol.* 141, 537–547.
Underwood, H. (1981b). Parietalectomy affects photoperiodic responsiveness in thermoregulating lizards (*Anolis carolinensis*). *Comp. Biochem. Physiol.* 69A, 575–578.
Underwood, H. (1981c). Effects of pinealectomy and melatonin on the photoperiodic gonadal response of the male lizard, *Anolis carolinensis. J. Exp. Zool.* 217, 417–422.
Underwood, H. (1983a). Circadian pacemakers in lizards: Phase-response curves and effects of pinealectomy. *Am. J. Physiol.* 244, R857–R864.
Underwood, H. (1983b). Circadian organization in the lizard *Anolis carolinensis:* a multioscillator system. *J. Comp. Physiol.* 152, 265–274.
Underwood, H. (1985a). Pineal melatonin rhythms in the lizard *Anolis carolinensis:* effects of light and temperature cycles. *J. Comp. Physiol.* A157, 57–65.
Underwood, H. (1985b). Extraretinal photoreception in the lizard *Sceloporus occidentalis:* phase response curve. *Am. J. Physiol.* 248, R407–R414.
Underwood, H. (1985c). Parietalectomy does not affect testicular growth in photoregulating lizards. *Comp. Biochem. Physiol.* 80A, 411–413.
Underwood, H. (1985d). Annual testicular cycle of the lizard *Anolis carolinensis:* effects of pinealectomy and melatonin. *J. Exp. Zool.* 233, 235–242.
Underwood, H. (1986a). Light at night cannot suppress pineal melatonin levels in the lizard *Anolis carolinensis. Comp. Biochem. Physiol.* 84A, 661–663.
Underwood, H. (1986b). Circadian rhythms in lizards: phase response curve for melatonin. *J. Pineal Res.* 3, 187–196.
Underwood, H. (1987). Circadian organization in lizards: perception, translation, and transduction of photic and thermal information. In *Processing of Environmental Information in Vertebrates* (M. H. Stetson, ed.). Springer-Verlag, New York, pp. 47–70.
Underwood, H. (1990). The pineal and melatonin: regulators of circadian function in lower vertebrates. *Experientia* 46, 120–128.
Underwood, H., and Calaban, M. (1987a). Pineal melatonin rhythms in the lizard *Anolis carolinensis* I. Response to light and temperature cycles. *J. Biol. Rhythms* 2, 179–193.
Underwood, H., and Calaban, M. (1987b). Pineal melatonin rhythms in the lizard *Anolis carolinensis* II. Photoreceptive inputs. *J. Biol. Rhythms* 2, 195–206.
Underwood, H., and Hall, D. (1982). Photoperiodic control of reproduction in the male lizard *Anolis carolinensis. J. Comp. Physiol.* 146, 485–492.
Underwood, H., and Harless, M. (1985). Entrainment of the circadian activity rhythm of a lizard to melatonin injections. *Physiol. Behav.* 35, 267–270.
Underwood, H., and Hyde, L. L. (1989a). The effect of daylength on the pineal melatonin rhythm of the lizard *Anolis carolinensis. Comp. Biochem. Physiol.* 94A, 53–56.
Underwood, H., and Hyde, L. L. (1989b). Daytime light intensity affects night-time pineal melatonin levels in the lizard (*Anolis carolinensis*). *Comp. Biochem. Physiol.* 94A, 467–469.
Underwood, H., and Hyde, L. L. (1990). A circadian clock measures photo-

periodic time in the male lizard *Anolis carolinensis*. *J. Comp. Physiol.* A167, 231–243.

Underwood, H., and Menaker, M. (1970). Extraretinal light perception: entrainment of the biological clock controlling lizard locomotor activity. *Science* (N.Y.) 170, 190–193.

Underwood, H., and Menaker, M. (1976). Extraretinal photoreception in lizards. *Photochem. Photobiol.* 23, 227–243.

Vivien-Roels, B. (1983). The pineal gland and the integration of environmental information: possible role of hydroxy- and methoxyindoles. *Molec. Physiol.* 4, 331–345.

Vivien-Roels, B. (1985). Interaction between photoperiod, temperature, pineal and seasonal reproduction in non-mammalian vertebrates. In *The Pineal Gland: Current State of Pineal Research* (B. Mess, C. Rúzsás, L. Tima, and P. Pévet, eds.). Elsevier, Amsterdam, pp. 187–209.

Vivien-Roels, B., and Arendt, J. (1979). Variations circadiennes et circannuelles de la mélatonine épiphysaire chez *Testudo hermanni* G. (Reptile-Chelonien) dans des conditions naturelles d'éclairement et de température. *Ann. Endocrinol.* (Paris) 40, 93–94.

Vivien-Roels, B., and Arendt, J. (1981). Relative roles of environmental factors, photoperiod and temperature in the control of serotonin and melatonin circadian variations in the pineal organ and plasma of the tortoise, *Testudo hermanni* Gmelin. In *Melatonin: Current Status and Perspectives* (N. Birau and W. Schloot, eds.). Pergamon Press, New York, pp. 401–406.

Vivien-Roels, B., and Arendt, J. (1983). How does the indoleamine production of the pineal gland respond to variations of the environment in a nonmammalian vertebrate, *Testudo hermanni* Gmelin? *Psychoneuroendocrinology* 8, 327–332.

Vivien-Roels, B., Arendt, J., and Bradtke, J. (1979). Circadian and circannual fluctuations of pineal indoleamines (serotonin and melatonin) in *Testudo hermanni* Gmelin (Reptilia, Chelonia). *Gen. Comp. Endocrinol.* 37, 197–210.

Vivien-Roels, B., Petit, A., and Lichté, C. (1971). Dosage spectrofluorimetrique de la sérotonine épiphysaire chez les reptiles: etude des variations du taux de sérotonine en fonction de la période d'éclairement. *C. R. Acad. Sc. Paris* 272, 459–461.

Vivien-Roels, B., Pévet, P., and Claustrat, B. (1988). Pineal and circulating melatonin rhythms in the box turtle, *Terrapene carolina triunguis:* effect of photoperiod, light pulse, and environmental temperature. *Gen. Comp. Endocrinol.* 69, 163–173.

Webb, G. J. W., and Witten, G. J. (1973). Critical thermal maxima of turtles: validity of body temperature. *Comp. Biochem. Physiol.* 45A, 829–832.

Whittier, J. M., Mason, R. T., and Crews, D. (1987). Role of light and temperature in the regulation of reproduction in the red-sided garter snake, *Thamnophis sirtalis parietalis*. *Can. J. Zool.* 65, 2090–2096.

Wood, S. C., Johansen, K., Glass, M. L., and Maloiy, G. M. O. (1978). Aerobic metabolism of the lizard *Varanus exanthematicus:* effects of activity, temperature, and size. *J. Comp. Physiol.* 127, 331–336.

Yokoyama, K., Oksche, A., Darden, T. R., and Farner, D. S. (1978). The sites

of encephalic photoreception in photoperiodic induction of the growth of the testes in the white-crowned sparrow, *Zonotrichia leucophrys gambelii*. *Cell Tiss. Res.* 189, 441–467.
Young, R. W. (1977). The daily rhythm of shedding and degradation of cone outer segment membranes in the lizard retina. *J. Ultrastructure Res.* 61, 172–185.

6

Reptilian Coloration and Behavior

William E. Cooper, Jr., and Neil Greenberg

CONTENTS

I. INTRODUCTION

A. General Significance of Coloration to Behavior

The coloration and color patterns of animals strongly influence numerous aspects of their behavior, especially social behavior, predatory and antipredatory behavior, and maintenance, including thermoregulation. Because the potentially optimal coloration may differ for each of several roles and may vary in time and space, the resultant coloration may in many cases best be viewed as an adaptive compromise between conflicting selective pressures exerted by social, predatory, antipredatory, thermoregulatory, and perhaps other demands. However, such compromises are more often mentioned than documented experimentally.

An example of compromise is seen in the bright display colors of many iguanid, lacertid, and teiid lizards. Many of these colors are located ventrally or ventrolaterally and may be revealed by display postures for social communication, but at other times remain concealed, thus reducing risk of predation (Endler, 1980). Similarly, the bright dewlap coloration of anoles is hidden except when it is revealed during brief behavioral displays. Presumably, predation pressure selects for phasic rather than tonic, continually visible, displays of color. Furthermore, some predators are cryptic to avoid detection by their prey. Alternatively, the efficiency of communication may be enhanced by restricting visibility of chromatic signals except in their specific functional contexts. Being subject to diverse selective pressures, coloration may reflect underlying homeostatic dynamics of multiple physiological systems. Finally, body color, like behavior, is an external manifestation of the internal state of the animal and is arrayed on a continuum of temporal stability from obligate morphological and developmental states to states responding rapidly to short-term environmental change.

B. The Multiple Roles of Color Patterns

Coloration has widespread social significance in vertebrates, being important in aggressive, sexual, and parental behavior. Color differences often support behavioral maintenance of reproductive isolation among closely related species. Intraspecific variation in color patterns may affect mate choice. In species having chromatic sexual dimorphism, the sex having lower parental investment (Trivers, 1972), usually the male, is the more brightly colored. The bright color in many such species is widely viewed as a product of sexual selection, which acts by having the members of the opposite sex preferably mating with brightly colored individuals (e.g., Darwin, 1871; Andersson, 1986).

Color is an element of various maintenance behaviors, including thermoregulation, maintenance of water balance, predation, defense, and possibly others, such as behavior associated with solar shielding or its absence. Interactive effects of color associated with several behaviors or indirect effects of coloration on behaviors also occur. A prime example of interactive effects is shown by cryptic coloration and behavior; the need to avoid detection by predators may markedly influence the use of coloration in social behavior (Endler, 1980).

Coloration that enhances the ability of prey to avoid predation and of predators to avoid detection by prey has been documented extensively (Cott, 1940; Wickler, 1968; Edmunds, 1974; see recent reviews of reptilian material by Arnold, 1988; Greene, 1988; Pough, 1988). In a wide range of vertebrate and invertebrate species, color patterns are rendered cryptic by mechanisms such as disruptive coloration, countershading, and background matching. The effectiveness of these mechanisms is attributable to altered contrast, which is a function of ambient illumination (i.e., spectral irradiation), the optical background, and body coloration (i.e., spectral reflectance) (Hailman, 1979). All of these are affected by the visual acuity of the perceiver. Some reptiles may enhance camouflage or crypsis by microhabitat selection, allowing background matching (Vetter and Brodie, 1977; Lillywhite et al., 1977; Gibbons and Lillywhite, 1981), and by physiological color change (Moermond, 1978). Lizards that are well camouflaged may allow closer approach by potential predators before initiating escape behavior than do more conspicuous ones (Heatwole, 1968). Accordingly, foraging behaviors and predator-avoidance behaviors are often adjusted to minimize detection.

The extent to which reptiles avoid predation by reliance on crypsis rather than by active defense or rapid locomotion strongly affects dorsal body coloration in snakes (Jackson et al., 1976; Sweet, 1985). Species that can flee rapidly but have minimal defensive ability tend to be unicolored, to be uniformly speckled, or to have longitudinal stripes. Such snakes, especially striped ones, tend to be fast and diurnal and to forage actively in the open. Species that have well-developed defensive ability, cannot flee quickly, and forage in moderate light intensity usually have irregularly banded or blotched-spotted dorsal coloration. Similar considerations may apply to lizard coloration. Lizard species that forage actively and rely on flight to avoid predation tend to be unicolored, speckled, or striped, whereas those that are ambush foragers and rely on concealment to avoid predation may have blotched patterns (Jackson et al., 1976).

Other species that are venomous, poisonous, or show some level of protection from predation may advertise their undesirable properties by bright warning coloration (i.e., aposematic coloration). Still other species, themselves desirable as prey, capitalize on the protection afforded noxious species by chromatic mimicry of aposematic patterns (Pough, 1988). Their partial protection leads many aposematically colored species to remain in place upon close approach of potential predators subsequent to being detected. Thus, many species that mimic the aposematic coloration of harmful models also resemble them behaviorally in this respect. Cryptic species commonly fail to move until detected but then flee when closely approached. Some other palatable reptiles employ bright coloration to deter predation attempts (Dial, 1986) or deflect attacks away from vulnerable body parts to the tail, especially if the latter can be autotomized. Some reptilian predators respond to aposematic coloration by avoidance, sometimes learned (e.g., Sexton, 1960, 1964; Gans, 1965; Sexton et al., 1966), but such responses in other taxa to color are beyond the scope of this review.

Body coloration and temperature regulation are intimately related among poikilothermic vertebrates because reflectance strongly affects the warming rates of animals exposed to insolation (e.g., Norris, 1967). Desert lizards are typically lighter at high temperatures, becoming darker only at very low body temperatures (Atsatt, 1939). Between extremes of body temperature, illumination strongly affects coloration, lizards being more lightly colored in the dark. The review of color adaptation in desert reptiles by Norris (1967) concludes that color matching between substrate and body is confined to the visible spectrum. This finding implies that the selective pressure for color matching is established by predation rather than thermoregulation; it does not rule out that coloration has an important role for thermoregulation. Indeed, infrared radiation is more important than visible light for thermal balance. Consequently, coloration, although constrained by defensive requirements, has important effects on thermoregulation.

Reflectance is a key physiological factor affecting body temperature (see review by Bartholomew, 1982). Interaction of changes of body color, body temperature, and subcutaneous blood flow complicate interpretation of observed changes. Control of heat gain by radiation is only one of several functions of body coloration. At relatively low body temperatures, enhanced subdermal circulation, attributable to increase of local temperature, facilitates heat transfer from the surface of the body to the core. At relatively high body temperatures, the skin may blanch, but circulatory heat transfer continues unabated. Thus, behavioral adjustments are more important than color change for thermoregulation at high body temperatures. Color change may be most important whenever behavioral mechanisms such as microhabitat selection and postural adjustment are potentially dangerous.

In species that alternately bask and withdraw to shade, behavior is necessarily associated with coloration. This is especially clear in those species that show temperature-dependent physiological color changes from a dark phase when cool to a light phase when hot (Norris, 1967; Bradshaw and Main, 1968; Porter and Norris, 1969; Heatwole, 1970; Hamilton, 1973; Cogger, 1974).

Body coloration may reflect underlying processes affecting water balance. For example, quinones used for chemical defense darken the exoskeleton of insects and reduce its permeability to water (Needham, 1974). Coloration may be affected by phenotypic matching to the background coloration of the habitat occurring over a period of weeks by pigment deposition in juveniles; the color pattern attained by juveniles may be fixed thereafter (Norris, 1967). Habitat selection may be influenced by the degree of matching to background coloration because natural selection would favor those individuals remaining in habitats spectrally similar to their body coloration. Additionally, the presence or absence of pigmentary ultraviolet shielding may affect the need for avoidance of strong sunlight by behavioral microhabitat selection. Correlative studies indicate that the black peritoneum of many species could be adaptive as ultraviolet shielding (Porter, 1967). However, melanin, although possessing some utility as a radiation shield, is a rather poor sunscreen due to its photosensitizing ability and the ease with which it transfers heat (Burtt, 1981). Also, a black peritoneum may serve to warm internal organs. Epidermal darkening in response to ultraviolet radiation appears in humans to be a nonspecific hyperplasia in response to trauma (Morison, 1985). No similar responses have been reported in reptiles, but the possibility of their occurrence in addition to the obvious dermal pigmentary responses cannot be excluded. Finally, there are aspects of color that have never been analyzed functionally, such as the color changes of *Candoia* (Hedges et al., 1989).

C. Reptilian Coloration and Behavior

This review attempts a comprehensive survey of relationships of body coloration to behavior in reptiles, focusing on morphological, physiological, and behavioral aspects. It was undertaken in the hope that the information will have some heuristic value. Body coloration is at a nexus of multiple causal mechanisms and adaptive needs, spanning levels of organization from the physiological to populational. It is uniquely suited to provide insights clarifying fundamental issues in the causation of behavior.

All of the relationships described for vertebrates apply to some reptiles. However, studies relating color patterns and behavior have focused largely on the social behavior of lizards because many species of lizards show dramatic color changes and display bright colors during social encounters. Possible influences of color on social behavior are poorly understood in turtles, crocodilians, rhynchocephalians, amphisbaenians, and snakes. Although color may be important to behavior in these groups, chromatic signals are much less frequent than in lizards. This could be in part due to limited utility in the aquatic habitats of most turtles and crocodilians and in the fossorial habitats of amphisbaenians and some snakes. The living representatives of Rhynchocephalia may be nocturnal, as are many snakes. Another factor that may make chromatic signals unimportant in snakes is their heavy reliance on the chemical senses in social behavior (see Chapter 4). Most lizards, on the other hand, are diurnal and terrestrial. Furthermore, chromatic signals are most highly developed in those lizard families in which visual cues are more critical than chemical ones to foraging behavior (Evans, 1961). Physiological color changes for thermoregulatory purposes are important in lizards and have been reported in several snakes (Section III.B). They too may be less important in the aquatic turtles and crocodilians and in the fossorial amphisbaenians, in which the habitat restricts such opportunities. However, some crocodilians appear to be capable of limited physiological lightening and darkening (Kirshner, 1985; Samuel S. Sweet, personal communication, 1990), as is the tuatara (Waring, 1963).

We lack comprehensive surveys of adaptive coloration such as background matching, countershading, and disruptive coloration for any reptilian order (but see Jackson et al., 1976, for snakes). Influences of color on nonsocial reptilian behaviors are very poorly known, especially from a physiological viewpoint. Relationships between body color and several nonsocial behaviors, including thermoregulatory behavior, hydroregulatory behavior, predation, and defense, are discussed in Section VIII.

D. Properties and Perceptions of Color

The physical and perceptual properties of color are crucial to chromatic roles in reptilian behavior. Nevertheless, physical aspects of light, although well defined, are often neglected in reptilian behavioral studies. An even more serious limitation that applies to most behavioral research involving color, including that on reptiles, is our lack of knowledge regarding color perception in nonhuman species (Hailman, 1977). Ethologists have long been aware that the perceptual worlds of different species can be remarkably dissimilar (von Uexküll, 1921). This led Tinbergen (1951) to stress sensory physiology as a starting point in ethology. Failure to consider physical properties of color and to determine differential spectral sensitivities have plagued most studies attempting to demonstrate the presence of color vision in reptiles. Only a few basic principles are mentioned here; the sources cited should be consulted for more complete treatments.

Color is a complex phenomenon that defies adequate definition on the basis of a single property (Gruber, 1979). For example, color could be formally represented as a particular range of wavelengths of light, but the perceived color of light of a constant wavelength changes when its physical intensity is altered. Three specific qualities of light must be taken into account in designing research: hue, brightness (lightness), and saturation. Hue is what is ordinarily meant by color, such as red or blue, and corresponds roughly to the dominant wavelength. Brightness refers to the relative reflectance of the stimulus, dark objects being highly absorbant, bright ones reflecting a high percentage of incident light. Saturation expresses the percentage of the stimulus consisting of light at the dominant wavelength. It is measured perceptually as the degree of departure from a gray (which has zero saturation) of comparable brightness (ASTM, 1968).

Many researchers have adopted the notation of perceptual attributes of color developed by A. H. Munsell. The Munsell system is much less complex than alternatives such as spectrophotometry and the international system (Committee Internationale de l'eclairage). It has thus become the principal system for organizing hue, brightness, and saturation (called hue, value, and chroma in the Munsell system). By comparing a specimen with a series of color chips under controlled lighting and viewing conditions, it is possible to obtain rapid, repeatable assessments having high interobserver reliability. However, even when the three chromatic characteristics of a stimulus are clearly defined, it is important to realize that the perceptual experience to a human experimenter may be affected by other colors in the visual field and by the perceiver's emotional state, state of color adaptation, color memory, and expectations. Worse yet, reptilian perception of the same stimulus may be quite different from that of the human observer.

Physics texts should be consulted for treatment of the physical properties of light, such as irradiation, reflection, and refraction. We note here that when light strikes the surface of a nonreflecting substance, it can be absorbed, transmitted, or reflected. Absorbed light transmits its energy to the absorbing substance; the color of the transmitted or reflected light represents those wavelengths (frequencies) remaining after absorption (Sustare, 1979).

II. CELLULAR AND SUBCELLULAR BASES OF COLORATION AND COLOR CHANGE

Skin coloration is produced by a combination of structural and pigmentary effects. Structural coloration is attributable to the physical

Table 6.1 The types, pigments, organelles, and visual properties of reptilian dermal chromatophores

Type	Pigments	Organelles	Visual properties
Melanophore	Melanin	Melanosomes	Black, brown, yellow, red
Xanthophore	Carotenoids	Carotenoid vesicles	Yellow, orange
	Pteridines	Pterinosomes	
Erythrophore	Carotenoids	Carotenoid vesicles	Red, orange
	Pteridines	Pterinosomes	
Iridophore (leucophore)	Purines	Reflecting platelets	Highly reflective, often iridescent, no inherent color
Mosaic	Potentially all pigments above	Potentially all organelles above	May vary within the cell and with relative amounts of the various organelles

properties of biological structures rather than to the reflective and absorptive properties of specific pigments. Three types of structural coloration are recognized: (1) iridescence due to interference phenomena, (2) iridescence attributable to diffraction by regularly arranged fibrils or lamellae, and (3) blue coloration due to Tyndall scattering, which results from differential scattering of light (Bagnara and Hadley, 1973).

Many vertebrate colors are produced by pigment molecules, notably melanins, guanine, adenine, hypoxanthine, uric acid, pteridines, and carotenoids. Among reptiles, these pigments are contained in specialized integumental cells called chromatophores (chromatocytes), which are embryologically derived from neural crest cells (Bagnara and Hadley, 1973).

Chromatophores are classified largely on the basis of colors of the pigments they contain (Bagnara, 1966; Table 6.1). The major types are melanophores, xanthophores, erythrophores, and iridophores. Melanophores contain melanosomes, the organelles containing the protein pigment melanin, which is usually brown or black, but may in some instances be yellow or red. Melanophores are unique in producing their own pigment. Deposition of a polymer of tyrosine metabolites on a small (0.5 μm) intermediate vesicle of Golgi origin generates a melanin granule (melanosome). This process is best known in mammals but also seems to occur in reptiles (Bagnara and Hadley, 1973). Melanophores are of two types, the relatively large dermal melanophore, which participates in rapid color change, and the spindle-shaped epidermal melanophore, which is responsible for slow deposition of melanin granules in epidermal cells. Dermal melanophores, which are most abundant in the dorsal integument, are very sensitive to pituitary hormones. Epidermal melanophores are most concentrated just superficial to the stratum germinativum. Internal, extraintegumentary melanophores are classified as dermal due to similar structure and hormonal sensitivity.

Xanthophores (yellow) and erythrophores (red) are closely related structurally and are distinguished by color. They include red, orange, or yellow light-filtering pigments, predominantly fat-soluble carotenoids obtained in the diet and pteridines (Obika and Bagnara, 1964). The organelles containing these pigments, the carotenoid vesicles and pterinosomes, are small (0.5 μm diameter) and of uncertain origin (Bagnara and Hadley, 1973). A single chromatophore may contain carotenoid vesicles, pterinosomes, or both (Bagnara and Hadley, 1973). Erythrophores and xanthophores occur primarily in the dermis but occasionally are found also in the epidermis.

Iridophores or leucophores are morphologically variable dermal cells that contain pigments based on purines (mainly guanine). The pigments are arranged as stacks of flat, reflecting platelets (Bagnara, 1966). They are highly reflective because their organelles may be oriented efficiently to reflect light. Iridophores frequently appear to be iridescent, but the reflecting organelles have no inherent color. They may appear yellow in reflected light or produce a metallic luster visible as flecks, stripes, or spots on larger surfaces, especially the venter. In hylid frogs, they may produce a blue color by Tyndall scattering (Bagnara et al., 1969).

In addition to the readily identified chromatophores, some mosaic or polychromatic "hybrid chromatophores" appear to possess organelles characteristic of other chromatophores, for example, melanophores containing pterinosomes and xanthophores containing melanosomes (Bagnara and Hadley, 1973). The significance of these mosaic chromatophores is uncertain. In the iguanid lizard *Phrynosoma modestum,* mosaic chromatophores simultaneously may contain organelles characteristic of melanophores, iridophores, and xanthophores (Sherbrooke, 1988; Sherbrooke and Frost, 1989; Fig. 6.1). The increased frequency of mosaic chromatophores under the influence of exogenous melanophore-stimulating hormone (MSH) indicates plasticity in cell types and suggests that they may be cells in transition between chromatophore types (Sherbrooke and Frost, 1989). Ontogenetically, melanophores appear first, followed by iridophores and finally xanthophores (Sherbrooke and Frost, 1989).

Vertebrate coloration depends on the arrangement of chromatophores, their pigments, pigment-bearing organelles, and on nonpigmentary structural effects. Two functional cellular associations are recognized, the dermal chromatophore unit and the epidermal mela-

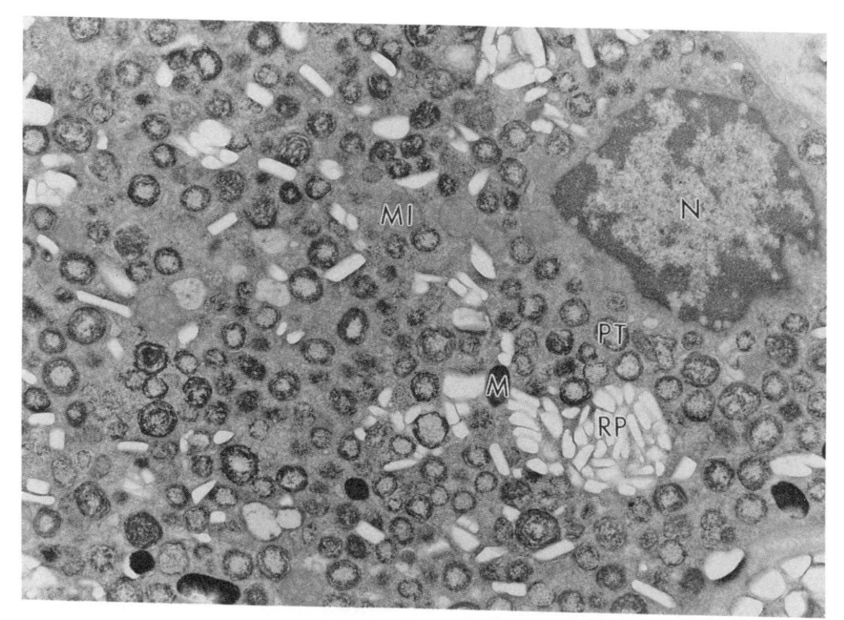

Fig. 6.1. *Phrynosoma modestum*. Electron micrograph of a mosaic chromatophore showing organelles typical of melanophores, iridophores, and xanthophores. M, melanosome; MI, mitochondrion; N, nucleus; PT, pterinosome; RP, reflecting platelet. From Sherbrooke and Frost, 1989.

nin unit. The dermal chromatophore unit in *Anolis carolinensis* (Bagnara et al., 1968; Taylor and Hadley, 1970) is a functional unit that includes an outer layer of light-filtering xanthophores located just below the basal lamina, an intermediate layer of light-scattering iridophores, and, deepest in the dermis, the light-absorbing dermal melanophores (Fig. 6.2). Although this arrangement in anoles has been considered representative for lizards, more species should be studied. In *Phrynosoma modestum* only two cell types are involved: melanophores are embedded in a thick layer of iridophores but have long processes that penetrate to the outer dermal surface. Whereas the regularity of reflecting platelets in the iridophores of *A. carolinensis* results in a perceived green color, the disorganized arrangement of reflecting platelets in *P. modestum* reflects a much wider spectrum of visible light (Sherbrooke and Frost, 1989).

Physiological color changes (i.e., rapid and reversible color changes taking place over seconds to hours) are effected by intracellular movements of the pigment-bearing organelles (Table 6.2) within the cells of the dermal chromatophore units. In amphibians, melanosomes are aggregated near the nuclei of the dermal melanophores in the light color phase (Bagnara et al., 1968). Because the nuclei lie deep to the iridophores, the dermal melanophores do not contribute much to light-phase coloration. Thus, the reflective iridophores remain fully exposed. Furthermore, the stimuli that induce aggregation of melanosomes simultaneously cause dispersion of the reflecting platelets in iridophores. During darkening, melanosomes disperse into the branching processes of the dermal melanophores that extend superficial to the iridophores, thereby covering the iridophores and producing a darker coloration. The xanthophores do not themselves show physiological color changes but contribute passively to the coloration by acting as yellow filters.

The dermal pigment system of the lizard *Anolis carolinensis* has been described in detail by Alexander and Fahrenbach (1969) and by Taylor and Hadley (1970), who extended and clarified von Geldern's (1921) extensive observations. Despite apparent differences from the dermal chromatophore unit (Alexander and Fahrenbach, 1969), Taylor and Hadley (1970) applied the amphibian terminology to anoles, a decision that has been accepted by subsequent workers. The dermal pigment system contains a superficial layer of xanthophores that contain membrane-bound inclusions filled with pteridines or carotenoids; deep to these is a two- to four-cell-thick intermediate layer of iridophores containing layers of guanine rodlets. Deep to the xanthophores and iridophores lies a basal stratum of relatively large melanophores, each of which shows vertically oriented dendritic processes that project between the overlying iridophores and xanthophores to terminate immediately above the xanthophores (Fig. 6.2).

Those melanophores of poikilotherms located on dark backgrounds generally contain dispersed melanosomes, producing a dark colora-

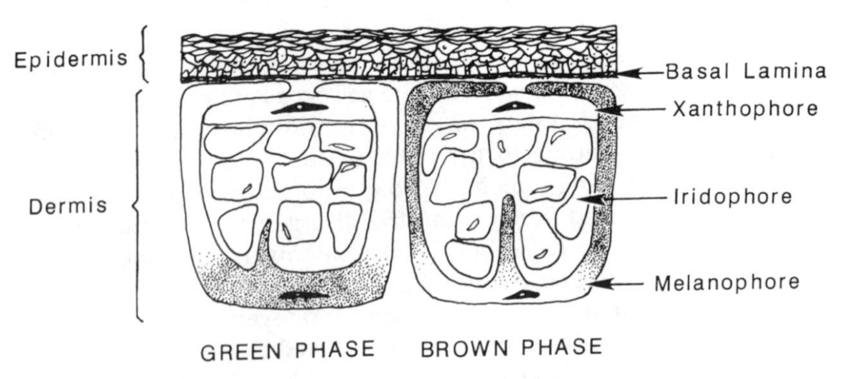

Fig. 6.2. *Anolis carolinensis.* The dermal chromatophore unit. In the green phase the melanosomes are aggregated below the xanthophores; in the brown phase the melanosomes are dispersed and many cover the xanthophores. (Modified and redrawn with permission from Taylor and Hadley, 1970).

Table 6.2 Agents producing physiological color changes and their effects in reptiles

Agent	Effect	Change	Species	Sources
MSH	Melanosome dispersion	Darkening	*Anolis carolinensis*	Kleinholz, 1938a, b; Horowitz, 1958
			Chamaeleo jacksonii	Canella, 1963
			Hemidactylus brookii	Noble and Bradley, 1933b
			Crotalus viridis	Rahn, 1941
ACTH	Melanosome dispersion	Darkening	*Chamaeleo jacksonii*	Canella, 1963
Epinephrine, norepinephrine	Melanosome aggregation (alpha receptors)	Lightening	*Anolis carolinensis*	Kleinholz, 1938b
			Phynosoma sp.	Redfield, 1916, 1918
	Melanosome dispersion (beta receptors)	Darkening	*Anolis carolinensis*	Kleinholz, 1938b, Redfield, 1916, 1918
Thyroxine, triiodothyronine	Blocks darkening by MSH	None	*Anolis carolinensis*	Bagnara and Hadley, 1973
Neural regulation		Wide range	*Chamaeleo sp.*	Brücke, 1852; Hogben and Mirvish, 1928

Abbreviations: MSH, melanophore stimulating hormone; ACTH, adrenocorticotropic hormone

tion; on light backgrounds the melanosomes aggregate in the perinuclear region, producing a lighter color. Physiological color changes are effected by dispersion and aggregation of melanosomes of dermal melanophores, alteration of the orientation of reflecting platelets of iridophores, or both. Color change attributable to migration of melanosomes within the dendritic processes of melanophores is exemplified by *Anolis carolinensis*, which changes from brown to green when melanosomes withdraw from the dendritic processes and aggregate in a perinuclear position, exposing the iridophores (see Fig. 6.2). Light of longer wavelengths, in the red and orange range, passes through the iridophores and is absorbed in the melanophore layer. The iridophores both scatter and reflect the remaining light, producing a blue color by Tyndall scattering. Pigments in the xanthophores remove the blue, leaving predominantly green light reflected from the surface of the lizard. Although the crystalline arrays in the anoline iridophore appear to be stable, those of other taxa (some anurans and fish) show changes in the crystalline structure; this suggests that such an effect could be responsible for color shifts in some reptiles (Rohrlich and Porter, 1972).

In the viperid snake *Bothriechis rowleyi*, as in *Anolis carolinensis*, the dermis includes an outer xanthophore layer, a middle layer of guanophores, and a basal layer of melanophores, but the melanophores lack the extensive processes that extend toward the epidermis in anoles (Gosner, 1989). The green skin color of *B. rowleyi* is produced as follows: incident light is filtered by the xanthophores, which absorb blue wavelengths; the guanophores transmit red and yellow wavelengths to the melanophore layer, where they are absorbed, but scatter and reflect green light to the outside. Thus, the green is produced by Tyndall scattering. If the melanophore layer is absent, as in yellow morphs of *Bothriechis schlegelii*, yellow and red are reflected, producing a yellow appearance (Gosner, 1989).

The bidirectional movements of melanin granules appear to be produced by macromolecular filaments. It is unclear whether microtubules, microfilaments, intermediate filaments, or combinations of them are involved. However, in *Anolis carolinensis* there is evidence that dispersion and aggregation are accomplished by different mechanisms. In vitro melanin does not disperse in response to MSH or photic stimulation upon treatment of the skin with cytochalasin-B, an actin inhibitor; this indicates a role for microfilaments in dispersion of previously aggregated granules. Colchicine treatment, which affects microtubules, has no effect on dispersion. Colchicine does affect aggregation, inhibiting it to different degrees (Vaughan, 1987a). Responses to both MSH and light are mediated by cyclic-AMP, but only the photic response can be blocked by a nonmetabolizable analogue of cyclic-GMP. Blocking of cell membrane calcium channels with lanthanum chloride inhibits chromatophore response to photic stimulation but not to MSH (Vaughan, 1987a). This suggests that signal transduction by chromatophores in response to light might be analogous to that of retinal rods (Vaughan, 1987a), which is consistent with the neural plate origin of both rods and melanophores. Indeed, even cells transplanted from at least the cephalic end of the ventral neural tube can differentiate into melanophores. Pharmacological evidence further suggests that physiologically, the melanophore is a "masked primitive neuron" (Burgers, 1966).

Epidermal melanophores, principal elements in morphological color change, cannot be presumed to be present in all adult reptiles and are sometimes present only at relatively early developmental stages (Luke, 1989). In some reptiles, extracellular epidermal melanin granules appear to originate in the dermis, but in others, adults have distinct epidermal melanophores. Among lizards examined (Luke, 1989), epidermal melanophores occur in adults in one of three anguid species (see also Schmidt, 1914; Maderson, 1965), 4 of 12 gekkonids,

5 of 6 lacertids (see also Maderson, 1965), 1 of 2 varanids, 4 of 4 agamids, and 1 of 1 chamaeleontids (see also Maderson, 1965). They are absent in the cordylids, iguanids (see also von Geldern, 1921; Klausewitz, 1954, Sherbrooke, 1988) and scincids studied. Reports regarding the presence or absence of epidermal melanophores in lizards are sometimes contradictory, as for *Anguis fragilis* (present: Schmidt, 1914; Maderson, 1965; absent: Batelli, 1880; Kerbert, 1876; Lange, 1931) and *Lacerta vivipara* (present: Maderson, 1965; absent: Klausewitz, 1953). Such discrepancies are the result of several factors (reviewed by Schmidt, 1917). Even when present, epidermal melanophores may be dispersed and have low density, making them difficult to observe (Schmidt, 1912). Furthermore, the presence of epidermal melanophores may vary with age, sex, location on the surface of the skin, and geographical location (Schmidt, 1914; Klausewitz, 1964; Luke, 1989). Finally, melanin granules may be deposited in the epidermis by dermal melanophores (Schmidt, 1913; Klausewitz, 1964) in the absence of epidermal melanophores.

The epidermal melanin unit of *Anolis carolinensis* includes the epidermal melanophore and associated Malpighian cells. The melanophore synthesizes melanin and transfers it to adjacent Malpighian cells. This arrangement can lead to great concentrations of melanin in the epidermis, causing the skin to be virtually black. The epidermal melanin unit participates in the gradual, long-lasting morphological color change, which often involves darkening or lightening due to changes in the amount of melanin in the epidermis (Bagnara and Hadley, 1973). Only dermal, not epidermal, melanophores are involved in rapid color change.

III. COLOR CHANGES AND THEIR REGULATION

A. General

Integumentary color changes are best known in lizards and snakes (e.g., Bagnara and Hadley, 1973; Rahn, 1941) but also have been reported in Testudines (e.g., Moll et al., 1981; Banks, 1986), Crocodilia (Kirshner, 1985), and Rynchocephalia (Waring, 1963). They are widely categorized as being physiological or morphological. Rapid shifts in color occurring in seconds, minutes, or hours constitute physiological color change. Among these are the quick color changes of chameleons and some agamid and iguanid lizards during aggressive behavior. The nature and course of physiological color change has been relatively well studied in some lizards, as have its physiological control and behavioral relationships. Physiological color changes usually, but not invariably, precede morphological ones. In addition to physiological and morphological changes in the chromatic properties of bodily surfaces, many reptiles are capable of behavioral color changes. These changes are brought about rapidly by postural adjustments and movements of scales to reveal surfaces differing in color from those currently in view. They do not require changes in the amount or dispersion of pigments. Mechanisms of physiological and morphological color changes are discussed in Sections III.B and III.C; behavioral color changes are considered in Sections VI.B, VI.C, VII, and VIII.

Although it is widely expressed in lizards, physiological color change otherwise has been described in only the single crocodilian, *Crocodylus porosus* (Kirshner, 1985) and a few snakes (*Candoia carinata*, Hedges et al., 1989; *Casarea dussumieri*, McAlpine, 1983; *Crotalus cerastes*, Klauber, 1930; *C. viridis*, Klauber, 1930, 1956; Rahn, 1941; Sweet, 1985; *Tropidophis canus*, *T. feicki*, and *T. greenwayi*, Rehák, 1987; *T. haetianus* and *T. melanurus*, Hedges et al., 1989) and suggested in a few others (Hedges et al., 1989). In all of these snakes except *C. cerastes* and *C. viridis*, the color changes follow a 24-hour cycle with the light phase occurring at night during the activity period (Fig. 6.3). Because color change is also associated with feeding and the color remains dark when inactive at night in *T. melanurus*, the color change may be directly linked to activity (Hedges et al., 1989).

Color changes taking days, weeks, or longer to effect constitute morphological color change. One widespread type of morphological color change in lizards is assumption of bright seasonal breeding colors (Table 6.3). Morphological change in sexual coloration also occurs seasonally in the snake *Vipera berus* (Rehák, 1987). Several turtles show morphological changes that are seasonal, sexually dichromatic, or both (Table 6.4; Auffenberg, 1964; reviewed by Moll et al., 1981). Some other reptiles may show seasonal color changes that apparently are unrelated to breeding. Such seasonal changes occur in a pygopodid lizard (*Ophidiocephalus taeniatus*; Ehmann, 1981) and a few elapid snakes, including *Pseudonaja nuchalis* (Banks, 1981), *Oxyuranus scutellatus* (Banks, 1981; Mirtschin and Davis, 1982), and *O. microlepidotus* (Mirtschin and Davis, 1982). Matching of background coloration by morphological color change has been reported in a wide variety of reptiles, including many lizards (e.g., Norris and Lowe, 1964; Norris, 1967; Luke, 1989), a few turtles (Wooley, 1957; Bellairs, 1970; Dunn, 1982; Banks, 1986), and two crocodilians (McIlhenny, 1935; Kleinholz, 1941; Kirshner, 1985). Such changes are often reversible (Luke, 1989). In other species, morphological color change is developmental or ontogenetic and permanent rather than seasonal. Among such changes are permanent bright sexually dimorphic coloration found only in adults and monomorphic ontogenetic changes in pattern between juveniles and adults.

All major forms of color change, physiological, morphological, and

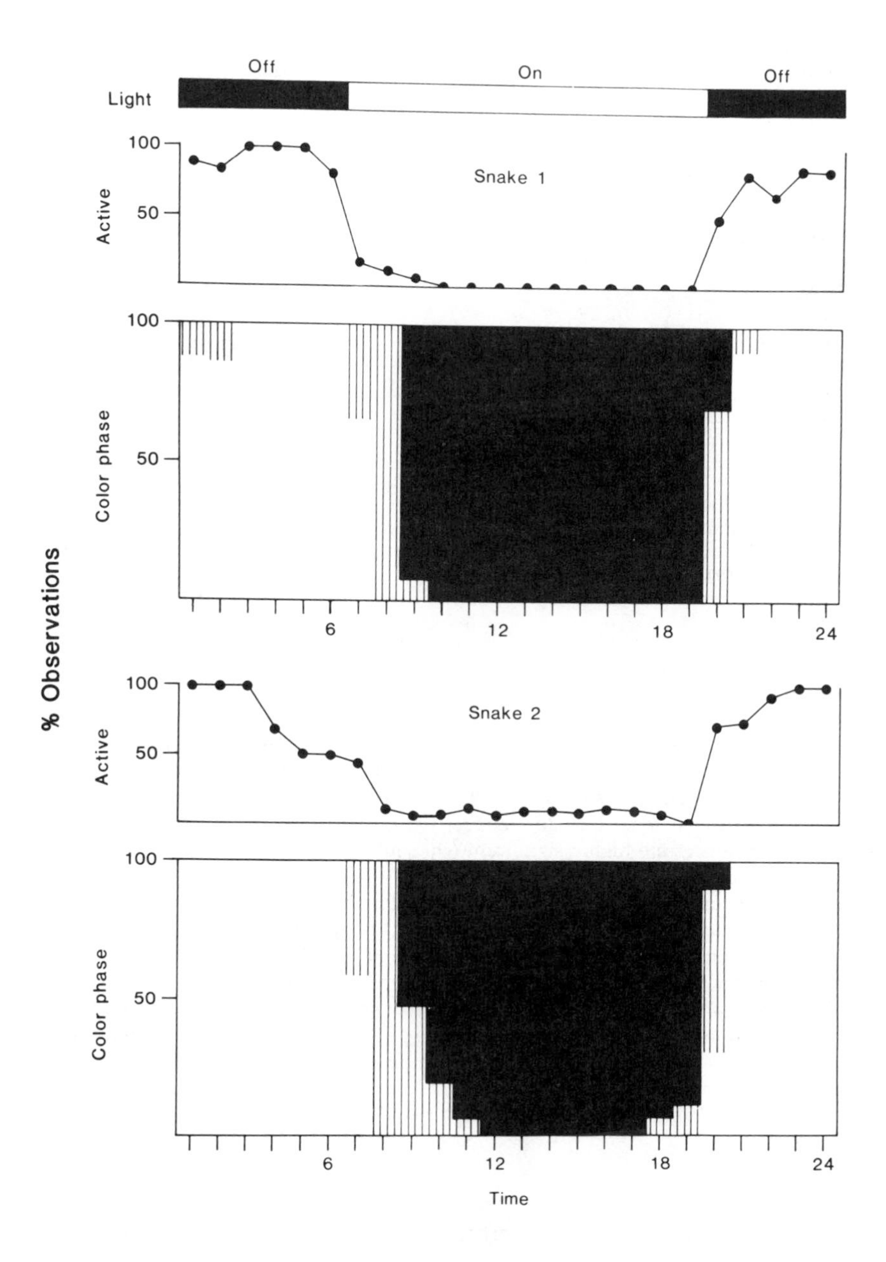

Fig. 6.3. *Tropidophis melanurus.* Color changes and activity in two individuals on a 13L:11D light cycle. The top bar indicates the light cycle. Values are averages for 330 observations in a 5-week period. For color phase: white, light phase; black, dark phase; vertical lines, transitional phase. (Reproduced with permission, Soc. Study Amphibs. and Reptiles, after S. B. Hedges, C. A. Hass, and T. K. Maugel, *J. Herpetol.* Vol. 23, No. 4, 1989.)

Table 6.3 Frequencies of chromatic sexual dimorphism in lizard families[a]

	Dimorphic		Total		Proportion	
Family	ssp.	sp.	ssp.	sp.	ssp.	sp.
Iguanidae	78	65	103	89	.76	.73
Agamidae	52	46	152	146	.34	.32
Gekkonidae	8(5)	8(5)	278	258	.03	.03
Pygopodidae	1	1	30	30	.03	.03
Scincidae	39	38	448	431	.09	.09
Lacertidae	28	27	77	72	.36	.38
Teiidae	9	8	18	17	.50	.47
Pooled Lacertidae-Teiidae	37	35	95	89	.39	.39
Cordylidae	22	18	60	41	.37	.44
Varanidae	0	0	33	33	.00	.00
Anguidae	3(1)	3(1)	19	19	.16	.16
Chamaeleonidae	0	0	16	16	.00	.00
Dibamidae	0	0	2	2	.00	.00
Helodermatidae	0	0	2	2	.00	.00
Xantusiidae	0	0	4	4	.00	.00
Xenosauridae	0	0	1	1	.00	.00
Pooled families lacking color dimorphism[b]	0	0	25	25	.00	.00
Totals	240	214	1,253	1,174	.19	.18

Note: To meet the assumptions of chi square tests for expected frequencies (Siegel, 1956), data were pooled as follows: (1) Lacertidae with Teiidae; (2) Annielidae, Dibamidae, Chamaeleonidae, Helodermatidae, Xantusiidae, and Xenosauridae. The frequency differences are highly significant (chi square = 274.09, df = 9, $p < 0.001$ for species and chi square = 305.95, df = 9, $p < 0.001$ if subspecies are included), indicating real and substantial variation in frequency of intraspecific sex differences in coloration among families. These results are robust. Although chi square values are altered by dropping rather than pooling data for those families not showing dimorphic colors, or even dropping data on teiids, the interpretations are not affected.

[a]Numbers of subspecies (ssp.) and species (sp.) showing dimorphism, total sample size, and proportions of total subspecies and species that are dimorphic. Numbers in parentheses for Gekkonidae and Anguidae indicate the numbers showing dimorphic colors that are bright.

[b]Annielidae, Dibamidae, Chamaeleonidae, Helodermatidae, Xantusiidae, and Xenosauridae.

behavioral, can and do co-occur in single species. Thus, body coloration is jointly determined by rapid, reversible physiological and behavioral changes and by slower morphological changes, which may be irreversible or only slowly reversible, in dermal and epidermal chromatophores. The processes of physiological and morphological color change are often intimately related. For example, melanotropic hormone stimulates both the rapid movement of melanin granules

Table 6.4 Sexual and seasonal dichromatism reported for chelonians

Family/Species	Sexual Dichromatism	Seasonal Dichromatism[a]	Source
Chelidae			
Phrynops dahli	+		Pritchard and Trebbau, 1984
Phrynops zuliae	+		Pritchard and Trebbau, 1984
Rheodytes leukops	+		Legler and Cann, 1980
Cheloniidae			
Caretta caretta	+		Pritchard and Trebbau, 1984
Chelonia mydas	+		Frazier, 1971
Dermochelyidae			
Dermochelys coriacea	+		Pritchard and Trebbau, 1984
Lepidochelys olivacea	+		Pritchard and Trebbau, 1984
Emydidae			
Batagur baska	+	M	Anderson, 1878; Moll, 1978, 1980
Callagur borneoensis	+	M	Moll et al., 1981
Chinemys reevesii	+		Sachsse, 1975; Lovich et al., 1985
Clemmys guttata	+		Ernst and Barbour, 1972
Clemmys insculpta	–	M, F	Harding and Bloomer, 1979
Cuora flavomarginata	+	M	Bartlett, 1981b
Emys orbicularis	+		Pritchard, 1966
Heosemys sylvatica	+		Groombridge et al., 1983; Moll et al., 1986
Kachuga kachuga	+	M	Anderson, 1878
Kachuga trivittata	+	M	Theobald, 1868
Pseudemys nelsoni	+		Kramer, 1986
Sacalia bealei	+		Sachsse, 1975
Terrapene carolina	+		Anderson, 1961; Ernst and Barbour, 1972
Terrapene ornata	+		Legler, 1960
Trachemys decorata	+		Pritchard, 1979
Trachemys decussata	+		Pritchard, 1979
Trachemys scripta	+		Viosca, 1933
Kinosternidae			
Kinosternon angustipons	+		Legler, 1965
Kinosternon scorpioides	+		Pritchard and Trebbau, 1984
Kinosternon sonoriense	+		Hulse, 1976
Pelomedusidae			
Podocnemis expansa	+		Pritchard and Trebbau, 1984
Podocnemis erythrocephala	+		Pritchard and Trebbau, 1984
Podocnemis unifilis	–		Pritchard, 1979
Podocnemis vogli	+		Pritchard and Trebbau, 1984
Peltocephalus dumerilianus	+		Pritchard and Trebbau, 1984
Platysternidae			
Platysternon megacephalum	–	M, F	Wirot, 1979
Testudinidae			
Indotestudo elongata	–	M, F	Smith, 1931; Pritchard, 1979
Indotestudo forstenii	–	M, F	Auffenberg, 1964
Trionychidae			
Trionyx muticus	+		Webb, 1962
Trionyx spiniferus	+		Webb, 1962

After Moll et al., 1981.
[a]M, male; F, female.

within dermal melanophores and the slower synthesis of melanin (Bagnara and Hadley, 1973).

Besides changing skin coloration, many reptiles can rapidly and reversibly alter their appearances behaviorally by revealing or concealing colored patches of skin or other surfaces. Behavioral color changes are effected by postural adjustments or other movements. Examples include dewlap extension by *Anolis carolinensis* (Greenberg and Noble, 1944) and many other species of iguanid lizards (Carpenter and Ferguson, 1977), elevation and lateral flattening of the torso to reveal ventrolateral color patches in many sceloporines (e.g., Noble, 1934; Clarke, 1965), and protrusion of the brightly colored tongue of *Tiliqua scincoides* (Carpenter and Murphy, 1978). The snake *Leptotyphlops dulcis* (Gehlbach et al., 1968) and probably *L. scutifrons* (Visser, 1966) change color rapidly as a defensive behavior by mechanically elevating their scales. *Dasypeltis* show mimicry patterns (also present on models) that are on the interscalar skin and only appear upon inflation of the body (Gans, 1961).

B. Physiological Color Change

1. Overview Among reptiles, physiological color change occurs widely among squamates but is most fully expressed in lizards. Physiological color change may have a fixed rhythm but usually occurs in response to proximal environmental stimuli, such as change in general illumination, background, and social milieu. Physiological color changes are controlled by hormonal or neural action on chromatophores among poikilothermic vertebrates, particularly in reptiles (see Table 6.2). They also may be brought about by direct action of light and temperature on chromatophores.

2. Hormonal Regulation

Several pituitary, gonadal, adrenal, and thyroid hormones, in addition to melatonin and perhaps catecholamines of neural origin, affect physiological color change in reptiles. Virtually all of our detailed information regarding cellular influences of these hormones pertains specifically to their hormonal effects on melanophores. Very little is known about the possible participation of erythrophores, xanthophores, and iridophores in reptilian physiological color change (Sand, 1935). However, in *Anolis carolinensis,* the organelles of these chromatophores are not affected by hormonal stimulation (Taylor and Hadley, 1970).

The pars intermedia of the pituitary synthesizes a proteinaceous hormone called melanotropin, melanophore-stimulating hormone (MSH), melanocyte-stimulating hormone, and intermedin. During synthesis the prohormone pro-opiomelanocotin (POMC) is converted to MSH, which contains 13 amino acids. For reviews of the nature and function of MSH and related substances, see Eberle (1988) and Hadley (1988a, b, c). In the few reptiles studied, MSH darkens the skin by inducing melanosome dispersion. Evidence regarding pituitary origin and function of MSH is available for the lizards *Anolis carolinensis* (Kleinholz, 1938a, b; Horowitz, 1958), *Anolis roquet* (May and Thillard, 1957), *Hemidactylus brookii* (Noble and Bradley, 1933b), and *Chamaeleo jacksonii* (Canella, 1963) and the snake *Crotalus viridis* (Rahn, 1941).

Factors regulating secretion of MSH are unclear and appear to vary among vertebrates. For example, adrenergic and cholinergic hypothalamic neurons terminating in the pars intermedia appear to be important in frogs (Bagnara and Hadley, 1973) but are absent in the lizard *Xantusia riversiana* (Rodriguez and La Pointe, 1970). An additional complicating factor is the presence of both inhibiting and releasing factors for MSH in the amphibian and mammalian hypothalamus (Bagnara and Hadley, 1973). Adaptation to a light background in *Anolis carolinensis* induces elevation in the MSH concentration in the pars intermedia in comparison to dark adaptation. Presumably, dark-adapted anoles secrete more MSH (Dores et al., 1987). Serotonin may be an MSH-releasing factor in *A. carolinensis* (Thornton and Geschwind, 1975). In vitro, serotonin induces release of MSH from the pituitary gland. When injected subcutaneously, it darkens the skin, but when placed on excised skin, it does not affect melanocytes. These findings suggest that the injected serotonin acts directly on the pars intermedia, eliciting darkening indirectly through its effect on MSH.

Adrenocorticotropic hormone, or corticotropin (ACTH), is a larger protein also derived in the pituitary gland from POMC. ACTH is structurally very similar to MSH, having an N-terminal region containing the same amino acid sequence as MSH. Because parts of the ACTH molecule are comparable to MSH a-molecule, pigmentary effects are reasonable; in humans, hyperpigmentation is observed in disease states (such as Cushing's disease) that are associated with excessive levels of ACTH (Hadley, 1984). ACTH is also capable of inducing darkening in other taxa, including the lizard *Chamaeleo jacksonii* (Canella, 1963).

The catecholamines epinephrine and norepinephrine induce aggregation of melanosomes in vitro in *Anolis* and *Phrynosoma* (Redfield, 1916, 1918; Kleinholz, 1938b) but may also cause dispersion (Kleinholz, 1938b; Horowitz, 1958). Whether darkening or lightening occurs depends on the type of adrenergic receptors in the chromatophores. Alpha- and beta-adrenergic receptors are cellular elements that respond to catecholamines, whether from neural or adrenomedullary sources. Cells may possess alpha receptors, beta receptors, or both. If both are present, alpha receptors are usually dominant. The ratios of epinephrine and norepinephrine, as well as of these two receptor types, are significant determinants of melanosome movement and consequent change in coloration. Alpha-adrenergic receptors in the dermal melanophores of *A. carolinensis* mediate melanosome aggregation in response to catecholamines (Goldman and Hadley, 1969). The lightening of *A. roquet* in response to adrenalin (May and Thillard, 1957) presumably is attributable to similar aggregation. Melanosome aggregation mediated by alpha-adrenergic receptors is responsible for the excitement pallor and similar chromatic responses of anoles and many other lizards (e.g., *Sceloporus occidentalis,* Cooper and Ferguson, 1973b). Beta-adrenergic receptors have the opposite effect, inducing excitement darkening by dispersion of melanosomes. In *A. carolinensis,* initially dark individuals pale in response to stress-induced catecholamines, but initially light-green individuals develop a mottled appearance. The dark spots, most notably the postorbital eyespots, occur whenever melanophores contain only beta receptors or the beta receptors dominate the alpha receptors; melanophores involved in excitement pallor contain alpha and beta receptors (Hadley and Goldman, 1969).

The blanching reaction of many poikilothermic vertebrates when placed in the dark, especially in some larval teleosts, anurans, and urodeles, appears to be controlled by the pineal. In darkness the pineal secretes increased amounts of melatonin, which effects aggregation of melanosomes in the dermal melanophores (Bagnara and Hadley, 1973). However, different mechanisms are necessary to account for blanching in other developmental stages and in other vertebrates, particularly reptiles. The eyes do not control the blanching response (Sand, 1935; Parker, 1938) in the several iguanid and chamaeleonid lizards studied. Neither does melatonin reliably induce blanching in *Anolis carolinensis.* In this species, the pituitary regulates blanching, hypothetically under the reflex control of integumental photoreceptors (Kleinholz, 1938a).

The thyroid hormones thyroxine and triiodothyronine block the in vitro darkening effects of MSH in *Anolis carolinensis* (Bagnara and Hadley, 1973). Thyroxine also reduces the number of melanophores and increases the number of iridophores in salmonid fish (Bagnara and Hadley, 1973).

Beyond their individual effects, hormones may interactively determine color changes. In *Anolis carolinensis,* the integumentary melanophores that regulate color respond most notably to epinephrine, norepinephrine, and MSH. Accordingly, anoles have catecholamine receptors (Goldman and Hadley, 1969) and MSH receptors that appear to be distinct from catecholamine receptors. The MSH receptors can override the effect of catecholamines on skin color and may be linked with both alpha-adrenergic receptors (Carter and Shuster, 1982) and beta-adrenergic receptors (Vaughan and Greenberg, 1987) through a common adenyl cyclase molecule to form a functional unit in the membrane of the melanophore. When beta-adrenergic receptors are blocked by propanolol, darkening is not induced as rapidly by epinephrine and MSH (Vaughan and Greenberg, 1987). Another potentially significant interaction occurs between MSH and beta endorphin. This endogenous opiate has no independent melanotropic effect but potentiates the effect of MSH on melanophores in *A. carolinensis* (Carter and Shuster, 1982).

3. Neural Regulation

Neural regulation is important in some species of lizards but appears to be absent in other species that display pronounced physiological color changes. It was first demonstrated in chameleons (Brücke, 1852; Hogben and Mirvish, 1928). Electrical stimulation of the roof of the mouth or of the cloaca elicits a general pallor that is completed in about 35 to 50 seconds. If the spinal cord is sectioned, stimulation of the mouth induces pallor only on skin surfaces innervated by spinal nerves anterior to the point of sectioning; cloacal stimulation induces blanching only posterior to the point of sectioning (Hogben and Mirvish, 1928; Fig. 6.4). Neural regulation of color change has also been observed in *Phrynosoma,* which shows both nervous and endocrine regulation of physiological color change (Redfield, 1916, 1918; Parker, 1938). In contrast, hormones alone control rapid color changes in *Anolis carolinensis* (Kleinholz, 1936, 1938a, b). The situation is less clear in chameleons. Hogben and Mirvish (1928) claimed no evidence for hormonal chromatophore regulation in *Bradypodion pumilis,* but their exclusion of MSH as a chromatophore effector may have been premature (Parker, 1938). The possible effectiveness of endocrines in *B. pumilis* is underscored by the demonstration that both MSH and ACTH induce darkening in the closely related *C. jacksonii* (Canella, 1963).

The neurotransmitter acetylcholine may affect chromatophores in several ways, producing either darkening or lightening indirectly by effects on pituitary release of MSH or adrenal release of catecholamines. However, in reptiles, acetylcholine has no known direct effects on chromatophores (Hadley and Goldman, 1969).

4. Environmental Factors

Environmental agents, such as light and temperature, may induce physiological color change directly by acting on chromatophores or indirectly through autonomic or humoral responses. Even humidity may affect pigment migration (Burgers, 1966). Any agent constituting an environmental stressor may produce such indirect effects. For example, reduced barometric pressure stimulates increased epinephrine release in *Anolis carolinensis* as indicated by color change (Rahn, 1956).

Reptilian skin is photosensitive (Weber, 1983). Light typically induces melanin dispersion in lizard skin (*Chamaeleo* sp.: Brücke, 1852; Zoond and Eyre, 1934; *Phrynosoma coronatum:* Redfield, 1916; Parker, 1938; *Anolis equestris:* Smith, 1929; Hadley, 1931; Kleinholz, 1938a). However, light induces aggregation of melanin in some amphibians (Weber, 1983). Skin of *A. carolinensis* exposed in vitro to visible light from a tungsten source (110 W m^{-2}) manifests a 30% to 50% reduction in reflectance within 2 to 4 minutes as it changes from green to brown. The response is completed in 3 minutes at light intensities as low as 50 W m^{-2} and is completely reversed within 2 minutes in darkness (Vaughan, 1987a). Maximum photic response is achieved at 50 W m^{-2}, which is only 0.5% as intense as midwinter sunshine in the temperate zone at Knoxville, Tennessee (Vaughan, 1987a). The skin is sensitive to light of intensities as low as 5.0 W m^{-2} (Vaughan, 1987a). There appear to be important differences between the mechanisms of light-induced and MSH-induced pigment dispersion (see Section II).

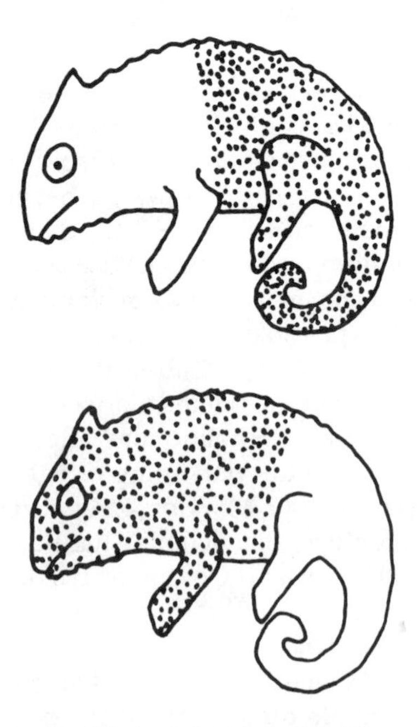

Fig. 6.4. *Bradypodion pumilis.* Effects of electrical stimulation on color following section of the spinal cord. Upper: Oral stimulation following section at the eighth vertebra. Lower: Cloacal stimulation following section at the twelfth vertebra. (From Hogben and Mirvish, 1928, with permission, Company of Biologists Ltd.)

Temperature may affect chromatophores directly and appears to be a significant variable in the expression of the dermal photic response of *Anolis carolinensis*. In the laboratory, darkening under illumination occurs only at temperatures below 13°C; the effect is extinguished above this level (Vaughan, 1987b). This phenomenon has also been observed near the northern edge of the geographic range during winter; here midwinter basking is commonly observed, and the intensity of visible light can be approximately 1000 W m^{-2}. Considering the increased absorption of light by darker animals, the thermal advantage conferred on a 5.0-g anole by being brown rather than green is up to 0.8 cal/min (Vaughan, 1987b), which confers a gain of approximately 0.21°C/min over thermoneutrality.

Cold slows, but does not abolish, the darkening effect of MSH on the skin of *Anolis carolinensis*. It abolishes the lightening effect of norepinephrine on MSH-darkened skin, indicating that the aggregate (retraction to a perinuclear position) of melanin granules has greater energetic requirements than does their dispersion in the distal processes of the chromatophore (Hadley and Goldman, 1969).

5. A Caveat

Physiological color change has been studied more thoroughly in *Anolis carolinensis* than in any other vertebrate, yet even in this species and its congeners our knowledge of its regulatory processes is far from complete. Early field observation on color changes in *A. carolinensis* revealed a wider range of chromatic variation than occurs in typical laboratory studies (von Geldern, 1921; Strecker, 1928). *A. coelestinus* shows a similar broad range of color patterns not easily accounted for by currently understood mechanisms of physiological color change (Moermond, 1978; Fig. 6.5). Some of the complexity of color changes in the field may be explicable in terms of direct effects of environmental factors, especially temperature (Vaughan, 1987b). Temperature may be especially important in regulating diel color cycles such as those of several island boas (Hedges et al., 1989) and the rattlesnake *Crotalus viridis* (Sweet, 1985).

C. Morphological Color Change

1. Regulation of Morphological Color Change Morphological color changes, those requiring hours or more to complete, may be induced by environmental stimuli such as increased or decreased exposure to light. Epidermal melanophores are responsible for much of such light-induced change, but dermal melanophores also participate. MSH induces increased melanin production in the melanophores, as well as melanosome dispersion in the dermal melanophores.

Change in morphological coloration in response to background coloration has been reported in crocodilians (Kirshner, 1985), turtles (Dunn, 1982), lizards (Luke, 1989), and snakes (Rehák, 1987). In addition, temperature may affect the development of coloration in crocodilians. The development of coloration may be affected by incubation temperatures in *Alligator mississippiensis*. Hatchlings from eggs incubated at elevated temperatures have an altered color pattern and have darker dorsal pigmentation (Deeming and Ferguson, 1989).

Seasonal breeding coloration in some lizard species is induced by gonadal and possibly adrenal steroid hormones (Cooper and Ferguson, 1972a, b, 1973a, b; Cooper and Clarke, 1982; Cooper and Crews, 1987; Cooper et al., 1987). Preliminary evidence indicates that seasonal dichromatic coloration is under androgenic control in the turtle *Callagur borneoensis* (Moll et al., 1981). Because these steroidally produced morphological color changes often involve orange and yellow pigments, they are likely to result from pigment synthesis or deposition in the xanthophores and erythrophores, as noted for other poikilothermic vertebrates (Bagnara and Hadley, 1973). However, androgens may induce morphological color change by other mechanisms. Androgens induce ventrolateral darkening in male *Sceloporus occidentalis* via effects on melanophores (Kimball and Erpino, 1971). It is not known whether androgens act directly on melanophores in this species or through stimulation of MSH secretion. In the turtle *Callagur borneoensis*, androgens appear to induce two color changes (Moll et al., 1981). First, the sides of the head lighten because the epidermis thickens enough to prevent dermal melanophores from influencing

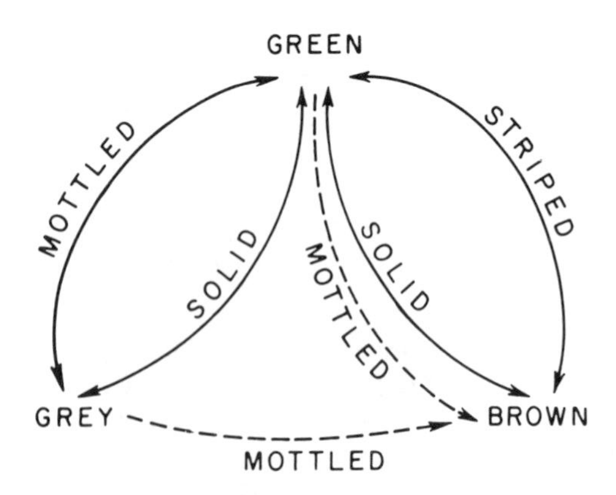

Fig. 6.5. *Anolis coelestinus.* Color changes among three basic background colors are augmented by variation among uniform, mottled, and striped patterns. Solid arrows represent observed changes. The broken arrow represents a suspected change. (Reproduced with permission, Soc. Study Amphibs. and Reptiles, after Moermond, 1978.)

surface coloration and melanosomes disappear from the basal epidermis. Second, a middorsal stripe on the head turns bright orange-red due to an increase in vascularization in the stratum laxum of the dermis.

Morphological color changes that involve quantitative change in integumentary pigment are relatively slow because they require synthesis or degradation of pigment granules. Melanic darkening may occur due to deposition of pigment particles in epidermal cells, their retention in dermal melanophores, or their presence in an increased number of cells. These changes are most conspicuous in, but are not limited to, melanophores. During the dark or light adaptation of some fish and amphibians, the guanine content of iridophores apparently changes in a manner reciprocal and complementary to changes in the melanophores (Bagnara and Hadley, 1973). Physiological color changes usually, but not invariably, precede morphological ones.

2. Ontogenetic Color Change

Ontogenetic changes in color pattern unrelated to reproduction are very widespread in reptiles. They are striking in many snakes, lizards, and turtles. Even some crocodilians show more intense color patterns as juveniles, the patterns fading with growth. Permanent developmental changes are sometimes classified as obligatory, whereas seasonal changes are said to be facultative (Hinton, 1976). Behavioral relationships of obligatory color changes have received very little attention in reptiles, and physiological ones almost none. Nevertheless, there are numerous hints that behavioral changes may correspond to developmental chromatic ones, suggesting avenues for profitable study. For example, habitat selection for background matching (see Section I.A) (*Anolis cuvieri*, Gorman, 1977), temperament (*Elaphe climacophora*, Hadley and Gans, 1972), and various other behaviors may change with ontogenetic color state.

Several types of ontogenetic color change may have behavioral significance in lizards. Juveniles of numerous species of skinks (*Eumeces* and others), some teiids, lacertids, and perhaps others have bright blue or orange tail coloration that disappears as the lizard grows. In some species, the bright tail coloration is combined with jet black body coloration with light longitudinal stripes. As such lizards grow, the bright tail coloration and, if present, the bold body coloration are replaced by cryptic color patterns. The common argument is that changes in the predator suite with increasing size are responsible for such changes (Cooper and Vitt, 1985); however, this has not been demonstrated. Hatchlings of one species, *Eumeces obsoletus*, raise their brightly colored tails perpendicular to the substrate when confronted by a human potential predator (personal observation, 1971). Adults, which lack the bright tail coloration, do not perform this behavior. It also has been suggested that hatchlings of this species attract prey by wriggling their brightly colored tails (Smith, 1946). Other striking ontogenetic color changes presumably related to defense mechanisms occur in a presumed lacertid mimic (*Eremias lugubris*) of a noxious beetle (Huey and Pianka, 1977) and a possible anguid mimic (*Diploglossus lessonae*) of a noxious millipede (Vanzolini, 1958). Neither the physiological nor the selective basis for ontogenetic changes in color and associated behaviors has been established experimentally for any of these species.

Ontogenetic color changes in many snakes may be related to change from the greater importance of crypsis in avoiding predators by juveniles to active defense or escape by fleeing in adults (Jackson et al., 1976). In several species of viperid snakes, juveniles have been reported to raise and wriggle conspicuously colored tails to attract prey. Caudal luring by tails having coloration contrasting with the general body coloration has been reported in seven species of crotalines (*Agkistrodon bilineatus*, Allen, 1949; Neill, 1960; *A. contortrix*, Neill, 1948; *Hypnale hypnale*, Henry, 1925; *A. piscivorus*, Pycraft, 1925; Wharton, 1960; *Bothrops asper*, Tryon, 1985; *B. atrox*, Pycraft, 1925; Burger and Smith, 1950; *Bothriopsis bilineata*, Greene and Campbell, 1972; *Calloselasma rhodostoma*, Schuett, 1984; *Crotalus lepidus*, Kauffield, 1943; *Sistrurus catenatus*, Schuett et al., 1984; and *Sistrurus miliarius barbouri*, Jackson and Martin, 1980) and two species of viperines (*Cerastes vipera*, Heatwole and Davison, 1976; *Vipera russellii*, Henderson, 1970). In two of these species, *B. bilineata* and *C. vipera*, conspicuous tail coloration is retained in adulthood; adults of these species continue caudal luring. Adults of the remaining species do not practice caudal luring, and most of them lose the conspicuous tail coloration ontogenetically. It is not known whether adult *V. russelli* in the population studied (Henderson, 1970) have conspicuous tail coloration. Thus, caudal luring is associated with conspicuous tail coloration and appears to be lost if the caudal coloration fades ontogenetically.

Less is known about caudal luring in other families of snakes. Caudal luring by adults having brightly colored tails occurs in the elapid *Acanthophis antarcticus* (Carpenter et al., 1978). A juvenile boid *Chondropython viridis* employed caudal luring in the presence of rodent prey, but the luring gradually diminished as the bright tail coloration (contrasting with that of the body) faded over a period of 28 months (Murphy et al., 1978). Caudal luring has been reported in two other boids, *Boa constrictor* (Radcliffe et al., 1980) and *Tropidophis canus curtus* (Bartlett, 1981a). In addition to the species mentioned, many crotalines, numerous boids, and the taxonomically enigmatic *Atractaspsis corpulenta* have been suggested as possible caudal lurers (Neill, 1960). In the boids, the conspicuous tail coloration of juveniles suggests that if caudal luring occurs, it may be lost ontogenetically; their bright tail coloration fades, as in several viperids. Physiological mechanisms underlying the ontogenetic changes in color and caudal luring are unknown.

IV. COLOR VISION

Reptiles' spectral sensitivities and their ability to discriminate among hues have been relatively well studied in turtles and lizards, with much less information available for other groups. A single study (Nickel, 1960) demonstrates color vision in crocodilians, but color vision is unreported in rhynchocephalians, amphisbaenians, burrowing lizards, and snakes. Nevertheless, the cell types and properties in reptilian retinas strongly suggest the possible importance of color vision in some groups. Behavioral data on color vision must be interpreted cautiously, even in the few species that have been studied, because most investigators have not controlled brightness, saturation, or both.

The morphology and pattern of the retinal cells are reviewed by Underwood (1970), the source for the cellular information that follows unless otherwise stated. In Testudines, the retina contains two types of single cones, double cones, and the less numerous rods. The cones contain oil droplets of various colors, predominantly reds and yellows; the rods lack oil droplets. The oil droplets appear to function as cutoff filters that eliminate light of short wavelengths (Granda, 1979). Oil droplet color is associated with cone type, red or pale green droplets being found in red-sensitive cones, yellow droplets in green-sensitive cones, and clear droplets in blue-sensitive cones in *Trachemys scripta* (Ohtsuka, 1985; Fig. 6.6). There are three types of single cones in this species, each containing a distinct visual pigment (Liebman and Granda, 1971). The peaks of psychophysically determined sensitivity spectra of a proposed three-receptor system correspond well to the wavelengths of maximum absorption for the three pigments (Granda et al., 1972). These studies of *T. scripta* provide the strongest evidence for color vision in any reptile. *Chelonia mydas* also has three cone pigments (Granda, 1979).

Retinal cell types and constituents differ somewhat in other reptiles (Underwood, 1970). Crocodilian retinas have single cones, double cones, and rods. However, all cones lack oil droplets, and the rods are the most numerous type. The rhynchocephalian retina includes major and minor single cells similar to the single cones of diurnal lizards and turtles but having colorless oil droplets. There are also double cells similar to lacertilian double cones but with colorless oil droplets and small droplet-free cells similar to very small cones. Lizards have all-cone retinas with two types of single cones and double cones, all types having oil droplets. The cones have oil droplets of various colors, primarily yellow in diurnal species. In some nocturnal lizards the oil droplets are colorless. Pygopodids and some geckos lack oil drop-

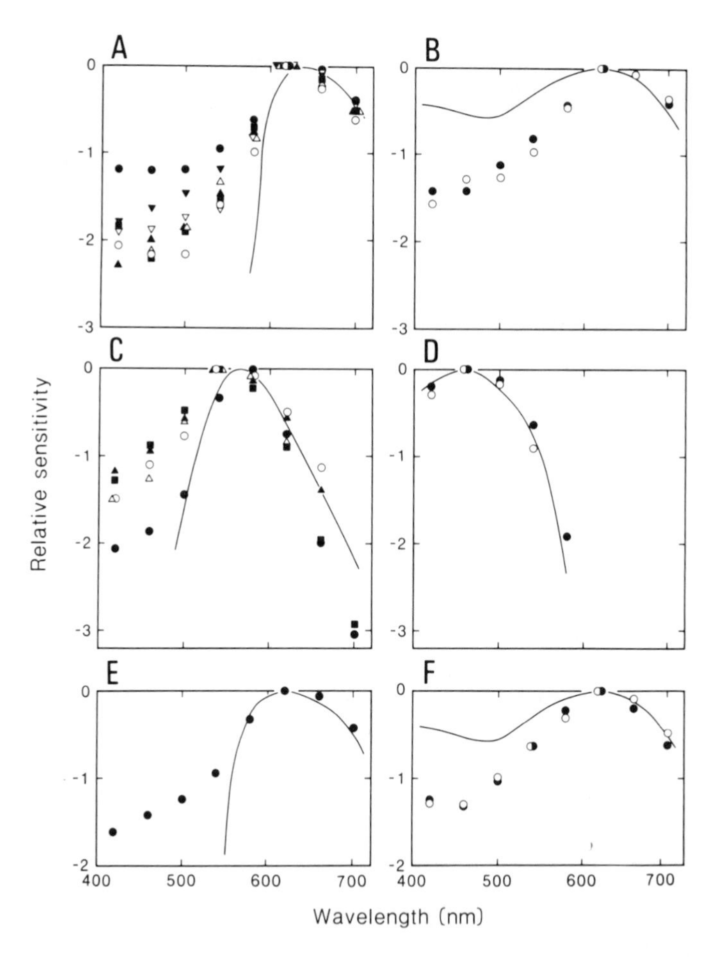

Fig. 6.6. *Trachemys scripta elegans.* Spectral sensitivity of six types of cone cells. A to D represent individual cones containing red, pale green, yellow, and clear oil droplets, respectively. E and F represent the principal and accessory members of double cones. Each symbol represents a different cone. Continuous lines A, C, and E are the calculated absorption spectra of the visual pigment-oil droplet combinations. Note the sharp cut-off in the shorter wavelength due to the colored oil droplet. Both pale green and clear oil droplets transmit the visible light, and therefore the absorption spectra of the red- (B and F) and blue-absorbing (D) pigments are illustrated by continuous lines. (From Ohtsuka, 1985, *Science* [N.Y.], 229, 874–877, © 1985 by the AAAS.)

lets. Modifications of the basic cell types occur in various geckos, all of which have a second type of double cone not found in other families. The retinas of amphisbaenians have small visual cells resembling embryonic cones of lizards, some containing small oil droplets. Snakes have variable retinal cell types but always lack oil droplets. Some forms have only rods, others rods and cones, still others only cones. Some of the cone cell types occur only in snakes.

Among turtles, operant conditioning and visual discrimination studies establish the presence of color vision. Two related species of turtles, *Chrysemys picta* and *Trachemys scripta,* show behavioral evidence of at least two types of visual receptor cells, one sensitive to red and dominant at high intensity, the other sensitive to blue and dominant at low intensity (Armington, 1954; Sokol and Muntz, 1966). *Mauremys caspica* readily discriminate among various wavelengths of controlled brightness, whether the light source is transmitted spectral light or light reflected from standardized pigmented papers (Wojtusiak, 1933); *Mauremys japonica* also discriminates red and other hues from blue in visual discrimination tests using colored paper (Kuroda, 1933; see discussion by Burghardt, 1977). Further evidence of abilities to discriminate among hues is available for *Caretta caretta* (Fehring, 1972), *Geochelone elephantopus* (Quaranta, 1952), *G. gigantea* (Quaranta, 1952), *Emys orbicularis* (Bartowiak, 1949), *Chinemys reevesii* (Bartowiak, 1949), *C. picta* (Graf, 1967; Muntz and Sokol, 1967; Muntz and Northmore, 1968), and *T. scripta* (Muntz and Sokol, 1967). Three species, *Emys orbicularis, Chinemys reevesii,* and *Cuora amboinensis,* have been shown to detect infrared light and discriminate it from visible light (Wojtusiak and Mlynarski, 1949).

Color vision thus appears to be widespread in turtles. In experimental tests requiring turtles to choose to enter a compartment having its entrance covered with red, yellow, green, or blue cellophane, turtles of four families differ in their responses (Ernst and Hamilton, 1969). The single chelydrid species studied, *Chelydra serpentina,* strongly prefer blue chambers (Fig. 6.7). Among kinosternid species, *Kinosternon subrubrum* prefers blue, but three species of *Sternotherus* differ, two apparently preferring yellow and one red. The strongest color preference is shown by the trionychid, *Trionyx spiniferus,* which prefers yellow chambers. Among emydid turtles, *C. picta, Terrapene carolina,* and *Pseudemys concinna* have no color preferences for chambers, but *P. floridana* and *T. scripta* prefer blue. The study of color vision by Ernst and Hamilton (1969) has been strongly criticized (Hailman and Jaeger, 1971) for failure to control important visual variables. Similar considerations apply to almost all attempts to study color vision in reptiles. All too often, neither brightness nor saturation are controlled. Only for one reptilian species, *T. scripta,* has both a physiological basis for color vision and a corresponding behavioral ability in psychophysical tests been established (Granda et al., 1972).

Little is known about color vision among crocodilians, although it occurs in the two species studied. *Caiman crocodilus* and *C. latirostris* discriminate among red, green, yellow, and blue papers matched for brightness (Nickel, 1960).

Lizards have long been presumed to have color vision because many species are brightly colored, and their brightly colored areas often are exposed or otherwise emphasized during social displays. Analysis of color shift electroretinograms reveals clear differential responses to red, yellow, and green by three species of diurnal iguanid lizards, *Sceloporus olivaceus, S. cyanogenys,* and *Crotaphytus collaris* (Forbes et al., 1964). The lizard's retinas are quite insensitive to blue. The plateau of response often occurs at 565 nm, which corresponds to the peak sensitivity of iodopsin and to the yellow color of the oil droplets in cones.

All species tested in discrimination studies have color vision. These include representatives of Agamidae, Iguanidae, Lacertidae, and Tei-

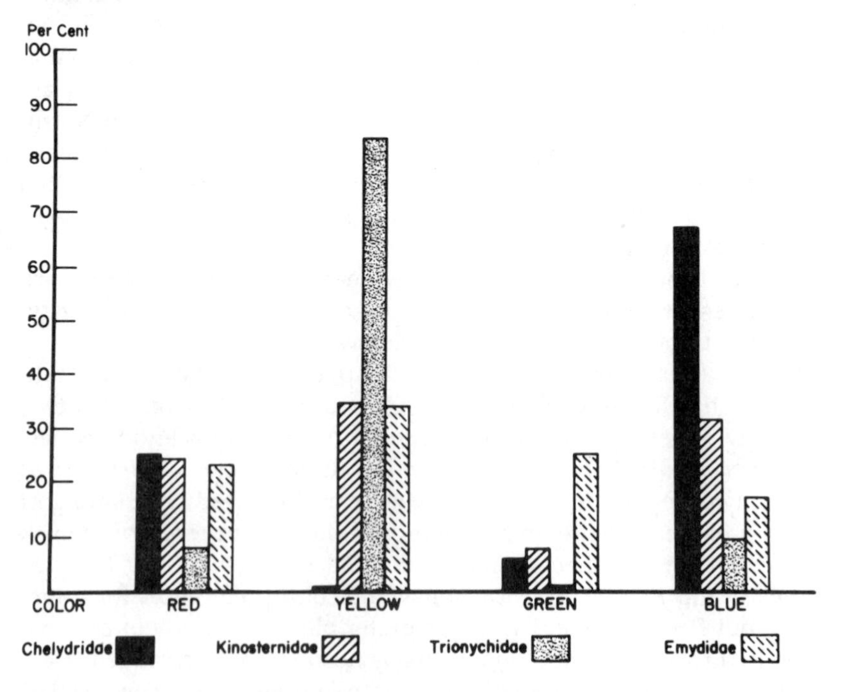

Fig. 6.7. Preferences of some North American turtles for entering chambers having entrances covered with cellophane of different colors. (Reproduced with permission, Soc. Study Amphibs. and Reptiles, after C. H. Ernst and H. F. Hamilton, *J. Herpetol.* Vol. 3, No. 3–4, 1969.)

idae. Lacertids have been most thoroughly studied. *Lacerta agilis* and *L. viridis* can discriminate at least eight hues (Wagner, 1933), but there is conflicting evidence about differential sensitivity. Wagner reported that these species are most sensitive in the red and blue wavelengths, least in green, but, in a more detailed work primarily on *L. agilis,* Swiezawska (1949) stated that these lizards are most sensitive in the yellow-green region, which corresponds to the body coloration, and least sensitive to red and violet. Additional work with this species shows that color is also an important prey characteristic (Svoboda, 1969). *L. viridis* is capable of red-green and yellow-blue discriminations (Dücker and Rensch, 1973). *Cnemidophorus tigris,* a member of the closely related family Teiidae, can discriminate single steps on the Munsell color scale (Benes, 1969).

Differential sensitivity to light of different wavelengths is known in five iguanids, *Anolis agassizi, Iguana iguana, Dipsosaurus dorsalis, Tropidurus delanonis,* and *Uta palmeri. I. iguana* can learn red-green and yellow-blue discriminations, the latter most rapidly (Rensch and Adrian-Hinsberg, 1963). *A. agassizi* exhibit stronger feeding responses to small orange and yellow objects than to red, green, or black ones, perhaps because the orange and yellow objects are more similar to the color of broken bird eggs that these lizards eat (Rand et al., 1975). Similarly, feeding responses of *U. palmeri* in the field are elicited more frequently by small red objects than by yellow, blue, or black objects. Adult females also respond strongly to yellow objects (Hews and Dickhaut, 1989). *Tropidurus delanonis* most frequently approach red cards, followed by yellow, but do not approach cards of other colors (Werner, 1978). Because brightness and saturation were not controlled in the three field studies just described, it is uncertain whether the lizards respond to hue or not. Sensitivity to near-ultraviolet wavelengths that may be significant in detection of pheromone deposits has recently been discovered in another iguanid, *D. dorsalis* (Alberts, 1989). Ultraviolet vision could be, but is not necessarily, responsible for the increased frequency of head-nodding in *Agama agama* and intraspecific threatening behavior of the iguanids, *D. dorsalis* and *Crotaphytus collaris,* upon exposure to ultraviolet light (Moehn, 1974). The possibility that reptiles may have socially significant color patterns that are detectable only at ultraviolet wavelengths should be investigated. The agamid *A. agama* can also discriminate between red and green and between yellow and blue stimuli (Dücker and Rensch, 1973).

For reptiles, there is little physiological information relating functional aspects of vision to body coloration, but three ways in which coloration may affect color vision have been suggested (Burtt, 1981, 1984, 1986). First, reflectance around the eye may interfere with vision; thus dark colors around the eyes can reduce reflected light. Second, eyelines may assist predators in visually tracking and striking moving prey. Finally, dark coloration of the iris may decrease the amount of light it transmits, thereby increasing resolution. In addition, the widespread use of blue as a display color in the iguanid lizard genus *Sceloporus*, despite the insensitivity of these lizards to blue wavelengths (Forbes et al., 1964), suggests that the display colors may be perceived as dark, which is consistent with the frequent juxtaposition of blue and black on surfaces displayed.

V. COLORATION, SOCIAL BEHAVIOR, AND CHROMATIC SEXUAL DIMORPHISM

A. Coloration and Social Behavior

Reptiles exhibit various morphological changes in pattern and coloration associated with growth, maturation, and seasonal reproduction; many lizards additionally show physiological color changes in some cases closely tied to social behavior. In many species of turtles, crocodilians, and snakes (e.g., Golder, 1987; Madsen, 1987), juvenile coloration is brighter than that of the adult. In some lizards and many snakes juveniles have color patterns totally different from those of adults (e.g., Conant, 1975; Jackson et al., 1976). For example, in the colubrid snake *Coluber c. constrictor*, juveniles are strongly patterned, having dark blotches on a gray background dorsally and small dark spots ventrally; adults are uniformly black except on the chin and throat, which usually have some white (Conant, 1975). The coloration of the lacertid lizard *Lacerta schreiberi* is light brown with light bars, or ocelli, on the flanks in juveniles but is darker brown or green with black spots or blotches on the dorsum in adults (Arnold and Burton, 1978). Such changes in color pattern are widespread. Other examples of ontogenetic color changes include melanization of males in certain turtles, such as *Trachemys scripta* (Mount, 1975; Lovich et al., 1990) and developmental acquisition of permanent sexual dichromatism in various lizards, especially in many species of sceloporine and anoline iguanids.

Seasonally developed breeding colors and permanent chromatic sexual dimorphism are most pronounced in lizards, occurring in numerous species of several families (Table 6.3). Bright breeding colors usually occur exclusively in males, but females of several species of lizards show bright colors distinct from those of males (Cooper, 1984). Although bright breeding colors are widely believed to provide important social cues, their behavioral significance has been studied in only a few species. The extent and distribution in lizards of sexual differences in bright coloration that potentially affect social behavior are presented in Section V.B.

Coloration can be used as a visual signal to convey socially important messages in several contexts. The importance of such cues is largely unknown except in lizards. However, in the turtle *Geochelone carbonaria,* head coloration may provide an important cue for sex recognition by conspecific males. Males with heads painted black do not elicit aggression from resident males, but those having heads painted to resemble the natural male head coloration (black with orange, white, or yellow markings) are challenged (Auffenberg, 1965). In *Trachemys scripta,* melanistic males are more aggressive and dominant than nonmelanistic males (Lardie, 1983); melanistic male *Pseudemys nelsoni* initiate most aggressive encounters (Kramer, 1986). Numerous species of turtles are known to have sexually dimorphic coloration (Table 6.4; see Moll et al., 1981; Gibbons and Lovich, 1990). For example, male *Terrapene carolina* have red eyes and females have brown ones (Ernst and Barbour, 1972). Such differences are potentially important as social signals, but their behavioral roles have not been studied. One obvious difference in sexual dichromatism between turtles and lizards is that turtles only rarely have dichromatic coloration as bright as that of many lizards. In one snake, *Vipera berus,* melanistic males have been hypothesized to have greater reproductive success than normally patterned males due to their larger size and consequent ability to win male-male combats over access to females (Andrén and Nilson, 1981; Madsen and Stille, 1988). Whether or not the darker coloration is an important social signal in either agonistic or courtship behavior is unknown.

In lizards, chromatic cues may support sex recognition and may play roles in sexual behavior and intraspecific agonistic behavior (Cooper, 1984, 1988; Greenberg et al., 1984; Zucker, 1989). Color differences among closely related syntopic species may also provide important cues for recognizing conspecifics and may thus participate in maintaining reproductive isolation (Losos, 1985). Whereas little is known about the physiological underpinnings of some behavioral roles of color, the relationships between color and social behavior are described in hopes of stimulating future research.

B. Chromatic Sexual Dimorphism

Sexual differences in coloration, especially those involving bright epigamic colors, are widespread in lizards, much rarer in snakes and turtles (Section III.A), and apparently absent in other reptiles. Lizard taxa show substantial variation in frequency of dimorphic color patterns. This variation has been surveyed at the familial level by compiling data from sources describing coloration in several major lizard faunas throughout the world (Smith, 1935; Fitzsimons, 1943; Donoso-Barros, 1966; Pienaar, 1966; Conant, 1975; Arnold and Burton, 1978;

Rivero, 1978; McCoy, 1980; del Toro, 1982; Cogger, 1983; Stebbins, 1985). Because some species covered in these sources are poorly known, dimorphism is likely present in some species for which is it not described. Thus, the frequencies of chromatic dimorphism found in this survey are minimal estimates. The proportion of species and subspecies showing dimorphic coloration differs among families (see Table 6.3). Variation in frequency of color dimorphism among genera is also pronounced within some families but has not been formally examined.

Iguanids, agamids, cordylids, teiids, and lacertids show the greatest incidence of chromatic dimorphism. In agamids, the observed frequency of color dimorphism is probably well below the true value because many of the descriptions are based on preserved specimens, whereas physiologically regulated bright colors, especially of males during social displays (e.g., Harris, 1964), may be lost in preservation or absent in small samples. Although most male and female chameleons have monomorphic cryptic coloration, both sexes of several species rapidly develop distinctive bright color patterns during sexual and aggressive behavior. These color patterns differ between sexes in some species, but only a few species have been studied in detail (Parcher, 1974). Thus, the extent of chromatic dimorphism is underrepresented to an unknown degree. Although most geckos are nocturnal, the few cases of color dimorphism occur primarily in diurnal forms; however, many nocturnal geckos can darken or lighten their skin and display intrinsic patterns. Chromatic dimorphism occurs in relatively few skinks, but is highly developed in some genera, especially *Eumeces*. Families generally lacking bright coloration may have higher frequencies of sexual dimorphism in coloration than suggested by the survey data because differences in less striking colors are less likely to be reported. This may be the case for anguids (e.g., Vial and Stewart, 1989).

VI. SEX RECOGNITION AND INTRASPECIFIC AGGRESSIVE AND SEXUAL BEHAVIORS

A. Overview

Coloration is believed to provide important cues allowing identification of sex by conspecifics in many lizards. As documented above, chromatic sexual dimorphism occurs extensively in certain families. That the agonistic and sexual display behaviors of numerous lizards in the families Agamidae, Chamaeleonidae, Iguanidae, Gekkonidae, Lacertidae, and Teiidae maximally expose brightly colored patches to conspecifics, or that those forms showing rapid color changes adopt the bright coloration during territorial and sexual behavior, provides circumstantial evidence for a role in sex recognition (e.g., Noble and

Bradley, 1933a; Kramer, 1937; Kitzler, 1941; Greenberg and Noble, 1944; Fitch, 1958; Clarke, 1965; Parcher, 1974; reviewed by Stamps, 1977, and Becker, 1984).

In many of these lizards, the colorful patches are hidden except during displays or appear only during displays by physiological color changes. For example, bright colors are often found on the dewlap of anoles (Rand and Williams, 1970), the gular regions of various other lizards (e.g., Clarke, 1965; Becker, 1984), and on the ventrolateral surfaces and on the ventral side of the tail in many species (e.g., Fitch, 1958; Clarke, 1965). While displaying, a lizard may flatten its body laterally and/or tilt its side toward the recipient of the display to reveal ventrolateral colors, tilt a tail upward to show its ventral pattern, and enlarge the throat by means of the hyoid apparatus, revealing the gular or dewlap colors. This distribution of pigment and behavior is consistent with the coverable badge hypothesis, which states that individuals benefit by being able to produce either bright signals of dominance or duller signals of submissiveness (Hansen and Rohwer, 1986). However, it is also consistent with the hypothesis that the bright coloration is important in social contexts such as sex identification but is hidden, except during display, to reduce the probability of detection, by predators, as in some aposematic systems, or by prey.

Despite its presumedly widespread importance, a role of bright color in sexual recognition and social behavior has been experimentally established in only a few species and families, specifically in the Agamidae, Iguanidae, Lacertidae, and Scincidae, and strongly suggested for several species in Chamaeleonidae. Suggestive evidence based on field observations is available for other species. For only three groups of lizards, the iguanid genera *Anolis* and *Sceloporus* and the scincid genus *Eumeces*, has a demonstrated role in social behavior of male dimorphic coloration been studied physiologically. However, the female reproductive coloration is related to aggressive behavior and has been induced hormonally in several iguanid lizards.

The following material on male coloration and behavior adopts a taxonomic approach at the familial level. Within sections on each family, behavioral and physiological information is presented sequentially for each taxon.

B. Male Coloration

1. General Male breeding colors may function as social releasers in a high proportion of the dimorphic species given in Table 6.3. However, experimental studies establishing this function have been reported for only the few species for which data are summarized in Table 6.5. In these, behavior is strongly influenced by the bright display colors of conspecific males.

Table 6.5 Evidence that bright male coloration affects or is affected by social behavior

Species	Response	Condition	Source
Anolis carolinensis	Females prefer males with red dewlap	Males with red or green dewlaps	Sigmund, 1983
	Subordinate males darker	Formation of dominance hierarchy	Greenberg et al., 1984
Sceloporus undulatus	Males attack	Females painted with male blue	Noble, 1934; Cooper and Burns, 1987
	Males court	Males painted as females	
Urosaurus ornatus	Green males more aggressive	Male throat color orange or green	Hover, 1985
	Male behavior affected by opponent's throat color	Throat color morphs reversed by painting	
Agama agama	Males attack	Model or female painted as male	Harris, 1964
Eumeces laticeps	Males attack	Females painted as male	Cooper and Vitt, 1988
Lacerta sp.	Males attack	Females painted as male	Kramer, 1937; Kitzler, 1941
	Males court	Males painted as female	

2. IGUANIDAE

Roles of male morphological color changes and bright reproductive coloration have been best studied in Iguanidae. Such differences are believed to be important in the aggressive behavior and sexual behavior of numerous species in genera including, but not limited to, *Anolis* (e.g., Sigmund, 1983; Greenberg et al., 1984; Losos, 1985), *Sauromalus* (Prieto and Ryan, 1978), *Sceloporus* (Cooper and Burns, 1987), *Urosaurus* (Hover, 1985; Zucker, 1989), and *Uta* (Ferguson, 1966). Evidence for some of these lizards is merely suggestive. In *Sauromalus obesus tumidus* sex recognition has been reported to be based in part on color differences, but no evidence for this assertion has been presented (Prieto and Ryan, 1978). In contrast, experimental evidence suggests that coloration and body form may be important in releasing courtship behavior and aggression in *Uta stansburiana stejnegeri* (Ferguson, 1966). However, no effect of coloration per se was established.

Throat coloration and aggressiveness are related in male *Urosaurus ornatus*. A New Mexican population shows three distinct morphs for throat color, that is, ventral gular coloration. Gular colors of these morphs are (1) centrally green bordered by orange, (2) uniformly orange or orange with a few scattered green scales, and (3) uniformly yellow. These throat colors are clearly visible to conspecifics during aggressive encounters, when the hyoid apparatus is extended ventrally. Orange males involved in natural agonistic encounters in the field are significantly larger than green males despite a lack of difference in mean size for the entire population. Because large size provides an advantage in many iguanid lizards (Greenberg and Noble, 1944; Rand, 1967; Ferner, 1974; Ruby, 1978; Stamps, 1978; Tokarz, 1985), the involvement of smaller green males in aggressive behavior suggests that they may be more aggressive than orange males (Hover, 1985).

Relative aggressiveness of the abundant green and orange morphs, but not of the rare yellow morph, was studied by simultaneously placing two nonresident males in an outdoor enclosure and observing their social interactions (Hover, 1985). The green males dominate orange males in over 80% of trials and have higher display rates than do orange males. However, further experiments involving color modification by painting show that green males having the throat painted orange have no advantage over orange males having the throat painted orange as a control for effects of painting. Neither is there any difference in frequency of dominance between such males and green males painted green. Orange males painted green dominate orange males painted orange in a marginally significant percentage of trials but do not differ in dominance from green males painted green.

One interpretation of these results is that they indicate that the behavior of orange males is affected by their perception of the throat coloration of the opponent (Hover, 1985). The effect of throat color perception on green males is ambiguous because green males painted green do not dominate either the orange males painted green or other green males painted green. Their loss of advantage over males colored orange demonstrates the importance of the opponent's throat color to green males.

An alternative interpretation of the data for *Urosaurus ornatus* is possible. The failure by green males to dominate green males painted orange may be due to the greater dominance potential of green males and have little to do with perception of color. Green males painted orange still behave like green males and win half their encounters with green opponents. This suggests that for the more dominant green males, the behavior of an opponent may be more important than its throat coloration in determining encounter outcomes. Physiological aspects of throat coloration have not been studied, but this fascinating chromatic and behavioral polymorphism should be studied further. Because throat coloration of *U. ornatus* is geographically variable, being monomorphic in some populations and polymorphic in others, the results could be specific to particular populations.

Variation in dewlap coloration is even more complex in other populations of *Urosaurus ornatus*. Dewlaps in one population have five distinct dewlap color morphs (Thompson and Moore, 1987). Some of the morphs are discrete, but continuous variation in color occurs among others. Males of one morph have orange dewlaps having centrally located blue spots of variable size. In laboratory studies of males matched by size, males having blue spots more than 10% larger than those of their opponents dominate in 45 of 48 trials as indicated by exclusion of the other male from the area around a heat lamp. Preliminary field observations suggest that males employ several reproductive strategies related to dewlap coloration (Thompson and Moore, 1987).

In some populations of *Urosaurus ornatus*, some males are dark brown, almost black, whereas others have a light gray-brown coloration similar to that of females (Zucker, 1989; Fig. 6.8). Dark males maintain territories that do not overlap with those of other dark males. Their territories usually contain several females. The dark males appear to be dominant over light males, which occupy home ranges that overlap somewhat with the territories of dark males. Dark coloration is not permanent but may be assumed when a light male displaces a former resident from a territory or occupies a newly vacated territory (Zucker, 1989). In an outdoor enclosure, the most dominant, or alpha, male among six develops dark dorsal coloration. When the alpha male is removed, the beta male becomes dominant and darkens (Carpenter and Grubitz, 1960). The dark coloration may be a signal of dominant status within a territory.

Blue ventrolateral coloration is widely distributed among males in the iguanid genus *Sceloporus*. Male *S. undulatus* reveal the bright blue colors on the throat and side to females or to other males during social displays. Displays by females also reveal the absence or lesser development of blue. Lizard social behavior can often be profitably studied by tethered introduction experiments, in which experimental lizards are introduced into the territories of free-living lizards. In the first such study, Noble (1934) painted females with the male blue pattern, causing them to be treated as male by conspecific males. Furthermore, females with other bright colors painted on their sides were also treated as male.

Unfortunately, insufficient details have been published to allow evaluation of any hypotheses about coloration. Cooper and Burns (1987) reported a statistically analyzable extension of the study by Noble (1934) in which several groups of tethered lizards were pre-

Fig. 6.8. *Urosaurus ornatus.* The dorsal coloration of dominant males (right) in some populations is much darker than the common dorsal color pattern of males and females (left). (Reproduced with permission, Soc. Study Amphibs. and Reptiles, after N. Zucker, *Journal of Herpetology* Vol. 23, No. 4, 1989.)

sented to resident territorial males. Unpainted males and females were presented to establish the normal responses of the residents. Other males and females were painted to resemble members of the opposite sex; these were the critical experimental subjects for establishing the importance of coloration in sex identification. Control for effects of painting was accomplished by presenting lizards painted the normal color for their sex. The resident males clearly respond to the ventrolateral coloration, reacting aggressively to females painted blue and courting males painted white. Painting has no effect, as indicated by the nearly identical response to unpainted lizards and paint controls of the same sex. Thus, male *Sceloporus undulatus* appear to identify lizards with blue ventrolateral patches as male and those lacking the blue as female.

The communicative significance of sexually dimorphic coloration in *Sceloporus* is underscored by a comparative study of aggression in *S. virgatus, S. undulatus consobrinus,* and *S. undulatus tristichus,* three taxa differing in the degree of sexual dichromatism (Vinegar, 1975). In *S. virgatus,* neither sex has blue belly patches, but both have blue throats. The blue throat coloration of females is either surrounded or replaced by orange during the breeding season. Male *S. u. tristichus* have blue belly and throat patches; females have both patches, but the belly patches are lighter than in males. Male coloration is brightest in *S. u. consobrinus,* in which males have blue belly and throat patches, but females have only the throat patch, lacking blue on the belly. Observations of social interaction following introduction of tethered stimulus lizards to free lizards in the field reveal that male *S. u. consobrinus* are more aggressive to conspecific males than are males of the other two taxa. The most brightly colored females, those of *S. u. tristichus,* were also the most aggressive to conspecifics. These results tentatively suggest that the degree of chromatic dimorphism may be adjusted to the intensity of intraspecific aggressiveness.

Chromatic displays are related to dominance in male *Sceloporus.* In *S. cyanogenys,* after stable dominance relationships have been established, an alpha male commonly tips its head at an acute angle upon approach by a subordinate male. This behavior displays the blue gular area of the dominant male, causing the subordinate male to arrest its movement and lower its body to the substrate (Greenberg, 1977b). In some populations, the adult male *S. u. erythrocheilus* show orange and yellow color morphs for throat coloration; other populations include adult males lacking either bright color (Rand, 1988, 1990). The bright color morphs appear to be discrete and to be fixed once they develop because there are no males having intermediate coloration and no males change from one morph to the other (Rand, 1990). They also vary in frequency among populations (Rand, 1990). Orange males always dominate yellow ones in aggressive encounters. They also spend more time with females and more time courting females than do yellow males (Rand, 1988). Nevertheless, orange and yellow males do not differ in size (Rand, 1990).

The ventrolateral regions of hatchling *Sceloporus* are white in both sexes, but adult males characteristically have blue patches bordered medially by black regions that are sometimes extensive in large individuals. Males approaching maturity develop the dark pigmentation by increasing the number of scales containing melanophores, the number of melanophores per scale, and the degree of melanosome dispersion. The black color in *S. occidentalis* is enhanced by exogenous androgen in juveniles of both sexes, but males show greater pigment dispersion (Kimball and Erpino, 1971).

The physiological mechanism for establishing blue coloration in male *Sceloporus* is unknown, but investigations suggest hormonal induction. Following castration, adult male melanosome dispersion decreases; melanosome dispersion does not decrease in castrated adult females but increases after treatment with androgen (Kimball and Erpino, 1971). Because adult male coloration does not show appreciable seasonal variation, the major hormonal influence on the dark coloration appears to be developmental, although low levels of androgen may be required for long-term maintenance of the black patches. This appears to account for the failure of testosterone pellets implanted in intact adult male *S. olivaceus* to produce detectable color change (Forbes, 1941). The lack of the "typical masculine coloration" in some adult males, and the lack of reference by Kimball and Erpino (1971) to the blue portion of male coloration, leads to the suggestion that the blue color may not be under direct androgenic control. However, a preliminary report suggests that the blue ventral coloration of *S. undulatus erythrocheilus* responds to exogenous testosterone more strongly in males than in females; orange throat coloration is induced by exogenous testosterone in both sexes (Rand, 1989). That the intensity of throat coloration covaries with the annual testicular cycle further suggests androgenic control (Rand, 1990).

In the related *Sceloporus occidentalis*, the dorsum of excited individuals has a different type of bright blue coloration. The blue color develops during intense aggressive bouts and also in lizards handled by experimenters. Preliminary study shows that injections of epinephrine induce this rapid color change in intact lizards (Cooper and Ferguson, 1973b). The color change is induced by aggregation of melanosomes within the melanophores, revealing a blue structural color presumably produced by Tyndall scattering from the iridophores (Bagnara and Hadley, 1973). Unfortunately, neither removal of adrenal glands nor measurements of circulating epinephrine levels were performed to establish the role of epinephrine. Brightening of some other iguanids during prolonged aggressive encounters might be similarly controlled.

The distinctive bright coloration of females in several iguanid species sometimes occurs in juvenile but not in adult males. In *Tropidurus delanonis*, similarity between coloration of juvenile males and adult females may reduce aggression by adult males (Werner, 1978). Juvenile male *Crotaphytus collaris* have a red-orange pattern of spots and bars similar to that of adult females. This coloration appears within a few weeks of hatching at approximately 45-mm snout-vent length and disappears before sexual maturity in the fall of hatching or early in the next activity season by a length of about 76 mm (Fitch, 1956). The adult female orange-red coloration has been hypothesized to be an inhibitory stimulus for courtship (Fitch, 1967) or aggressive behavior (Clarke, 1965). A further hypothesis is that the orange-red pattern of juvenile males has the same function as that of adult females (Fitch, 1967); the juvenile coloration of male *T. delanonis* and *C. collaris* may thus be considered a form of female mimicry (Rand, 1986).

Studies of the hormonal basis of such juvenile coloration have been inconclusive. Progesterone, which induces the bright female coloration in *Crotaphytus collaris* (Cooper and Ferguson, 1972b), fails to induce the orange-red pattern in intact juvenile males; corticosterone injections also fail (Rand, 1986). Circulating concentrations of progesterone and androgen, both of which induce the female orange spots when administered exogenously (Cooper and Ferguson, 1972b), did not differ between those males having the orange-red color and those lacking it (Rand, 1986). However, the radioimmunoassay data for androgen reported by Rand (1986) give the total concentration of both testosterone and dihydrotestosterone but do not separate the two. Rand (1986) hypothesized that dihydrotestosterone might circulate in early stages of sexual maturation and specifically that it might stimulate synthesis of the orange pigment. Subsequent radioimmunoassays prove this hypothesis to be incorrect. Tests specific for the two androgens reveal an elevated titer only of testosterone (Rand, 1986). An alternative hypothesis is that the responsiveness of integumentary androgen receptors involved in pigment synthesis may change ontogenetically (Rand, 1986).

The large iguanid genus *Anolis* shows at least three aspects of coloration with possible behavioral significance: dewlap coloration, sexually dimorphic body coloration, and body coloration that is modifiable by physiological color change. The dewlap is a flap of skin in the gular region that lies flat against the body in resting position but may be extended by the hyoid apparatus. When extended, the dewlap is a thin, dorsoventrally flattened structure having a large surface area on either side. It is extended repeatedly during both courtship and aggressive encounters, in many species revealing a bright coloration.

Color appears to be very important to anoline behavior. Because the dewlap coloration is prominently displayed in intraspecific social behavior, it is believed to have some intraspecific signal function. Males usually have larger and more brightly colored dewlaps, suggesting that these features may allow recognition of sex and perhaps play a role in sexual selection. Nevertheless, brightly colored dewlaps may be present in both sexes. Additionally, variation in dewlap coloration among sympatric species of anoles has suggested a role in maintaining reproductive isolation. Only evidence regarding intraspecific behavioral roles of dewlap coloration is treated here; the interspecific relationships between dewlap coloration and behavior are discussed in a separate section below. A second behaviorally important aspect of color in anoles is that the overall body coloration in some anoline species undergoes rapid color transformations related to social behavior.

Most of the information on possible intraspecific functions of anole dewlap color has been obtained from studies of *Anolis carolinensis*, in which the dewlap is bright red in most populations. However, in a region of southwestern Florida these anoles have dewlap coloration ranging from gray to magenta (Duellman and Schwartz, 1958; Christman, 1980). Because demonstrating an intraspecific role for dewlap coloration has proven to be experimentally difficult, a brief history of landmarks in its study is given. Greenberg and Noble (1944) performed the first series of experiments. First they observed that resident males respond aggressively to females having attached artificial red dewlaps or having the red-painted throat distended by saline injection. However, males also behave aggressively toward females with distended but unpainted throats. This indicates that the presence of an enlarged dewlap evokes aggression and thus provides an operational mechanism allowing identification of a stranger as a male.

Greenberg and Noble (1944) were unable to demonstrate any role for the red dewlap color, although they believed that it advertises the presence of territorial males to intruders. They also studied the potential importance of dewlap color in mate choice by females. In these experiments females were allowed to approach and remain with three categories of males: unaltered males, males having dewlaps painted green, and males having dewlaps artificially prevented from being extended. No evidence for any preference for red dewlap coloration was obtained, whether or not new females were released between the established territories of males, new males were released in the vicinity of an established female, or established females were tested with established males. However, the results were inconclusive because only 11 trials were conducted for all conditions combined.

The effects of dewlap color in courting males on sexual receptivity by females have been tested (Crews, 1975a). Whenever courted (at different times) by males of *Anolis carolinensis* having natural red dewlap coloration and males having dewlaps made blue by injection of India ink, those females receptive to males with red dewlaps also tended to be receptive to males with dewlaps made blue, and the behavioral responses did not differ. A study of long-term physiological effects of the dewlap color of the male on female reproductive condition has indicated that exposure to males having either red or blue dewlaps induces ovarian recrudescence earlier than does exposure to castrated males or males having immobilized dewlaps. The rate and degree of ovarian recrudescence do not differ between females exposed to males with red dewlaps and those exposed to males with blue dewlaps.

There is evidence that female *Anolis carolinensis* prefer males with red dewlaps to those with green dewlaps (Sigmund, 1983). In the experimental design used, females known to be sexually receptive are placed between two males matched for size, behavior (unspecified), and head-bobbing. At the start of each trial a female occupies a terrarium midway between two terraria situated 4 m apart, each containing a single male having either a red dewlap or a dewlap painted green. The most distant glass wall of the terrarium at each end is covered with red or green paper. A female is considered to have made a choice when it moves from its initial central perch to a perch at one end of the central terrarium and remains there for 5 minutes or if it tries to crawl through the end glass. Trials in which one male displays more than the other are aborted and repeated on another day. Twenty-seven different females responded in six experiments representing all combinations of background and dewlap colors. Eighteen trials were conducted per experiment.

Female preference for color is not affected by background coloration. However, with end terraria having a green background, females approach males with red dewlaps significantly more often than males with green dewlaps. Females show no preference for the red background. Twice as many females chose males with red dewlaps on green backgrounds as males with green dewlaps on red backgrounds; however, this result proved to be not statistically significant, perhaps due to small sample size. Males with red dewlaps on red backgrounds are approached significantly more frequently than males with green dewlaps on green backgrounds.

These data indicate that the red coloration of dewlaps is an important intraspecific social signal and suggest that dewlap coloration might have been sexually selected by differential choice of females for mates with bright red dewlaps. Sigmund (1983) has suggested that previous experiments do not demonstrate any effects of dewlap coloration because the lizards have been placed too close to each other. Another possible explanation for the apparent absence of any effect of dewlap coloration in the study by Crews (1975a) is that the physiology of the eye may favor the red and blue ends of the spectrum. Gull chicks show a strong tendency to peck both the naturally colored red model bills and blue model bills, which do not occur naturally (Hailman, 1964, 1967). The colors of the models most frequently pecked match the peaks of spectral sensitivity of the gulls (Hailman, 1967, 1969). Thus a preference for the natural red dewlap coloration in *Anolis carolinensis* could have been obscured by a strong response to the artificial blue dewlap coloration. From a distance, a flash of dewlap coloration during dewlap extension may attract the attention of a female, identifying a potential mate. Additional factors are presumably important to mate selection at close range. Although the study (Sigmund, 1983) shows that the red color of the male dewlap affects female behavior, the tests for overall effects of dewlap color, background color, and contrast between dewlap and background are

Table 6.6 Pteridines in the dewlaps and dewlap coloration of ten species of *Anolis*

Species	Dewlap color	DRO	ISO	XAN	SP	AHP	BIO
Series 1							
Grass anoles:							
A. pulchellus	Red	+	+	+	−	+	+
A. krugi	Orange	+	+	+	−	−	+
A. poncensis	White	−	+	−	−	+	+
Tree anoles:	Red, green	+	+	−	+	−	−
A. gundlachi	Brown	+	+	+	+	+	+
Series 2							
A. stratulus	Red	+	+	+	+	+	+
A. evermanni	Yellow	−	+	+	+	+	+
A. acutus	Yellow	−	+	+	+	+	+
A. cuvieri	Yellow	−	+	+	+	+	+
A. carolinensis	Red	+	+	+	+	+	+

Source: Data reprinted with permission from Ortiz and Maldonado, 1966.
Abbreviations: DRO, drosopterins; ISO, isoxanthopterin; XAN, xanthopterin; SP, sepiapteridine; AHP, 2-amino-4-hydroxypteridine; BIO, biopterin. + indicates presence in the dewlap; − indicates absence.

based on data pooled from all experiments. They are questionable due to lack of independence among observations.

Dewlap color in anoles appears to be determined largely by pteridines, although carotenoids may participate. *Anolis carolinensis* has yellow pigments in erythrophores and red granules deeper in the dewlap (von Geldern, 1921). The red and orange dewlap pigments of the Puerto Rican *A. pulchellus* include three pteridines tentatively identified as neodrosopterin, drosopterin, and isodrosopterin, with drosopterin predominant. The same three pigments occur in the dewlaps of *A. krugi, A. cristatellus,* and *A. stratulus,* but in smaller amounts. All of these species have red or orange on at least part of the dewlap. Yellow carotenoids also occur in the dewlaps of these species and in *A. evermanni,* but red and orange carotenoids are absent. Several species that lack orange or red dewlaps, including *A. evermanni, A. poncencis,* and *A. cuvieri,* also lack these pteridines (Ortiz et al., 1963). The three drosopterins are present in all five species studied with red or orange dewlaps, including *A. carolinensis,* and are absent in four species having yellow or white dewlaps (Table 6.6). *Anolis gundlachi* has a brown dewlap but only small amounts of drosopterins that are restricted to the dewlap's margin (Ortiz and Maldonado, 1966).

Dewlap coloration is revealed when the dewlap is erected by the retrobasal process of the hyoid apparatus under the control of the nucleus ambiguus of the brainstem through the branchiomeric components of the vagal and glossopharyngeal nerves (Font et al., 1986; Font, 1988). Dewlap extension is sometimes confused with throat extension (gular inflation), an engorgement of the buccal floor that may be attributable to a different pattern of hyoid activation. However, extension of throat and dewlap occur in different social contexts; the dewlap is extended primarily during assertion and challenge displays, whereas the throat is extended at close quarters during intense agonistic bouts and is often a prelude to biting (Cooper, 1977; Greenberg, 1977a).

Body size and predation may be important variables affecting the importance of dewlap coloration and body coloration as social signals. A few species of small anoles show greatly reduced dewlaps. At least one, *Anolis bahorucoensis,* is chromatically dimorphic in body coloration, with the male being much more brightly colored than the female. The head-bobbing displays are infrequently performed by males and are less vigorous than in species having larger dewlaps. Only the head and neck move in *A. bahorucoensis;* the dewlap is not pulsed. Intense predation by congeners on small anoles, coupled with the increased probability of being detected while performing dewlap displays, may account for reduced dewlap size (Fitch and Henderson, 1987). The bright body coloration of males may substitute for bright dewlap coloration during close-range social encounters with conspecifics; it is cryptic from a greater distance, reducing vulnerability to predation (Fitch and Henderson, 1987).

In addition to dewlap coloration, body coloration is behaviorally significant in some anoles. In *Anolis carolinensis,* physiological changes in body coloration are related to aggressive behavior and dominance. These changes affect both the overall skin coloration and the color in specific regions. During intensely aggressive bouts, likely to lead to actual fighting, a small patch on the side of the head posterior to the eye turns very dark in both sexes, an effect also produced by handling (Greenberg and Noble, 1944). This eyepatch provides a valuable indicator of circulating epinephrine following adrenal (chromaffin) activation because it is devoid of alpha-adrenergic receptors (Hadley and Goldman, 1969). The darkening effect of epinephrine due to its action on beta-adrenergic receptors can be overridden by the response of alpha-adrenergic receptors. Consequently, body coloration can be green both at very high and at very low levels of circulating epinephrine and brown at intermediate levels. However, a visible dark eyepatch is a positive indication of high levels of epinephrine. Fig. 6.9 illustrates these relationships among coloration, stress, and circulating levels of hormones.

Skin color changes during formation of dominance hierarchies by adult male *Anolis carolinensis* leave dominant individuals greener than subordinates, which tend to be darker (Fig. 6.10; Greenberg et al., 1984; Greenberg and Crews, 1990). In the field, green males are significantly larger than brown ones, but brown and green females do not differ in size. This suggests that body coloration may imperfectly signal both sex and dominance status in males (Medvin, 1990). The

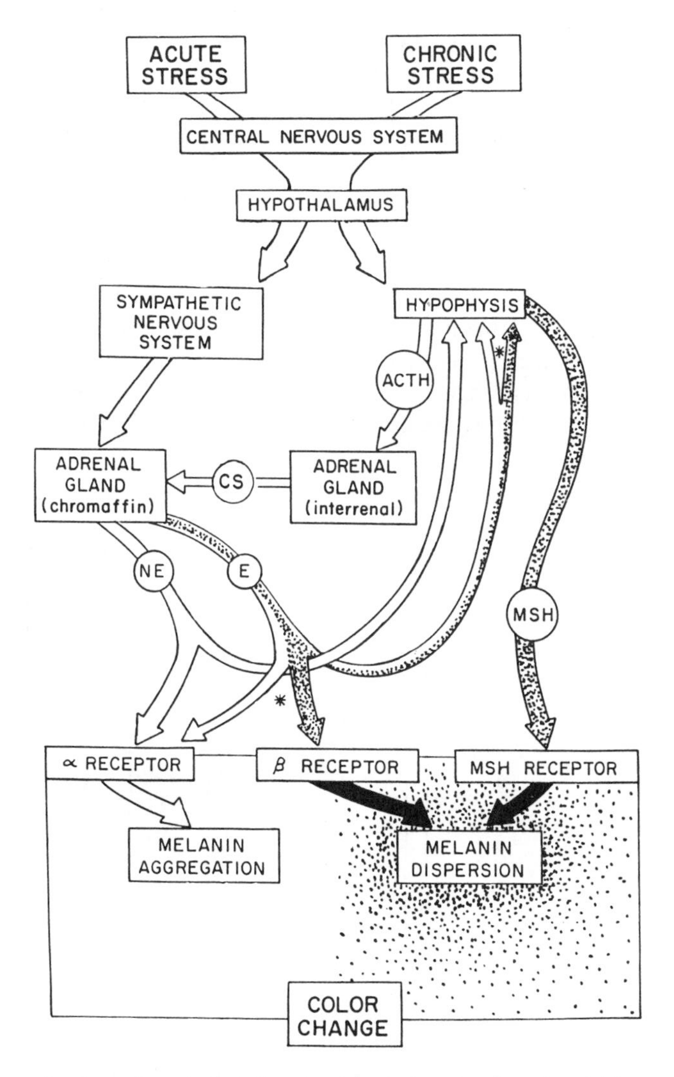

Fig. 6.9. *Anolis carolinensis*. Putative physiological mechanisms by which stress affects color change complexly. ACTH, adrenocorticotropic hormone; CS, corticosterone; E, epinephrine; MSH, melanophore-stimulating hormone; NE, norepinephrine. CS stimulates additional synthesis of E by enzymatic conversion from NE. Asterisk indicates that E biphasically causes darkening or lightening of body color. (From Greenberg and Crews, 1983.)

skin color of castrated subordinate *A. carolinensis* does not differ from that of isolated males, suggesting that they are not substantially stressed (Greenberg et al., 1984). Circulating corticosterone levels were higher in intact subordinates than in dominant lizards, isolated lizards, or in castrated subordinates in this long-term study, but were not elevated in a short-term study (Greenberg and Crews, 1990). This indicates that an adrenocortical stress response occurs in intact subordinate males (confirming and extending observations on *Dipsosaurus dorsalis* by Callard et al., 1973, and on *Cnemidophorus sexlineatus* by Brackin, 1978). Further, androgens appear to play a role in the stress response because castrates that fight and subsequently become subordinate do not change body color or have elevated corticosterone levels (Greenberg et al., 1984).

Chronic stress induces increased release of MSH in *Anolis carolinensis* (Greenberg et al., 1986), as in mammals (mice, Sandman et al., 1973). Because ACTH contains an amino acid sequence similar to that of the smaller MSH molecule, corticotropin-releasing factor may cause simultaneous release of ACTH and MSH (Proulx-Ferland et al., 1982, in rats). For the same reason, darkening due to stress could be

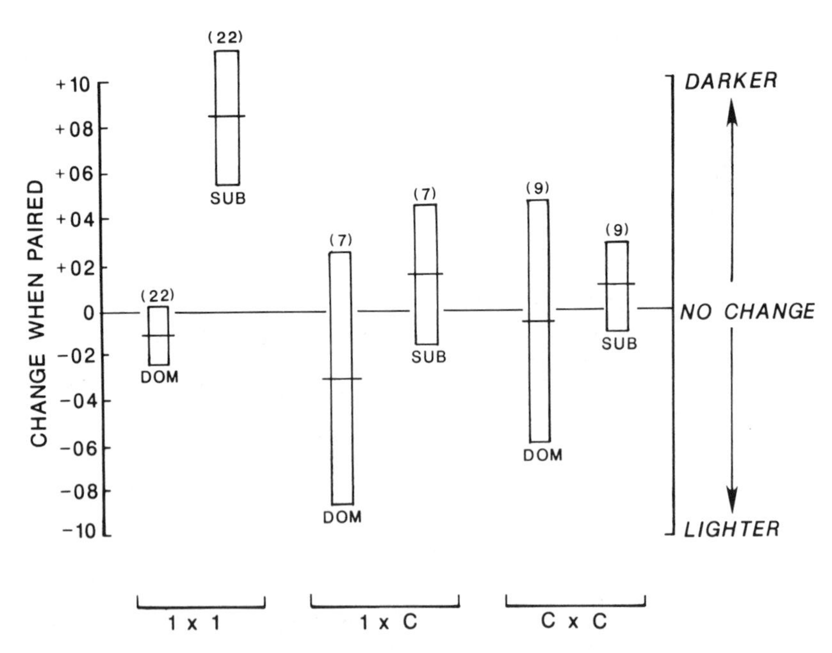

Fig. 6.10. *Anolis carolinensis*. Mean changes in body color following pairing. Horizontal bars show group means, and vertical bars represent 2 standard errors. 1, intact; C, castrate. In pairings of intact lizards the subordinates became significantly darker. (From Greenberg et al., 1984.)

induced jointly by MSH and ACTH. In humans, injections of ACTH darken the skin (Lerner and McGuire, 1964). Corticosterone may additionally provide indirect effects on body color in *A. carolinensis* by facilitating conversion of norepinephrine to epinephrine by enzymatic methylation (Wurtman and Axelrod, 1965; Wurtman et al., 1967). Furthermore, subcutaneous grafts of pars intermedia tissue will release MSH in stressed *A. carolinensis,* inducing darkening (Muerling et al., 1974). Further study will be required to determine the roles of MSH, corticosterone, and ACTH in stress-induced color change.

The possibility that MSH and sex steroid hormones may interact in regulating physiological color change has not been investigated but cannot be discounted. Subordinate male *Anolis carolinensis* court at reduced frequency and have reduced circulating concentrations of sex steroid hormones (Greenberg and Crews, 1990); these lizards are characteristically a light brown, indicative of altered melanotropin level or mild adrenal activation with attendant epinephrine secretion. In principle, either combination may cause melanophore dispersion by stimulating MSH receptors or beta-adrenergic receptors, respectively.

The absence of any chronic stress disease suggests that the light brown coloration of subordinate males is more likely to result from an increased level of circulating MSH than from epinephrine. Indeed, elevated circulating MSH levels have been detected in *Anolis carolinensis* subjected to chronic social stress (Greenberg et al., 1986). In addition to the effect of increased MSH due to stress, subordinate animals may be darker because of changes in responsiveness to stimuli activating release of adrenomedullary catecholamines. Alternatively, darker coloration in subordinates could result from increased density of beta-adrenergic or MSH receptors. Apparently, social stress may cause either acute depression of circulating MSH levels or chronic increase of MSH levels (Greenberg et al., 1986). Whatever the mechanism, castration ameliorates the melanotropic response to social stress.

It may be significant that the hormones MSH, ACTH, corticosterone, and epinephrine are often associated with the behavioral and color patterns manifested by *Anolis carolinensis* during aggressive interactions and in dominance relationships. Of particular interest is reduced aggressiveness (Patterson et al., 1980) and enhanced retention of conditioned avoidance behavior (DeWied, 1966; DeWied and Bohus, 1966). These and other behavioral effects seem related by a general enhancement of attentiveness (LaHost et al., 1980; Datta and King, 1982), an idea supported by observations that treatment with exogenous MSH reduces tonic immobility in *A. carolinensis* (Stratton and Kastin, 1976). This observation suggests that MSH may modulate responsiveness to some external stimuli. Further study will be required to determine the contributions of stress-sensitive hormones to acute and chronic color changes manifested by *A. carolinensis* and the manner in which their actions are integrated with body coloration and behavior.

3. AGAMIDAE

During the breeding season, males of many agamid species have bright reproductive coloration, but conspecific females lack it. Although the bright colors are characteristic of reproductively active males and occur during courtship and territorial advertisement (Harris, 1964), they are subject to rapid physiological changes. Examples of species having such changeable colors are *Agama agama* (Harris, 1964), *Calotes versicolor* (Lydekker, 1901), and *C. mystaceus* (Smith, 1915). Such changes are considered below with other physiological color changes.

Male *Agama agama* from different geographic regions have a bright orange, vermillion, or yellow head with a black or blue body during the reproductive color phase. Males presented in the field with cardboard model lizards painted with the male pattern often respond aggressively, even though the models are immobile (Harris, 1964). Wooden model lizards effectively evoke challenge displays and fighting responses from males in the field, but only whenever the models are painted with orange or vermillion head and body of dark color (black, brown, dark green, gray, and blue). Combined with light blue, pink, yellow, or buff body coloration, orange and vermillion head coloration does not induce aggression. With blue or black body color, only orange-, vermillion-, and dark yellow–headed models are challenged or attacked. In addition, a single living female painted with orange head and blue body was repeatedly challenged by a male (Harris, 1964). The bright orange head coloration is likely an important releaser of aggressive behavior by male *A. agama*, but it is ineffective alone. Thus, the color pattern appears to act as a gestalt, that is, its effects depend on the overall combination of chromatic features rather than solely on independently acting chromatic components.

Female *Agama agama* also appear to respond to male breeding coloration. They display with arched back and elevated tail, directing this only to males in reproductive color phase; however, evidence from model experiments and naturalistic observation is inconclusive (Harris, 1964). The behavioral importance of reproductive color patterns in *A. agama* is supported by highly suggestive evidence. However, this must be interpreted with caution because no statistical analyses and no information on sample sizes have been published (Harris, 1964).

In *Agama agama* physiological color changes associated with behavior begin within 15 to 20 seconds and are completed within minutes. Courting males are virtually always in bright reproductive color phase unless they begin courting immediately after fighting. In this event they may be in fear color phase (so described by Harris, 1964); males then develop a pale mottled body coloration, and the formerly orange head and middle section of the tail turn dark brown. Males retain bright breeding colors throughout courtship and copulation. Only dominant males in bright reproductive color phase head-bob early in courtship. Whereas males almost always begin aggressive interactions in reproductive color phase, both combatants rapidly switch to fear color phase once the aggressive activity becomes intense. Lizards with the dark reticulate color phase show a dark ground color overlain with a reticulate pattern. Males and females closely resemble each other in this color state, which appears when the lizards are dark and cool, as at night. The physiological basis for assumption of bright coloration by dominant males is unknown, but castration causes the coloration to fade and exogenous testosterone restores it in castrates (Inoué and Inoué, 1977).

Male coloration is closely related to dominance. *Agama agama* usually form small groups of about a dozen individuals, of which only the dominant male displays the reproductive color phase. Any other adult males in the territory of the dominant male remain in reticulate phase and are challenged if they change to reproductive color phase. Such adult males, which are denied reproductive opportunities by the dominant male, tend to be too small to establish their own territories. In the challenge display the head and body are bobbed up and down, whereas the gular fold is extended. A dominant male directs challenge displays only to other males in reproductive color phase. Juvenile males and adult females share a color pattern that is distinct from that of adult males. Both are tolerated by adult males, which appear to distinguish males in juvenile color phase from adult females by behavioral differences (Harris, 1964). Dominant male coloration is brighter when the male is near the center of its territory rather than at its periphery; subordinate coloration brightens as distance from a dominant male increases (Madsen and Loman, 1987; Table 6.7).

In encounters between pairs of adult male *Agama agama* staged in wire cages placed in outdoor enclosures, the dominant members of pairs have significantly brighter head coloration and assume dark blue-black body coloration. The coloration of subordinate males tends

Table 6.7 Effect of distance from the nearest dominant (cock) male on coloration of the nearest subordinate adult male *Agama agama*

Color of subordinate adult males	Distance to nearest cock males (m)	SD	*n*
Gray/brown	8.5	4.7	87
Green spotted head	7.8	4.5	88
Red cheeks	10.1	4.2	105
Red	11.3	5.1	24
Cock color	13.7	5.8	6

Source: T. Madsen and J. Loman, 1987.

toward a duller orange to brown head and a gray to brown body. Aggressive behavior occurs until one of the lizards turns pale but ceases as soon as the subordinate coloration is adopted. When the head and gular fold of the dominant member of a pair are painted green, the coloration of subordinates shows a nonsignificant brightening (Inoué and Inoué, 1977). This effect may be a real one that is masked by small sample size (10 pairs, brightening occurring in 7). After painting, the body coloration of dominant males fades significantly (Inoué and Inoué, 1977). In the absence of controls for effects of painting, it is impossible to state whether this color change represents a response to painting or to altered behaviors by the subordinates.

In the Australian agamid *Amphibolurus maculosus*, the expression of display coloration is related to dominance status (Mitchell, 1973). Males that display and fight develop bright coloration consisting of orange-yellow ventrolateral markings that intensify to reddish-orange ventrally with a pale midventral area. Subordinate males that do not display or fight either retreat from dominant males or, if attacked, perform a subordinate gesture by rolling over onto their backs. These subordinate males develop only faint yellow ventrolateral coloration.

4. CHAMAELEONIDAE

Lizards of the family Chamaeleonidae are widely known for their ability to undergo physiological color change, but very little is known about the relationships of these changes to social behavior. What little is known about the physiological mechanisms producing these changes indicates that rapid color changes may be produced both hormonally and neurally (see Section III). Bright color patterns limited to the reproductive season occur in some male chameleons, such as *Chamaeleo dilepis* (Brain, 1961), *C. hoehnelii* (Bustard, 1965), *C. jacksonii* (Stamps, 1977), *C. pardalis* (Bourgart, 1970), and *Bradypodian pumilis* (Burrage, 1973; Kästle, 1967). Males of some species, including

Brookesia superciliaris (Parcher, 1974), *C. brevicornis* (Parcher, 1974), *C. dilepis* (Brain, 1961), *C. gastrotaenia* (Parcher, 1974), *C. gracilis* (Bustard, 1967), *C. nasatus* (Parcher, 1974), *C. parsonii* (Parcher, 1974), *B. pumilis* (Burrage, 1973; Kästle, 1967), and *C. willsii,* have bright coloration during aggressive behavior or courtship but are cryptically colored at other times. Dominant males of some species such as *C. hoehnelii* (Bustard, 1965) and *C. jacksonii* (Stamps, 1977) display the bright pattern for long intervals. The bright color pattern differs from that of females and subordinate males; a dominant male may lose the distinctive pattern if defeated. At least one species, *C. namaquensis,* is consistently sexually dimorphic in body coloration, but neither sex is brightly colored and no bright male breeding coloration has been reported (Robinson, 1978). These variations in coloration and mechanisms of color change imply considerable intrafamilial diversity in physiological control.

Color changes occur reliably during social encounters of several chameleons. In these, both sexes may change colors in aggressive situations, and males may assume a distinctive color pattern that is identical during assertion displays and courtship. *Chamaeleo brevicornis* has brown nondisplay coloration, but the rostral process, centers of the occipital lobes, and the top of the head of males turn white during agonistic encounters. During courtship or assertion display males assume red coloration of the rostral process and centers of the occipital lobes, yellow on the eye coverings and margins of the occipital lobes, and dark reticulation on the flanks (Parcher,1974). Similar findings of two distinct bright color patterns associated respectively with courtship and aggressive behavior, and each different than the cryptic pattern, have been reported in *C. nasatus, C. willsii, C. gastrotaenia,* and *C. parsonii* (Parcher, 1974).

An attempt to demonstrate experimentally an effect of bright female display coloration on courtship by males was made by applying green blocking pigments over female body parts involved in the chromatic display green, thereby masking possible color changes (Parcher, 1974). The percentage of time spent in courtship did not change significantly during staged encounters using stained and unstained females both for *C. brevicornis* and *C. nasatus.* The staining treatment produces no effect in *C. brevicornis,* but males of *C. nasatus* are less likely to court stained than unstained females. Although insufficient details of the experimental design were presented to determine the appropriate analysis, Fisher exact tests of the data presented by Parcher (Siegel, 1956) give good approximations of the null probabilities. For *C. brevicornis,* the Fisher probabilities exceed 0.10. However, males are significantly more likely to court nonstained females than females with head and body stained ($p = 0.005$), head only stained ($p = 0.001$), or body only stained ($p = 0.001$). These results suggest that the bright female coloration is used in sex recognition by conspecific males and should be confirmed by future study allowing rigorous statistical analysis and controlling for possible effects of staining per se on behavior.

5. Gekkonidae

Many gekkonid lizards, especially nocturnal species, use vocal signals. However, diurnal species of several genera have brightly colored males. Male *Gonatodes vittatus* have permanent bright coloration that is lacking in females (Quesnel, 1957). Males of several species of *Phelsuma* and *Lygodactylus* undergo physiological color changes associated with social behavior. In these species, males assume bright coloration as territory holders or when dominant in cages, but these colors are rapidly lost if the male loses an aggressive encounter (Loveridge, 1947; Kästle, 1964; Pienaar, 1966; Greer, 1967). A previously subordinate male *Lygodactylus* assumes the bright color phase after defeating a formerly dominant male (Kästle, 1964). Males of *Phelsuma abbotti, P. dubia,* and *P. lineata* show these changes, but the most striking change occurs in *P. lineata,* in which the color shifts from brilliant blue to olive. This species is the most diurnal and aggressive of the three and defends the largest territory (Kästle, 1964).

6. Scincidae

Males of many species in the genus *Eumeces* develop a bright orange head coloration during the spring breeding season. This color is absent or is weakly expressed in females. After the breeding season the color fades until the following spring. Male *E. laticeps* emerging from hibernation in March in South Carolina have a dull coloration but turn bright orange in April. This color is maintained through May and early June but then begins to fade. Males are highly aggressive to conspecific males and court females while their heads are brightly colored but do not court and are not aggressive at other times (Vitt and Cooper, 1985). Similar relationships among head coloration, behavior, and season occur in *E. fasciatus* (Fitch, 1954) and *E. inexpectatus* (Vitt and Cooper, 1986). In all three species the heads of adult males are wider and longer than those of females of the same snout-vent length. However, the head sizes of juvenile males and females do not differ. At sexual maturity the head of a male begins to grow more rapidly than the head of a female, producing a greater difference between the sexes with increasing size. Large males have heads much broadened in the temporal region (Vitt and Cooper, 1985, 1986). Presumably the increase in head size is accompanied by increased bite strength. Large males inflict damaging injuries on each other by biting during fights and may kill smaller males (Cooper and Vitt, 1987a, c). Thus, the orange head coloration in these species emphasizes the primary weapon used by males in intraspecific fighting.

The orange head coloration of *Eumeces* has social significance. Male *E. laticeps* respond much more aggressively to females having heads painted orange (Cooper and Vitt, 1988). Upon introduction of normal and of modified females into the home cages of the males, males perform aggressive head-tilt displays or bite more frequently in tests with females having orange heads than with those painted the normal brown color or unpainted. Thus the orange head color elicits aggression and identifies the lizard as male. However, only about half the males are aggressive to orange-headed females and the aggressive behaviors in all cases are performed before chemosensory investigation of the female. After a male tongue-flicked a female, it performs no further aggressive acts. Males that show no aggression toward orange-headed females had tongue-flicked either the female directly or the substrate where the female had been before initiating any social behavior. It therefore appears that the orange pigmentation is an important social cue, but that pheromones take precedence in sex identification when the chromatic and chemical cues conflict (Cooper and Vitt, 1988).

Early observations suggested that orange coloration in male *Eumeces* is under androgenic control. Castration of *E. fasciatus* during the breeding season decreases the intensity of male head color (Reynolds, 1943) and testosterone injections induce deposition of orange pigment in females, which normally lack brightly colored heads (Edgren, 1959). A study of *E. laticeps* shows that the head color of males castrated in early June fades to a dull orange by early July, at which time the head color of intact controls has also faded (Cooper et al., 1987). Castrates implanted in early June with Silastic capsules containing testosterone propionate develop fully brightened orange heads within 1 month, whereas the heads of intact controls remain brown. This study demonstrates sequential changes in head coloration within lizards following androgen treatment, but effects of castration and hormone treatment are confounded due to lack of castrated controls. However, another part of the same study shows that bright orange head coloration is induced in castrated males with dull head coloration following implantation of testosterone propionate, but that the dull head color does not change in castrated controls (Cooper et al., 1987). Exogenous testosterone propionate, administered as subcutaneous pellets or implanted intraperitoneally in Silastic capsules, reliably induces bright orange head coloration in intact male *E. fasciatus*, *E. inexpectatus*, and *E. laticeps* (Cooper and Vitt, 1987b, c).

Aggressive and sexual behavior in male *Eumeces* are also regulated by androgen. Male *E. fasciatus* castrated during the breeding season do not court sexually active females (Reynolds, 1943). Males of *E. fasciatus, E. inexpectatus,* and *E. laticeps* do not normally court and are not aggressive toward conspecific males after the breeding season but become aggressive and court conspecific females once the head brightens following androgen treatment (Cooper and Vitt, 1987b, c). Male *E. laticeps* castrated late in the breeding season in early June no longer respond to conspecific males aggressively by early July; neither do they court females rendered sexually receptive by estrogen treatments (Cooper et al., 1986, 1987). After 1 week of exogenous androgen treatment, no males court or are aggressive; after 1 month of implantation, all males court females and react aggressively to males. These results exactly parallel those obtained for orange head coloration. Males are sexually and aggressively active only when their heads are bright orange. Comparisons of the behavior of castrated males following androgen replacement with that of castrated controls reveal that the testosterone-treated males are significantly more likely to court females and perform head-tilt displays or bite males (Cooper et al., 1987). Thus androgens induce orange head coloration and activate sexual and aggressive behavior in *Eumeces* (Fig. 6.11). The temporal and endocrine links between the bright color and reproduc-

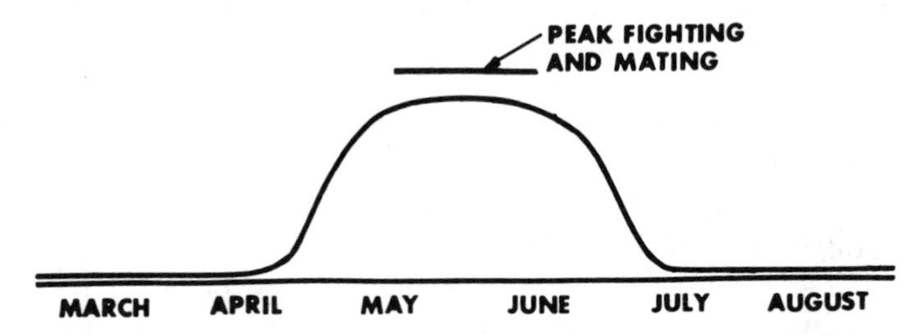

		CASTRATE	ANDROGEN
BEFORE	COLOR	DULL	DULL
	AGGRESSION	NONE	NONE
	COURTSHIP	NONE	NONE
AFTER	COLOR	DULL	BRIGHT
	AGGRESSION	NONE	INTENSE
	COURTSHIP	NONE	ACTIVE

Fig. 6.11. *Eumeces laticeps.* Seasonal occurrence of social behavior by males and the induction by exogenous androgen of social behaviors and orange head coloration in castrates.

tively important behaviors and the responsiveness of males to orange head coloration suggest that the head coloration of males is a product of sexual selection.

7. Lacertidae

In many lacertid species, males are more brightly colored than females. The importance of this sexual coloration has been demonstrated for several species by experiments in which lizards were painted with colors of the opposite sex (Kramer, 1937; Kitzler, 1941). Conspecific males attack rather than court females painted to resemble males; males court other males painted with the female pattern. Even artificial model lizards and dead lizards painted with male coloration are treated as male. However, olfactory stimuli are also clearly involved in recognition of sex because males that have responded aggressively to females painted with male colors stop behaving aggressively after tongue-flicking them (Kramer, 1937; Kitzler, 1941).

C. Female Coloration

1. Introduction Female coloration appears to be socially important in some species of lizards that lack distinctive bright coloration. Of special interest is the bright female secondary sexual coloration of several species of iguanid, agamid, and chamaeleonid lizards. Females of a single species may have bright display coloration that differs in saturation, hue, and distribution. These bright colors of females are of evolutionary interest because bright female coloration is relatively rare in vertebrates. In certain birds it has been associated with female emancipation from maternal care and in dendrobatid frogs and syganthid fish with male parental behavior (Fiedler, 1954; Lack, 1968; Trivers, 1972).

In these other vertebrates, the two major correlates of bright female coloration are highly developed paternal behavior and in some cases courtship by females. Brightly colored iguanid females, in contrast, do not actively court males; males in these species vigorously court females but do not perform any parental care; their sexual advances are often rejected by females (Noble and Bradley, 1933a; Greenberg and Noble, 1944; Cooper, 1985). Bright male secondary sexual coloration is usually interpreted as being sexually selected by female mating preference, whereas the relatively drab coloration of females in the same species is attributed to selection for crypticity, especially while females are associated with the young. However, iguanids lack either parental behavior or female courtship. Thus, the selective prerequisites for sexual selection of bright female coloration through male choice are absent. Because they lack maternal behavior after oviposition, female iguanids are relatively free to evolve bright coloration, as are female birds emancipated from maternal care (Lack, 1968).

Despite such arguments, it remains possible that male choice based on some unidentified reproductive correlate of female coloration could have selected for bright female color patterns. For example, greater fertility by bright females due to superior selection of oviposition sites or enhanced survival of bright females between copulation and oviposition could make male choice of bright females adaptive. Similarly, male choice would be adaptive if the brightness communicated some adaptive attribute of the eggs. Even a slight initial advantage in fertility by bright females could stimulate intensification of the brightness of female coloration by runaway selection (Fisher, 1930).

In some iguanid species, the bright female coloration is found only in mature females during the breeding season, especially in gravid ones, and is associated with aggressive displays, aggressive courtship rejection behaviors, and actual aggressive attack of conspecific males (e.g., Fitch, 1956, Clarke, 1965, Cooper et al., 1983; Fig. 6.12). Behaviors performed by brightly colored gravid females are back-arching, in which the back is arched upward, and sidlehopping (Carpenter, 1962), in which the female hops on stiffly extended legs, usually while oriented laterally to an adjacent male. Although the several hypothe-

Fig. 6.12. *Holbrookia propinqua.* The rejection posture of the gravid female (upper) to the courting male (lower). (From Clarke, 1965.)

sized functions of bright coloration have not been studied in most species, sex recognition appears to be a common denominator in the few species examined for this.

2. Sex Recognition

Female coloration could be used to identify sex among conspecifics in any species having sexual dichromatism, but only those cases in which sex recognition has been demonstrated experimentally will be discussed. In species lacking bright female secondary sexual colors, dull female coloration presumably allows sexual recognition by conspecific males, especially if the males are brightly or otherwise distinctively colored, unless the sexes are chromatically monomorphic. This is the case in the sexually dichromatic lacertid species *Lacerta vivipara* (Bauwens et al., 1987). Males of this species court conspecific males painted in the female color pattern. Because no males painted in the male color pattern were tested, it is possible that the responses of males were to paint rather than to hue, but this is unlikely. No discernible effects of paint were detected in similar studies employing paint controls (Cooper, 1986; Cooper and Burns, 1987).

Much more attention has been devoted to roles of bright female coloration in sex recognition. Among the species studied are *Holbrookia propinqua* (Cooper, 1984, 1986, 1988; Cooper and Crews, 1987), *Sceloporus virgatus* (Vinegar, 1972), and *Tropidurus delanonis* (Werner, 1978). For all three species, the primary experimental technique has been field observation of responses to lizards having color patterns altered by painting.

During the breeding season, female *Sceloporus virgatus* develop orange throat coloration that surrounds or replaces the blue throat patches that are present throughout the remainder of the activity season in females and continuously in adult males. The orange pigmentation is most intense in gravid females but usually appears before ovulation, when the ovarian follicles are enlarged. Although males court females with blue or orange throat coloration, they more frequently court males with throats painted orange than blue (Vinegar, 1972). They also frequently misidentify the sex of females with throats painted blue, responding to them with threat displays rather than courtship. Gravid females respond aggressively to introduced lizards of both sexes, arching the back and sidlehopping (Table 6.8). Nongravid females having blue to pale orange throats in spring or fading orange toward the end of the breeding season perform aggressive threat displays similar to those by conspecific males toward introduced lizards but do not sidlehop. Some gravid females having orange throat coloration sidlehop, but none emit male-typical threat

Table 6.8 Response of female *Sceloporus virgatus* to introduced lizards

Dates	1 May–15 May	30 May–8 July	17 July–19 August
Reproductive condition of residents	Oviductal eggs	After oviposition	Ovarian eggs
Range of throat color of residents	Blue–pale orange	Light orange, blue surrounded by orange, rusty orange, blue surrounded by orange	Fading orange–blue
Range of throat color of female intruders	Blue	Blue surrounded by orange	Fading orange–blue surrounded by orange
Response to introduced females:			
Threat	5	0	15
Sidlehopping	0	5	0
Other	7	3	19
Response to introduced males:			
Threat	3	0	4
Sidlehopping	0	8	0
Chasing	0	0	3
Other	9	4	6

Source: From Vinegar, 1972.
Note: All males had blue throat coloration.

displays. These findings indicate that orange throat coloration is an important cue for sex recognition by males and that the social behavior of females changes during gravidity when the orange throat coloration is most intense. Only gravid females sidlehop, which is primarily a courtship rejection behavior (Vinegar, 1972).

Adult female *Tropidurus delanonis* have a uniformly dark red color on the sides of the head and neck and on the chin, throat, and chest, whereas males are black, light blue, and yellow in the same regions (Werner, 1978). The red coloration brightens in gravid females and fades noticeably within several days after oviposition. When the red pigment brightens, females copulate; this occurs about 10 to 16 days before oviposition. Before copulation, females arch their backs in response to males; beginning after copulation and lasting until after oviposition, back-arching represents a rejection display to courting males until after oviposition. Juveniles of both sexes have red pigmentation similar to that of adult females. Juvenile males having red coloration respond to adult males with a femalelike display and withdrawal. Once the red pigmentation is lost, the young males no longer act like females.

The red color is a social signal sufficient for recognition of females, as demonstrated by color-modification experiments (Werner, 1978). Territorial resident males painted to resemble females are initially treated as females by other males and females but are subsequently recognized as male by their behavior responses. Males initially attack and often bite females painted with the patterns of adult or young males. In addition to facilitating sex recognition, the bright female coloration may advertise the location of the female to males. Males and females approach 25-mm red disks in color vision tests. Also, juveniles are associated with the territories of adult females. This has led to the speculation that the red female coloration functions to attract juveniles to the territories of females, in which they are assured of a good food supply and may eventually establish territories of their own (Werner, 1978). This proposed function cannot be accepted without further evidence. No details or analysis of the color preference tests have been presented. Even if the lizards approach red disks more frequently than disks of other hues, juvenile lizards of either sex might not approach and remain near other lizards having red anterior coloration. Furthermore, no data have been given to establish any advantage accruing to juveniles associated with female territories, either in feeding opportunity or subsequent establishment of territories.

In *Holbrookia propinqua*, immature females and adult females before the first reproductive cycle of the breeding season have predominantly light brown dorsal and white ventral coloration and a pair of dark ventrolateral oblique stripes, the latter being more prominent in adult males. During the breeding season females develop a bright orange pigmentation on the ventrolateral area between the forelegs and hindlegs, on posterior surfaces of the thighs, the sides of the tail base, and on the ventral portion of all but a short proximal segment of the tail. The ventrolateral dark stripes fade in brightly colored females; in some females they become almost completely obscured. As the orange coloration develops, a bright yellow pigment is deposited on the sides of the head and neck, and a more subtle yellow suffusion appears on the dorsolateral skin. Males lack the orange pigment, but many males have yellow on the sides of the head and neck. However, the presence of numerous dorsal and lateral light spots immediately distinguishes males with yellow pigment from the females.

When rejecting courtship, gravid female *Holbrookia propinqua* display the bright orange area to males by orienting laterally to them and arching the back while lowering the head. The sides are compressed and tilted slightly away from the male, increasing the apparent size and maximally exposing the lateral orange stripe. The hindlimb closer to the male is oriented slightly toward the male and the tail is simultaneously elevated and tilted away, revealing the orange areas on the posterior side of the thigh and ventral tail surface. Brightly colored gravid females also reject courtship by sidlehopping (Clarke, 1965; Cooper, 1984, 1986, 1988). These aggressive rejection behaviors differ greatly from the responses of plainly colored females, which never back-arch or sidlehop. Plainly colored females retreat from males or lash their tails laterally, sometimes both, but do not direct aggressive behaviors to males (Cooper, 1988).

The bright female color pattern of *Holbrookia propinqua* allows sex recognition by both sexes. Males painted with the orange and yellow color components are courted by males (Cooper, 1984, 1986); females of both color patterns react to them with aggressive behavior, retreat, or neutral responses typically directed to other females (Cooper, 1988). The responses to males painted to resemble brightly colored females are statistically distinct from those exhibited to unpainted males and to paint controls consisting of males painted with the typical male color pattern. In response to such males, females usually perform behaviors typical of their responses to other females, rather than the aggressive rejection displays usually directed to males. The orange and yellow color components both contribute to sex recognition. Almost all lizards showing only the yellow color, that is, males painted with only the yellow component and females with the orange component obscured, are courted by males (Cooper, 1986); brightly colored females respond to them with aggression, retreat, or neutrality (Cooper, 1988).

Both color components contribute to sex recognition, but apparently unequally. The yellow component alone is as effective for identification of sex as the entire pattern, there being no significant differences in response frequencies by either males or females to lizards bearing the entire bright color pattern or only its yellow portion (Cooper, 1986, 1988). Responses to males bearing only the orange component are mixed. About half of males painted with only the orange color component are courted by males and treated with aggression by brightly colored females; the other half are subjected to aggression by males or courtship rejection displays by females (Cooper, 1986, 1988). These results may indicate that for sex recognition, the orange component is less important than the yellow, that the conflicting cues provided by orange ventrolateral coloration and the lightly spotted dorsolateral coloration of the males result in confusion, or that the orange component is not as effective as the yellow simply because it is less visible due to its distribution. The third possibility may be dismissed because the orange paint is clearly visible from the side. Further experiments with plainly colored females having orange paint added should allow resolution of the remaining hypotheses. Because a paint mixture matching the dorsal ground coloration of females has not been attained, no experiments have been performed in which the yellow coloration of females is obscured, leaving only the orange.

The dark ventrolateral markings do not appear to be important for sex recognition. Females painted to resemble males by enhancing the dark ventrolateral bars and males with the ventrolateral bars obscured are correctly identified by conspecific males (Clarke, 1965) and females (Cooper, 1988). Although the markings have no known intraspecific importance, an interspecific signal function cannot be excluded.

In some chamaeleonid females, physiological color change during social encounters rapidly produces bright colors. The effect of bright female display coloration on courtship by males has been tested by staining green those body parts of females involved in the chromatic display, thus masking color changes (Parcher, 1974). The percentage of time spent in courtship during staged encounters between stained and unstained females does not change in *Chamaeleo brevicornis* or *C. nasatus*. Inspection of the data on proportion of males courting suggests that the staining treatment produces no effect in *C. brevicornis*, but that male *C. nasatus* are less likely to court stained than unstained females.

Although insufficient details of the experimental design have been presented by Parcher (1974) to determine the appropriate analysis, Fisher exact tests give good approximations of the null probabilities. For *Chamaeleo brevicornis*, the Fisher exact probabilities exceed 0.10. However, male *C. nasatus* are significantly more likely to court nonstained females than they are to court females with head and body stained ($p = 0.005$), head only stained ($p = 0.001$), or body only stained ($p = 0.001$). These results suggest that bright female coloration is used in sex recognition by conspecific males. The suggestion should be confirmed by future study designed to include controls for effects of staining and to allow rigorous statistical analysis.

3. Color, Reproductive Condition, and Sex Steroid Hormones

In a number of species, female coloration is brightest whenever the females are gravid (Table 6.9). The relationship between color intensity and reproductive condition has been studied in some detail in a few of these species. In *Crotaphytus collaris*, the orange lateral spots brighten from a faint to intense orange within 12 hours. Brightening occurs during the final stages of vitellogenesis, reaching maximum intensity shortly before ovulation and 10 days before oviposition (Fig. 6.13). The orange spots fade as the corpora lutea degenerate following oviposition (Ferguson, 1976). Females of the related species *Gam-*

Table 6.9 Lizard species showing aggressive courtship rejection displays, including back-arching, sidlehopping, and similar behaviors and bright female coloration

Species	Aggressive rejection	Time brightest	Source
Iguanidae			
Amblyrhynchus cristatus	+	G	Eibl-Eibesfeldt, 1966; Trillmich, 1983
Callisaurus draconoides	+	G	Clarke, 1965
Cophosaurus texanus	+	G	Clarke, 1965
Crotaphytus collaris	+	G	Fitch, 1956; Ferguson, 1976
Crotaphytus insularis	?	G	Stebbins, 1985
Gambelia silus	+	G	Montanucci, 1965
Gambelia wislizeni	+	G	Montanucci, 1967; Medica et al., 1973
Holbrookia lacerata	+	G	Clarke, 1965
Holbrookia maculata	+	G	Clarke, 1965
Holbrookia propinqua	+	G	Clarke, 1965; Cooper et al., 1983
Petrosaurus mearnsi	?	G	Stebbins, 1985
Petrosaurus thallassinus	?	N	Stebbins, 1985
Sator angustus	?	N	Clarke, R. F., unpublished
Sator grandaevus	?	N	Clarke, R. F., unpublished
Sceloporus graciosus	+	N	Stebbins and Robinson, 1946; Stebbins, 1985
Sceloporus virgatus	+	G	Vinegar, 1972
Sceloporus woodi	?	N	Cooper, unpublished
Tropidurus albemarlensis	+	G	Stebbins et al., 1967
Tropidurus bivittatus	+	N	Carpenter, 1966; Van Denburgh and Slevin, 1913; Werner, 1978
Tropidurus delanonis	+	G	Werner, 1978
Tropidurus duncanensis	+	N	Carpenter, 1966; Van Denburgh and Slevin, 1913; Werner, 1978
Tropidurus habeli	+	N	Carpenter, 1966; Van Denburgh and Slevin, 1913; Werner, 1978
Tropidurus occipitalis koepckeorum	?	N	Dixon and Wright, 1975
Tropidurus occipitalis	?	N	Dixon and Wright, 1975
Tropidurus pacificus	+	N	Carpenter, 1966; Van Denburgh and Slevin, 1913; Werner, 1978
Tropidurus peruvianus	?	N	Dixon and Wright, 1975
Tropidurus stolzmani	?	N	Dixon and Wright, 1975
Uma notata	+	G	Carpenter, 1963; Stebbins, 1985
Urosaurus ornatus	+	G	Zucker and Boecklen, 1990; Zucker, personal communication
Urosaurus bicarinatus	?	G	del Toro, 1982
Uta palmeri	+	G	Wilcox, 1980
Agamidae			
Agama agama	+	N	Harris, 1964
Amphibolurus maculosus	+	G	Mitchell, 1973
Chamaeleonidae			
Chamaeleo brevicornis	+	N	Parcher, 1974
Chamaeleo nasatus	+	N	Parcher, 1974
Chamaeleo willsii	+	N	Parcher, 1974

Note: In the *Aggressive rejection* column, + indicates presence, − indicates absence, and ? indicates lack of information. In the *Time brightest* column, G indicates that coloration is brightest when the female is gravid, N that the species develops bright coloration, but the timing of maximum brightness is not known.

belia wislizeni develop orange-red spots shortly before ovulation when the ovarian follicles reach 12 mm (Turner et al., 1969). The spots fade after oviposition but brighten again if a second clutch is produced (Medica et al., 1973). Orange throat coloration appears before ovulation in *Sceloporus virgatus* (see Table 6.8), at which time the ovarian follicles are enlarged; it is most intense after ovulation and fades following oviposition (Vinegar, 1972). Female *Holbrookia propinqua* emerge from hibernation plainly colored and develop the yellow and orange components of bright coloration simultaneously in late spring. The pigments are deposited rapidly during the terminal stages of vitellogenesis; the ovarian follicles then are at least 5 mm in diameter. The pigments attain full brightness before ovulation. All gravid females are brightly colored (Cooper et al., 1983). In *Uta palmeri,* the most brightly colored females are those with oviductal eggs and/or corpora lutea (Fig. 6.14). The bright pink-red stripe appears before ovulation, when the ova reach a diameter greater than 5 mm; the

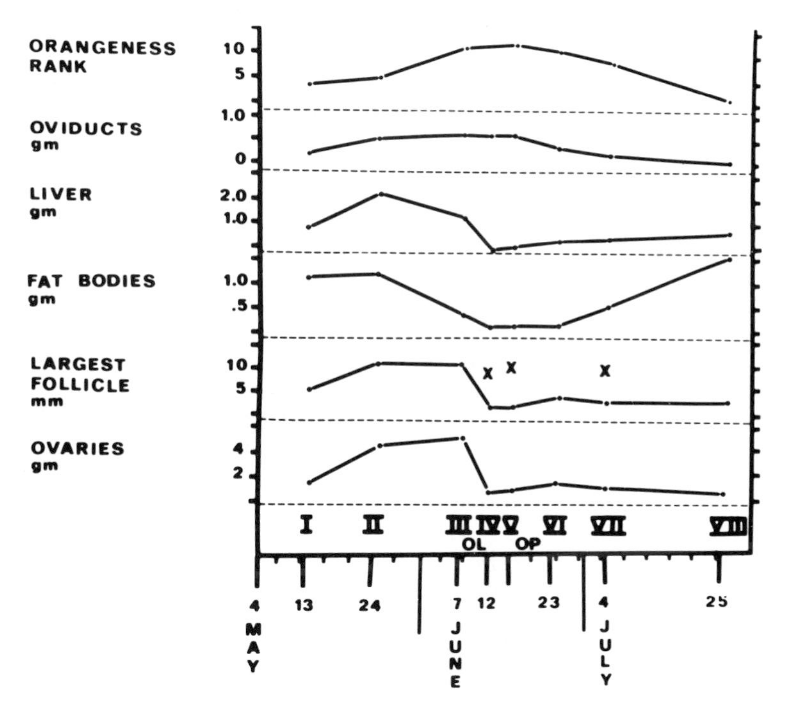

Fig. 6.13. *Crotaphytus collaris.* Brightening of the orange spots of females occurs just before ovulation (OL). The orange color is brightest while the females are gravid and fades following oviposition (OP). Roman numerals represent reproductive stages: I, early vitellogenic; II, late vitellogenic; III, preovulatory; IV, early gravid; V, late gravid; VI, VII, and VIII, postgravid. (From Ferguson, 1976.)

stripe is most intense in gravid females (Wilcox, 1980). It is unknown whether the bright reproductive coloration fades appreciably in seasonally iteroparous species with short interclutch intervals, such as *H. propinqua* and *U. palmeri*.

The relationships between female coloration and reproduction are complex in *Urosaurus ornatus*, differing between populations in New Mexico (Zucker and Boecklen, 1990). In Aguirre Spring, female throat coloration is yellow, gold, or orange. Early in the activity season, the throat is yellow, gold, or pale orange. Later, its color shifts to a darker

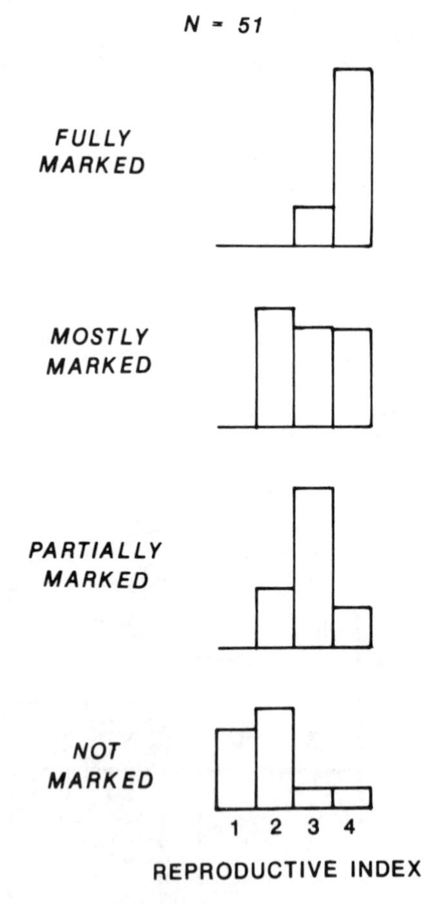

Fig. 6.14. *Uta palmeri*. Brightness of female coloration is greatest in gravid females and those with large ovarian eggs. Reproductive index: 1, ovarian follicles ≤ 5 mm; 2, ovarian eggs > 5 mm; 3, oviductal eggs; 4, corpora lutea. Breeding coloration: not marked—no red pigment; partially marked—red pigment on lateral trunk extending partially up the tail; fully marked—red pigment on lateral trunk and fully covering the tail. (From Wilcox, 1980.)

orange hue of longer dominant wavelength. The population of *U. ornatus* at Dona Ana has orange throat coloration in all females. As in the other species studied, female coloration is related to reproductive cyclicity. As the first of two seasonal clutches develops at Aguirre Spring, females having nonvitellogenic and small vitellogenic eggs (< 3.4 mm) have the lightest, least saturated, most yellow throat coloration. Females having large ovarian follicles (3.5 to 6.5 mm) tend to shift toward a darker orange of greater saturation, but the darkest, most saturated orange of longest dominant wavelength occurs in gravid females. As vitellogenesis for the second clutch begins, females in both populations have darker throats with longer dominant wavelength than they had at the same stage for the first clutch. At Aguirre Spring, females with yellow throats have larger clutches than do females with orange throats. This effect is independent of body size because hue and snout-vent length are uncorrelated.

Repeated recaptures of a few female *Urosaurus ornatus* establish that throat coloration changes in individuals, shifting to a darker, more saturated, orange hue of longer dominant wavelength (Zucker and Boecklen, 1990). Data for a single female suggest that throat coloration may become lighter and that the dominant wavelength may decrease between successive clutches (Zucker and Boecklen, 1990). Although relationships between female throat coloration and behavior have not been studied in *U. ornatus,* the pattern of color changes is concordant with those observed in other species in which female coloration is related to sex recognition (see Section VI.C.2) and aggressive rejection of courtship (see Section VI.C.4).

In the few species studied, bright female coloration may be induced readily by exogenous hormones. Orange spots appear in ovariectomized *Crotaphytus collaris* following injections of progesterone or testosterone propionate, but progesterone produces more intense coloration (Cooper and Ferguson, 1972a, b). Estradiol injections produce no color change but prime the response to progesterone (Fig. 6.15). That is, detectable color change in response to progesterone appears after 6 to 8 days, but this interval is shortened to 4 days after pretreatment with estradiol. Once the color begins to change, this change proceeds at the same rate whether or not estrogen has been administered. Thus, there is no synergistic effect on the rate of pigment deposition or on the ultimate intensity of orange coloration between estradiol and progesterone (Cooper and Ferguson, 1973a, b). Stress may interfere with development of bright coloration, because females given concurrent corticosterone and progesterone fail to develop the orange pigment induced by similar doses of progesterone alone (Rand, 1986). Similar hormonal mechanisms regulate the orange-red breeding coloration of female *Gambelia wislizeni.* Progesterone, but not estrogen, induces bright coloration in ovariectomized females and estrogen hastens the response to progesterone. In addition, the orange pigment is deposited by intact females in response to administration of mammalian follicle-stimulating hormone (Medica et al., 1973).

In iguanid lizards, hormonal regulation of female reproductive coloration has been most extensively studied in *Holbrookia propinqua*. The sex steroid hormones testosterone, progesterone, and estradiol all participate in development of bright coloration (Fig. 6.15). Progesterone and testosterone induce the bright color pattern in ovariectomized (Cooper and Crews, 1987) or intact females (Cooper and Clarke, 1982, for orange only). Watson (1974) reported that testosterone, but not progesterone, induced brightening, although progesterone prevented fading of bright coloration. However, a reanalysis of Watson's data reveals significant brightening of both colors in response to progesterone (Cooper and Crews, 1987).

Studies of the effects of estradiol on female coloration in *Holbrookia propinqua* have yielded contradictory results. Cooper and Clarke (1982) report that estradiol injections do not induce brightening of the orange component of intact females, but over half the females died within 8 days. In an attempt to reduce the circulating concentrations of estrogen to sublethal doses, ovariectomized females have been im-

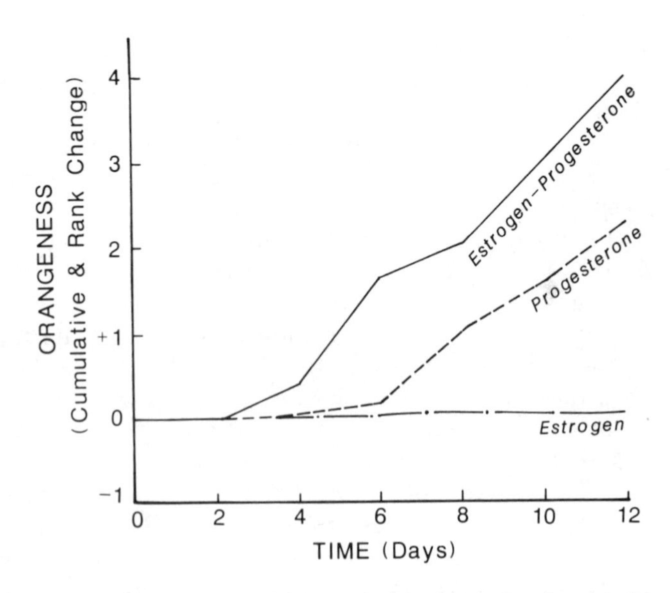

Fig. 6.15. *Crotaphytus collaris.* Progesterone induces bright female coloration. Estradiol does not induce the bright coloration, but pretreatment with estrogen allows more rapid deposition of pigment in response to progesterone. (From Cooper and Ferguson, 1973b.)

planted with Silastic capsules containing estradiol mixed with cholesterol. Such females show a significant increase in orange pigmentation (Cooper and Crews, 1988). However, the intensity of the orange coloration induced by estradiol is much less than that induced by either progesterone or testosterone (Fig. 6.16). Estradiol has no effect on the yellow color component. No priming or synergistic effects of estradiol occur in groups treated simultaneously or sequentially with estradiol and progesterone. Watson (1974) reports that estradiol induces incomplete but statistically significant brightening of both color components.

Reasons for the discrepant results of the three studies are unclear. They could include differential stress effects blocking color production as in *Crotaphytus collaris* (Rand, 1986) and differences in dosage and method of administration. Cooper and Clarke (1982) and Cooper and Crews (1987) report that females suffer high mortality rates whether estradiol was administered in low continuous dosage through Silastic capsules or by daily injections of 1 μg dissolved in 0.01 ml peanut oil. Watson (1974) reports no mortality in 32 females injected every other day with 5 μg estradiol in aqueous suspension (volume not reported). Another possible explanation of the disparate results on the yellow component is that its incomplete brightening is a pharmacological effect. Injection of large doses in aqueous suspension might be expected to elevate estradiol concentrations transiently to levels considerably higher than those produced by the continuous diffusion of small quantities from the Silastic capsules. Thus the continuous exposure to lower doses of estradiol might prove fatal, whereas brief exposure to higher levels may increase survival and pharmacologically elicit the yellow color component. Another possibility is that the studies may have differed in the availability of pigments, such as carotenoids or their precursors, which are not synthesized but acquired by the lizards from invertebrates in their diet. The effect of physiological doses of estradiol on the yellow pigmentation needs further study.

Plasma concentrations of sex steroid hormones in female *Holbrookia propinqua* vary with reproductive condition and coloration in a manner consistent with results of the color induction studies. Females having follicles of a size greater than 5.0 mm or oviductal eggs (and corpora lutea) are almost all brightly colored and have higher concentrations of circulating progesterone (Fig. 6.17) and androgen than do plainly colored females, almost all of which lack follicles as large as 5.0 mm (Cooper and Crews, 1988). Brightly colored females in a laboratory sample had higher circulating levels of estradiol, progesterone, and androgens than did plainly colored females. In a field sample of brightly colored females, only progesterone and androgens

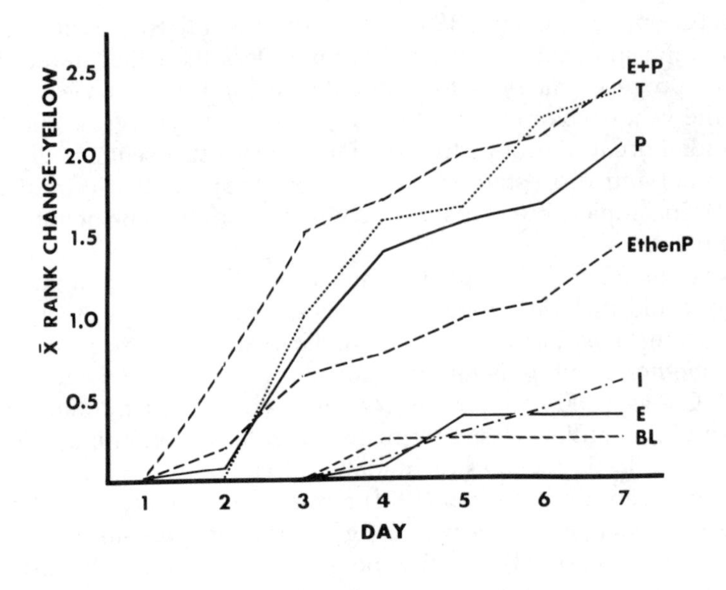

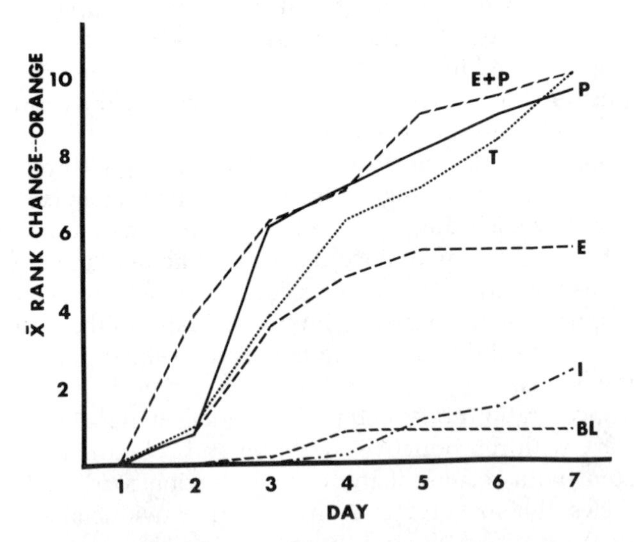

Fig. 6.16. *Holbrookia propinqua.* Bright colors are induced hormonally in females. BL, blank control; E, estrogen; E + P, simultaneous estrogen and progesterone; EthenP, sequential estrogen and progesterone; I, intact control; P, progesterone; T, testosterone. (From Cooper and Crews, 1987.)

were elevated. The difference in results with estradiol may simply reflect more recent brightening of females in the laboratory sample, several of which brightened shortly before being bled (Cooper and Crews, 1988). Because the pigments are deposited during the terminal stages of vitellogenesis, estradiol levels are presumably elevated during brightening.

The extent of brightening that occurs before ovulation suggests that levels of progesterone and perhaps androgen are also elevated at this time. Following ovulation, estradiol levels presumably fall; however, high progesterone levels are maintained in gravid lizards by the corpora lutea and possibly by increased adrenal synthesis (as suggested for *Lacerta vivipara*, Dauphin-Villemant and Xavier, 1985). Progesterone appears to be the primary agent responsible for inducing and maintaining bright female coloration. Mean concentrations of plasma progesterone in brightly colored females are 47 ng/ml for the laboratory samples and 38 ng/ml for the field samples, several times higher than the respective androgen concentrations of 7 ng/ml and 3 ng/ml (Cooper and Crews, 1988). Nevertheless, androgens undoubtedly contribute to the bright reproductive coloration. High estrogen levels before ovulation may stimulate partial brightening of the orange color component.

The source of circulating androgens in female *Holbrookia propinqua* is unknown, but androgens are known to be synthesized by the ovaries and are present in the plasma of some female lizards (Chan and

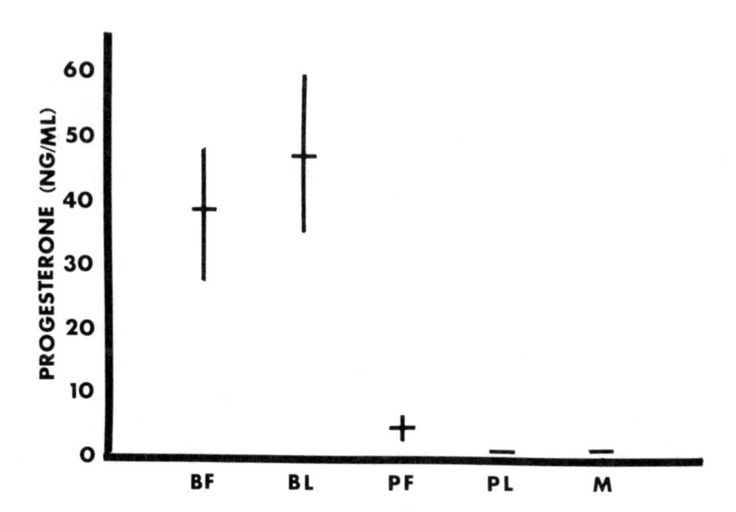

Fig. 6.17. *Holbrookia propinqua*. Circulating progesterone concentrations are elevated in brightly colored females. Groups are BF, bright field females; BL, bright laboratory females; PF, plain field females; PL, plain laboratory females; and M, males. (From Cooper and Crews, 1988.)

Callard, 1974; Judd et al., 1976; Arslan et al., 1978). The adrenal gland synthesizes androgens in the snake *Naja naja* (Tam et al., 1972), but the adrenal of the lizard *Lacerta vivipara* does not do so (Dauphin-Villemant and Xavier, 1985). Whatever their sources, plasma androgen concentrations are far lower in females than the 25 ml/ng mean for males (Cooper and Crews, 1988).

In a pilot study conducted in July, 1983 (Cooper and Vitt, unpublished), intact female *Sceloporus virgatus* having faded throat coloration were injected with progesterone, testosterone, estradiol, or estrogen simultaneously with progesterone or testosterone. All failed to develop a bright orange throat coloration; however, at least one of the females showed no orange despite being gravid. Whether or not sex steroids induce the orange throat coloration in this lizard should be further investigated outside of the breeding season or in ovariectomized females lacking the bright coloration. If sex steroids do induce the orange throat coloration, the experimental results may possibly be explained by blockage by glucocorticoids (Rand, 1986) or by absence of sufficient pigments or their precursors in the diet.

4. Color and Aggressive Courtship Rejection

Aggressive rejection of courtship by sidlehopping, back-arching, and similar behaviors occurs in many iguanids, including many species known to have bright secondary sexual coloration in females (Table 6.9). Of the 30 iguanid species known to have bright secondary sexual coloration in females, 19 perform distinct aggressive rejection displays and another becomes aggressive while brightly colored (Trillmich, 1983, for *Amblyrhynchus cristatus*). The courtship rejection behaviors of the remaining species are unknown. Thus bright female coloration and aggressive rejection are strongly associated. In those species for which the timing of bright coloration is known, maximal color intensity develops while the female is gravid ($n = 18$, $p < 0.001$ under the assumption that a female is equally likely to be most intensely colored when gravid or nongravid). In the 14 species for which females are known to be brightest while gravid and which are known to display aggressive courtship rejection, the aggressive rejection displays are limited to the brightly colored gravid females ($p < 0.001$). It should not be inferred that bright coloration and aggressive courtship rejection evolved simultaneously. On the contrary, aggressive rejection occurs in numerous iguanids that have been omitted from Table 6.9 because they lack bright female coloration.

Aggressive courtship rejection is so widespread that it may be primitive for Iguanidae or some of its subfamilies; its expression may be constrained by the degree of sexual dimorphism. Its possibly primitive nature is supported by the occurrence of aggressive courtship rejection behaviors in the closely related families Agamidae and Chamaeleonidae (Table 6.9). Among iguanids, aggressive courtship rejection behaviors have been reported in sceloporines, tropidurines, crotaphytines (Table 6.9), iguanines (e.g., Dugan, 1982; Werner, 1982), and anolines (Carpenter and Ferguson, 1977). However, no distinctive courtship rejection displays limited to, or most intensely expressed by, gravid females have been described for iguanines. Sidlehopping was recorded for only one anoline species, *Anolis carolinensis,* in an extensive literature review (Carpenter and Ferguson, 1977). However, we have not observed this behavior, nor has D. Crews (personal communication, 1990), despite extensive study of this species.

The taxonomic distribution of bright female coloration is similar to that of the aggressive rejection of courtship by sidlehopping and back-arching. Among iguanids, crotaphytines, sceloporines, and tropidurines are almost the only groups having bright secondary sexual coloration of females. *Amblyrhynchus cristatus* is the only iguanine reported to develop bright gravid coloration (Eibl-Eibesfeldt, 1966); it is not known to have distinctive courtship rejection displays such as sidlehopping or back-arching. In all of the species listed in Table 6.9, except *A. cristatus,* the bright female color pattern differs from that of conspecific males. Gravid *A. cristatus* develop bright coloration similar to that of males and aggressively compete for limited oviposition sites (Eibl-Eibesfeldt, 1966). Because the bright colors of the remaining species are associated with aggressive courtship rejection, they appear to differ adaptively from those of *A. cristatus.* No anolines are known to develop bright gravid coloration.

In those crotaphytines, sceloporines, and tropidurines having bright female secondary sexual coloration, the behavior of females toward males changes dramatically at the time that bright coloration appears. Before its appearance, nonreceptive females in many species reject courtship primarily by avoidance, but brightly colored gravid females become highly aggressive, performing the distinct rejective displays described above (e.g., Fitch, 1956; Clarke, 1965; Vinegar, 1972; Werner, 1978; Cooper, 1984, 1986, 1988).

A similar association of bright female display coloration and aggressive courtship rejection may occur in Agamidae and Chamaeleonidae, but only fragmentary information is available. The two agamids listed in Table 6.9 have bright female morphological coloration distinct from that of the male and distinct rejection displays; the three chameleons have physiologically controlled chromatic displays associated with female rejection of courtship. Sidlehopping and back-arching have not been reported, but the former would be maladaptive given the digital adaptation for grasping branches, and the latter might be undetectable in lizards having permanently arched backs.

Female chameleons threaten courting males by opening their mouths; they bite if approached too closely (Parcher, 1974). As do iguanid females, female agamids and chamaeleonids of species having bright female coloration reject male courtship most vigorously when gravid (Harris, 1964; Mitchell, 1973; Parcher, 1974). In the agamid *Amphibolurus maculosus,* the behavior of females toward conspecific males changes when the female develops bright breeding coloration (Mitchell, 1973). Females switch from circumduction to a display with the head elevated. If a male approaches rapidly, a female may roll onto her back (a submissive behavior among males), then right herself and display the bright ventrolateral coloration to the male. After displaying, the female advances upon the male in a stiff-legged display walk. After being fertilized, females become highly aggressive toward males.

Bright female coloration and its placement are believed to reflect an adaptive balance between social costs and benefits of visual communication and costs due to predation. By visually communicating nonreceptivity to conspecific males, females may reduce the time spent in courtship and therefore decrease the probability of attracting predators. Chromatic signaling may only be adaptive if the species also has effective means of aggressive courtship rejection. This would explain the limitation of bright gravid female coloration to species having aggressive rejection behaviors. For example, most iguanine species may lack bright secondary sexual coloration because their gravid females lack distinctive, aggressive courtship rejection displays. Because they are slowed by the increased weight, especially in species having large ratios of clutch mass to body mass, gravid females are more vulnerable to predators (Shine, 1980). This may explain the greater intensity of courtship rejection, including distinctive displays, in gravid female iguanids, agamids, and chameleons. Anoles have a fixed clutch size of one and a low relative clutch mass. The absence or rarity in anolines of aggressive courtship rejection displays and bright female coloration in Anolinae may reflect a lesser increase in the vulnerability of gravid females to predation as they become gravid. Other factors must account for the absence of aggressive courtship rejection displays in lizards such as iguanines.

Any intraspecific benefit from reduction of courtship would be at least partially offset if the bright coloration increased the detectability of females to predators. Although gravid female iguanids have continuous bright colors for long intervals, in many species crypticity is maintained by ventral or ventrolateral placement of the bright pigments. The coloration of *Holbrookia propinqua* is bright at close range, but the orange component is largely hidden except during rejection displays; the yellow surfaces remain visible, but from a distance blend with the sandy substrate on sunny days (Cooper, 1986). All iguanid and agamid species known to have bright female coloration are essentially terrestrial, although they may be good climbers. Bright ventral coloration may be more detectable or disrupt crypticity to aerial predators having keen color vision. This may help to explain the lack of bright female coloration in anoles. In the chameleons listed in Table 6.9, specific bright color patterns appear rapidly whenever the female threatens the male. Physiological control of color patterns signaling aggressive courtship rejection allows chameleons to be cryptically colored except during actual courtship.

5. Sex Steroid Hormones, Color, and Behavior

Relationships between female coloration, aggressive courtship rejection, and endocrines have been studied only in *Holbrookia propinqua*. In courtship encounters staged in the laboratory, ovariectomized plainly colored females implanted with progesterone or testosterone in Silastic capsules exhibit aggressive rejective behaviors (Fig. 6.18) after they become brightly colored (Cooper and Crews, 1988). These behaviors are similar to those of intact, brightly colored females and are not performed by intact plainly colored females or by ovariectomized plainly colored controls implanted with empty capsules.

Estradiol does not induce aggressive rejection of courtship (Fig. 6.18); however, females implanted simultaneously with estradiol and progesterone more frequently behave aggressively toward males than do intact plainly colored females, plainly colored controls having

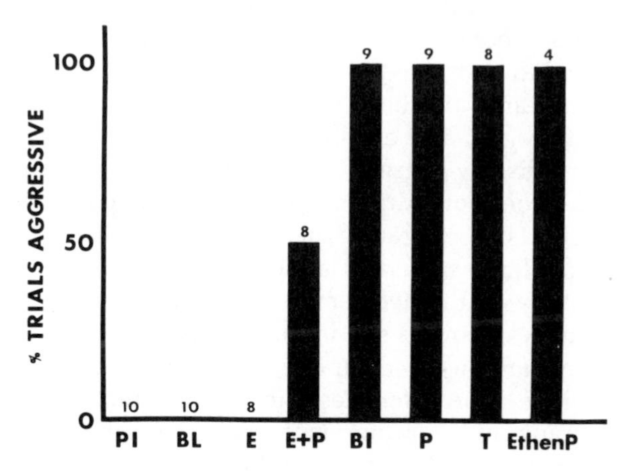

Fig. 6.18. *Holbrookia propinqua*. Progesterone and testosterone activate aggressive rejection of courtship by females; estradiol may reduce female aggressiveness toward males. PI, plainly colored intact; BL, blank control; E, estradiol; BI, brightly colored intact; E + P, simultaneous estradiol and progesterone; P, progesterone; T, testosterone; EthenP, sequential estradiol and progesterone. (From Cooper and Crews, 1987.)

empty Silastic capsules and females given only estradiol. Such females react aggressively toward males less frequently than do females given progesterone alone, testosterone, or estrogen followed by progesterone (Cooper and Crews, 1987). The frequency of aggression by females treated simultaneously with estradiol and progesterone does not differ significantly from that of any other group. However, the differences approach significance as closely as possible given the small sample sizes and an intermediate degree of aggressive rejection by the group given estradiol and progesterone simultaneously. The probability that frequency of aggression was the same as that by females given estradiol and progesterone simultaneously was 0.06 for brightly colored intact females; for ovariectomized females, it was 0.06 for those receiving progesterone, 0.08 for those receiving testosterone, and 0.14 for those initially receiving estrogen that was later replaced by progesterone. The differences would probably be significant were the sample sizes larger, allowing reasonably high statistical power. Indeed, if data for the groups receiving progesterone alone and for those receiving progesterone replacing estradiol are pooled, the females in the pooled sample show a significantly higher frequency of aggressive rejection than do females simultaneously treated with estrogen and progesterone (Fisher exact test, $p = 0.014$). If data for testosterone-treated females are added to those for these two groups, the probability becomes 0.003. Thus, estradiol may partially inhibit aggressive rejection.

The relationships of circulating steroid levels to female behavior are consistent with the results of sex steroid hormone implantations. Responses in the laboratory of intact females to introduced males are directly related to hormone levels. Brightly colored intact females sidlehop and back-arch and have higher plasma concentrations of progesterone, androgen, and estradiol than do plainly colored females, none of which display aggressive courtship rejection. Because exogenous estradiol does not induce aggressive courtship rejection and has only minor effects on coloration, its correlation with aggressive rejection does not indicate a causal relationship (Cooper and Crews, 1988).

Although fully brightened female *Holbrookia propinqua* are aggressive toward males, females still in the process of brightening, at least when brightening is nearly complete, are no more aggressive to males than are plainly colored females and perform neither back-arching nor sidlehopping. In laboratory encounters, only brightening females are sexually receptive. However, within a day after copulating, females begin to display aggressive rejection to males (Cooper and Crews, 1988). The mechanisms that terminate receptivity and initiate aggressive rejection behavior are not known, but possibilities include induction by copulation and temporal changes in the hormonal milieu of brightening females independent of copulation.

Sex steroid hormones profoundly alter the behavior of females toward conspecific males; they induce back-arching, sidlehopping, and biting. The same steroid hormones that induce bright female coloration activate the aggressive courtship rejection. Estradiol, which induces partial brightening only of the orange color component but does not stimulate aggressive courtship rejection, is the sole exception. These findings are corroborated by the elevation of circulating levels of progesterone and androgen in brightly colored, aggressive females. Their shared steroidal regulation of bright female coloration and aggressive courtship rejection strengthens the hypothesis that these phenomena are adaptively linked.

The above data permit presentation of a hypothetical scenario of the endocrine regulation of bright female coloration and behavior in *Holbrookia propinqua* as a stimulus for further investigation. The available evidence suggest that females, which are plainly colored upon emergence from hibernation and before maturation, at these times have low levels of circulating estrogen, progesterone, and androgen. As the ovarian follicles enlarge during vitellogenesis, levels of estradiol increase, but the color does not change appreciably until the terminal phase of vitellogenesis, at which time levels of progesterone and, perhaps, androgen begin to increase.

Sexual receptivity in brightening females may be stimulated by estradiol alone, which is effective in *Anolis carolinensis* (Noble and Greenberg, 1941; Greenberg and Noble, 1944; Crews, 1975b; Valenstein and Crews, 1977) and *Eumeces laticeps* (Cooper et al., 1986), or by a synergism between estradiol and progesterone, as established for *A. carolinensis* (McNicol and Crews, 1979; Wu et al., 1985). Androgens may possibly participate because exogenous androgens induce receptivity in female *A. carolinensis* (Mason and Adkins, 1976; Adkins and Schlesinger, 1979; Crews and Morgentaler, 1979). High levels of estrogen may inhibit aggressive courtship rejection despite increasing levels of progesterone in brightening females.

Subsequently, high levels of progesterone and androgen induce complete brightening and activate aggressive rejection of courtship. Estrogen levels presumably decline in gravid females, as suggested by the elevation of estradiol in lizards that had recently brightened while being maintained in the laboratory but not in animals collected in the field, which presumably had a longer mean duration of bright coloration (Cooper and Crews, 1988). Decreased estrogen would terminate sexual receptivity. Bright coloration and aggressive rejection behavior appear to be maintained in gravid females by high concentrations of progesterone and androgen, progesterone levels being maintained by luteal activity and androgen levels by ovarian or perhaps adrenal synthesis, as discussed previously.

At the end of the breeding season and after oviposition between successive clutches, levels of progesterone and possibly androgen decline below those necessary to support aggressive courtship rejection and bright coloration. For *Holbrookia propinqua,* it is not known whether the bright colors fade between successive clutches within the breeding season. Presumably some degree of brightness is retained through the remainder of the breeding season. Evidence suggesting retention of bright coloration is that (1) all adult females are brightly colored in the second half of August (Cooper, unpublished observations, 1990) and (2) there may be insufficient time for complete fading to occur between successive clutches because clutches are produced at intervals of approximately 30 days (Judd, 1976). Furthermore, estradiol, progesterone, and testosterone, which may be present in peak concentrations at different times during the cycle, are all effective in keeping bright coloration from fading (Watson, 1974). Increasing estradiol at a time of lowered levels of progesterone and androgen presumably induces sexual receptivity near the time of ovulation in later clutches. Females store sperm in seminal receptacles of the middle and anterior vagina up to a maximum of 75 days or more. However, few females store sperm this long and then only in small quantities (Adams and Cooper, 1988). Without at least one more copulation after the initial reproductive cycle, it is unlikely that fully fertile clutches could be produced throughout the reproductive season. Further study is needed to determine whether the bright coloration fades and intensifies cyclically throughout the reproductive season, whether there are corresponding shifts in behavioral response to males, and how these changes in color and behavior are related to hormonal events.

6. Adaptive Role of Bright Female Coloration

Although the bright coloration of female *Holbrookia propinqua* (Cooper, 1984, 1986), *Sceloporus virgatus* (Vinegar, 1972), and *Tropidurus delanonis* (Werner, 1978) provides cues sufficient for sex recognition, it is highly improbable that sex recognition alone could account for the evolution of bright female coloration. In most iguanid species that lack bright female coloration, females are recognized by both sexes. Females may be distinguished by lack of bright or otherwise distinct male coloration, by sex differences in agonistic and courtship behaviors, and sometimes by sexually dimorphic anatomical traits (see Ferguson and Carpenter, 1977; Stamps, 1977). Male *H. propinqua,* for example, have bright coloration distinct from that of both female color phases, are larger than females, more aggressive, and perform vigorous courtship displays. If sex recognition were its sole function and if it were necessary for accurate identification of sex, females of many more species might be expected to be brightly colored. Whereas bright female coloration might improve the efficiency of sex recognition, it need not be essential for it. Consequently, a balance of selection between increased risk of predation due to conspicuousness and increased efficiency of signaling could favor bright female coloration in some species and prevent it in others. Furthermore, some lizards can see ultraviolet light (Alberts, 1989); it is possible that females lacking bright coloration in the visible spectrum could be marked in the ultraviolet range.

Bright pigmentation presumably exacts a metabolic cost due to the need to acquire and synthesize the pigments and an additional cost due to increased detectability to predators. These costs must be outweighed by some selective benefit for brightness to evolve or be maintained. In males of many vertebrates, dimorphic traits that seemingly have adverse effects on survival, including bright coloration, exaggerated ornamentation, and extremely enlarged weaponry, are attributed to sexual selection (Darwin, 1871). Bright male secondary sexual coloration may often evolve under intersexual selection based on mate choice, but bright female coloration is unlikely to do so. The scarcity of bright female coloration in female iguanid lizards and the presence of other structural and behavioral cues provided by these females suggest that mate selection is not the primary selective factor for bright female coloration.

Several other functional hypotheses have been advanced as causes for bright female coloration (Table 6.10). Little or no information is available about many of these for most species. Indeed, it is likely that the adaptive basis for bright coloration in females may differ among species. Distinct hypotheses may pertain in different species; therefore, negative evidence about a hypothesis in one species does not invalidate the hypothesis for other species. Furthermore, the hypotheses need not be mutually exclusive.

According to the female signal hypothesis, the bright female secondary sexual coloration directs signals to conspecific females. Crucial evidence on the tenability of this hypothesis is lacking for many species, but preliminary evidence suggests that brightly colored female *Gambelia wislizeni* may be more tolerant of other bright females than of males or less brightly colored conspecific females (Moore, 1983). This observation may not be merely chance because the female signal hypothesis was first suggested for the closely related *Crotaphytus collaris* (Yedlin and Ferguson, 1973). However, the hypothesis is not tenable for *Holbrookia propinqua*. Brightly and plainly colored fe-

Table 6.10 Hypotheses regarding adaptive roles of bright female coloration

Hypothesis	Description
Sex recognition:	Bright female coloration evolved to identify females to conspecifics. Pro: Bright female coloration allows sex recognition. Con: Cannot explain evolution of bright female coloration because: 1) females are recognized in species lacking it by behavioral, pheromonal, or other cues (yet it could be a contributor); 2) male *Holbrookia propinqua* have a color pattern distinct from both female color phases; and 3) all adult females should be brightly colored if sex recognition were the sole function of the bright color.
Female signal:	Bright female coloration is a social signal to conspecific females. Pro: Preliminary results suggest that bright female *Gambelia wislizeni* are more tolerant of other bright females than of males or plain females. Con: In *H. propinqua* 1) bright and plain females are equally aggressive to females of either color phase; and 2) bright females perform rejection displays only to males and only bright females perform these displays.
Aggression avoidance:	Bright coloration helps aggressive gravid females prevent or reduce reciprocation of aggressive behavior by males. Pro: 1) Nongravid, plainly colored females do not need bright coloration because they are not aggressive to males. 2) Males are less aggressive to males painted in either the yellow or orange female color component than to males lacking these colors. Con: 1) Males are not consistently aggressive to sidlehopping females, even in species lacking bright female colors. 2) Female aggressive rejection is distinct from male aggressive behavior and could itself identify females, thereby reducing aggression by males.
Maturity:	Bright colors advertise sex and reproductive competence. Pro: Bright female coloration allows sex recognition, appears during the final phases of vitellogenesis, and is retained by females while gravid. Con: 1) Colors fade between clutches in *G. wislizeni*. 2) Colors are bright at times when fertilization would not occur (complicated by sperm storage) and when courtship is rejected. 3) Females of most species lack such coloration but are courted by males.
Courtship rejection:	Aggressive rejection by gravid females could be reliably signaled by bright coloration, thereby avoiding increased predation risk associated with fruitless courtship and/or copulation. Pro: 1) Bright females reject courtship and retain bright colors while gravid. 2) Bright females display the orange areas on the side, rear of thigh, and tail during rejection. 3) Plain females do not reject courtship aggressively. Con: 1) Newly brightened female *H. propinqua* do not reject courtship.

continued

Table 6.10 *(continued)*

	2) Males of several species court and occasionally copulate with brightly colored females. 3) Bright female coloration is lacking in many species having aggressive courtship rejection, but males do not appear to be more aggressive toward females of these species.
Courtship stimulation:	Bright coloration signals that the female is in a condition favorable for fertilization. Pro: 1) Females brighten in the final stages of vitellogenesis. 2) Females are sexually receptive near the completion of brightening. Con: 1) Bright coloration remains long after the female becomes nonreceptive. 2) Brightly colored gravid females reject courtship. 3) Plainly colored females are receptive occasionally.
Conditional signal:	Being readily identifiable from a distance, brightly colored females may avoid challenges by males patrolling their territories yet advertise sexual receptivity by rapid brightening when fertilizable. By monitoring coloration of females on their territories, males could detect brightening, indicating impending receptivity, and recognize fully brightened females that are likely to aggressively reject courtship. Pro: 1) Females become receptive when brightening but are aggressive for a long period thereafter. 2) All species having bright female colors and aggressive rejection are territorial. 3) Males appear to recognize resident bright females. Con: Nothing yet.

males of that species are equally aggressive to females of either color phase. Brightly colored females direct sidlehopping and back-arching displays only to males, and only brightly colored females perform these behaviors (Cooper, 1988). Further study is needed to evaluate the female signal hypothesis for crotaphytines.

Another proposed function is that it permits gravid females to reduce or avoid aggression from males despite aggressively rejecting courtship (Clarke, 1965). Some evidence is consistent with such a role in *Holbrookia propinqua*. According to this hypothesis, plainly colored females do not need bright coloration because they are not aggressive to males and are therefore not subjected to aggression by males. Males are less aggressive to other males painted with either or both of the female yellow or orange components than to males that are normally colored (Cooper, 1984, 1986). However, this may reflect misidentification of sex and the resulting increase in courtship rather than a reduction in aggression to females due to bright coloration.

The latter interpretation is favored by a reduction in aggressive rejection by brightly colored females of males painted with the yellow or orange component. The prominent display of orange surfaces by gravid females, which almost exclusively occurs while these are aggressively rejecting courtship (Clarke, 1965; Cooper, 1988), is the strongest point favoring the aggression-avoidance hypothesis. Other considerations suggest that bright coloration is unnecessary to reduce male aggression. In *H. propinqua* (Clarke, 1965; Cooper, 1984, 1988) and in *S. virgatus* (Vinegar, 1972) female aggressive rejection behavior is distinct from male aggressive behavior. However, males court plainly colored females in the former species, suggesting that they recognize them in the absence of chromatic signals. Even in species lacking bright female color patterns, males are not consistently aggressive to sidlehopping females (e.g., *S. undulatus*, Cooper and Burns, 1987).

An empirical test based on published and previously unpublished data on responses of males to introduced tethered females allows the aggression-avoidance hypothesis to be rejected for *Holbrookia propinqua* (Cooper, 1984, and W. E. Cooper, unpublished observations, 1983). Males court, with equal probability, normal brightly colored females and those that are painted to resemble plainly colored females, even though the painted females perform aggressive rejection displays. Males court all brightly colored females whether or not painted to resemble brightly colored females. Although males are aggressive to 2 of 10 of the brightly colored females painted to resemble plainly colored females, 1 of 10 males is also aggressive to a plainly colored female painted with the bright female pattern (data from Cooper, 1984, and unpublished, 1983). None of these frequencies of aggressive behavior differs significantly (Fisher's exact test, $p > 0.10$). Furthermore, males are occasionally aggressive toward unrestrained, unpainted females in the field (W. E. Cooper, unpublished observations, 1983, 1985). Therefore, there is no evidence for *H. propinqua* that bright coloration reduces aggression by males.

The maturity hypothesis states that bright female coloration advertises sex and reproductive competence. Although only mature females are brightly colored (with the same reservations regarding possible ultraviolet color patterns and selective balance between efficiency of signaling and increased risk of predation), this hypothesis may be rejected. As in the sex recognition hypothesis, even mature females of most iguanid species lack bright pigmentation. Furthermore, the colors are brightest at times when fertilization would not occur and when courtship is rejected. The most important negative evidence is that adult females are plainly colored much of the time even in species having bright female secondary sexual coloration. For example, the bright colors fade between clutches in *Gambelia wislizeni* (Medica et al., 1973) and perhaps other seasonally iteroparous species.

Rather than simply indicating maturity, the bright colors might specifically stimulate courtship by males at times at which the female is in a condition favorable for fertilization (Fitch, 1956). Support for the courtship stimulation hypothesis is furnished by rapid brightening during the final stages of vitellogenesis observed in several species (Ferguson, 1976; Werner, 1978; Wilcox, 1980; Cooper et al., 1983). Even more suggestive is the sexual receptivity of female *Holbrookia propinqua* near the completion of brightening (Cooper and Crews, 1988). Despite the coincidence of brightening and receptivity, stimulation of courtship cannot be the entire explanation for bright female coloration. The brilliant pigments remain long after the female becomes nonreceptive, being characteristic of gravid females that reject courtship. The possibility that females are constrained to a formerly adaptive peak, with bright colors being retained long after their usefulness in stimulating courtship has ended, seems remote in view of their presumed cost in terms of detectability to predators. Finally, males court females lacking bright coloration. Male *H. propinqua* court females of both color phases; also, plainly colored females are sometimes receptive shortly before brightening (Cooper, 1984; W. E. Cooper, unpublished observations, 1985).

The performance of aggressive rejection displays and biting of courting males by brightly colored females suggests that bright female coloration could be a chromatic signal that courtship will be rejected aggressively. Gravid females would benefit by avoiding predation risks associated with fruitless courtship, copulation, or both. Responsiveness to the rejection signal would enable the male to avoid aggressive rejection, increased risk of predation associated with courtship and rejection, and waste of time and energy. Plain female *Holbrookia propinqua* do not perform the aggressive rejection behaviors; bright females perform them when gravid and retain brightness at least as long as they are gravid (Clarke, 1965; Cooper, 1984, 1988; Cooper and Crews, 1988). The display of orange areas on the sides, rear of thighs, and tail during rejection displays suggests that the orange component is related specifically to courtship rejection, as in the related aggression avoidance hypothesis.

A primary objection to the courtship rejection hypothesis has been that males of several species court and sometimes copulate with brightly colored females (Fitch, 1956; Clarke, 1965). Given the sexual receptivity of newly brightened female *Holbrookia propinqua* (Cooper and Crews, 1988), this is to be expected. However, sexually receptive, newly brightened females do not display the partially concealed bright orange areas or perform any of the aggressive rejection behaviors. Thus, although the bright yellow component is continuously visible, the orange patches are revealed primarily during aggressive rejection. It might be argued that chromatic displays are not needed in species having aggressive rejection behaviors, but their joint occurrence suggests that bright female coloration and aggressive rejection displays may not be entirely redundant. Alternatively, their redundancy might reduce the time required for males to cease courting, possibly through a learned association between the display of orange areas and subsequent refusal to copulate and actual aggression.

Many species, including some members of the genera *Crotaphytus, Gambelia, Holbrookia, Petrosaurus, Sator,* and *Tropidurus,* have continuously visible bright female coloration. In some cases, the coloration appears to have a function other than advertisement of courtship rejection. It has been proposed that bright females of territorial species such as *H. propinqua* and *T. delanonis* may avoid challenge by resident males (Werner, 1978; Cooper, 1986). All species having bright female coloration and aggressive courtship rejection are territorial, having males that regularly patrol their territories, aggressively challenging any unidentified lizards. Because females occupy smaller territories, either within or overlapping male territories, males may become familiar with expected female locations. The yellow color component in *H. propinqua* is a static signal visible at all times and identifying a female, as opposed to the orange component, which is emphasized only during display. Females become sexually receptive during brightening of the clearly visible yellow pigmentation but remain aggressive for long periods thereafter. Therefore, males could monitor females for color change, courting intensely when the females are rapidly brightening and more perfunctorily at other times. That males may have the capacity for such sophisticated behavior is suggested by their ability to distinguish females residing within their territories from other females (Cooper, 1985). Similarly, male *Urosaurus ornatus* might gauge female reproductive state by observing throat coloration (Zucker and Boecklen, 1990). Nevertheless, it must be emphasized that the function of the continuously visible bright female colors is poorly understood and may vary with species. Also, there is but limited information on ancillary cues. Functional evaluation of the bright female coloration will require data on the social structure and mating systems of several species and of the relative intensity of courtship by territorial males to plainly colored, brightening, and fully brightened females.

VII. COLORATION AND INTERSPECIFIC SOCIAL INTERACTION

Although it has long been suspected that interspecific differences in coloration contribute to maintenance of ethological isolation among closely related lizards, little empirical evidence is available on this important issue. Many iguanids have species-typical displays involving pushups, head-bobbing, and, in some species, dewlap pulsing. These displays often differ sufficiently to serve as ethological isolating mechanisms (Ferguson, 1977; Ferguson and Carpenter, 1977; Jenssen, 1977); the displays have been implicated in maintaining reproductive isolation in the *torquatus* species group of *Sceloporus* (Hunsaker, 1962). Only in anoles has serious consideration been given to the possible significance of interspecific color differences. Because the bright dewlap coloration in anoles is exposed only during displays involving dewlap extension, interspecific differences in coloration and display movements may be mutually reinforcing.

Dewlap coloration varies greatly among species in *Anolis*. Color and pattern of dewlaps are often quite similar among species in limited anole faunas, being—with a few exceptions—yellow to yellow-orange throughout the Lesser Antilles. Variation of dewlap colors is much greater in larger anole assemblages such as those from Hispaniola and Cuba (Williams and Rand, 1977). In an assemblage of eight sympatric anoles from La Palma, Hispaniola, the dewlap of each species has a different color (Table 6.11), so that response to dewlap coloration could by itself allow accurate identification of every species (Rand and Williams, 1970). Dewlap coloration alone does not separate all species in other anole faunas (e.g., in Cuba), supporting the contention that multiple cues may be important in species recognition (Rand and Williams, 1970; Williams and Rand, 1977). Species-specific dewlap coloration often differs markedly among sympatric or parapatric anoles, which is consistent with use of dewlap coloration in species recognition (Williams and Rand, 1977). However, dewlap color varies geographically in other species (Webster and Burns, 1973; Crews and Williams, 1977; Case and Williams, 1984; Williams and Case, 1986) and sometimes within single populations (Case and Williams, 1984).

The Haitian populations of the sibling species *Anolis brevirostris* and *A. distichus* vary geographically in dewlap color (Webster and Burns, 1973). Dewlap color in *A. brevirostris* is discontinuous in contact zones with *A. distichus*. However, subsequent study has revealed complex relationships between dewlap color and proximity of populations in these species. In different areas, these species may be sympatric and have differing dewlap colors, be syntopic and have similar dewlap

Table 6.11 Dewlap color in a local assemblage of anoles from La Palma, Hispaniola

	Species Present							
Color	ins	ric	chr	eth	dis	cyb	chl	coc
Black	...	...	...	...	...	...	X	...
Gray	...	...	...	X	...	...	...	...
White	X	...	X	X	...	...	...	X
Red	...	...	...	...	X	...	...	...
Pink	...	X	...	...	...	...	...	...
Orange	X	...	...	...	...	...	...	...
Yellow	...	X	...	...	X	X	...	...
Blue	...	...	...	...	...	...	X	...
Blue-gray	X	X	...	...	...	...	...	...
Gray-violet	...	...	X	...	...	...	...	...
Violet	...	X	...	...	...	...	...	...

Source: Data from Rand and Williams, 1970, *American Naturalist*.
Species: ins, *insolitus*; ric, *ricordii*; chr, *christophei*; eth, *etheridgei*; dis, *distichus*; cyb, *cybotes*; chl, *chlorocyanus*; and coc, *cochranae*. The last two species are suspected but not confirmed in La Palma.

colors, and parapatric and have similar dewlap colors (Williams and Case, 1986). Thus it is clear that dewlap coloration cannot by itself play an important role in maintaining reproductive isolation between all pairs of adjacent populations. Another, as yet untested, hypothesis is that variable dewlap colors may act as signals allowing assortative mating among individuals having primarily physiological adaptations to local conditions (Crews and Williams, 1977).

The role of dewlap coloration in species recognition has been studied directly in only a single case, that of the morphologically similar sympatric species *Anolis marcanoi* and *A. cybotes* (Losos, 1985). In color modification experiments in which the normally red dewlaps of *A. marcanoi* and white dewlaps of *A. cybotes* were painted to resemble those of the other species, male *A. marcanoi* are significantly more aggressive to male *A. cybotes* painted to resemble *A. marcanoi* than to unpainted ones. The reduced aggressiveness to painted heterospecific males compared to conspecific males suggests that other factors are involved. For instance, the form of the species-typical assertion and challenge display may also affect aggressiveness. However, the aggressiveness of male *A. marcanoi* to male *A. cybotes* having dewlaps painted red is slightly greater than to conspecific males having dewlaps painted white; this perhaps indicates primacy of color in determining intensity of aggression (Losos, 1985). Another possible explanation for the reduced aggressiveness to heterospecific males bearing conspecific dewlap coloration is that in paint-modification trials the male *A. cybotes* are less aggressive than conspecific males toward *A. marcanoi* despite their artificial white dewlap coloration in these trials. Also, male *A. marcanoi* may respond to a hierarchical structure lacking in the aggressive repertoire of *A. cybotes*.

Although this color modification study very likely produced robust results, the results are not conclusive due to two features of the experimental protocol. Several lizards in each experimental condition were tested more than once, but there is no rationale for choosing the individuals to be retested. Mean number of trials per individual varied substantially with treatment, having values of 1.4, 1.3, 1.5, and 2.8. No information has been presented about the numbers of trials in which each retested individual participated. Nonrandom selection of individuals to be retested could have biased results. A potentially more serious problem is that results of different treatments were compared by Mann-Whitney U tests, which demand independent observations, whereas the same individuals were in many cases tested in both treatments being compared. The comparisons between groups were thus based on a mixture of correlated and independent observations, with trial sequence and its likely effects unspecified for correlated observations. Thus the significance levels reported must be considered to be approximations. Because this intriguing and otherwise well-designed study gives the only experimental evidence for an interspecific function of dewlap coloration, such function must be regarded as highly probable but unproven pending further study.

The Anolinae represent an excellent group for studying possible interspecific roles of color, but numerous other visually oriented reptiles show pronounced interspecific color variation. Behavioral responses to conspecific and heterospecific color patterns should be examined for these reptiles, including some emydid turtles and other lizard taxa such as other iguanids, agamids, and chamaeleonids having marked interspecific differences in color patterns. For example, the social displays of chameleons appear to include species-typical chromatic components subject to rapid physiological change (Parcher, 1974), which could be involved in recognition of conspecifics.

VIII. BACKGROUND COLOR MATCHING

A. General

Coloration matching the background in natural habitats is important because it enables reptiles to avoid detection by predators. The importance of matching is reflected in the diversity of reptiles, including snakes, lizards, crocodilians, and turtles that have phenotypic plasticity in coloration allowing them to match the predominant background coloration (e.g., Redfield, 1916; Norris and Lowe, 1964; Kirshner,

1985; Banks, 1986; Luke, 1989; Section III.A). The match between body coloration and background coloration can be attained in several ways. Animals having a particular body coloration may behaviorally select habitats or microhabitats in which they are inconspicuous against the background. In some species capable of rapid color changes, such as chameleons, the animal may change color to match the current background (Zoond and Eyre, 1934). Background matching by physiological color change is also believed to occur in *Anolis carolinensis* (von Geldern, 1921; Medvin, 1990), but its extent in nature is unknown.

Morphological color changes may also gradually shift body coloration toward inconspicuousness on the predominant background in some lizards, turtles, and crocodilians (Redfield, 1916; Kirshner, 1985; Banks, 1986; Luke, 1989) and perhaps in snakes. Such changes may be reversible throughout life or may consist of irreversible changes that occur during growth. At least one lizard, the lacertid *Podarcis taurica,* changes color seasonally, being a bright green in spring and early summer and becoming dark or olive green to brown or olive brown in late summer. The changes in body coloration correspond to changes in the dominant coloration of the local vegetation—the green body coloration persists only as long as the grass remains green (Chondropoulos and Lykakis, 1983). Due to its effects on thermoregulation and the importance of body temperature to efficiency of many behaviors, background matching can potentially affect diverse behaviors (Luke, 1989). Reliance on crypsis through background matching may also favor decreased frequency of movement, especially in the presence of predators. Many squamates undergo physiological color changes with temperature (Section III.B), whether or not their body coloration matches that of the local substrate. These physiological changes may also affect behavioral efficiencies.

B. Body Coloration, Color Matching, Thermoregulation, and Behavior

Despite the great importance of crypsis in avoiding detection by predators, scant attention has been given to behavioral selection of backgrounds appropriate for crypsis by reptiles. However, such behavioral selection has been studied experimentally in a few lizards. The gecko *Ptychozoon kuhli* from Thailand chooses a brown background resembling its body coloration over tan or white backgrounds (Vetter and Brodie, 1977). In the chapparal of southern California, periodic fires blacken the stalks of shrubs for several years. After the charred surface has worn off the stalks, the stalks are lighter in color. The darkly colored iguanid lizard *Sceloporus occidentalis* perches selectively on the dark stalks. After the stalks lighten and the lizards are no longer cryptic, they switch perch preference to rock outcrops. In the laboratory, the lizards strongly preferred to perch on dark branches rather than light ones (Lillywhite et al., 1977). Two closely related species of Australian agamid lizards are sympatric on rocky outcrops, being ecologically segregated on rocks of differing color: *Amphibolurus decresii* occurs on pinkish yellow rocks and *A. vadnappa* on dark reddish brown rocks. Both species are camouflaged by body coloration that matches that of the substrate. In substrate selection experiments, both species strongly prefer to perch on rocks matching their camouflage color (Gibbons and Lillywhite, 1981).

The presumed explanation for color matching is that crypsis lessens the likelihood of predation. Two lines of evidence favor this hypothesis. Lizards (*Uta stansburiana*) that have coloration contrasting with that of the substrate are more likely to be killed by avian predators than are substrate-matched lizards (laboratory tests, Luke, 1989). Similarly, lizards (*Sceloporus undulatus*) differing from background coloration have higher frequencies of tail breaks in the field (Gillis, 1989), suggesting that they are subject to greater predation pressure.

Influences of the coloration of an individual reptile on its own behavioral performance are very poorly understood. Coloration may affect heating rates in lizards, especially during basking (e.g., Cole, 1943; Norris, 1967; Porter and Norris, 1969; Porter et al., 1973; Pearson, 1977; Porter and James, 1979; Rice and Bradshaw, 1980; Porter and Tracy, 1983; Luke, 1989). Coloration therefore has two potentially important effects on behavior. First, individuals having coloration that allows more rapid warming to preferred body temperature may have more time available for numerous activities, both social and nonsocial (Fig. 6.19). Second, because the performance curves of many behaviors are thermally dependent in ectotherms (e.g., in lizards, probability of feeding, Waldschmidt et al., 1985; tongue-flicking rate, Cooper and Vitt, 1986), the performance efficiencies of the behaviors may vary with coloration (Luke, 1989).

The thermoregulatory, antipredatory, and social functions of color may conflict. For example, dark body coloration could confer advantage due to higher rates of heating and permit achievement of higher body temperatures (for lizards, Cowles and Bogert, 1944; Pearson, 1977; Rice and Bradshaw, 1980; Porter and Tracy, 1983; for snakes, Gibson and Falls, 1979; Andrén and Nilson, 1981), although crypsis might demand light body coloration.

The adjustment of coloration to match local substrates may influence thermoregulation and thus affect many behaviors indirectly (Luke, 1989). Populations of the iguanid lizard *Uta stansburiana* living on light alluvial soil are much lighter in color than those on darker lava substrates (Norris and Lowe, 1964; Norris, 1967). From experi-

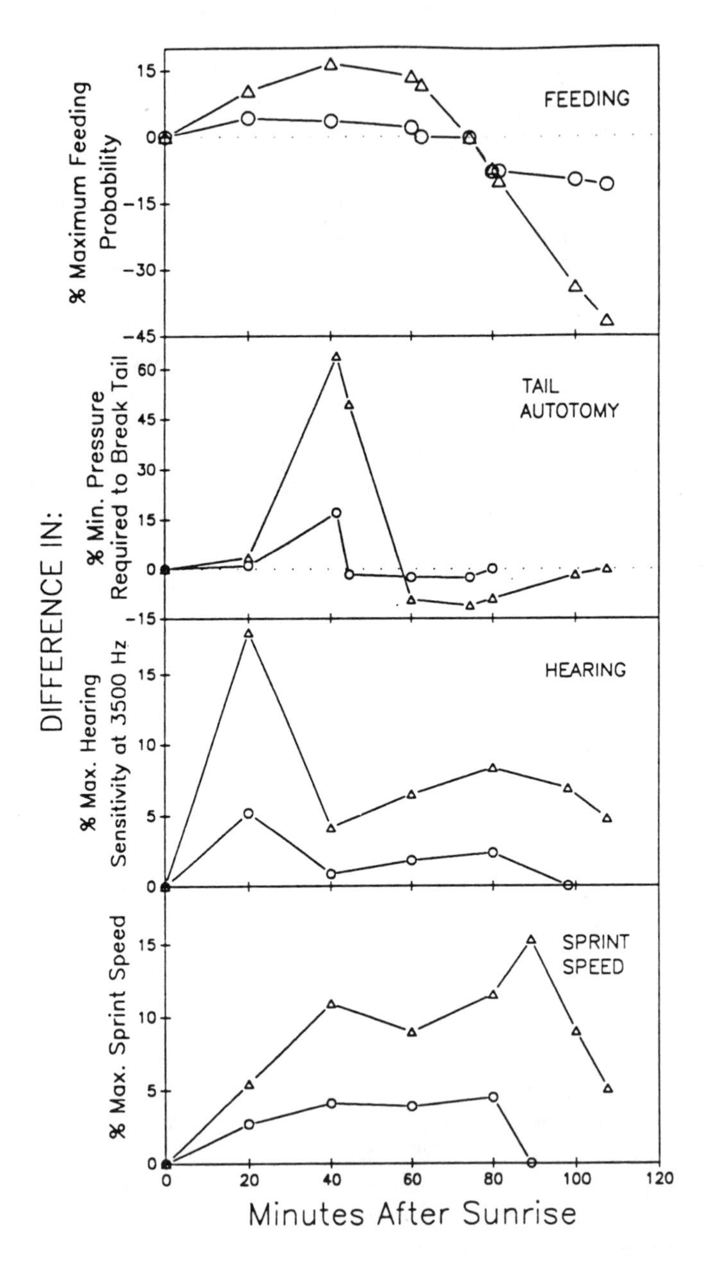

Fig. 6.19. *Uta stansburiana.* Differences in percent maximum performance curves between dark and light lizards for four behavioral and physiological processes are plotted by time of day. Temperature differences assumed between dark and light lizards are 3.2°C (triangles) and 0.8°C (circles). (From Luke, 1989.)

ments with model lizards and calculations based on absorbance, it is estimated that darkly colored individuals may be 0.8 to 3.2°C warmer than lightly colored individuals. They may thus be somewhat better able than lightly colored individuals to perform various activities during the entire basking interval (Luke, 1989). Placement of model lizards in basking positions at angles of 30° to the sun and sheltered from wind shows darker models warming to optimal temperatures 5 minutes faster than lighter ones. Estimates of this difference from models in several different conditions ranged from 5 to 14 minutes. Calculated estimates were larger still (up to 32 minutes). Only the morning activity period has been studied, and the total time gained may be greater when afternoon activity is considered as well (Luke, 1989).

The importance of coloration varies with insolation, dark coloration reducing the basking time required to reach preferred body temperature slightly more in clear than in hazy conditions (Luke, 1989). Dark coloration may be more important for two populations of *Uta stansburiana* in winter than in summer (Norris, 1967; Luke, 1989) because (1) there is less haze in winter, (2) the cool temperatures of winter preclude effective delay of daily emergence until the ambient temperatures allow rapid warming, and (3) the time available for activity, both in hours per day and days per month, is less in winter than in summer. The reduction in activity time may be important because any time gained by rapid warming to activity temperature by dark lizards constitutes a greater portion of the total activity time (Luke, 1989).

The relative effects of light and dark body coloration on a behavior are determined by the heating rates, especially during basking, and the relationship between body temperature and behavioral performance (Luke, 1989; Fig. 6.19). Sprint speed in dark *Uta stansburiana* may be substantially greater than that of light ones throughout the basking interval. The greater speed reflects the warmer state of darker lizards at any given time after basking begins. It also reflects the increase in sprint speed with temperature up to the preferred body temperature of 36°C, with peak speed maintained up to 38°C. On the other hand, the probability of feeding in the presence of a prey item is slightly greater for dark than light individuals for about the first 80 minutes of basking. After that time, it is less because the probability of feeding is highest at 32°C and declines below the preferred body temperature (Luke, 1989). Because the efficiency of various behaviors may have substantially different thermal relationships, determination of the effects of color on behavior will demand detailed quantitative study of both heating curves and behavioral performance curves.

Behavioral influences of physiological color changes through thermoregulatory effects are potentially important, but have not been studied directly. In some populations of *Uta stansburiana,* the lizards are dark below 25°C but blanch at and above that temperature. Because they emerge at 25°C, any possible thermoregulatory benefits during the activity period are lost (Luke, 1989). However, many other species of snakes and lizards begin basking while dark and blanch as body temperature rises (e.g., Norris, 1967; Pearson, 1977; Hoppe, 1979; Vaughan, 1987b). In such species, having dark body coloration at low temperatures may decrease the time required to reach behaviorally optimal body temperatures and thus increase available time for activity and feeding (Porter and James, 1979; Porter and Tracy, 1983). By elevating the body temperature during basking, dark body coloration may also increase behavioral efficiencies at any given time after the onset of basking. Because blanching may be a continuous process as in the iguanid lizard *Liolaemus multiformis* (Pearson, 1977), the relative advantage may be less than that observed for species such as *U. stansburiana* (Luke, 1989), which maintain constant dark or light body coloration during the entire basking interval.

IX. METHODS AND PROSPECTS FOR FUTURE STUDIES

Techniques for studying the physiology of pigmentation and color vision are well developed (see Bagnara and Hadley, 1973; Burghardt, 1977); however, they generally have not been applied to most reptilian groups. For example, a basic assumption of studies on social effects of lizard coloration has been that lizards have color vision; however, neither absorption spectra of the visual pigments nor psychophysical sensitivity spectra have been determined for any species of lizard. Chromatic stimuli could be important due to brightness rather than hue in colorblind species, but responsiveness to light of different wavelengths should be determined. Much of our knowledge of reptilian pigmentation and physiological color change derives from studies of the single species, *Anolis carolinensis,* which may well be an atypical representative of lizards in several respects. Recent work on *Uta stansburiana* (Luke, 1989) and *Phrynosoma modestum* (Sherbrooke and Frost, 1989) reveals important differences and points the way to a broader perspective. Comparative studies of reptilian pigments, mechanisms of color change, and color vision are long overdue. One area in great need of study is the neural regulation of color change. The evidence for neural regulation in reptiles is quite old, and the studies antedate modern knowledge of hormonal regulation. The early evidence cited above favoring neural regulation in chameleons and *Phrynosoma* should be reevaluated by modern methods.

A variety of techniques for studying behavioral responses to chromatic social stimuli is available, but the easiest technique to apply and most successful to date has been observation of behavioral changes induced by modification of colors in living animals. Color modification can be combined with introduction of tethered or unrestrained individuals in the field. It can be used in a variety of experimental paradigms in the laboratory, such as simultaneous and successive discrimination experiments. Other techniques having potential for studying social responses to coloration include presentation of standardized stimuli on film loops or videotape. Editing capacity gives videotape presentations great flexibility and use of electronic filters permits controlled change of color. Use of painted model animals has been attempted many times with limited success, but no one has reported studies using reasonably realistic movable mechanical models. Controls for the effects of painting or other color modification should always be included. Experimental design and analysis should be considered thoroughly before starting to collect data. There has been an unfortunate tendency to publish results that are inconclusive due to insufficient sample size or lack of statistical independence among observations. For instance, several of the studies here reviewed applied statistical tests requiring single observations per individual to data consisting of a mixture of single and multiple observations per individual.

Some topics that were noted in Section I.A have not been discussed further due to lack of information (predatory behavior, hydroregulation) or to recent coverage elsewhere in *Biology of the Reptilia* (Vol. 16, ch. 1; defensive behavior, Greene, 1988). Some thermoregulatory behaviors are associated with color in forms that show physiological color change. We know that predation selects for color matching (Norris, 1967); does this imply that the necessity for color matching constrains the thermoregulatory behavior of species that change color with temperature? That could explain why *Uta stansburiana* blanches at emergence temperature in populations living on light backgrounds (Luke, 1989). A new approach may make it possible to study complex effects of coloration, through its influence on thermoregulation, on performance of various behaviors (Luke, 1989).

Virtually nothing is known regarding the possible physiological regulation of behavioral aspects of the thermoregulatory role of body coloration. However, mammalian studies provide fascinating hints that melanophore-stimulating hormone (MSH) might be important in this regard. MSH may be a neurotransmitter and is found in the septum and anterior hypothalamus, regions of the brain associated with thermoregulation in mammals (O'Donohue and Dorsa, 1982). During fever, its concentration rises in the septum of rabbits and falls in the arcuate nucleus (Samson et al., 1981). It and ACTH are the most potent known fever reducers, perhaps 25,000 times more potent than acetaminophen (Murphy et al., 1983). Injected into the central nervous system or peripherally, MSH and ACTH reduce fever at doses that do not affect normal temperature (Lipton and Clark, 1986), possibly by competing with leukocyte pyrogen for receptor sites in the brain (Murphy et al., 1983).

Many reptiles are cryptically colored in their natural habitats (e.g., Norris, 1967; Jackson et al., 1976). How does current coloration affect the choice of background color in species having and in those lacking the ability to physiologically change color? Studies of microhabitat selection, especially with respect to coloration, are needed for a variety of taxa. Corresponding studies might determine whether the behavioral selection of substrates with a color matching body coloration reduces the probability of predation. In a chromatically complex environment offering backgrounds with multiple colors, there is potential for interaction of physiological color change and background selection. The complexity of the relationship between background color and body color is seen in *Anolis carolinensis;* in the field, green males and brown females match their backgrounds above chance levels, but brown males and green females do not (Medvin, 1990). Finally, although coloration may affect hydroregulation in amphibians (Hoppe, 1979), it is unlikely to do so in reptiles, and nothing is known regarding reptilian hydroregulatory behavior related to color.

Many intriguing aspects of reptilian coloration remain virtually unstudied. It is important to know how relationships between coloration and behavior are affected by the combined adaptive requirements of selective pressures for social behavior, defense, predation, and thermoregulation. A common hypothesis is that many lizards have evolved ventral and lateral display colors and behavioral displays to expose them because conspicuous dorsal coloration would increase detectability to predators (e.g., Cooper, 1986). Unfortunately, there is no information regarding relative predation intensity, types of predators, or other ecological factors relevant to the presence of chromatic signals or the location of bright pigments on the body. Sexually dimorphic coloration in lizards often plays a role in sex recognition, but the influence of such coloration on mate choice has been little studied (Section VII.A). Thus little is known about sexual selection of coloration in lizards and virtually nothing in other reptiles.

Intraspecific and interspecific behavioral functions of chromatic signals are only beginning to be described. Despite the occurrence of sexual dichromatism in diverse reptilian taxa, the possible signal function of color has been studied in only a few lizards. Social functions of color should also be studied in dichromatic turtles, other families of lizards, and snakes. Most controlled studies on lizards have been carried out on iguanids; the behavioral significance of bright static breeding colors of many lacertids, teiids, cordylids, and fewer members of other families have not been investigated. The rapid color changes associated with social behavior in many agamids and chameleons should provide new insights into the functions of color patterns. In these families the bright display colors may be expected to appear in social contexts similar to those eliciting the displays of otherwise hidden color patches by many iguanids.

The evolutionary history and function of bright breeding colors in males and females are controversial. Investigation of roles of male colors in mate choice and of female colors in courtship rejection combined with measurement of the intensity of sexual or natural selection on coloration where variation exists could resolve the major issues. How various chromatic signals are related to fitness is unknown. Sexual selection of chromatic social signals is strongly suspected in many lizards, but studies demonstrating its action and modes are needed. Evaluation of the roles of interspecific differences in coloration among syntopic congeners requires much more work, not only for anoles, but for many sceloporine iguanids, lacertids, and other taxa. Study of the influences of social and environmental stimuli on color in reptiles requires systematic observational and experimental regimes, such as those described for *Anolis carolinensis* (Greenberg et al., 1984). Too much of our current information in this area is anecdotal.

Chromatic phenomena in reptiles exist in a profusion of striking and subtle manifestations having adaptive significance in social and other behaviors. We have studied few aspects of these, usually in a cursory manner. Thus opportunities abound for clarifying the contributions of color to fitness, microevolutionary change in coloration, and the physiological control of static and labile color patterns in reptiles. A basic consideration for studies of color and behavior is that to be behaviorally important, chromatic properties such as wavelength and saturation must permit an animal to respond in an appropriate manner at the appropriate time. In social systems, coordination demands that the signal and receiver evolve in parallel.

Although numerous important topics have been essentially ignored, this is particularly astonishing for two topics widely recognized to be potentially important: behavioral relationships of ontogenetic color change and social roles of interspecific differences in color. First, very little is known regarding ontogenetic color changes involving major alteration of pattern in lizards, snakes, and, to a lesser extent, turtles. Defensive behaviors and caudal luring associated with conspicuous tail coloration in juvenile lizards and snakes and behavioral changes associated with the loss of mimetic coloration in *Eremias lugubris* and *Diploglossus lessonae* are obvious topics for future work. Such studies should correlate the timing of behavioral and gross chromatic changes with underlying events in the integument. The relationships of widespread differences in the basic juvenile and adult color patterns in squamates (excluding adult breeding patterns) to microhabitat selection and predation deserve more attention.

A second topic for fruitful investigation is the potential role of interspecific color differences in maintaining reproductive isolation and reducing interspecific aggressive behavior. Results of one study (Losos, 1985) suggest that differences in dewlap coloration may reduce interspecific aggression in two sympatric anoles, but the results are not conclusive. There is no other experimental evidence on interspecific social influences of coloration. In many other lizard genera, sexually dimorphic color patterns, some known to be androgenically controlled, differ among species in ways that could have an important effect on interspecific social behavior.

One of the most exciting dimensions in the study of reptilian coloration is the prospect for better understanding of the reciprocal and complementary aspects of internal state and behavior in relation to the causes, corollaries, and consequences of coloration. Physiological bases of coloration have for the most part been investigated independently of any knowledge about possible behavioral functions of the pigments or color changes. Standard methods of extraction, separation, identification and assay of pigments, gonadal steroid hormones, catecholamines, and MSH have been employed in such work. Ideally, evidence on hormonal induction of color changes is based on surgical or physiological removal of the source followed by administration of exogenous hormones. Measurements of endocrine levels of intact individuals of differing color states can provide important corroborative evidence and reveal fluctuations of hormone levels as colors change.

Nontrophic effects of MSH that may affect behavior (while coloration is altered trophically) are of great potential importance but are poorly understood. In all taxa studied, MSH has psychoactive effects, as well as numerous pigmentary and correlated somatic ones. The multiplicity of biological roles of melanotropic peptides has been emphasized in a recent exhaustive survey (Hadley, 1988a, b, c). Of particular interest are the associations of MSH with a variety of behavioral phenomena, including learning, memory, and attention (Beckwith, 1988; Datta and King, 1982), sexual behavior (Thody et al., 1979; Thody and Wilson, 1988), and thermoregulation (Lipton, 1988). The observations that socially subordinate *Anolis carolinensis* are darker in body color than dominants, are nonaggressive, and display manifest reduced reproductive activity are interesting in light of mammalian studies associating elevated MSH with decreased aggressiveness (Patterson et al., 1980). MSH has been localized by immunochemistry in the central nervous system of the lizards *Podarcis muralis* (Vallarino, 1984) and *A. carolinensis* (Khachaturian et al., 1984). The mechanism of action of MSH is poorly understood, although at least some of its effects probably involve central dopaminergic mechanisms (Monnett and Lichtensteiger, 1980).

A major gap in our knowledge of those color changes regulated by gonadal steroid hormones is the lack of data on effects of reptilian gonadotropins. Interestingly, there is evidence that MSH may integrate gonadotropin release with environmental stressors in mammals (Khorram et al., 1985; Reid et al., 1981). It is possible, although unlikely, that gonadotropins might directly induce color change. Because luteinizing hormone increases aggressiveness in the male starling and because starlings are unusual in that castration increases dominance (Mathewson, 1961), possible effects of gonadotropins on color should be investigated. Clarification of the nature of reptilian gonadotropins should be coupled with determination of their effects on coloration. Few studies have considered the relationship of color to social status or the effects of the color of one individual on the physiology of other individuals; such studies should be emphasized now that the signal function of color has been established in several species. In addition to the hormones already mentioned, it may be anticipated that adrenal glucocorticosteroids and pituitary ACTH will be measured frequently in studies of social dominance and stress. Effects of social chromatic signals on the reproductive tract should be investigated.

An intriguing topic that is unstudied in reptiles and poorly understood in other animals is the relationship between display coloration and the color of the background occupied. Only a beginning has been made in attempts to predict what colors an animal should use to display in different environmental circumstances (e.g., Hailman, 1977, 1979; Burtt, 1986). Visual signals would be most conspicuous whenever their brightness (luminance) and saturation differ most markedly from background. Hailman (1979) also makes the apparently reasonable assumption that the greatest contrast in hue can be obtained at the complementary dominant wavelength.

Determination of irradiance spectra in specific microhabitats permits conclusions about colors that would render animals conspicuous. For example, spectra from several microhabitats in southeastern pine forests and ecotones of the United States suggest that (1) dark colors produce the most contrast against a blue or gray sky, (2) blue-green against loblolly pine trunks, (3) dark blue against sweet gum trunks, leaf litter from hardwoods, sand dunes, and grass, and (4) light blue against green leaves (Hailman, 1979). This could account for the high frequency of blue display coloration in many sceloporine iguanid lizards found primarily on tan and brown backgrounds and for conspicuous blue tail coloration that draws attention of predators to the autonomous tails (Arnold, 1988) in many small scincid and teiid lizards. It may also explain why dominant males in some populations of *Urosaurus ornatus* (Zucker, 1989) assume a very dark body coloration. Their darkness makes them very conspicuous against the light background. As they perch on boulders, they may be viewed from above against the light gray-brown of the boulders or from below against the sky. In coniferous forests, maximum contrast with the habitat as background may be obtained by white stimuli, whereas yellow and orange provide better contrast against a background of broad leaves (Burtt, 1986). It might be possible to relate such predictions to variation in chromatic signals among species of anoles.

The prospects for improving our ability to predict the dominant wavelength, saturation, and brightness are exciting, but many problems remain that will require further conceptual and empirical advances. A major stumbling block is that the predictions made to date are based on the assumption that visual perception is similar in humans and animals. For example, a cubic space of perceived colors based on dominant frequency, excitation purity, and relative luminance (Burtt, 1986) depends on a number of assumptions regarding the relationship of luminance to brightness, the relationship of excitation purity to perceptual saturation, and variation in discriminability of hues among wavelengths. Unfortunately, we do not know whether these assumptions are true for any reptile. For example, lizards seeing ultraviolet light (Alberts, 1989) might perceive different wavelengths as complementary than do human beings. Detailed psychophysical studies are needed to determine the chromatic perceptual spaces of representative reptiles. Another current problem is that the yellows, oranges, and reds, which are widely used as reptilian display colors, are not predicted to provide maximum contrast on any background studied by Hailman (1979). This could indicate that there are conceptual difficulties with the approach, possibly based on interspecific differences in perception of complementary colors or on the fact that the backgrounds yet studied are inappropriate. One important consideration for predicting animal coloration is its determination by the balance of multiple selective forces. In addition to providing clear intraspecific signals, coloration must often provide crypsis, warning, or other antipredatory features, may be subject to thermoregulatory limitation, may serve as interspecific signals to competitors, and may be constrained by current or historical selection for reproductive isolation from related sympatric species. Signals to competing species may be chromatically convergent to ensure correct interpretation (Baylis, 1979).

Other aspects of reptilian chromatic signals also need attention. Do reptiles respond to graded chromatic stimuli or respond at maximum intensity above some unknown threshold levels of stimulation? Graded responses to continuously varying intensity of socially significant color patterns have been discovered in birds that maintain dominance hierarchies (reviewed by Rohwer, 1982), but the possibility that similar responses occur in reptiles has not been investigated. How specifically do reptiles respond to chromatic properties of stimuli? What are the relative importances of brightness, saturation, and hue? The importance of such questions may be seen by the observation that male *Sceloporus undulatus* will act aggressively not only to females painted blue in the ventrolateral region normally colored blue in males but also to those painted several other bright colors (Noble, 1934; Cooper and Burns, 1987). This suggests that brightness or saturation may be as important as hue, perhaps more so. Opportunities for study of the effects of such variables and their interactions on behavior abound.

Having completed a review of reptilian coloration and its relationships to behavioral physiology, we are struck by both the mass of information already accumulated and the huge gaps in our knowledge and understanding. Progress is being made in several laboratories in which the physiology and behavior of reptilian coloration are being concurrently studied. We hope that this chapter will be a useful reference for our colleagues, and that it will interest new researchers and stimulate interest in some new lines of research. Unraveling the complex determinants of reptilian coloration remains a fascinating and rewarding task.

X. SUMMARY

Many reptiles, especially many lizards, have color vision and the ability to produce chromatic social signals. These signals may be based on sexual differences in pigment synthesis in mature individuals and may be permanent or seasonal, fluctuating with the ovarian cycle. Other lizards, especially agamids and chameleons, but also some iguanids, show rapid color changes having social import in aggressive and reproductive contexts; unfortunately, study of the signal functions of such changes is insufficient for generalization.

Sexually dimorphic coloration is far more common in lizards than in other extant reptiles but also occurs in a few turtles and snakes. Among the lizards, it is most widespread in the Iguanidae, Agamidae, Lacertidae, Teiidae, and Cordylidae. It is found in fewer species of several other families, most notably Scincidae, and even in a few diurnal members of Gekkonidae. Bright male colors are exhibited primarily during sexual and aggressive displays in iguanids, agamids, chamaeleonids, gekkonids, lacertids, and teiids. In many cases, the colors are evident only during social encounters because they are located ventrally, or are only transiently apparent under physiological control. Bright male coloration is a sufficient cue for sex recognition in some lacertids, agamids, iguanids, and scincids. In species capable of physiological color changes, color often reflects dominance status, as in anoles and some agamids. Polymorphic display colors are related to behavioral dominance in at least one iguanid.

Static dimorphic colors of males are under androgenic control in the two species studied, an iguanid and a scincid. The linkage of dimorphic coloration and social behavior is exemplified by the induction of bright male head coloration and activation of sexual and aggressive behavior in male skinks (*Eumeces*) by testosterone. Melanins and pteridines are known to be involved in male dimorphic coloration, but carotenoids may be important as well. Some mechanisms of physiological color change have been documented in some detail for *Anolis carolinensis,* but other mechanisms need to be adduced to explain the rapid color changes even of some other anoles and to determine the influences of ACTH and adrenocortical stress on the darkening response of subordinate individuals. Roles of catecholamines, MSH, and the nervous system in the displays of agamids and chamaeleonids are uncertain.

Female color patterns allow sex recognition in at least some iguanid and lacertid lizards. Males of *Lacerta vivipara* recognize females in part by their comparatively drab coloration. Bright secondary sexual coloration of females is a sufficient cue for sex recognition by both sexes in some iguanid lizards. It is unlikely to have evolved to attract males into performing parental care as in some other vertebrates; rather, it appears to signal courtship rejection in several species and is limited to species having distinctive aggressive courtship rejection displays. Bright female coloration may have additional adaptive functions or effects, such as letting females be identified from a distance by territorial males as suggested for *Tropidurus delanonis* and *Holbrookia propinqua.* In all species studied, the bright colors appear or intensify slightly before ovulation, reach maximal intensity while the female is gravid, and fade after oviposition or at the end of the breeding season.

Both the bright coloration and the courtship rejection behavior are induced by sex steroids. In the three species studied, bright female pigmentation may be induced by exogenous progesterone and testosterone, both of which have elevated plasma concentrations in *Holbrookia propinqua,* the only species examined. However, progesterone may be the primary agent of color change because progesterone concentrations are much higher than those of androgen. The effects of estrogen appear to vary among species, having either no independent effect or inducing partial brightening of at least one color component. It has a priming effect on the response to progesterone in at least one species, but not in the other species studied. Exogenous progesterone and testosterone but not estradiol activate the aggressive courtship rejection behaviors of *H. propinqua.* Estradiol does not activate these behaviors, but decreases the probability that a female treated with progesterone will act aggressively toward males. Given also that female *H. propinqua* are sexually receptive while brightening just before ovulation, it appears that receptivity may occur when estrogen levels are high and perhaps when progesterone levels are rising. When circulating progesterone concentrations become sufficiently high, receptivity may be replaced by aggressive rejection of courtship.

Interspecific differences in color pattern have long been claimed to provide cues for species recognition, especially in assemblages of closely related species of lizards. However, only one behavioral study of effects of interspecific differences in coloration has been published. That study suggests that interspecific differences in dewlap coloration may allow selective responsiveness to conspecifics by anoles. Variation in dewlap coloration increases in large assemblages of anoles, and the dewlap color of each species is usually distinct from that of each syntopic anole; hence, dewlap coloration may be the key cue for species recognition. The limited available data are consistent with the classic interpretation that the bright colors of anole dewlaps may have evolved under selection against interspecific mating, that is, as an ethological isolating mechanism.

Relatively little is known about the relationships among body coloration, color change, substrate color matching, thermoregulation, and behavior. Some lizards behaviorally select perches having coloration that makes the lizards inconspicuous. In laboratory trials, lizards that occupy substrates contrasting with their body coloration have higher tail break rates and are more likely to be killed by avian predators in laboratory trials than are color-matched individuals. However, there have been no simultaneous demonstrations of behavioral selection of matching background and resultant decrease in the likelihood of predation.

Because darkly colored lizards warm more rapidly during basking than lightly colored ones, darkly colored lizards may have more time to perform numerous important activities once preferred body temperatures have been attained. Because they have higher body temperatures between the onset and termination of basking than do lightly colored individuals, darkly colored lizards may more effectively feed, escape predators, or perform other activities having performance curves that improve as temperature increases and have optima near preferred body temperature.

ACKNOWLEDGMENTS

Much of this portion of the manuscript was written by WEC while he was a faculty research participant at the Savannah River Ecology Laboratory (SREL) and was partially supported by grant DE-AC09-76SR00819 from the U.S. Department of Energy to the University of Georgia through SREL. WEC is grateful to Justin Congdon, Whit Gibbons, and Laurie Vitt for interesting discussions about roles of color in reptilian behavior and especially to his co-workers on coloration and behavior in lizards: Caroline Adams, Bob Clarke, Nina Burns, David Crews, Jim Dobie, Gary Ferguson, Lou Guillette, and Laurie Vitt. NG thanks Gerry Vaughan for interesting discussions, encouragement, and collaboration relating to color changes in anoles. We owe a debt of gratitude to the editors, the reviewers, and to numerous individuals who provided help in various ways.

REFERENCES

Adams, C. S., and Cooper, W. E., Jr. (1988). Oviductal morphology and sperm storage in the keeled earless lizard, *Holbrookia propinqua*. *Herpetologica* 44, 190–197.

Adkins, E., and Schlesinger, L. (1979). Androgens and the social behavior of male and female lizards, *Anolis carolinensis*. *Horm. Behav.* 13, 139–152.

Alberts, A. (1989). Ultraviolet visual sensitivity in desert iguanas: implications for pheromone detection. *Anim. Behav.* 38, 129–137.

Alexander, N. J., and Fahrenbach, W. H. (1969). The dermal chromatophores of *Anolis carolinensis* (Reptiles, Iguanidae). *Am. J. Anat.* 126, 41–56.

Allen, E. R. (1949). Observations on the feeding habits of the juvenile cantil. *Copeia* 1949, 225–226.

Anderson, J. (1878). *Anatomical and Zoological Researches and Zoological Results of the Yunnan Expeditions,* Vol. I. Government Printing Office, Calcutta, India.

Anderson, P. K. (1961). *The Reptiles of Missouri.* University of Missouri Press, Columbia.

Andersson, M. (1986). Evolution of condition-dependent sex ornaments and mating preferences: sexual selection based on viability differences. *Evolution* 40, 804–816.

Andrén, C., and Nilson, G. (1981). Reproductive success and risk of predation in normal and melanistic colour morphs of the adder, *Vipera berus. Biol. J. Linn. Soc.* 15, 235–245.

Armington, J. (1954). Spectral sensitivity of the turtle, *Pseudemys. J. Comp. Physiol. Psychol.* 47, 1–6.

Arnold, E. N. (1988). Caudal autotomy as a defense. In *Biology of the Reptilia,* Vol. 16 (C. Gans and R. B. Huey, eds.). Alan R. Liss, New York, pp. 235–273.

Arnold, E. N., and Burton, J. A. (1978). *A Field Guide to the Reptiles and Amphibians of Britain and Europe.* William Collins Sons & Co., London.

Arslan, M., Lobo, J., Zaidi, A. A., and Quazi, M. H. (1978). Steroid levels in preovulatory and gravid lizards (*Uromastix hardwicki*). *Gen. Comp. Endocrinol.* 34, 300–303.

ASTM (American Society for Testing and Materials) (1968). Standard method of specifying color by the Munsell system. *Annual Book of ASTM Standards,* Philadelphia, pp. 236–257.

Atsatt, S. R. (1939). Color changes as controlled by color and light in the lizards of the desert regions of southern California. *Publ. Univ. Calif. Los Angeles, Biol. Sci.* 1, 237–276.

Auffenberg, W. (1964). A first record of breeding colour changes in a tortoise. *J. Bombay Nat. Hist. Soc.* 61, 191–192.

Auffenberg, W. (1965). Sex and species discrimination in two sympatric South American tortoises. *Copeia* 1965, 335–342.

Bagnara, J. T. (1966). Cytology and cytophysiology of non-melanophore pigment cells. *Intern. Rev. Cytol.* 20, 173–205.

Bagnara, J. T., and Hadley, M. E. (1973). *Chromatophores and Color Change: the Comparative Physiology of Animal Pigmentation.* Prentice-Hall, Englewood Cliffs, N.J.

Bagnara, J. T., Hadley, M. E., and Taylor, J. D. (1969). Regulation of bright-colored pigmentation of amphibians. *Gen. Comp. Endocrinol.* Suppl., 2, 425–438.

Bagnara, J. T., Taylor, J. D., and Hadley, M. E. (1968). The dermal chromatophore unit. *J. Cell Biol.* 38, 67–79.

Banks, C. (1981). Notes on seasonal colour change in a western brown snake. *Herpetofauna* 13, 29–30.

Banks, C. B. (1986). Induced colour change in a Krafft's tortoise. *Victorian Nat.* 103, 158–159.

Bartholomew, G. A. (1982). Physiological control of body temperature. In *Biology of the Reptilia,* Vol. 12 (C. Gans and F. H. Pough, eds.). Academic Press, London, pp. 167–211.

Bartlett, R. D. (1981a). An incidence of caudal luring by the Bimini dwarf boa, *Tropidophis canus curtus. Noah* 8, 12–13.

Bartlett, R. D. (1981b). Observations on the Asian box turtle, *Cuora flavimarginata,* in captivity. *North Ohio Assoc. Herpetol. Newslett.* 8, 4–6.

Bartowiak, W. (1949). The ability of tortoises to discriminate colour saturations. *Bull. Int. Acad. Pol. Sci. Let., Ser. B. Sci., Nat.* 1949, 22–58.

Batelli, A. (1880). Beiträge zur Kenntnis des Baues der Reptilienhaut. *Archiv. Mikr. Anat.* 18, 346–361.

Bauwens, D., Nuitjen, K., van Wezel, H., and Verheyen, R. F. (1987). Sex recognition by males of the lizard *Lacerta vivipara:* an introductory study. *Amphibia-Reptilia* 8, 49–57.

Baylis, J. R. (1979). Environmental light and conspicuous colors. In *The Behavioral Significance of Color* (E. H. Burtt, Jr., ed.). Garland STPM Press, New York, pp. 289–358.

Becker, C. (1984). Some notes on the gestural communication system during the mating season of the green lizard (*Lacerta viridis*). *Herptile* 9, 91–97.

Beckwith, B. E. (1988). The melanotropins: learning and memory. In *The Melanotropic Peptides,* Vol. II: *Biological Roles* (M. E. Hadley, ed.). CRC Press, Boca Raton, Fla., pp. 43–72.

Bellairs, A. (1970). *The Life of Reptiles.* Universe Books, New York.

Benes, E. S. (1969). Behavioral evidence for color discrimination by the whiptail lizard, *Cnemidophorus tigris. Copeia* 1969, 707–722.

Bourgart, R. M. (1970). Recherches écologique et biologiques sur le *Chamaeleo pardalis* Cuvier 1829. De l'ile de la Réunion et de Madagascar. *Bull. Soc. Zool. Fr.* 95, 259–268.

Brackin, M. F. (1978). The relation of rank to physiological state in *Cnemidophorus sexlineatus* dominance hierarchies. *Herpetologica* 34, 185–191.

Bradshaw, S. D., and Main, A. R. (1968). Behavioral attitudes and regulation of temperature in *Amphibolurus* lizards. *Comp. Biochem. Physiol.* 20, 855–865.

Brain, C. K. (1961). *Chamaeleo dilepis*—a study on its biology and behaviour. *J. Herp. Soc. Rhodesia* 15, 15–20.

Brücke, E. (1852). Untersuchungen bei dem Farbwechsel des afrikanischen Chameleons. *Denkschr. Akad. Wiss. Wien.* 4, 179–210.

Burger, W. L., and Smith, P. W. (1950). The coloration of the tail tip of young fer-de-lance: sexual dimorphism rather than adaptive coloration. *Science* (N.Y.) 112, 431–433.

Burgers, A. J. C. (1966). Biological aspects of pigment cell research. In *Structure and Control of the Melanocyte* (G. Della-Porta and O. Muhlbock, eds.). Springer-Verlag, New York, pp. 6–15.

Burghardt, G. M. (1977). Learning processes in reptiles. In *Biology of the Reptilia,* Vol. 7 (C. Gans and D. W. Tinkle, eds.). Academic Press, London, pp. 555–681.

Burrage, B. R. (1973). Comparative ecology and behaviour of *Chamaeleo pumilis pumilis* (Gmelin) and *C. namaquensis* A. Smith (Sauria: Chamaeleonidae). *Ann. S. Afr. Mus.* 51, 1–158.

Burtt, E. H., Jr. (1981). The adaptiveness of animal colors. *Bioscience* 31, 723–729.

Burtt, E. H., Jr. (1984). Colour of the upper mandible: an adaptation to reduce reflectance. *Anim. Behav.* 32, 652–658.

Burtt, E. H., Jr. (1986). An analysis of physical, physiological, and optical aspects of avian coloration with emphasis on wood-warblers. *Ornithol. Monogr.* 38, 1–126.

Bustard, H. R. (1965). Observations on the life history and behavior of *Chameleo hoehnelii* (Steindachner). *Copeia* 1965, 401–410.

Bustard, H. R. (1967). The comparative behavior of chameleons: fight behavior in *Chameleo gracilis* Hallowell. *Herpetologica* 23, 44–50.

Callard, I. P., Chan, S. W. C., and Callard, G. V. (1973). Hypothalamic-pituitary-adrenal relationships in reptiles. In *Brain-pituitary-adrenal relationships* (A. Brodish and E. S. Redgate, eds.). Karger, Basel, pp. 270–312.

Canella, M. F. (1963). Note di fisiologia dei cromatofori dei vertebrati pecilotermi, particolarmente dei lacertili. *Monit. Zool. Ital.* 71, 430–480.

Carpenter, C. C. (1962). Patterns of behavior in two Oklahoma lizards. *Am. Midl. Nat.* 67, 132–151.

Carpenter, C. C. (1963). Patterns of behavior in three forms of the fringe-toed lizards (*Uma*—Iguanidae). *Copeia* 1963, 406–412.

Carpenter, C. C. (1966). Comparative behavior of the Galapagos lava lizards (*Tropidurus*). In *The Galapagos: Proceedings of the Galapagos International Scientific Project* (R. I. Bowman, ed.). University of California Press, Berkeley, pp. 269–273.

Carpenter, C. C., and Ferguson, G. W. (1977). Variation and evolution of stereotyped behavior in reptiles. Part I. A Survey of stereotyped reptilian behavioral patterns. In *Biology of the Reptilia,* Vol. 7 (C. Gans and D. W. Tinkle, eds.). Academic Press, New York, pp. 355–554.

Carpenter, C. C., and Grubitz, G., III. (1960). Dominance shifts in the tree lizard (*Urosaurus ornatus*—Iguanidae). *Southwest. Nat.* 5, 123–128.

Carpenter, C. C., and Murphy, J. B. (1978). Tongue display by the common bluetongue (*Tiliqua scincoides,* Reptilia, Lacertilia, Scincidae). *J. Herpetol.* 12, 428–429.

Carpenter, C. C., Murphy, J. B., and Carpenter, G. C. (1978). Tail luring in the death adder, *Acanthophis antarcticus* (Reptilia, Serpentes, Elapidae). *J. Herpetol.* 12, 547–577.

Carter, R. J., and Shuster, S. (1982). The association between the melanocyte-stimulating hormone receptor and the alpha$_2$-adrenoceptor on the *Anolis* melanophore. *Br. J. Pharmacol.* 75, 169–176.

Case, S. M., and Williams, E. E. (1984). Study of a contact zone in the *Anolis distichus* complex in the central Dominican Republic. *Herpetologica* 40, 118–137.

Chan, S. W. C., and Callard, I. P. (1974). Reptilian ovarian steroidogenesis and the influence of mammalian gonadotrophins (follicle-stimulating hormone and luteinizing hormone) in vitro. *J. Endocrinol.* 62, 267–275.

Chondropoulos, B. P., and Lykakis, J. J. (1983). Ecology of the Balkan wall lizard, *Podarcis taurica ionica* (Sauria: Lacertidae) from Greece. *Copeia* 1983, 991–1001.

Christman, S. P. (1980). Preliminary observations on the gray-throated form of *Anolis carolinensis* (Reptilia: Iguanidae). *Florida Field Nat.* 8, 11–16.

Clarke, R. F. (1965). An ethological study of the iguanid lizard genera *Callisaurus, Cophosaurus,* and *Holbrookia*. *Emporia State Res. Stud.* 13(4), 1–66.

Cogger, H. G. (1974). Thermal relations of the Mallee Dragon *Amphibolurus fordi* (Lacertilia: Agamidae). *Austr. J. Zool.* 22, 319–339.

Cogger, H. G. (1983). *Reptiles and Amphibians of Australia*, 3rd ed. Reed Books Pty. Ltd., Frenchs Forest, New South Wales.

Cole, L. C. (1943). Experiments on toleration of high temperatures in lizards with reference to adaptive coloration. *Ecology* 24, 94–108.

Conant, R. (1975). *A Field Guide to the Reptiles and Amphibians of Eastern and Central North America*, 2nd ed. Houghton Mifflin Co., Boston.

Cooper, W. E., Jr. (1977). Information analysis of agonistic behavioral sequences in male iguanid lizards, *Anolis carolinensis*. *Copeia* 1977, 721–735.

Cooper, W. E., Jr. (1984). Female secondary sexual coloration and sex recognition in the keeled earless lizard, *Holbrookia propinqua*. *Anim. Behav.* 32, 1142–1150.

Cooper, W. E,. Jr. (1985). Female residency, individual recognition, and courtship intensity in a territorial earless lizard, *Holbrookia propinqua*. *Amphibia-Reptilia* 6, 63–69.

Cooper, W. E., Jr. (1986). Chromatic components of female secondary sexual coloration: Influence on social behavior of male keeled earless lizards (*Holbrookia propinqua*). *Copeia* 1986, 980–986.

Cooper, W. E., Jr. (1988). Aggressive behavior and courtship rejection in brightly and plainly colored female keeled earless lizards (*Holbrookia propinqua*). *Ethology* 77, 265–278.

Cooper, W. E., Jr., and Burns, N. (1987). Social significance of ventrolateral coloration in the fence lizard, *Sceloporus undulatus*. *Anim. Behav.* 35, 526–532.

Cooper, W. E., Jr., and Clarke, R. F. (1982). Steroidal induction of female reproductive coloration in the keeled earless lizard, *Holbrookia propinqua*. *Herpetologica* 38, 425–429.

Cooper, W. E., Jr., and Crews, D. (1987). Hormonal induction of secondary sexual coloration and rejection behaviour in female keeled earless lizards (*Holbrookia propinqua*). *Anim. Behav.* 35, 1177–1187.

Cooper, W. E., Jr., and Crews, D. (1988). Sexual coloration, plasma concentrations of sex steroid hormones, and responses to courtship in the female keeled earless lizard (*Holbrookia propinqua*). *Horm. Behav.* 22, 12–25.

Cooper, W. E., Jr., and Ferguson, G. W. (1972a). Relative effectiveness of progesterone and testosterone as inductors of orange spotting in female collared lizards. *Herpetologica* 28, 64–65.

Cooper, W. E., Jr., and Ferguson, G. W. (1972b). Steroids and color change during gravidity in the lizard, *Crotaphytus collaris*. *Gen. Comp. Endocrinol.* 18, 69–72.

Cooper, W. E., Jr., and Ferguson, G. W. (1973a). Estrogenic priming of color change induced by progesterone in the collared lizard, *Crotaphytus collaris*. *Herpetologica* 29, 107–110.

Cooper, W. E., Jr., and Ferguson, G. W. (1973b). Induction of physiological color change in male *Sceloporus occidentalis* by epinephrine. *Copeia* 1973, 341–342.

Cooper, W. E., Jr., and Vitt, L. J. (1985). Blue tails and autotomy: enhancement of predation avoidance in juvenile skinks. *Z. Tierpsychol.* 70, 265–276.

Cooper, W. E., Jr., and Vitt, L. J. (1986). Thermal dependence of tongue-flicking and comments on use of tongue-flicking as an index of squamate behavior. *Ethology* 71, 177–186.
Cooper, W. E., Jr., and Vitt, L. J. (1987a). Deferred agonistic behavior in a long-lived scincid lizard *Eumeces laticeps:* field and laboratory data on the roles of body size and residence in agonistic strategy. *Oecologia* 72, 321–326.
Cooper, W. E., Jr., and Vitt, L. J. (1987b). Ethological isolation, sexual behavior and pheromones in the fasciatus species group of the lizard genus *Eumeces. Ethology* 75, 328–336.
Cooper, W. E., Jr., and Vitt, L. J. (1987c). Intraspecific and interspecific aggression in lizards of the scincid genus *Eumeces:* chemical detection of conspecific sexual competitors. *Herpetologica* 43, 7–14.
Cooper, W. E., Jr., and Vitt, L. J. (1988). Orange head coloration of the male broad-headed skink (*Eumeces laticeps*), a sexually selected social cue. *Copeia* 1988, 1–6.
Cooper, W. E., Jr., Adams, C. S., and Dobie, J. L. (1983). Female color change in the keeled earless lizard, *Holbrookia propinqua:* relationship to the reproductive cycle. *Southwest. Nat.* 28, 275–280.
Cooper, W. E., Jr., Mendonca, M. T., and Vitt, L. J. (1986). Induction of sexual receptivity in the female broad-headed skink, *Eumeces laticeps,* by estradiol-17β. *Horm. Behav.* 20, 235–242.
Cooper, W. E., Jr., Mendonca, M. T., and Vitt, L. J. (1987). Induction of orange head coloration and activation of courtship and aggression by testosterone in the male broad-headed skink (*Eumeces laticeps*). *J. Herpetol.* 21, 96–101.
Cott, H. B. (1940). *Adaptive Coloration in Animals.* Methuen & Co, London.
Cowles, R. B., and Bogert, C. M. (1944). A preliminary study of the thermal requirements of desert reptiles. *Bull. Am. Mus. Nat. Hist.* 83, 261–296.
Crews, D. (1975a). Effects of different components of male courtship behavior on environmentally induced recrudescence and mating preferences in the lizard, *Anolis carolinensis. Anim. Behav.* 23, 349–356.
Crews, D. (1975b). Psychobiology of reptilian reproduction. *Science* (N.Y.) 189, 1059–1065.
Crews, D., and Morgentaler, A. (1979). Effects of intracranial implantation of oestradiol and dihydrotestosterone on the sexual behaviour of the lizard *Anolis carolinensis. Endocrinology* 82, 373–381.
Crews, D., and Williams, E. E. (1977). Hormones, reproductive behavior, and speciation. *Am. Zool.* 17, 271–286.
Darwin, C. (1871). *The Descent of Man, and Selection in Relation to Sex.* John Murray, London.
Datta, P. C., and King, M. G. (1982). Alpha-melanocyte-stimulating hormone and behavior. *Neurosci. Biobehav. Rev.* 6, 297–310.
Dauphin-Villemant, C., and Xavier, F. (1985). *In vitro* steroid biosynthesis by the adrenal gland of the female *Lacerta vivipara* Jacquin: the metabolism of exogenous precursors. *Gen. Comp. Endocrinol.* 58, 1–9.
Deeming, D. C., and Ferguson, M. W. J. (1989). The mechanism of temperature dependent sex determination in crocodilians: a hypothesis. *Am. Zool.* 29, 973–985.

Denburgh, J. van, and Slevin, J. R. (1919). The Galapagoan lizards of the genus *Tropidurus;* with notes on the iguanas of the genera *Conolophus* and *Amblyrhynchus. Proc. Calif. Acad. Sci.* 2, 133–202.
DeWied, D. (1966). Inhibitory effect of ACTH and related peptides on extinction of conditioned avoidance behavior in rats. *Proc. Soc. Exp. Biol. Med.* 122, 29–32.
DeWied, D., and Bohus, B. (1966). Long-term and short-term effects on retention of a conditioned response in rats by treatment with long-acting pitressin and alpha-MSH. *Nature* (London) 212, 1484–1486.
Dial, B. E. (1986). Tail display in two species of iguanid lizards: a test of the "predator signal" hypothesis. *Am. Nat.* 127, 103–111.
Dixon, J. R., and Wright, J. W. (1975). A review of the lizards of the genus *Tropidurus* in Peru. *Contrib. Sci., Nat. Hist. Mus. Los Angeles Co.* 217, 1–39.
Donoso-Barros, R. (1966). *Reptiles de Chile.* University of Chile Press, Santiago.
Dores, R. M., Wilhelm, M. W., and Sandoval, D. M. (1987). Steady state analysis of alpha-melanotropin in the pars intermedia of *Anolis carolinensis:* effect of background adaptation. *Gen. Comp. Endocrinol.* 68, 153–160.
Dücker, G. V., and Rensch, B. (1973). Die visuelle Lernkapazität von *Lacerta viridis* und *Agama agama. Z. Tierpsychol.* 32, 209–214.
Duellman, W. E., and Schwartz, A. (1958). Amphibians and reptiles of southern Florida. *Bull. Florida State Mus. Biol. Sci.* 3, 181–324.
Dugan, B. (1982). The mating behavior of the green iguana, *Iguana iguana.* In *Iguanas of the World: Their Behavior, Ecology, and Conservation* (G. M. Burghardt and A. S. Rand, eds.). Noyes Publications, Park Ridge, N.J., pp. 320–341.
Dunn, R. W. (1982). Effects of environment upon tortoise pigmentation. *Austr. Zool.* 21, 105–107.
Eberle, A. N. (1988). *The Melanotropins: Chemistry, Physiology and Mechanisms of Action.* Karger, Basel, Switzerland.
Edgren, R. A. (1959). Hormonal control of red head coloration in the five-lined skink, *Eumeces fasciatus* Linnaeus. *Herpetologica* 15, 155–157.
Edmunds, M. (1974). *Defence in Animals.* Longman, Harlow, Essex, England.
Ehmann, H. (1981). The natural history and conservation of the bronzeback (*Ophidiocephalus taeniatus* Lucas and Frost) (Lacertilia, Pygopodidae). In *The Proceedings of the Melbourne Herpetological Symposium* (C. B. Banks and A. A. Martin, eds.). Zoological Board of Victoria, Melbourne, pp. 7–13.
Eibl-Eibesfeldt, I. (1966). Das Verteidigen der Eiablageplätze bei der Hood-Meerechse (*Amblyrhynchus cristatus venustissimus*). *Z. Tierpsychol.* 23, 627–631.
Endler, J. A. (1980). Natural selection on color patterns in *Poecilia reticulata. Evolution* 34, 76–91.
Ernst, C. H., and Barbour, R. W. (1972). *Turtles of the United States.* University of Kentucky Press, Lexington.
Ernst, C. H., and Hamilton, H. F. (1969). Color preferences of some North American turtles. *J. Herpetol.* 3, 176–180.
Evans, L. T. (1961). Structure as related to behavior in the organization of

populations of reptiles. In *Vertebrate Speciation* (W. F. Blair, ed.). University of Texas Press, Houston, pp. 148–178.

Fehring, W. K. (1972). Hue discrimination in hatchling loggerhead turtles (*Caretta caretta*). *Anim. Behav.* 20, 632–636.

Ferguson, G. W. (1966). Releasers of courtship and territorial behaviour in the side blotched lizard *Uta stansburiana*. *Anim. Behav.* 14, 89–92.

Ferguson, G. W. (1976). Color change and reproductive cycling in female collared lizards (*Crotaphytus collaris*). *Copeia* 1976, 491–494.

Ferguson, G. W. (1977). Display and communications in reptiles: a historical perspective. *Am. Zool.* 17, 167–176.

Ferguson, G. W., and Carpenter, C. C. (1977). Social displays of Reptiles: communications value, ultimate causes of variation, taxonomic significance and heritability of population differences. In *Biology of the Reptilia,* Vol. 7 (C. Gans and D. W. Tinkle, eds.). Academic Press, New York, pp. 404–554.

Ferner, J. W. (1974). Home-range size and overlap in *Sceloporus undulatus erythrocheilus* (Reptilia: Iguanidae). *Copeia* 1974, 332–337.

Fiedler, K. (1954). Vergleichende Verhaltensstudien an Seenadeln, Schlangennaden und Seepferdchen (Syngnathidae). *Z. Tierpsychol.* 11, 358–416.

Fisher, R. A. (1930). *The Genetical Theory of Natural Selection.* Clarendon Press, Oxford.

Fitch, H. S. (1954). Life history and ecology of the five-lined skink, *Eumeces fasciatus*. *Univ. Kans. Publ., Mus. Nat. Hist.* 8, 1–154.

Fitch, H. S. (1956). An ecological study of the collared lizard (*Crotaphytus collaris*). *Univ. Kans. Publ., Mus. Nat. Hist.* 8, 213–274.

Fitch, H. S. (1958). Natural history of the six-lined racerunner (*Cnemidophorus sexlineatus*). *Univ. Kans. Publ. Mus. Nat. Hist.* 11, 1–62.

Fitch, H. S. (1967). Discussion. In *Lizard Ecology: a Symposium* (W. W. Milstead, ed.). University of Missouri Press, Columbia, p. 75.

Fitch, H. S., and Henderson, R. W. (1987). Ecological and ethological parameters in *Anolis bahorucoensis,* a species having rudimentary development of the dewlap. *Amphibia-Reptilia* 8, 69–80.

Fitzsimons, V. F. (1943). The lizards of South Africa. *Mem. Transvaal Mus.* (Pretoria) 1, 1–528 + xxiv pl.

Font, E. (1988). "Functional Morphology and Central Control of Dewlap Extension in the Displays of *Anolis equestris* (Sauria, Iguanidae)." Ph.D. dissertation, University of Tennessee, Knoxville. Dissertation Abstract number DA8911723.

Font, E., Greenberg, N., and Switzer, R. C. (1986). Brainstem origins of motoneurons controlling the hyoid contribution to the dewlap display of *Anolis* lizards. *Soc. Neuroscience Abstracts* 12, 498 (abstract).

Forbes, A., Fox, S., McCarthy, E., and Yamashita, E. (1964). Quantitative response to color shift in diurnal lizards. *Proc. U.S. Natl. Acad. Sci.* 52, 667–672.

Forbes, T. R. (1941). Observations on the urogenital anatomy of the adult male lizard, *Sceloporus,* and on the action of implanted pellets of testosterone and of estrone. *J. Morphol.* 68, 31–69.

Frazier, J. (1971). Observations on sea turtles at Aldabra Atoll. *Philos. Trans. Roy. Soc.* (London) B260, 373–410.

Gans, C. (1961). Mimicry in the procryptically colored snakes of the genus *Dasypeltis. Evolution* 15, 72–91.

Gans, C. (1965). Empathic learning and the mimicry of African snakes. *Evolution* 18, 705.

Gehlbach, F. R., Watkins, J. F., II, and Reno, H. W. (1968). Blind snake defensive behavior elicited by ant attacks. *Bioscience* 18, 784–785.

Geldern, C. E. von. (1921). Color changes and structure of the skin of *Anolis carolinensis. Proc. Calif. Acad. Sci.* 10, 77–117.

Gibbons, J. R. H., and Lillywhite, H. B. (1981). Ecological segregation, color matching, and speciation in lizards of the *Amphibolurus decresii* species complex (Lacertilia: Agamidae). *Ecology* 62, 1573–1584.

Gibbons, J. W., and Lovich, J. E. (1990). Sexual dimorphism in turtles with emphasis on the slider turtle (*Trachemys scripta*) . *Herp. Monogr.* 4, 1–29.

Gibson, R., and Falls, J. B. (1979). Thermal biology of the common garter snake *Thamnophis sirtalis* (L.). II. The effects of melanism. *Oecologia* 43, 99–109.

Gillis, J. E. (1989). Selection for substrate reflectance-matching in two populations of red-chinned lizards (*Sceloporus undulatus erythrocheilus*) from Colorado. *Am. Midl. Nat.* 121, 197–200.

Golder, F. (1987). Zur Haltung und Fortpflanzung von *Boiga cyanea* (Duméril & Bibron, 1854) und Angeben zum Farbdimorphismus zwischen juvenilen und adulten Tieren. *Salamandra* 23, 78–83.

Goldman, J. M., and Hadley, M. E. (1969). *In vitro* demonstration of adrenergic receptors controlling melanophore responses of the lizard, *Anolis carolinensis. J. Pharmacol. Exp. Ther.* 166, 1–9.

Gorman, G. C. (1977). Comments on ontogenetic color change in *Anolis cuvieri* (Reptilia, Lacertilia, Iguanidae). *J. Herpetol.* 11, 211.

Gosner, K. L. (1989). Histological notes on the green coloration of arboreal pit vipers; genus *Bothrops. J. Herpetol.* 23, 318–320.

Graf, V. A. (1967). A spectral sensitivity curve and wavelength discrimination for the turtle, *Chrysemys picta picta. Vis. Res.* 7, 915–928.

Granda, A. M. (1979). Eyes and their sensitivity to light of differing wavelengths. In *Turtles: Perspectives and Research* (M. Harless and H. Morlock, eds.). John Wiley and Sons, New York, pp. 247–266.

Granda, A. M., Maxwell, J. H., and Zwick, H. (1972). The temporal course of dark adaptation in the turtle, *Pseudemys,* using a behavioral paradigm. *Vis. Res.* 12, 653–672.

Greenberg, B., and Noble, G. K. (1944). Social behavior of the American chameleon (*Anolis carolinensis* Voigt). *Physiol. Zool.* 17, 392–439.

Greenberg, N. (1977a). A neuroethological study of display behavior in the lizard *Anolis carolinensis* (Reptilia, Lacertilia, Iguanidae). *Am. Zool.* 17, 191–201.

Greenberg, N. (1977b). An ethogram of the blue spiny lizard, *Sceloporus cyanogenys* (Reptilia, Lacertilia, Iguanidae). *J. Herpetol.* 11, 177–195.

Greenberg, N., and Crews, D. (1983). Physiological ethology of aggression in amphibians and reptiles. In *Hormones and Aggressive Behavior* (B. Svare, ed.). Plenum Press, New York, pp. 469–506.

Greenberg, N., and Crews, D. (1990). Endocrine and behavioral responses to aggression and social dominance in the green anole lizard, *Anolis carolinensis. Gen. Comp. Endocrinol.* 77, 246–255.

Greenberg, N., Chen, T., and Crews, D. (1984). Social status, gonadal state, and the adrenal stress response in the lizard, *Anolis carolinensis. Horm. Behav.* 18, 1–11.

Greenberg, N., Chen, T., and Vaughan, G. L. (1986). Melanotropin levels are altered by acute and chronic social stress in lizards. *Soc. Neurosci.* (abstract) 12, 834.

Greene, H. W. (1988). Antipredator mechanisms in reptiles. In *Biology of the Reptilia,* Vol. 16 (C. Gans and R. B. Huey, eds.). Alan R. Liss, New York, pp. 1–152.

Greene, H. W., and Campbell, J. A. (1972). Notes on the use of caudal lures by arboreal green pit vipers. *Herpetologica* 28, 32–34.

Greer, A. E. (1967). Ecology and behavior of two sympatric *Lygodactylus* geckos. *Breviora, Mus. Comp. Zool.* 268, 1–20.

Groombridge, B., Moll, E. O., and Vijaya, J. (1983). Rediscovery of a rare Indian turtle. *Oryx* 17, 130–134.

Gruber, S. H. (1979). Mechanisms of color vision, an ethologist's primer. In *The Behavioral Significance of Color* (E. H. Burtt, Jr., ed.). Garland STPM Press, New York, pp. 184–236.

Hadley, C. E. (1931). Color changes in excised and intact reptilian skin. *J. Exper. Zool.* 58, 321–331.

Hadley, M. E. (1984). *Endocrinology.* Prentice-Hall, Englewood Cliffs, N.J.

Hadley, M. E. (1988a). *The Melanotropic Peptides,* Vol. I, *Source, Synthesis, Chemistry, Secretion, Circulation, and Metabolism.* CRC Press, Boca Raton, Fla.

Hadley, M. E. (1988b). *The Melanotropic Peptides,* Vol. II: *Biological Roles.* CRC Press, Boca Raton, Fla.

Hadley, M. E. (1988c). *The Melanotropic Peptides,* Vol. III: *Mechanisms of Action and Biomedical Applications.* CRC Press, Boca Raton, Fla.

Hadley, M. E., and Goldman, J. M. (1969). Physiological color changes in reptiles. *Am. Zool.* 9, 489–504.

Hadley, W. F., and Gans, C. (1972). Convergent ontogenetic change of color pattern in *Elaphe climacophora* (Colubridae: Reptilia). *J. Herpetol.* 6, 75–78.

Hailman, J. P. (1964). Coding of the colour preference of the gull chick. *Nature* (London) 204, 710.

Hailman, J. P. (1967). Spectral discrimination: an important correction. *J. Optical Soc. Am.* 57, 281–282.

Hailman, J. P. (1969). Spectral pecking preference in gull chicks: possible resolution of a species difference. *J. Comp. Physiol. Psychol.* 67, 465–467.

Hailman, J. P. (1977). *Optical Signals: Animal Communication and Light.* Indiana University Press, Bloomington.

Hailman, J. P. (1979). Environmental light and conspicuous colors. In *The Behavioral Significance of Color* (E. H. Burtt, Jr., ed.). Garland STPM Press, New York, pp. 289–358.

Hailman, J. P., and Jaeger, R. G. (1971). On criteria for color preferences in turtles. *J. Herpetol.* 5, 83–85.
Hamilton, W. J., III. (1973). *Life's Color Code.* McGraw-Hill, New York.
Hansen, A. J., and Rohwer, S. (1986). Coverable badges and resource defence in birds. *Anim. Behav.* 34, 69–76.
Harding, J. H., and Bloomer, T. J. (1979). The wood turtle, *Clemmys insculpta:* a natural history. *Bull. N. Y. Herpetol. Soc.* 15, 9–26.
Harris, V. A. (1964). *The Life of the Rainbow Lizard.* Hutchinson & Co., London.
Heatwole, H. (1968). Relationship of escape behavior and camouflage in anoline lizards. *Copeia* 1968, 109–113.
Heatwole, H. (1970). Thermal ecology of the desert dragon *Amphibolurus inermis. Ecol. Monogr.* 40, 425–457.
Heatwole, H., and Davison, E. (1976). A review of caudal luring in snakes with notes on its occurrence in the saharan sand viper, *Cerastes vipera. Herpetologica* 32, 332–336.
Hedges, S. B., Hass, C. A., and Maugel, T. K. (1989). Physiological color changes in snakes. *J. Herpetol.* 23, 450–455.
Henderson, R. W. (1970). Caudal luring in a juvenile Russell's viper. *Herpetologica* 26, 276–277.
Henry, G. M. (1925). Notes on *Ancistrodon hypnale,* the hump-nosed viper. *Ceylon J. Sci. Sect B. (Spolia Zeylanica),* 13, 257–258.
Hews, D. K., and Dickhaut, J. C. (1989). Differential response to colors by the lizard *Uta palmeri* (Reptilia: Iguanidae). *Ethology* 82, 134–140.
Hinton, H. E. (1976). Colour changes. In *Environmental Physiology of Animals* (J. Bligh, J. L. Cloudsley-Thompson, and A. G. MacDonald, eds.). John Wiley, New York, pp. 390–412.
Hogben, L. T., and Mirvish, L. (1928). The pigmentary effector system. V. The nervous control of excitement pallor in reptiles. *J. Exp. Biol.* 5, 295–308.
Hoppe, D. M. (1979). The influence of color on behavioral thermoregulation and hydroregulation. In *The Behavioral Significance of Color* (E. H. Burtt, Jr., ed.). Garland STPM Press, New York, pp. 35–62.
Horowitz, S. B. (1958). The energy requirements of melanin granule aggregation and dispersion in the melanophores of *Anolis carolinensis. J. Cell. Comp. Physiol.* 51, 341–357.
Hover, E. L. (1985). Differences in aggressive behavior between two throat color morphs in a lizard, *Urosaurus ornatus. Copeia* 1985, 933–940.
Huey, R. B., and Pianka, E. R. (1977). Natural selection for juvenile lizards mimicking noxious beetles. *Science* (N.Y.) 195, 201–203.
Hulse, A. C. (1976). Growth and morphometrics of *Kinosternon sonoriense* (Reptilia, Testudines, Kinosternidae). *J. Herpetol.* 10, 341–348.
Hunsaker, D., II. (1962). Ethological isolating mechanisms in the *Sceloporus torquatus* group of lizards. *Evolution* 16, 62–74.
Inoué, S., and Inoué, Z. (1977). Colour changes induced by pairing and painting in the male rainbow lizard, *Agama agama agama. Experientia* 33, 1443–1444.
Jackson, J. F., Ingram, W., III, and Campbell, H. W. (1976). The dorsal pigmentation pattern of snakes as an antipredator strategy: a multivariate approach. *Am. Nat.* 110, 1029–1053.

Jackson, J. F., and Martin, D. L. (1980). Caudal luring in the dusky pygmy rattlesnake, *Sistrurus millarius barbouri*. *Copeia* 1980, 926–927.

Jenssen, T. A. (1977). Evolution of anoline lizard display behavior. *Am. Zool.* 17, 203–215.

Judd, F. W. (1976). Demography of a barrier island population of the keeled earless lizard, *Holbrookia propinqua*. *Occas. Pap. Mus. Texas Tech Univ.* 44, 1–45.

Judd, H. L., Laughlin, G. A., Bacon, J. P., and Benirschke, K. (1976). Circulating androgen and estrogen concentrations in lizards (*Iguana iguana*). *Gen. Comp. Endocrinol.* 30, 341–345.

Kästle, V. W. (1964). Verhaltenstudien an Taggeckonen der Gattungen *Lygodactylus* und *Phelsuma*. *Z. Tierpsychol.* 21, 486–507.

Kästle, V. W. (1967). Soziale Verhaltenweisen von Chamaeleonen aus der *pumilis*- und *bitaeniatus*-Gruppe. *Z. Tierpsychol.* 24, 313–341.

Kauffield, C. F. (1943). Growth and feeding of newborn Price's and green rock rattlesnakes. *Am. Midl. Nat.* 29, 607–614.

Kerbert, C. (1876). Ueber die Haut der Reptilien und anderer Wirbelthiere. *Arch. Mikr. Anat.* 13, 205–262.

Khachaturian, H., Dores, R. M., Watson, S. J., and Akil, H. (1984). β-endorphin/ACTH immunocytochemistry in the CNS of the lizard *Anolis carolinensis:* evidence for a major mesencephalic cell group. *J. Comp. Neurol.* 229, 576–584.

Khorram, O., Bedran de Castro, J. C., and McCann, S. M. (1985). Stress-induced secretion of α-melanocyte stimulating hormone and its physiological role in modulating the secretion of prolactin and luteinizing hormone in the female rat. *Endocrinology* 117, 2483–2489.

Kimball, F. A., and Erpino, M. J. (1971). Hormonal control of pigmentary sexual dimorphism in *Sceloporus occidentalis*. *Gen. Comp. Endocrinol.* 16, 375–384.

Kirshner, D. (1985). Environmental effects on dorsal colouration in the saltwater crocodile, *Crocodylus porosus*. In *Biology of Australasian Frogs and Reptiles* (G. Grigg, R. Shine, and H. Ehmann, eds.). Surrey Beatty and Sons, Pty Limited, Chipping Norton, New South Wales, Australia, pp. 397–402.

Kitzler, G. (1941). Die Paarungsbiologie einiger Eidechsen. *Z. Tierpsychol.* 4, 353–402.

Klauber, L. M. (1930). New and renamed subspecies of *Crotalus confluentus* Say, with remarks on related species. *Trans. San Diego Soc. Nat Hist.* 6, 95–144.

Klauber, L. M. (1956). *Rattlesnakes, Their Habits, Life Histories, and Influences on Mankind.* University of California Press, Berkeley.

Klausewitz, W. (1953). Histologische Untersuchen über das Farbkleid des mittelamerikanschen Zaunleguans *Sceloporus m. malachiticus*. *Zool. Anz.* 152, 199–206.

Klausewitz, W. (1954). Histologische Untersuchen über das Farbkleid der Zauneidechse *Lacerta a. agilis*. *Senck. Biol.* 45, 425–444.

Kleinholz, L. H. (1936). Studies on reptilian color changes. I. A preliminary report. *Proc. Natl. Acad. Sci.* 22, 454–456.

Kleinholz, L. H. (1938a). Studies on reptilian colour changes. II. The pituitary and adrenal glands in the regulation of the melanophores of *Anolis carolinensis*. *J. Exp. Biol.* 15, 474–491.

Kleinholz, L. H. (1938b). Studies on reptilian colour changes. III. Control of the light phase and behaviour of isolated skin. *J. Exp. Biol.* 15, 492–499.

Kleinholz, L. H. (1941). Behaviour of melanophores in the alligator. *Anat. Rec.* 81 (suppl), 121.

Kramer, G. (1937). Beobachtungen über Paarungsbiologie und soziales Verhalten von Mauereidechsen. *Z. Morphol. Okol. Tiere.* 32, 752–783.

Kramer, M. (1986). Field studies on a freshwater Florida turtle, *Pseudemys nelsoni*. In *Behavioral Ecology and Population Biology* (L. C. Drickamer, ed.). Privat, I.E.C., Toulouse, France.

Kuroda, R. (1933). Studies on visual discrimination in the tortoise *Clemmys japonica*. *Acta Psychol. Keijo* 2(2), 31–59.

Lack, D. (1968). *Ecological Adaptations for Breeding in Birds.* Methuen, London.

LaHost, G. J., Olson, G. A., Kastin, A. J., and Olson, R. D. (1980). Behavioral effects of melanocyte stimulating hormone. *Neurosci. Biobehav. Rev.* 4, 9–16.

Lange, B. (1931). Integument der Sauropsiden. *Vergl. Anat.* 1, 375–448.

Lardie, R. L. (1983). Aggressive interactions among melanistic males of the red-eared slider, *Pseudemys scripta elegans* (Wied). *Bull. Oklahoma Herp. Soc.* 8, 105–117.

Legler, J. M. (1960). Natural history of the ornate box turtle, *Terrapene ornata ornata* Agassiz. *Univ. Kansas Publ. Mus. Nat. Hist.* 11, 527–669.

Legler, J. M. (1965). A new species of turtle, genus *Kinosternon* from Central America. *Univ. Kansas Misc. Publ. Nat. Hist.* 15, 615–625.

Legler, J. M., and Cann, J. (1980). A new genus and species of chelid turtle from Queensland, Australia. *Contrib. Sci., Los Angeles Co. Mus.* 324, 1–18.

Lerner, A. B., and McGuire, J. S. (1964). Melanocyte-stimulating hormone and adrenocorticotrophic hormone. *N. Engl. J. Med.* 270, 539–546.

Liebman, P. A., and Granda, A. M. (1971). Microspectrophotometric measurements of visual pigments in two species of turtle, *Pseudemys scripta* and *Chelonia mydas*. *Vis. Res.* 11, 105–114.

Lillywhite, H. B., Friedman, G., and Ford, N. (1977). Color matching and perch selection by lizards in recently burned chaparral. *Copeia* 1977, 115–121.

Lipton, J. M. (1988). MSH in CNS control of fever and its influence on inflammation/immune responses. In *The Melanotropic Peptides,* Vol. 2: *Biological Roles* (M. E. Hadley, ed.). CRC Press, Boca Raton, Fla., pp. 97–113.

Lipton, J. M., and Clark, W. G. (1986). Neurotransmitters in temperature control. *Ann. Rev. Physiol.* 46, 613–623.

Losos, J. B. (1985). An experimental demonstration of the species-recognition role of *Anolis* dewlap color. *Copeia* 1985, 905–910.

Loveridge, A. (1947). Revision of the African lizards of the family Gekkonidae. *Bull. Mus. Comp. Zool. Harvard Coll.* 98, 1–469.

Lovich, J. E., Ernst, C. H., and Gotte, S. W. (1985). Geographic variation in the Asiatic turtle *Chinemys reevesii* (Gray) and the status of *Geoclemys grangeri* Schmidt. *J. Herpetol.* 19, 238–245.

Lovich, J. E., McCoy, C. J., and Garstka, W. R. (1990). The development and significance of melanism in the slider turtle. In *Life History and Ecology of the Slider Turtle* (J. W. Gibbons, ed.). Smithsonian Institution Press, Washington, D.C., pp. 233–254.

Luke, C. A. (1989). "Color as a Phenotypically Plastic Character in the Side-blotched Lizard, *Uta stansburiana*." Ph.D. dissertation, University of California, Berkeley.

Lydekker, R. (1901). Reptiles. In *Library of Natural History*, Vol. 5. Saalfield Publishing Co., New York, pp. I–XV, 2357–2621.

Maderson, P. F. A. (1965). The embryonic development of the squamate integument. *Acta Zool.* 46, 275–295.

Madsen, T. (1987). Are juvenile grass snakes, *Natrix natrix*, aposematically coloured? *Oikos* 48, 265–267.

Madsen, T., and Loman, J. (1987). On the role of colour display in the social and spatial organization of male rainbow lizards (*Agama agama*). *Amphibia-Reptilia* 8, 365–372.

Madsen, T., and Stille, B. (1988). The effect of size dependent mortality on colour morphs in male adders, *Vipera berus*. *Oikos* 52, 73–78.

Mason, P., and Adkins, E. K. (1976). Hormones and social behavior in the lizard, *Anolis carolinensis*. *Horm. Behav.* 7, 75–86.

Mathewson, S. F. (1961). Gonadotropic control of aggressive behavior in starlings. *Science* (N.Y.) 134, 1522–1523.

May, R., and Thillard, M. (1957). Sur les changements de coloration du lézard *Anolis roquet*. *Bull. Biol. France Belgique* 91, 186–202.

McAlpine, D. F. (1983). Correlated physiological color change and activity patterns in an Indian Ocean boa. *J. Herpetol.* 17, 198–210.

McCoy, M. (1980). *Reptiles of the Solomon Islands*. Wau Ecology Institute, Papua, New Guinea.

McIlhenny, E. A. (1935). *The Alligator's Life History*. Christopher Publishing House, Boston.

McNicol, D., Jr., and Crews, D. (1979). Estrogen/progesterone synergy in the control of female sexual receptivity in the lizard, *Anolis carolinensis*. *Gen. Comp. Endocrinol.* 38, 68–74.

Medica, P. A., Turner, F. B., and Smith, D. D. (1973). Hormonal induction of color change in female leopard lizards, *Crotaphytus wislizenii*. *Copeia* 1973, 658–661.

Medvin, M. B. (1990). Sex differences in coloration and optical signalling in the lizard, *Anolis carolinensis* (Reptilia, Lacertilia, Iguanidae). *Anim. Behav.* 39, 192–193.

Mirtschin, P., and Davis, R. (1982). *Dangerous Snakes of Australia*, Rev. ed. Rigby Publishers, Adelaide, Australia.

Mitchell, F. J. (1973). Studies on the ecology of the agamid lizard *Amphibolurus maculosus* (Mitchell). *Trans. Roy. Soc. S. Austral.* 97, 47–76.

Moehn, L. D. (1974). The effect of quality of light on agonistic behavior of iguanid and agamid lizards. *J. Herpetol.* 8, 175–183.

Moermond, T. C. (1978). Complex color changes in *Anolis* (Reptilia, Lacertilia, Iguanidae). *J. Herpetol.* 12, 319–325.

Moll, E. O. (1978). Drumming along the Perak. *Nat. Hist.* 87, 36–43.
Moll, E. O. (1980). Natural history of the river terrapin, *Batagur baska* (Gray) in Malaysia (Testudines: Emydidae). *Malaysian J. Sci.* 6(A), 23–61.
Moll, E. O., Groombridge, B., and Vijaya, J. (1986). Redescription of the cane turtle with notes on its natural history and classification. *J. Bombay Nat. Hist. Soc.* (suppl) 83, 112–126.
Moll, E. O., Mattson, K. E., and Krehbiel, E. B. (1981). Sexual and seasonal dichromatism in the Asian river turtle *Callagur borneoensis*. *Herpetologica* 37, 181–194.
Monnett, F., and Lichtensteiger, W. (1980). Neuroendocrine regulation and central dopamine (DA) systems in physical and psychological stress. *J. Physiol.* (Paris) 76, 273–275.
Montanucci, R. R. (1965). Observations on the San Joaquin leopard lizard, *Crotaphytus wislizenii silus* Stejneger. *Herpetologica* 21, 270–283.
Montanucci, R. R. (1967). Further studies on leopard lizards, *Crotaphytus wislizenii*. *Herpetologica* 23, 119–126.
Moore, E. M. (1983). "The Function of Orange Breeding Coloration in the Social Behavior of the Long-nosed Leopard Lizard (*Gambelia wislizenii*)." Master's thesis, Oregon State University, Corvallis.
Morison, W. L. (1985). What is the function of melanin? *Arch. Dermatol.* 121, 1160–1163.
Mount, R. (1975). *The Reptiles and Amphibians of Alabama.* Auburn University Agricultural Experiment Station, Auburn, Ala.
Muerling, P., Klefbohm, B., and Larsson, L. (1974). Transplantation of the pars intermedia in an elasmobranch and a lizard. *Gen. Comp. Endocrinol.* 22, 347.
Muntz, W. R. A., and Northmore, D. P. M. (1968). Background light, temperature, and visual noise in the turtle. *Vis. Res.* 8, 787–800.
Muntz, W. R. A., and Sokol, S. (1967). Psychophysical thresholds to different wavelengths in light adapted turtles. *Vis. Res.* 7, 729–741.
Murphy, J. B., Carpenter, C. C., and Gillingham, J. C. (1978). Caudal luring in the green tree python, *Chondropython viridis* (Reptilia, Serpentes, Boidae). *J. Herpetol.* 12, 117–119.
Murphy, M. T., Richards, D. B., and Lipton, J. M. (1983). Antipyretic potency of centrally administered a-melanocyte stimulating hormone. *Science* (N.Y.) 221, 192–193.
Needham, A. E. (1974). *The Significance of Zoochromes.* Springer-Verlag, New York.
Neill, W. T. (1948). The yellow tail of juvenile copperheads. *Herpetologica* 4, 161.
Neill, W. T. (1960). The caudal lure of various juvenile snakes. *Q. J. Florida Acad. Sci.* 23, 173–200.
Nickel, E. (1960). Untersuchungen über den Farbensinn junger Alligatoren. *Z. vergl. Physiol.* 43, 37–47.
Noble, G. K. (1934). Experimenting with the courtship of lizards. *Nat. Hist.* 32, 3–15.
Noble, G. K., and Bradley, H. T. (1933a). The mating behavior of lizards; its

bearing on the theory of sexual selection. *Ann. New York Acad. Sci.* 35, 25–100.
Noble, G. K., and Bradley, H. T. (1933b). The relation of the thyroid and the hypophysis to the moulting process in the lizard, *Hemidactylus brookii. Biol. Bull.* 64, 289–298.
Noble, G. K., and Greenberg, B. (1941). Effects of seasons, castration and crystalline sex hormones upon the urogenital system and sexual behavior of the lizard (*Anolis carolinensis*). *J. Exp. Zool.* 88, 451–479.
Norris, K. S. (1967). Color adaptation in desert reptiles and its thermal relationships. In *Lizard Ecology: a Symposium* (W. W. Milstead, ed.). University of Missouri Press, Columbia, pp. 162–229.
Norris, K. S., and Lowe, C. H. (1964). An analysis of background color-matching in amphibians and reptiles. *Ecology* 45, 565–580.
Obika, M., and Bagnara, J. T. (1964). Pteridines as pigments in amphibians. *Science* (N.Y.) 143, 485–487.
O'Donohue, T. L., and Dorsa, D. M. (1982). The opiomelanotopinergic neuronal and endocrine systems. *Peptides* 3, 353–395.
Ohtsuka, T. (1985). Relation of spectral types to oil droplets in cones of turtle retina. *Science* (N.Y.) 229, 874–877.
Ortiz, E., Bachli, E., Price, D., and Williams-Ashman, H. G. (1963). Red pteridine pigments in the dewlaps of some anoles. *Physiol. Zool.* 36, 97–103.
Ortiz, E., and Maldonado, A. A. (1966). Pteridine accumulation in lizards of the genus *Anolis. Carib. J. Sci.* 6, 9–13.
Parcher, S. R. (1974). Observations on the natural histories of six Malagasy Chamaeleontidae. *Z. Tierpsychol.* 34, 500–523.
Parker, G. H. (1938). The colour changes in lizards, particularly in *Phrynosoma. J. Exp. Biol.* 15, 48–73.
Patterson, A. T., Rickerby, J., Simpson, J., and Vickers, C. (1980). Possible interaction of melanocyte-stimulating hormone and the pineal in the control of territorial aggression in mice. *Physiol. Behav.* 24, 843–848.
Pearson, O. P. (1977). The effect of substrate and of skin color on thermoregulation of a lizard. *Comp. Biochem. Physiol.* 58A, 353–358.
Pienaar, U. de V. (1966). The reptile fauna of the Kruger National Park. *Koedoe* 1, 1–223.
Porter, W. P. (1967). Solar radiation through the living body walls of vertebrates with emphasis on desert reptiles. *Ecol. Monogr.* 37, 273–296.
Porter, W. P., and James, F. C. (1979). Behavioral implications of mechanistic ecology. II. The African rainbow lizard, *Agama agama. Copeia* 1979, 594–619.
Porter, W. P., Mitchell, J. W., Beckman, W. A., and DeWitt, C. B. (1973). Behavioral implications of mechanistic ecology. Thermal and behavioral modeling of desert ectotherms and their microenvironment. *Oecologia* 13, 1–54.
Porter, W. P., and Norris, K. S. (1969). Lizard reflectivity change and its effect on light transmission through body wall. *Science* (N.Y.) 163, 482–484.
Porter, W. P., and Tracy, C. R. (1983). Biophysical analyses of energetics, time-space utilization, and distributional limits. In *Lizard Ecology: Studies of a Model Organism* (R. B. Huey, E. R. Pianka, and T. W. Schoener, eds.) Harvard University Press, Cambridge, Mass., pp. 55–83.

Pough, F. H. (1988). Mimicry and related phenomena. In *Biology of the Reptilia,* Vol. 16 (C. Gans and R. B. Huey, eds.). Alan R. Liss, New York, pp. 153–234.

Prieto, A. A., and Ryan, M. J. (1978). Some observations of the social behavior of the Arizona chuckwalla, *Sauromalus obesus tumidus* (Reptilia, Lacertilia, Iguanidae). *J. Herpetol.* 12, 327–336.

Pritchard, P. C. H. (1966). Notes on Persian turtles. *Br. J. Herpetol.* 3, 271–273.

Pritchard, P. C. H. (1979). *Encyclopedia of Turtles.* TFH Publ., Neptune, N.J.

Pritchard, P. C. H., and Trebbau, P. (1984). *The Turtles of Venezuela.* Soc. Study Amphibs. Reptiles, Lawrence, Kan.

Proulx-Ferland, L., Labreie, F., Dumont, D., and Cote, J. (1982). Corticotropin-releasing factor stimulates secretion of melanocyte-stimulating hormone from the rat pituitary. *Science* (N.Y.) 217, 62–64.

Pycraft, W. P. (1925). *Camouflage in Nature.* Hutchinson and Co., London.

Quaranta, J. V. (1952). An experimental study of color vision of the giant tortoise. *Zoologica, N.Y.* 37, 295–311.

Quesnel, V. C. (1957). The life history of the streak lizard, *Gonatodes vittatus* (Licht). *J. Trinidad Field Nat.* 1957, 5–14.

Radcliffe, C. W., Chiszar, D., and Smith, H. M. (1980). Prey induced caudal movements in *Boa constrictor* with comments on the evolution of caudal luring. *Bull. Md. Herp. Soc.* 16, 19–22.

Rahn, H. (1941). The pituitary regulation of melanophores in the rattlesnake. *Biol. Bull.* 80, 228–237.

Rahn, H. (1956). The relationship between hypoxia, temperature, adrenalin release and melanophore expansion in the lizard, *Anolis carolinensis. Copeia* 1956, 214–217.

Rand, A. S. (1967). Ecology and social organization in the iguanid lizard *Anolis lineatopus. Proc. U. S. Natl. Mus.* 122, 1–79.

Rand, A. S., Gorman, G. C., and Rand, W. M. (1975). Natural history, behavior, and ecology of *Anolis agassizi. Smithson. Contrib. Zool.* 176, 27–38.

Rand, A. S., and Williams, E. E. (1970). An estimation of redundancy and information content of anole dewlaps. *Am. Nat.* 104, 99–103.

Rand, M. S. (1986). "Histological, Hormonal, and Chromatic Correlates of Sexual Maturation in the Male Lizard, *Crotaphytus collaris.*" Master's thesis, Wichita State University, Wichita, Kan.

Rand, M. S. (1988). Courtship and aggressive behavior in male lizards exhibiting two different sexual colorations. *Am. Zool.* 28, 153A (abstract).

Rand, M. S. (1989). Androgen organization and activation of three sexually dimorphic characters in a species of lizard. *Am. Zool.* 29, 19A (abstract).

Rand, M. S. (1990). Polymorphic sexual coloration in male lizards (*Sceloporus undulatus erythrocheilus*). *Am. Midl. Nat.* 124, 352–359.

Redfield, A. C. (1916). The coordination of chromatophores by hormones. *Science* (N.Y.) 43, 580–581.

Redfield, A. C. (1918). The physiology of the melanophores of the horned toad *Phrynosoma. J. Exp. Zool.* 26, 275–333.

Rehák, I. (1987). Color change in the snake *Tropidophis feicki* (Reptilia:Squamata:Tropidophidae). *Vestnik Ceskoslovenske Spolecnosti Zoologicke* 51, 300–303.

Reid, L., Ling, N., and Yen, S. S. C. (1981). Melanocyte stimulating hormone induces gonadotropin release. *J. Clin. Endocrinol. Metab.* 52, 159–161.

Rensch, B., and Adrian-Hinsberg, Ch. (1963). Die visuelle Lernkapazität von Leguanen. *Z. Tierpsychol.* 19, 34–42.

Reynolds, A. E. (1943). The normal seasonal reproductive cycle in the male *Eumeces fasciatus* together with some observations on the effects of castration and hormone administration. *J. Morphol.* 72, 331–377.

Rice, G. E., and Bradshaw, S. D. (1980). Changes in dermal reflectance and vascularity and their effects on thermoregulation in *Amphibolurus nuchalis* (Reptilia:Agamidae). *J. Comp. Physiol.* 135, 139–146.

Rivero, J. A. (1978). *Los Anfibios y Reptiles de Puerto Rico.* University of Puerto Rico Editorial Universitaria, Mayaguez.

Robinson, M. D. (1978). Sexual dichromatism in the Namaqua chamaeleon, *Chamaeleo namaquensis. Madoqua* 11, 81–83.

Rodriguez, E. M., and La Pointe, J. (1970). Light and electron microscopic study of the pars intermedia of the lizard, *Klauberina riversiana. Z. Zellforsch.* 104, 1–13.

Rohrlich, S. T., and Porter, K. R. (1972). Fine structural observations relating to the production of color by the iridophores of a lizard, *Anolis carolinensis. J. Cell Biol.* 53, 38–52.

Rohwer, S. (1982). The evolution of reliable and unreliable badges of fighting ability. *Am. Zool.* 22, 531–546.

Ruby, D. E. (1978). Seasonal changes in the territorial behavior of the iguanid lizard *Sceloporus jarrovi. Copeia* 1978, 430–438.

Sachsse, W. (1975). *Chinemys reevesii* var. *unicolor* and *Clemmys bealei* var. *quadriocellata*-Auspragungen von Sexualdimorphismus der beiden "Nominatiformen." *Salamandra* 11, 20–26.

Samson, W. K., Lipton, J. M., Zimmer, J. A., and Glyn, J. R. (1981). The effect of fever on central a-MSH concentrations in the rabbit. *Peptides* 2, 419–423.

Sand, A. (1935). The comparative physiology of colour response in reptiles and fishes. *Biol. Rev.* 10, 361–382.

Sandman, C. A., Miller, L. H., Kastin, A. J., Shally, A. V., and Kendall, J. W. (1973). Neuroendocrine responses to physical and psychological stress. *J. Comp. Physiol. Psychol.* 34, 386–390.

Schmidt, W. J. (1912). Studien am Integument der Reptilien. I. Die Haut der Geckoniden. *Z. Wiss. Zool.* 101, 139–258.

Schmidt, W. J. (1913). Studien am Integument der Reptilien. IV. *Uroplatus fimbriatus* und die Geckoniden. *Zool. Jahrb. (Anat.)* 36, 377–464.

Schmidt, W. J. (1914). Studien am Integument der Reptilien. V. Anguiden. *Zool. Jahrb. (Anat.)* 38, 1–102.

Schmidt, W. J. (1917). Die Chromatophoren der Reptilienhaut. *Arch. mikr. Anat.* I, 90, 98–259 + ix plates.

Schuett, G. W. (1984). *Calloselasma rhodostoma* (Malayan pit viper). Feeding mimicry. *Herp. Rev.* 15, 112.

Schuett, G. W., Clark, D. L., and Krause, F. (1984). Feeding mimicry in the rattlesnake *Sistrurus catenatus,* with comments on the evolution of the rattle. *Anim. Behav.* 32, 625–626.

Sexton, O. J. (1960). Experimental studies of artificial Batesian mimics. *Behaviour* 3–4, 244–252.

Sexton, O. J. (1964). Differential predation by the lizard, *Anolis carolinensis,* upon unicoloured and polycoloured insects after an interval of no contact. *Anim. Behav.* 12, 101–110.

Sexton, O. J., Hoger, C., and Ortleb, E. (1966). *Anolis carolinensis:* effects of feeding on reaction to aposematic prey. *Science* (N.Y.) 153, 1140.

Sherbrooke, W. C. (1988). "Integumental Biology of Horned Lizards (*Phrynosoma*)." Ph.D. dissertation, University of Arizona, Tucson.

Sherbrooke, W. C., and Frost, S. K. (1989). Integumental chromatophores of a color-change, thermoregulating lizard, *Phrynosoma modestum* (Iguanidae; Reptilia). *Am. Mus. Novit.* 2943, 1–14.

Shine, R. (1980). Costs of reproduction in reptiles. *Oecologia* 46, 92–100.

Siegel, S. (1956). *Nonparametric Statistics for the Behavioral Sciences.* McGraw-Hill, New York.

Sigmund, W. R. (1983). Female preference for *Anolis carolinensis* males as a function of dewlap color and background coloration. *J. Herpetol.* 17, 137–143.

Smith, D. C. (1929). The direct effect of temperature changes upon the melanophores of the lizard *Anolis equestris. Proc. Natl. Acad. Sci. USA* 15, 48–56.

Smith, H. M. (1946). *Handbook of Lizards: Lizards of the United States and Canada.* Cornell University Press, Ithaca, N.Y.

Smith, M. (1915). On the breeding habits and colour changes in the lizard *Calotes mystaceus. J. Nat. Hist. Soc. Siam.* 1, 256–257.

Smith, M. A. (1931). *The Fauna of British India, including Ceylon and Burma: Reptilia and Amphibia,* Vol. 1: *Loricata, Testudines.* Taylor and Francis, London.

Smith, M. A. (1935). *The Fauna of British India, including Ceylon and Burma: Reptilia and Amphibia,* Vol. 2: *Sauria.* Taylor and Francis, London.

Sokol, S., and Muntz, W. R. A. (1966). The spectral sensitivity of the turtle, *Chrysemys picta picta. Vis. Res.* 6, 285–292.

Stamps, J. A. (1977). Social behavior and spacing patterns in lizards. In *Biology of the Reptilia,* Vol. 7 (C. Gans and D. W. Tinkle, eds.). Academic Press, London, pp. 265–334.

Stamps, J. A. (1978). A field study of the ontogeny of social behavior in the lizard *Anolis aeneus. Behaviour* 66, 1–31.

Stebbins, R. C. (1985). *A Field Guide to Western Reptiles and Amphibians,* 2nd ed. Houghton Mifflin Co., Boston.

Stebbins, R. C., and Robinson, H. B. (1946). Further analysis of a population of the lizard *Sceloporus graciosus gracilis. Univ. Calif. Publ. Zool.* 48, 149–168.

Stebbins, R. C., Lowenstein, J. M., and Cohen, N. W. (1967). A field study of the lava lizard (*Tropidurus albemarlensis*) in the Galapagos Islands. *Ecology* 48, 839–851.

Stratton, L. O., and Kastin, A. J. (1976). Melanocyte stimulating hormone and MSH/ACTH$_{4-10}$ reduce tonic immobility in the lizard. *Physiol. Behav.* 16, 771–774.

Strecker, J. K. (1928). Field observations on the color changes of *Anolis carolinensis* Voigt. *Contr. Baylor Univ. Mus.* 13, 3–9.

Sustare, B. D. (1979). Physics of light: an introduction for light-minded ethologists. In *The Behavioral Significance of Color* (E. H. Burtt, Jr., ed.). Garland STPM Press, New York, pp. 4–27.

Svoboda, B. (1969). "Die Bedeutung von Farb-, Form-, und Geruchsmerkmalen für das Vermeiden-lernen von Beuteobjekten bei *Lacerta agilis* L." Dissertation, University of Vienna, Vienna.

Sweet, S. S. (1985). Geographic variation, convergent crypsis, and mimicry in gopher snakes (*Pituophis melanoleucus*) and western rattlesnakes (*Crotalus viridis*). *J. Herpetol.* 19, 55–67.

Swiezawska, K. (1949). Colour-discrimination of the sand lizard, *Lacerta agilis* L. *Bull. Int. Acad. Pol. Sci. Let., Ser. B., Sci. Nat.* 1949, 1–20.

Tam, W. H., Phillips, J. G., and Lofts, B. (1972). Seasonal changes in the secretory activity of the adrenal gland of the cobra (*Naja naja* L.). *Gen. Comp. Endocrinol.* 19, 218–224.

Taylor, J. D., and Hadley, M. E. (1970). Chromatophores and color change in the lizard, *Anolis carolinensis. Z. Zellforsch.* 104, 282–294.

Theobald, W. (1868). Catalogue of the reptiles of British Burma, embracing the provinces of Pegu, Martaban, and Tenasserim, with descriptions of new or little known species. *J. Linn. Soc. Zool.* 10, 4–67.

Thody, A. J., and Wilson, C. A. (1988). The role of melanotropins in sexual behavior. In *The Melanotropic Peptides,* Vol. 2: *Biological Roles* (M. E. Hadley, ed.). CRC Press, Boca Raton, Fla., pp. 131–144.

Thody, A. J., Wilson, C. A., and Everard, D. (1979). Facilitation and inhibition of sexual receptivity in the female rat by α-MSH. *Physiol. Behav.* 22, 447–450.

Thompson, C. W., and Moore, M. E. (1987). Status signalling in male tree lizards. *Am. Zool.* 27, 50A.

Thornton, V. F., and Geschwind, I. I. (1975). Evidence that serotonin may be a melanocyte stimulating hormone releasing factor in the lizard, *Anolis carolinensis. Gen. Comp. Endocrinol.* 26, 346–353.

Tinbergen, N. (1951). *The Study of Instinct.* Clarendon Press, Oxford.

Tokarz, R. R. (1985). Body size as a factor determining dominance in staged agonistic encounters between male brown anoles (*Anolis sagrei*). *Anim. Behav.* 33, 746–753.

Toro, M. A. del. (1982). *Los Reptiles de Chiapas,* 3rd ed. Instituto de Historia Natural, Tuxtla Gutierrez, Chiapas, Mexico.

Trillmich, K. G. K. (1983). The mating system of the marine iguana (*Amblyrhynchus cristatus*). *Z. Tierpsychol.* 63, 141–172.

Trivers, R. L. (1972). Parental investment and sexual selection. In *Sexual Selection and the Descent of Man* (B. Campbell, ed.). Aldine-Atherton, Chicago, pp. 136–179.

Tryon, B. W. (1985). *Bothrops asper* (Terciopelo). Caudal luring. *Herp. Rev.* 16, 28.

Turner, F. B., Lannom, J. R., Medica, P. A., and Hoddenbach, G. A. (1969). Density and composition of fenced populations of leopard lizards *Crotaphytus wislizenii* in southern Nevada. *Herpetologica* 25, 247–257.

Uexküll, J. von. (1921). *Umwelt und Innenwelt der Tiere,* 2nd ed. Springer-Verlag, Berlin.

Underwood, G. L. (1970). The eye. In *Biology of the Reptilia,* Vol. 2 (C. Gans and T. S. Parsons, eds.). Academic Press, London, pp. 1–97.

Valenstein, P., and Crews, D. (1977). Mating induced termination of behavioral estrus in the female lizard, *Anolis carolinensis. Horm. Behav.* 9, 362–370.

Vallarino, M. (1984). Immunocytochemical localization of α-melanocyte-stimulating hormone in the brain of the lizard, *Lacerta muralis. J. Cell Tissue Res.* 237, 521–524.

Vanzolini, P. E. (1958). Sobre *Diploglossus lessonae,* com notas biometricas e sobre evolução ontogenetica do padrão de colorido (Sauria, Anguidae). *Papeis Avulsos Zool., Sao Paulo* 13, 179–211.

Vaughan, G. L. (1987a). Photosensitivity in the skin of the lizard, *Anolis carolinensis. Photochem. Photobiol.* 46, 109–114.

Vaughan, G. L. (1987b). Dermal photic response in *Anolis carolinensis:* the advantage to being brown in winter. *Proc. Joint Ann. Meeting Soc. Study Amphibians and Reptiles and Herpetologist's League* 1987:154 (abstract).

Vaughan, G. L., and Greenberg, N. (1987). Propanolol, a beta-adrenergic antagonist, retards response to MSH in skin of *Anolis carolinensis. Physiol. Behav.* 40, 555–558.

Vetter, R. S., and Brodie, E. D., Jr. (1977). Background color selection and antipredator behavior of the flying gecko, *Ptychozoon kuhli. Herpetologica* 33, 464–467.

Vial, J. L., and Stewart, J. R. (1989). The manifestation and significance of sexual dimorphism in anguid lizards: a case study of *Barisia monticola. Can. J. Zool.* 67, 68–72.

Vinegar, M. B. (1972). The function of breeding coloration in the lizard, *Sceloporus virgatus. Copeia* 1972, 660–664.

Vinegar, M. B. (1975). Comparative aggression in *Sceloporus virgatus, S. undulatus consobrinus,* and *S. u. tristichus* (Sauria: Iguanidae). *Anim. Behav.* 23, 279–286.

Viosca, P. (1933). The *Pseudemys troostii-elegans* complex, a case of sexual dimorphism. *Copeia* 1933, 208–210.

Visser, J. (1966). Colour change in *Leptotyphlops scutifrons* (Peters) and notes on its defensive behavior. *Zoologica Africana* 2:123–125.

Vitt, L. J., and Cooper, W. E., Jr. (1985). The evolution of sexual dimorphism in the skink *Eumeces laticeps:* an example of sexual selection. *Can. J. Zool.* 63, 995–1002.

Vitt, L. J., and Cooper, W. E., Jr. (1986). Skink reproduction and sexual dimorphism: *Eumeces fasciatus* in the southeastern United States, with notes on *Eumeces inexpectatus. J. Herpetol.* 20, 65–76.

Wagner, H. (1933). Ueber den Farbensinn der Eidechsen. *Z. vergl. Physiol.* 18, 378–392.

Waldschmidt, S. R., Jones, W. M., and Porter, W. P. (1985). The effect of body temperature and feeding regime on activity, passage time, and digestive coefficient in the lizard *Uta stansburiana. Physiol. Zool.* 59, 376–383.

Waring, H. (1963). *Colour Change Mechanisms of Cold-blooded Vertebrates.* Academic Press, London.

Watson, J. T. (1974). "Endocrine Control of Breeding Coloration and Reproduction in the Lizard *Holbrookia propinqua.*" Ph.D. dissertation, University of Texas Health Science Center at Houston, Graduate School of Biomedical Sciences, Houston.

Webb, R. G. (1962). North American recent soft-shelled turtles (family Trionychidae). *Univ. Kansas Publ. Mus. Nat. Hist.* 13, 429–611.

Weber, W. (1983). Photosensitivity of chromatophores. *Am. Zool.* 23, 495–506.

Webster, T. P., and Burns, J. M. (1973). Dewlap color variation and electrophoretically detected sibling species in a Haitian lizard, *Anolis brevirostris. Evolution* 27, 368–377.

Werner, D. I. (1978). On the biology of *Tropidurus delanonis,* Baur (Iguanidae). *Z. Tierpsychol.* 47, 337–395.

Werner, D. I. (1982). Social organization and ecology of land iguanas, *Conolophus subcristatus,* on Isla Fernandina, Galapagos. In *Iguanas of the World: Their Behavior, Ecology, and Conservation* (G. M. Burghardt and A. S. Rand, eds.). Noyes Publications, Park Ridge, N.J., pp. 342–365.

Wharton, C. H. (1960). Birth and behavior of a brood of cottonmouths, *Agkistrodon piscivorus piscivorus* with notes on tail-luring. *Herpetologica* 16, 125–129.

Wickler, W. (1968). *Mimicry in Plants and Animals.* McGraw-Hill, New York.

Wilcox, B. A. (1980). "Aspects of the Biogeography and Evolutionary Ecology of some Island Vertebrates." Ph.D. dissertation, University of California, San Diego.

Williams, E. E., and Case, S. M. (1986). Interactions among members of the *Anolis distichus* complex in and near Sierra de Baoruco, Dominican Republic. *J. Herpetol.* 20, 535–546.

Williams, E. E., and Rand, A. S. (1977). Species recognition, dewlap function and faunal size. *Am. Zool.* 17, 261–270.

Wirot, N. (1979). *The Turtles of Thailand.* Siam Farm Zoological Garden, Bangkok.

Wojtusiak, R. J. (1933). Ueber den Farbensinn der Schildkröten. *Z. Vergl. Physiol.* 18, 393–436.

Wojtusiak, R. J., and Młynarski, M. (1949). Investigations on the vision of infrared in animals. II. Further experience on water tortoises. *Bull. Acad. Pol. Sci. Let., Ser. B., Sci. Nat.* 1949, 95–120.

Wooley, P. (1957). Colour change in a chelonian. *Nature* (London) 179, 1255.

Wu, J., Whittier, J. M., and Crews, D. (1985). Role of progesterone in the control of female sexual receptivity in *Anolis carolinensis. Gen Comp. Endocrinol.* 58, 402–406.

Wurtman, R. J., and Axelrod, J. (1965). Adrenalin synthesis: control by the pituitary gland and adrenal glucocorticoids. *Science* (N.Y.) 150, 1464–1465.

Wurtman, R. J., Axelrod, J., and Tramezzani, J. (1967). Distribution of the adrenalin forming enzyme in the adrenal gland of a snake, *Xenodon merremii. Nature* (London) 215, 879–880.

Yedlin, I. N., and Ferguson, G. W. (1973). Variations in aggressiveness of free-living male and female collared lizards, *Crotaphytus collaris. Herpetologica* 29, 268–275.

Zoond, A., and Eyre, J. (1934). Studies in reptilian colour response. *Phil. Trans. Roy. Soc. London* B223, 27–35.
Zucker, N. (1989). Dorsal darkening and territoriality in a wild population of the tree lizard, (*Urosaurus ornatus*). *J. Herpetol.* 23, 389–398.
Zucker, N., and Boecklen, W. (1990). Variation in female throat coloration in the tree lizard (*Urosaurus ornautus*): relation to reproductive cycle and fecundity. *Herpetologica* 46, 387–394.

7

Nasal Chemical Senses in Reptiles: Structure and Function

Mimi Halpern

CONTENTS

I. INTRODUCTION

Reptiles use their nasal chemical senses in environmental exploration, prey detection, and a variety of species-typical social behaviors that include aggregation and mate recognition (Burghardt, 1970a, review). Of the several nasal chemical senses of reptiles, including possibly the trigeminal, nervus terminalis, and septal organ sense, the olfactory and vomeronasal systems have been most thoroughly investigated, and this review will be limited to those two systems. Contrasting these two systems morphologically and functionally in reptiles has provided general insights into their functional domains that extend well beyond this group.

Although it is generally accepted that the term *olfaction* refers to the sense mediated by the main olfactory system and that volatile odorants are the substances sensed by that system, there has been no generally accepted similar terminology for the vomeronasal sense. Recently, *vomerolfaction* has been suggested to refer to the sense mediated by the vomeronasal organ and *vomodor* has been suggested to refer to substances sensed by the vomeronasal organ (Cooper and Burghardt, 1990a). It remains to be seen whether this terminology will become generally accepted.

II. REVIEWS

The most complete and comprehensive review of the literature on chemical senses in reptiles is that of Burghardt (1970a). This extraordinarily complete and critical review of the literature to that time deals with the morphology of the vomeronasal and main olfactory systems in reptiles, the role of chemical senses in orientation, intraspecific aggregation, alarm, territorial marking, courtship, maternal behavior, predator recognition, and feeding behavior. In addition, it provides an especially detailed review of the literature on chemical perception in newborn snakes.

A somewhat later and more restricted review (Madison, 1977) discusses chemical communication in amphibians and reptiles, including scent production and scent reception. The review is limited to intraspecific interactions in the major living taxa, and therefore does not discuss feeding interactions, defense secretions, and alarm substances.

Several reviews of vertebrate chemoreception contain extensive discussions of reptilian nasal chemical senses (Halpern, 1980a; Moulton and Beidler, 1967). Moulton and Beidler review the peripheral olfactory and vomeronasal systems in a variety of vertebrates with particular attention to the functional significance of the vomeronasal system of reptiles. As part of a broad review of the use of chemical senses in interspecific communication in terrestrial vertebrates, Halpern reviews the literature on the use of chemical signals in prey detection, predator recognition, and defense.

The vomeronasal system of vertebrates has been reviewed recently, and several of these reviews have included discussion of the reptilian vomeronasal sense (Halpern, 1987; Wysocki and Meredith, 1987). The broad presence of the vomeronasal system among reptiles and its probable early appearance in their evolution have led to the speculation that the vomeronasal system was present in mammal-like (synapsid) reptiles and mediated response to sociochemicals involved in avoidance, aggregation, and eventually lactation (Duvall, 1986; Graves and Duvall, 1983a).

The excellent reviews of the morphology of the peripheral nasal chemosensory apparatus of reptiles (Gabe and Saint Girons, 1976; Parsons, 1967, 1970) are discussed in more detail below. These reviews are comprehensive and particularly well illustrated, providing the reader with graphic orientation to the major structures that comprise the peripheral sensory apparatus.

The role of chemical senses in stereotyped behavior of reptiles including olfactory releasing mechanisms for aggression, courtship, and mating is discussed by Carpenter and Ferguson (1977). There are summaries of the development of behavior (Burghardt, 1978) and learning processes in reptiles (Burghardt, 1977a), including discussions of the chemical senses in feeding, odor habituation, odor conditioning, and odor cues used in maze learning.

A number of recent reviews have been restricted to a single group of reptiles. For turtles, the chemical senses have been reviewed with particular emphasis on the olfactory system (Scott, 1979). In addition to the treatment of the gross structure of the nose, receptor anatomy (olfactory epithelium, vomeronasal epithelium), olfactory nerve, olfactory bulb, and central connections, the review includes an excellent discussion of olfactory electrophysiology, including the physiology of the receptors and olfactory bulb.

The review of the use of chemical cues in behavior of turtles (Manton, 1979) includes information on odor production, olfactory behavior, odor discrimination, orientation to odors, and imprinting to odors. Odor is produced by the musk, chin, and cloacal glands, particularly whenever turtles are stressed as well as during courtship and aggressive encounters. Responses to these odors include a variety of behaviors such as head-bobbing, exploration, and ramming. Odor discrimination and olfactory learning have been demonstrated. A review of turtle social behavior (Harless, 1979) discusses the role of olfactory stimuli. For the pheromones of turtles see Chapter 4.

Among lizards, the development and use of the nasal chemical senses varies enormously; this variation appears related to taxonomic group. The ascalabotans (Chamaeleonidae, Agamidae, Iguanidae, Gekkonidae, and Xantusiidae) appear to be more visual, whereas the autarchoglossans (Lacertidae, Scincidae, Teiidae, Cordylidae, Helodermatidae, Varanidae, and all other lizard families) appear to rely more on chemical communication as well as visual stimulation. A review of lizard chemoreception (Simon, 1983) discusses these differences and particularly emphasizes the use of the vomeronasal system and the use of tongue-flicking as a monitor of vomeronasal activation. It also emphasizes the multiple uses of tongue extrusions in lizard behavior and cautions against the uncritical acceptance of tongue-flicking activity as indicative of vomeronasal stimulation. However, changes in tongue-flick behavior suggest that the vomeronasal system of lizards may be involved in detection of conspecifics, species identification, territoriality and spacing, aggregation and hibernacula location, general exploration and responses to novel stimuli, food-seeking, predator detection, sex discrimination, courtship, and maternal care.

Recently, the chemical senses of snakes have received more attention than those of any other reptilian group. Several reviews deal with the anatomy and functional aspects of the nasal systems (Halpern, 1983; Halpern and Kubie, 1984). The roles of the chemical senses in snake foraging (Mushinsky, 1987), social behavior such as courtship, combat and aggregation (Gillingham, 1987), and spatial behavior including spatial patterns, home ranges, navigation, and aggregation (Gregory et al., 1987) have recently been summarized. The literature on conspecific trailing behavior (Ford, 1986) and strike-induced chemosensory searching (SICS) (Chiszar and Scudder, 1980; Chiszar et al., 1983a) also has been extensively reviewed.

The following survey of the literature on the main and accessory (vomeronasal) systems of reptiles focuses on reports published since 1970. It first discusses the anatomy, development, and physiology of the two systems. Next it discusses the mechanism of odor access to the vomeronasal system, followed by a review of the literature on chemical signals in reptilian behavior.

III. ANATOMY OF NASAL CHEMICAL SYSTEMS

A. Overview

The anatomy of the nasal cavities of reptiles has been described by numerous authors and comprehensively reviewed by Parsons (1959a, 1967, 1970). The section below is based on these references, and the reader is referred to them for more detail.

The nasal organs consist of a vestibule, the nasal cavity proper (including conchae), the vomeronasal (Jacobson's) organ, and the nasopharyngeal duct, which connects the nasal cavity proper to the internal naris (choana). The vestibule includes the tubular portion of the nasal cavity extending from external naris to nasal cavity proper. This discussion is restricted to the vomeronasal organ and the sensory portions of the nasal cavity (Figs. 7.1 to 7.3).

B. Main Olfactory Organ

The basic structure of the nasal cavity is fairly constant across reptilian groups, although the shape and proportions of the cavities and turbinals differ (Allison, 1953). The cavity is lined by a nonsensory respiratory epithelium and a posteriorly placed sensory epithelium (Fig. 7.4).

The pseudostratified, columnar sensory epithelia are generally similar in structure and component elements and differ only in degree of development and relative numbers of those elements (Gabe and Saint Girons, 1976). The epithelium proper contains three types of cells: supporting cells, bipolar neurons, and basal, undifferentiated cells (e.g., Ferri et al., 1982; Iwahori, 1984; Liquori et al., 1982). Distinguishing features of this epithelium include cilia on the dendritic enlargements of the bipolar neurons (Fig. 7.5). Scanning electron microscopy indicates that the surface morphology of the cilia may differ among lizards (Wang and Halpern, 1983). For example, the olfactory cilia of the alligator lizard (*Elgaria coerulea coerulea*) are smooth, slender, and of uniform diameter, whereas those of the tegu (*Tupinambis teguixin*) contain fine, granulated protrusions on their surfaces, and ciliary surfaces of *Gekko gecko* contain numerous large varicosities. Because the cilia are presumed to contain receptor proteins on or in their membranes, the surface modifications may reflect functional differences in these receptor cells.

The nuclei of the three types of cells of the epithelium occupy different positions. The nuclei of the basal cells lie closest to the basal lamina, the nuclei of the bipolar neurons are intermediate in position, and the oval nuclei of supporting cells are most superficial (e.g., Iwahori et al., 1987). Bowman's glands lie below the epithelium, but their ducts penetrate it and their contents are discharged into the lumen of the nasal cavity. The secretions of Bowman's glands and the support-

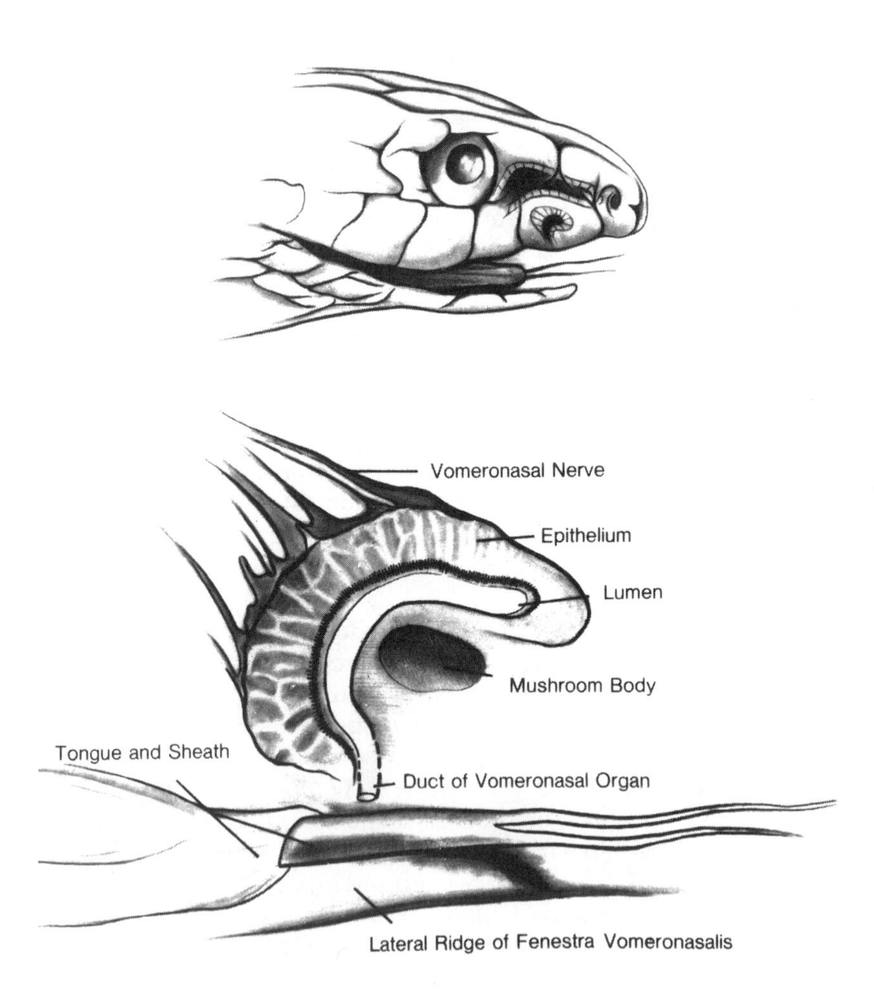

Fig. 7.1. *Thamnophis sirtalis.* Schematic drawing of lateral view of head illustrating relative position of vomeronasal organ and nasal cavity (above). Enlargement (below) illustrates the position of the principal features of the vomeronasal organ and the position of the tongue in relation to the lateral ridge of the fenestra vomeronasalis and the vomeronasal duct. (From Halpern, 1983.)

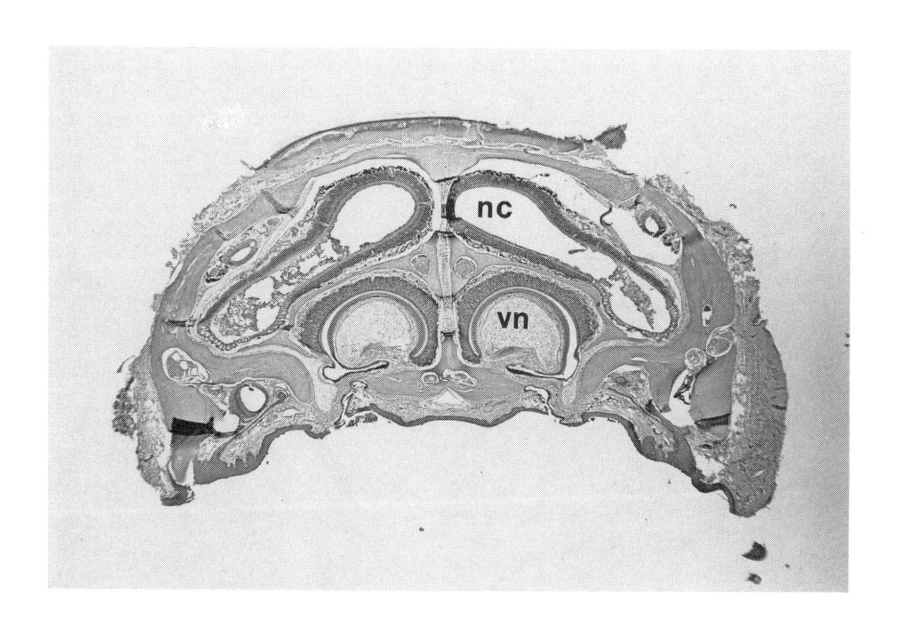

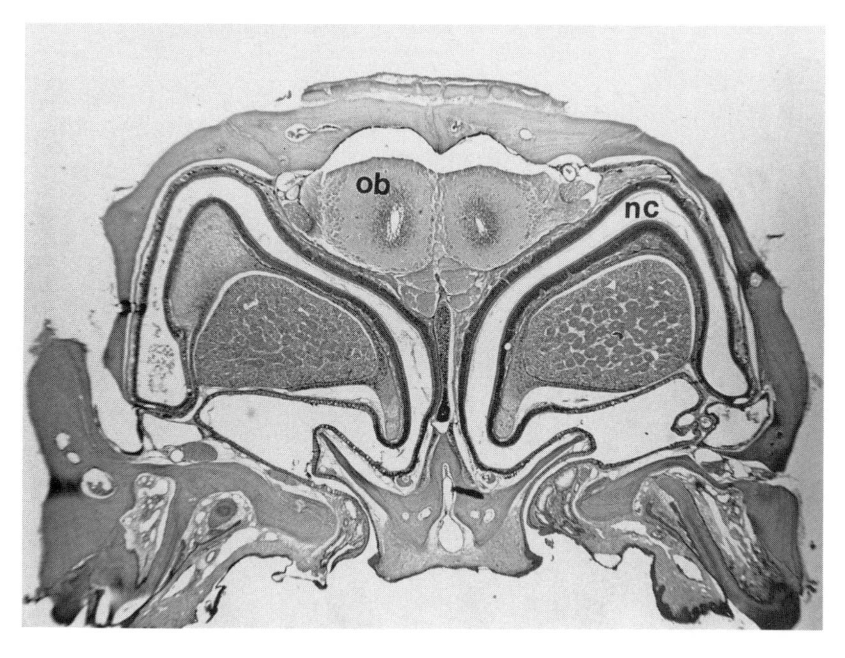

Fig. 7.2. *Chalcides ocellatus.* Top: Coronal section through head at level of vomeronasal organ (vn) and anterior portion of nasal cavity (nc). Bottom: Coronal section through head at level of olfactory bulb (ob) and caudal portion of nasal cavity (nc). (Courtesy, Brent Graves.)

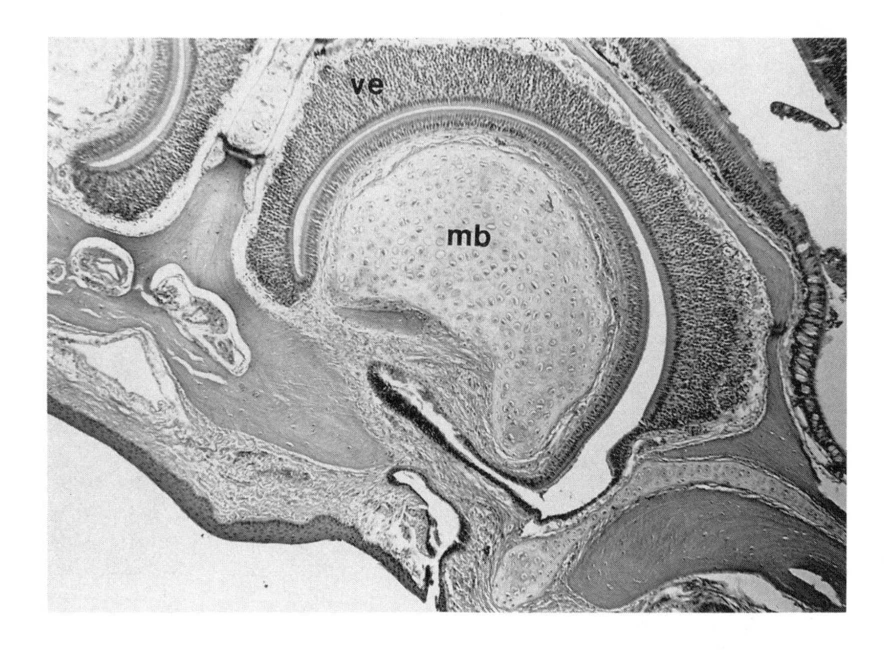

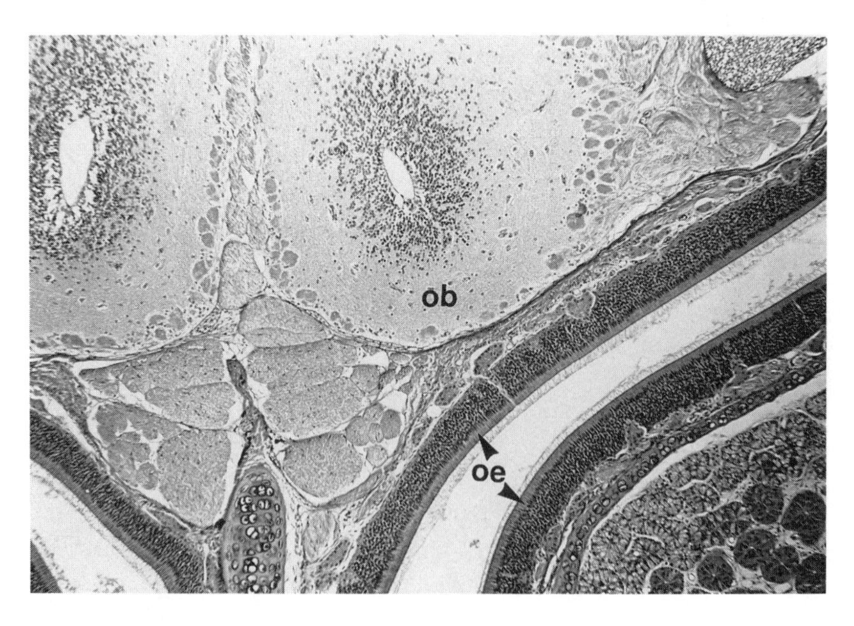

Fig. 7.3. *Chalcides ocellatus.* Top: Vomeronasal organ with vomeronasal epithelium (ve) and mushroom body (mb). Bottom: Olfactory epithelium (oe) and main olfactory bulb (ob). (Courtesy, Brent Graves.)

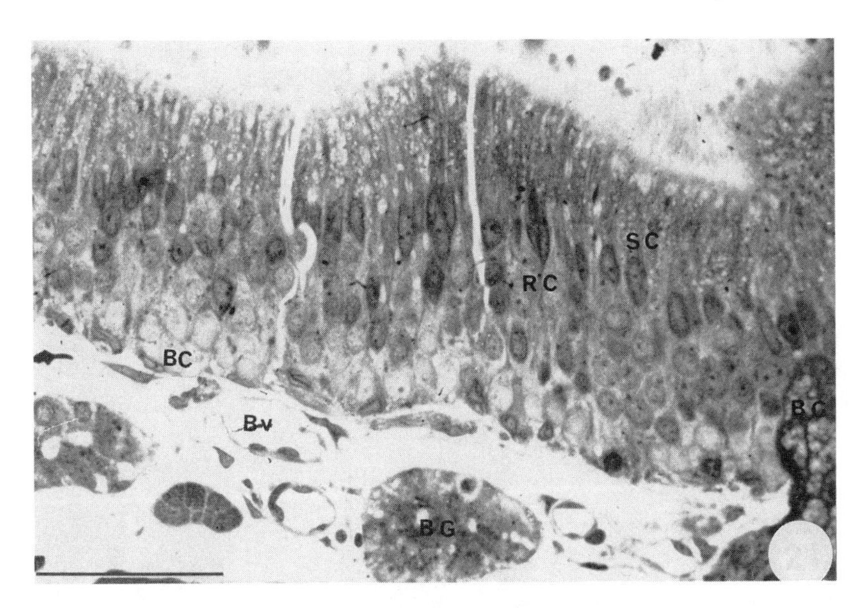

Fig. 7.4. *Thamnophis sirtalis.* Photomicrograph of toluidine blue O–stained section of olfactory epithelium. Lu, Lumen of nasal cavity; SC, supporting cells; RC, receptor cells; BC, basal cells; Bv, blood vessel; BG, Bowman's gland. Bar = 50 μm. (From Wang and Halpern, 1980a.)

ing cells are believed to be the source of the mucus that covers the apical surface of the epithelium. In *Testudo hermanni,* supporting cells in the olfactory epithelium contain paracrystalline, membrane-bound aggregations that may be related to secretory function (Delfino et al., 1986).

The neurons of the olfactory epithelium (and of the vomeronasal epithelium as well) are unique in being generated throughout the lifetime of the organism. This potency is a general vertebrate property. New neurons derive from cell populations along the base of the epithelium; transitional stages in morphological development are seen between the precursors in the basal cell layer and the mature neurons in the lower parts of the bipolar cell layer (e.g., Liquori et al., 1982).

The axons of the bipolar neurons of the epithelium pierce its basal lamina, collect to form the fila olfactoria, pass into the cranial vault, and terminate in the glomerular layer of the main olfactory bulb.

C. Vomeronasal Organ

The presence, or absence, and development when present, of a vomeronasal organ varies considerably among reptiles (Bertmar, 1981; Ne-

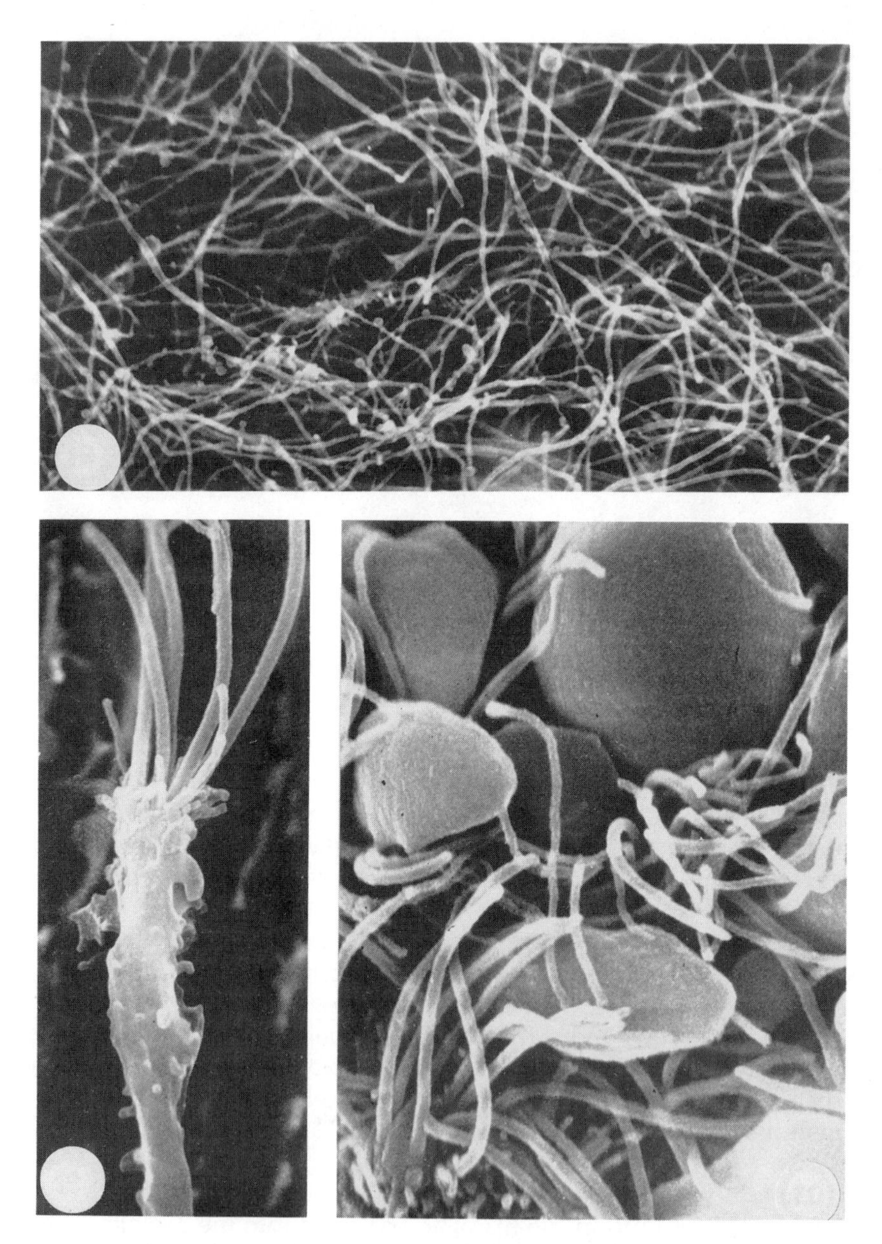

Fig. 7.5. *Thamnophis sirtalis*. Olfactory epithelium, scanning electron micrographs. Top: Heavy meshwork of cilia covering luminal surface of the olfactory epithelium. Bottom left: Isolated dendrite of an olfactory bipolar neuron illustrating the presence of dendritic spines and cilia. Bottom right: Cilia projecting between adjacent protuberances of supporting cells along the luminal surface of the olfactory epithelium. Distal segments of cilia were removed during preparation. (From Wang and Halpern, 1980b.)

gus, 1958; Saint Girons, 1975). Species of reptiles from aquatic and arboreal habitats tend to have diminished or vestigial vomeronasal organs (Bertmar, 1981).

Turtles lack a separate vomeronasal organ and are said to lack it entirely (Northcutt, 1970; Parsons, 1959a, 1970). Currently a structure in turtles in the ventral portion of the nasal cavity is thought to be homologous to the vomeronasal organ of other reptiles (Bertmar, 1981; Graziadei and Tucker, 1970; Hatanaka and Hanada, 1986; Hatanaka et al., 1982, 1988; Parsons, 1959a). The organ of turtles is histologically similar to other vomeronasal epithelia; that is, it lacks Bowman's glands and ciliated receptor cells and the axons of some of its neurons terminate in an accessory olfactory bulb.

Adult crocodilians lack a vomeronasal organ, but young crocodilians retain it (Parsons, 1959b).

In *Sphenodon,* unlike the squamates, the shape of the vomeronasal organ is tubular, the mushroom body is absent, and the organ is connected with both oral and nasal cavities (Parsons, 1970).

The vomeronasal organ is best developed in squamates. In general, it is well developed in all genera of snakes. The vomeronasal organs of lizards differ (Gabe and Saint Girons, 1976). In *Chamaeleo chamaeleon* the vomeronasal organ is extremely reduced, as are the olfactory nerves and olfactory epithelium (Haas, 1937). However, the vomeronasal organ is present and well developed in embryonic and neonate chameleons (Haas, 1947). Otherwise the vomeronasal organ may be absent (*Brookesia*), small with no mushroom body and poorly differentiated (*Anolis*), relatively small (*Chalarodon, Uma,* Agamidae, *Uta, phrynosoma*), moderately developed (*Iguana,* Cordylidae, *Casarea*), well developed (Gekkonidae, *Ameiva,* Scincoidea, *Crotalus*), or very well developed (Lacertidae, *Cnemidophorus,* Anguimorpha, most snakes) (Gabe and Saint Girons, 1976).

In all adult squamates the vomeronasal organ is separated from the nose by the secondary palate, which is an extension of the primary palate. The organ does not communicate with the nasal cavity (e.g., Labate et al., 1982) but is connected to the oral cavity by way of a narrow vomeronasal duct. The organ is hemispherical, with a thick sensory epithelium lining its dorsal, posterior aspect (Fig. 7.6). The base of the hemisphere is invaginated ventrally and rostrally by the mushroom body, a cartilaginous structure lined with nonsensory epithelium. Between the mushroom body and sensory epithelium intervenes a narrow lumen that is connected to the vomeronasal duct. The Harderian gland, located posteriorly and medially to the eye, empties into the lacrymal duct in snakes (Smith and Bellairs, 1947). This duct in turn joins the vomeronasal duct and probably provides the vomeronasal organ with the fluid that fills it. Secretions of the Harderian

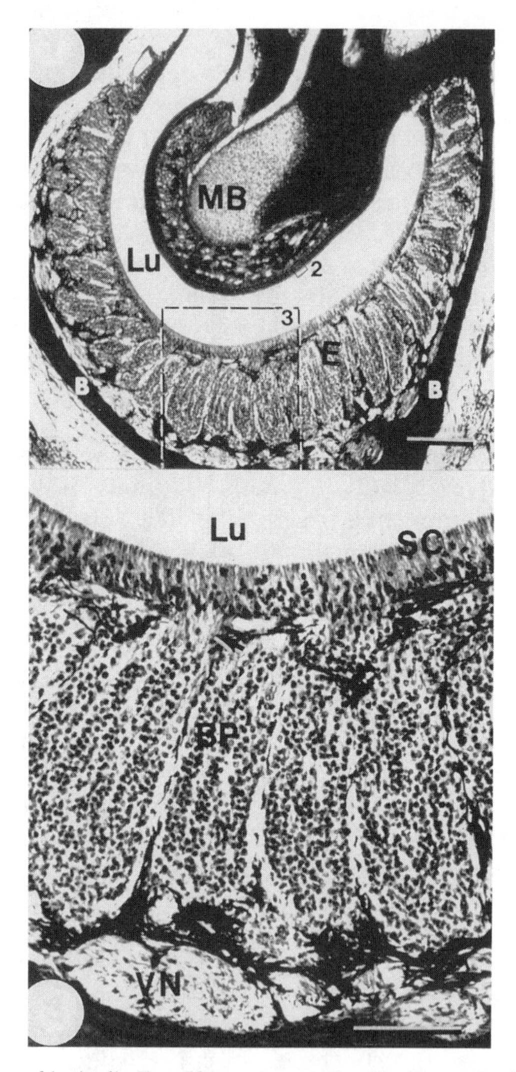

Fig. 7.6. *Thamnophis sirtalis.* Top: Photomicrograph of Bodian-stained horizontal section through the vomeronasal organ. Top of figure is rostral; right is medial. B, bony capsule of organ; E, sensory epithelium; Lu, lumen of organ; MB, mushroom body. Large rectangle encloses area enlarged in bottom figure. Bar = 200 μm. Bottom: Photomicrograph of sensory epithelium of vomeronasal organ illustrating the supporting cell layer (Sc) along the luminal surface and the columnar organization of the deeper layers of the epithelium. Dendrites (arrow) of bipolar neurons (BP) are seen at the apices of the columns. Axons aggregate at the bases of the columns to form the vomeronasal nerve (VN). Bar = 100 μm. (From Wang and Halpern, 1980a).

gland may be important for the sensory functions of the vomeronasal organ (Bellairs and Boyd, 1950).

As mentioned above, vomeronasal organ development varies greatly among reptiles. However, when present, its pseudostratified columnar sensory epithelium contains the same cellular elements as the olfactory epithelium (e.g., Iwahori, 1984). The axons of bipolar neurons form the vomeronasal nerves, which terminate in the accessory olfactory bulb. In contrast to the olfactory epithelium, the dendrites of the bipolar neurons are not ciliated but are covered with microvilli (Altner and Muller, 1968; Altner et al., 1970; Bannister, 1968; Kratzing, 1975). Bowman's glands are absent, but the supporting cells of the sensory epithelium contain secretory granules, as do cells in the mushroom body or along the junction between the sensory epithelium and that of the mushroom body.

The vomeronasal organ and its connections to the accessory olfactory bulb in *Trachyemys scripta* and *Chinemys reevesii* have been described in detail (Hatanaka and Hanada, 1986; Hatanaka et al., 1982, 1988). These turtles definitely have a vomeronasal organ. The axons of the receptor cells form a vomeronasal nerve that terminates in an accessory olfactory bulb. Approximately 1,300,000 receptor cells per side occur in the nasal cavity of *Chinemys,* whereas about 1,800,000 receptor cells per side occur in the vomeronasal organ per side of *Chinemys* (Hatanaka et al., 1982). The vomeronasal nerve is thicker than the main olfactory nerve (1.8 to 1), and the accessory olfactory bulb has a larger volume than the main olfactory bulb (1.2 to 1) (Hatanaka et al., 1988).

The relative development of the vomeronasal organ of squamates may be established by contrasting the number of sensory cells in the epithelium as a percentage of total cells in the organ (% RC) or as a ratio of sensory to supporting cells (RC/SC) (Table 7.1). Squamates show a correlation among the size of the vomeronasal organ, the development of the sensory epithelium, and the proportion of receptor cells in the epithelium (Gabe and Saint Girons, 1976); however, this relationship differs for *Sphenodon,* in which a small proportion of sensory cells is paired with a relatively large dorsal (sensory) epithelial surface.

Recently there has been particular interest in the relative importance of the main olfactory and vomeronasal systems in the ecology of various reptiles. The relative functional significance of a system should correlate with its relative morphological development. Reptiles fall into four main groups based on the general development of the chemical senses: each of these groups may be divided into subgroups based on the relative development of main olfactory epi-

Table 7.1 Vomeronasal receptor cells as a percentage of total epithelial cells (%RC) and receptor cells as a ratio of supporting cells (RC/SC) in groups of reptiles

	%RC	RC/SC
1. *Sphenodon*	23.5	0.4
2. *Anolis*	47.5	1.2
3. Agamidae	50.7–60.1	1.5–2.2
4. *Chalarodon*	64.4–65.2	2.4–2.5
Uma		
Phrynosoma		
5. Gekkonidae	74–82.6	4.4–8.3
Iguana		
Lacertidae		
Cordylidae		
Scincidae		
Casarea		
Crotalus		
6. Teiidae	83.3–92	7.6–16.6
Anguidae		
Varanus		
Amphisbaenia		
Most snakes		

Source: Gabe and Saint Girons, 1976.

thelia and vomeronasal organs (Table 7.2). Group 1 contains neither olfactory nor vomeronasal epithelia. Group 2 has these chemical senses poorly developed. Group 3 has both chemical senses well developed, usually with some inequality between them, and Group 4 has one chemical sense clearly predominating over the other. Chemical sense epithelia tend to be reduced among the Iguania, and this feature is particularly striking in arboreal species (Gabe and Saint Girons, 1976). The vomeronasal organ appears more regressed than the olfactory epithelium in *Anolis* and most Agamidae, whereas the olfactory epithelium is markedly regressed in aquatic snakes.

In reptiles with very well-developed vomeronasal organs, the basal lamina is indented, frequently separating the epithelium into subdivisions (Fig. 7.7). In snakes, these extensions of the basal lamina and the capillaries that accompany them extend from the base of the epithelium to just below the supporting cell layer. The subdivisions form regular columns perpendicular to the supporting cell layer that contain the bipolar neurons (Fig. 7.8) and basal, undifferentiated cells. The nuclei of bipolar cells are round and form a thick layer, two-thirds to three-quarters of the epithelial thickness (Iwahori, 1984; Takami and Hirosawa, 1987, 1990; Wang and Halpern, 1980a, b). An extensive vascular plexus lies at the base of the vomeronasal organ deep to the supporting cell layer and between adjacent columns of bipolar and basal cells.

Table 7.2 Relative development of epithelia of vomeronasal organ (VNO) and main olfactory organ (MO) based on percentage of sensory cells in VNO compared to percentage of sensory cells in MO

1. Neither present:
   Chamaeleonidae, completely atrophied
2. Chemical senses poorly developed (*Sphenodon*) or more or less atrophied (various lizards):
   a. VNO less developed than MO (VNO/MO = 0.48): *Sphenodon*
   b. More or less equal: *Anolis* and most Agamidae (VNO/MO = .90 to 1.27
   c. More developed VNO (VNO/MO = 1.13 to 1.27): Iguanidae other than *Anolis*; *Physignathus*
3. Chemical senses well developed:
   a. Percentage of sensory cells about equal (VNO/MO = .95 to 1.07): Gekkonidae, *Xantusia*, Lacertidae, Amphisbaenia
   b. Slightly greater percentage of sensory cells in VNO (VNO/MO = 1.06 to 1.19): *Lialis*, *Gerrhosaurus*, Scincidae, *Anniella*, *Typhlops*, *Xenopeltis*, *Eryx*, *Casarea*, *Oligodon*
   c. Percentage of sensory cells clearly greater in VNO (VNO/MO = 1.21 to 1.45): *Crotalus*, *Cordylus*, Teiidae, Anguidae, *Cylindrophis*, all nonburrowing colubrid snakes examined
4. Chemical senses very unequally developed (VNO/MO > 1.50):
   a. Extensive olfactory epithelium: *Varanus*
   b. Restricted olfactory epithelium: Acrochordidae

Source: Gabe and Saint Girons, 1976.

Although the neurons of both vomeronasal and olfactory epithelia are generated throughout the lifetime of vertebrates, neurogenesis has been most extensively studied in the vomeronasal system of *Thamnophis* (Wang and Halpern, 1982a, b, 1988). Electronmicroscopic studies (Wang and Halpern, 1980a) demonstrate that the undifferentiated basal cells of the vomeronasal epithelium undergo mitosis (Fig. 7.9) and that there is a morphological progression of cells with cytological features indicative of increasing neuronal maturation from basal to apical regions of the bipolar cell layer. Previous studies of *Thamnophis sirtalis* and *T. radix* (Kubie et al., 1978b) also demonstrate that the bipolar cell layers degenerate if the main olfactory and vomeronasal nerves respectively are sectioned. Light and electron microscopic studies of the vomeronasal epithelium of adult *Thamnophis* show that following unilateral vomeronasal nerve section, the bipolar neuron layer degenerates. Maximal degeneration occurs 2 weeks following denervation. The supporting layer exhibits some morphological changes, consonant with the loss of the processes of bipolar neurons that normally traverse it, but does not undergo mitosis or necrosis. The undifferentiated basal cell layer undergoes active cell proliferation following nerve section (Wang and Halpern, 1982a). The proliferating, basal undifferentiated cells appear to be the source of new bipolar neurons that replace the degenerated ones (Wang and

Halpern, 1982b, 1988). By 8 weeks following axotomy the receptor cell column is repopulated and new neuronal connections are established with the accessory olfactory bulb.

D. Olfactory and Vomeronasal Nerves

The main olfactory epithelium and vomeronasal epithelium are connected to the brain by the olfactory and vomeronasal nerves respectively, which have been recently described for a number of reptiles, for example, *Chinemys reevesii* and *Trachyemys scripta* (Hatanaka et al., 1988), *Chalcides ocellatus* and *Tarentola mauritanica* (Soliman, 1974), *Elaphe obsoleta quadrivittata* and *Thamnophis ordinoides* (Auen and Langebartel, 1977), *Naja n. naja* (Agarwal and Sharma, 1979), and three species of sea snakes—*Lapemis hardwickii, Hydrophis melanocephalus* and *Aipysurus eydouxi* (Young, 1987). Ultrastructural analysis of the olfactory nerves shows that *Testudo horsfieldii* has more than six times more axons than *Chinemys reevesii* (Matsuzaki and Shibuya, 1978; Matsuzaki et al., 1980).

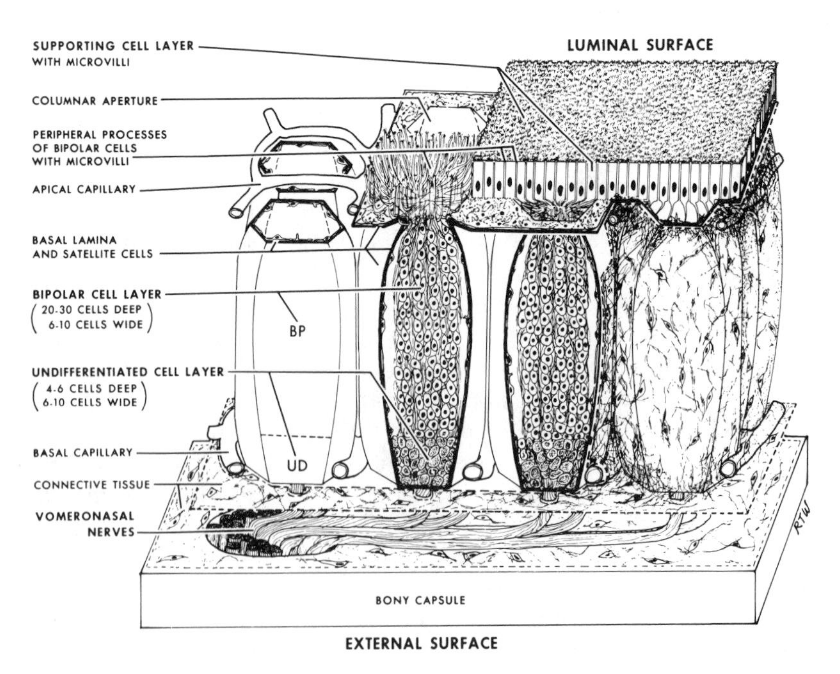

Fig. 7.7. *Thamnophis sirtalis.* Schematic three-dimensional diagram of the vomeronasal sensory epithelium as reconstructed from light transmission and scanning electron microscopy. (From Wang and Halpern, 1980b).

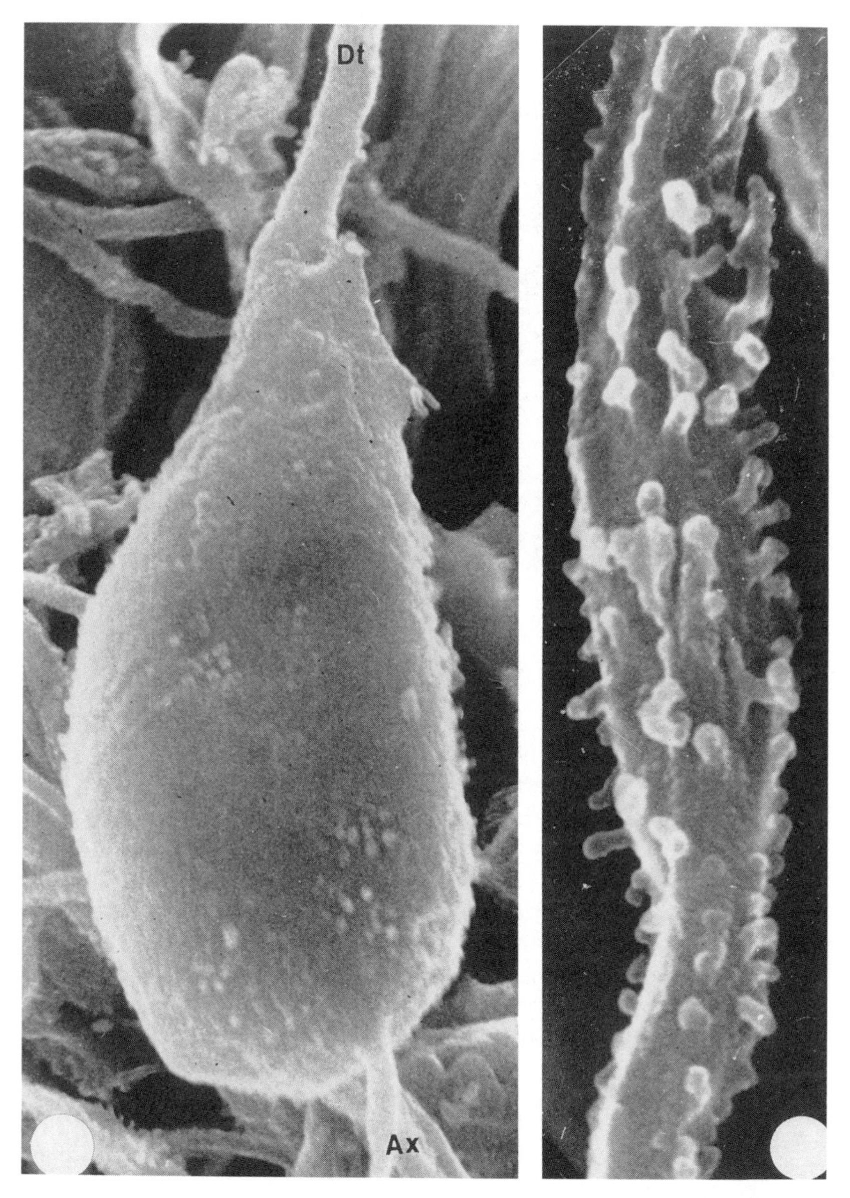

Fig. 7.8. *Thamnophis sirtalis.* Left: A vomeronasal bipolar neuron illustrating a dendrite (Dt) emanating from its apical pole and an axon (Ax) emanating from its basal pole. Right: A segment of a dendrite from a vomeronasal bipolar neuron illustrating the presence of fine dendritic spines on its surface. (From Wang and Halpern, 1980b.)

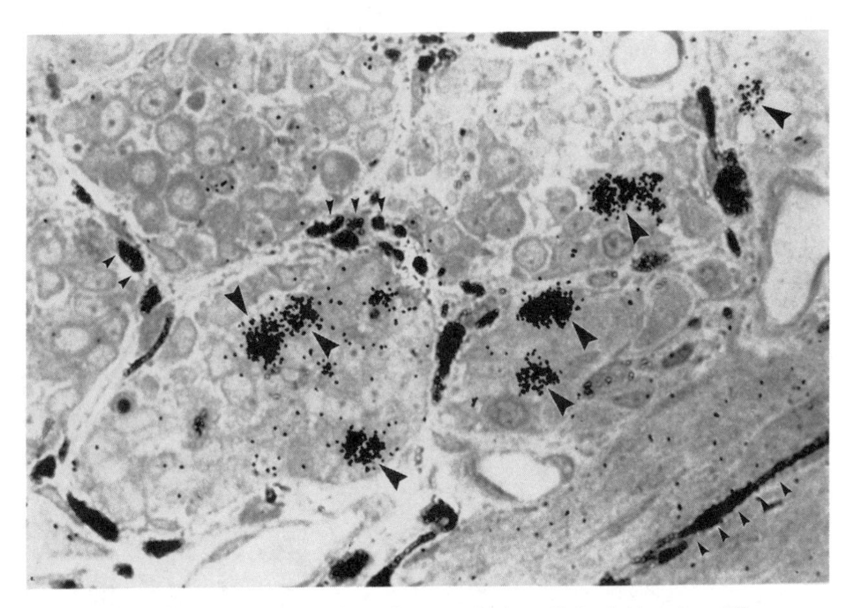

Fig. 7.9. *Thamnophis sirtalis.* ^{3}H-thymidine uptake by cells in the basal undifferentiated layer of the vomeronasal epithelium of a garter snake. Large arrowheads point to reduced silver grains overlying thymidine-labeled cells. Small arrowheads point to pigment. (From Wang and Halpern, 1988.)

E. Main and Accessory Olfactory Bulbs

The olfactory bulbs and the telencephalic structures to which they project have been described in a number of reptiles (Goldby and Gamble, 1957; Crosby and Humphrey, 1939). The size of the olfactory bulbs varies considerably (Crosby and Humphrey, 1939; Allison, 1953). Turtles and alligators have large olfactory bulbs, whereas those of lizards vary considerably in size; for example, those of *Anolis* are very small. In general, the size of the olfactory bulbs can be predicted from the level of development of the corresponding epithelia. The position of the bulbs also varies among reptiles. Bulbs may be separated from the rest of the telencephalon by a long stalk, as in alligators, or lie close to the cerebral hemispheres, as in snakes and turtles (Allison, 1953). However, the length of the olfactory stalk, and therefore the distance between the olfactory bulb and the rest of the telencephalic hemisphere, may increase in ontogeny as it does in *Caiman crocodilus* (Gans, 1980).

The main and accessory olfactory bulbs show parallel laminar patterns (Crosby, 1917; Crosby and Humphrey, 1939; García-Verdugo et al., 1986; Halpern, 1976, 1980b; Hatanaka et al., 1988; Iwahori et al., 1989a, b; Llahi and García-Verdugo, 1986, 1989a, b). The axons of bipolar neurons comprise the nerve layers and terminate in their respective glomerular layers. Glomeruli are composed of synaptic neuropil in which the axons of bipolar neurons synapse on the primary dendrites of mitral (and tufted) cells. Surrounding the glomeruli are the cell bodies of periglomerular cells with axons and dendrites that contribute to the formation of each glomerulus. An external granular layer may occur between the glomeruli and external plexiform layer. The mitral cell layer is not a monolayer, as is typically described in mammals, but is more diffusely organized. Internal to it is a broad internal plexiform layer. A relatively narrow granular layer lies adjacent to the olfactory ventricle, representing a rostral continuation of the lateral ventricle.

Differences and similarities in the olfactory bulbs of a number of reptiles have been described (Rudin, 1974). Among the reptilian orders, the architecture of the mitral cell and internal granular layers is relatively constant, but the other layers differ considerably. The differences among turtles, crocodiles, and squamates are easily observed; those between Sauria and Serpentes are less clear. An accessory olfactory bulb may be absent; its presence and size correlate with the presence and morphological development of a vomeronasal organ. Rudin reports that turtles and crocodiles lack an accessory olfactory bulb; however, recent studies dispute this for turtles (Rudin, 1974). The development of the accessory bulb is also correlated with the development of the nucleus sphericus, its major target area in the posterior telencephalon (Rudin, 1974). Among lizards, the variation that exists at the periphery is reflected in the development of olfactory structures in the telencephalon (Northcutt, 1978). Arboreal agamids, such as *Calotes* and *Draco,* show severe reduction of the nucleus sphericus and the lateral cortex, a principal target of the main olfactory bulb. In *Anolis,* parts of the lateral cortex are reduced to a nonneuronal ependymal layer. Agamids, chamaeleonids, and some iguanids have a very small nucleus sphericus, reduced to the point at which it can no longer be discerned. Ulinski (1983) describes a moderately well-developed nucleus sphericus in the desert iguana *Dipsosaurus dorsalis.* Teiids, helodermatids, and varanids have very well-developed accessory olfactory bulbs and a well-defined nucleus sphericus.

Whereas the major afferents to the main and accessory olfactory bulbs are the axons of the bipolar neurons of the olfactory and vomeronasal epithelia, cells of other telencephalic structures also project to the bulbs (see below). In *Gekko gecko,* both main and accessory olfactory bulbs receive afferents from the dorsal cortex (Hoogland and Vermeulen-Van der Zee, 1989), with the fibers traveling through the anterior olfactory nucleus and olfactory peduncle to reach their targets. As all other telencephalic connections of the main and accessory olfactory systems of reptiles remain segregated, it will be important to determine if this projection from the dorsal cortex to the bulbs originates from distinct cell groups.

F. Main Olfactory Bulb Projections

There are several relatively recent descriptions of the telencephalic hemispheres, including the olfactory receiving areas for *Chrysemys picta belli* (Northcutt, 1970), *Coluber c. constrictor* (Carey, 1967; Halpern, 1980b), *Boa c. constrictor, Elaphe obsoleta rossalleni, Nerodia sipedon, Pituophis catenifer, Thamnophis elegans, T. parietalis, T. radix, T. s. sirtalis, Typhlops punctata* (Halpern, 1980b), *Tupinambis teguixin* (Ebbeson and Voneida, 1969), *Iguana iguana* (Northcutt, 1967), and 17 lizard genera (Northcutt, 1978). Earlier descriptions of the telencephalon and olfactory areas of other reptiles have also been reviewed by Halpern (1980b).

The efferent connections of the main and accessory olfactory bulbs have been investigated in recent years with experimental degeneration techniques, anterograde or retrograde transport of enzymes such as horseradish peroxidase or lectins, or the anterograde transport of labeled proteins. The main projections of the olfactory bulb in *Testudo graeca* (Gamble, 1956), *Chrysemys picta belli* (Northcutt, 1970), *Trachemys scripta elegans*, and *Graptemys pseudogeographica* (Balaban, 1977) were observed using degeneration techniques and described as traveling in medial, intermediate, and lateral olfactory tracts. The projections in the medial and intermediate tracts are poorly developed and restricted to the anterior pole of the telencephalon. The lateral olfactory tract distributes olfactory bulb efferents to the anterior olfactory nucleus, olfactory tubercle, lateral (pyriform) cortex, and amygdaloid nuclei. The fibers in the lateral olfactory tract continue caudally to join the stria medullaris, cross in the habenular commissure, and distribute to contralateral pyriform cortex, amygdaloid nuclei, and paraolfactory areas.

The olfactory bulb projections of *Trionyx s. spiniferus* (Rolon and Skeen, 1980; Skeen and Rolon, 1982; Skeen et al., 1984) and *Trachemys scripta* (Reiner and Karten, 1985) have been examined using axonal transport techniques. These studies basically confirm the major findings previously reported on the basis of degeneration techniques.

The olfactory bulb projections of *Caiman crocodilus* have been investigated by the use of experimental degeneration techniques (Scalia et al., 1969). As in turtles, the olfactory bulbs project to the ipsilateral anterior olfactory nucleus, lateral pallium, olfactory tubercle, and part of the amygdaloid complex. Axons destined for the contralateral telencephalon enter the stria medullaris and cross in the habenular commissure.

The pattern of olfactory projections in *Tupinambis teguixin* (Heimer,

1969) and *Dipsosaurus dorsalis* (Ulinski and Peterson, 1981) has been investigated by experimentally induced degeneration. The main olfactory bulb projects to the ipsilateral rostral continuation of the medial cortex and septum, olfactory tubercle, and rostral lateral cortex. Intermediate and lateral olfactory tracts merge to form the stria medullaris, the axons of which cross in the habenular commissure and terminate in the contralateral lateral cortex, olfactory tubercle, retrobulbar formation, and septum. Similar results have been reported using axoplasmic transport techniques in *Podarcis hispanica* (Amiguet et al., 1986; Martinez-García et al., 1985) and *Gekko gecko* (Lohman et al., 1988; Hoogland and Vermeulen-Van der Zee, 1988).

The olfactory projections of *Thamnophis sirtalis, T. radix* (Halpern, 1976), and *Lampropeltis getulus floridana* (Kowell, 1974) have been examined by experimentally induced degeneration. The main olfactory bulb of these snakes projects to the ipsilateral anterior olfactory nucleus, olfactory tubercle, and lateral pallium. Axons destined for the contralateral hemisphere enter the stria medullaris thalami, cross in the habenular commissure, and terminate in the contralateral lateral pallium.

In all reptiles studied, the projections from the olfactory bulb to the lateral cortex terminate in the outermost portion of the molecular layer.

G. Projections of the Accessory Olfactory Bulb

The projections of the accessory olfactory bulb have been described for *Tupinambis teguixin* (Heimer, 1969), *Gekko gecko* (Lohman et al., 1988), *Lampropeltis getulus floridana* (Kowell, 1974), *Thamnophis radix*, and *T. sirtalis* (Halpern, 1976). The mitral cells of the accessory olfactory bulb always send their axons through the accessory olfactory tract to terminate in the core of the ipsilateral nucleus sphericus. Golgi and electron microscopic studies (Heimer, 1969; Ulinski and Kanarek, 1973) indicate that these axons terminate on the dendrites of cells of the mural layer of the nucleus sphericus. Projections from the accessory olfactory bulb do not cross the midline, and the areas of termination for the main and accessory olfactory bulb efferents do not appear to overlap in reptiles.

H. Lateral Pallium (Cortical) Connections

The architecture of the lateral cortex of the snakes *Boa constrictor, Nerodia sipedon,* and *Thamnophis sirtalis* has been described in detail (Ulinski and Rainey, 1980). Most afferents to the rostral lateral pallium derive from the projection from the main olfactory bulb. As described above, this projection is bilateral. Cells of the lateral pallium of *Thamnophis radix* send axons to the main olfactory bulb that terminate in the internal granular layer and adjacent portions of the internal plexiform layer (Halpern, 1980b).

The tertiary connections of the olfactory system have been studied most extensively in lizards and snakes. There is general agreement that the lateral cortex does not project out of the telencephalon. In *Tupinambis teguixin* (Lohman and Mentink, 1972; Lohman and Van Woerden-Verkley, 1976), *Podarcis hispanica* (Martinez-García et al., 1986a; Martinez-García and Olucha, 1988), *Gekko gecko* (Hoogland and Stoll, 1984), *Gallotia stehlinii* (Martinez-García and Olucha, 1988), *Nerodia sipedon* and *Thamnophis sirtalis* (Ulinski, 1976), and *T. radix* (Halpern, 1980b), the lateral cortex projects to the small celled part of the medial (or dorsomedial cortex). An exception to this finding is the report of Belekhova and Kenigfest (1983), who did not note retrogradely labeled cells in the lateral cortex of *Ophisaurus apodus* following injections of horseradish peroxidase into the mediodorsal cortex. *Podarcis hispanica* has an efferent projection to the contralateral telencephalic hemisphere (Martinez-García et al., 1986a, b). Some additional differences do exist in reports of the efferent projections. For example, whereas the reports for snakes limit the projections of the lateral cortex to the outer one-third of the molecular layer of the medial cortex (Ulinski, 1976), those for *Gekko* also note terminations on the distal parts of the basal dendritic tree (Hoogland and Stoll, 1984; Hoogland and Vermeulen–Van der Zee, 1988; Hoogland et al., 1985). These reports could reflect interspecific differences or those of technique.

The lateral cortex of *Gekko gecko* and *Iguana iguana* has been reported to receive a projection from the nucleus dorsalis anterior pars magnocellularis of the thalamus (Bruce and Butler, 1984). No similar observations based on experimental studies have been reported for other reptiles, so the generality of this finding remains to be demonstrated.

I. Connections of Nucleus Sphericus

In addition to the projection of the accessory olfactory bulb to the nucleus sphericus or to its homologue in all reptiles studied, additional projections to the nucleus have been described. In *Podarcis hispanica* and *Gallotia stehlinii,* the dorsal cortex projects to the nucleus sphericus and to the accessory olfactory bulb (Olucha et al., 1986). In *Tupinambis teguixin,* the anterior dorsal ventricular ridge projects to the nucleus sphericus (Voneida and Sligar, 1979). Similar studies are needed in snakes to test for the generality of these findings among squamates.

Anterograde transport of labeled proteins has been used to describe the efferent projections of nucleus sphericus of *Tupinambis te-*

guixin (Voneida and Sligar, 1979). Projections pass to the nucleus accumbens, the lateral amygdaloid nucleus, the nucleus of the lateral olfactory tract, the olfactory tubercle, anterior olfactory nucleus, granular layer of the olfactory bulb, and to the shell surrounding the ventromedial nucleus of the hypothalamus. In *Thamnophis radix* and *T. sirtalis* degeneration techniques document the efferent projections of the nucleus sphericus rostrally to the ipsilateral accessory olfactory bulb, to terminate in the internal granular layer and adjacent internal plexiform layer; the projections cross in the anterior commissure and terminate contralaterally in the marginal and mural layers of the nucleus sphericus. Caudally, the projections pass through the bed nucleus of the stria terminalis, preoptic area, and hypothalamus to terminate in the ventromedial nucleus of the hypothalamus. The described discrepancies in results regarding tertiary projections of the accessory olfactory system will require future clarification to determine if they reflect interspecific differences, differences in the techniques used, or interpretation of the observations.

J. Chemoarchitecture

The chemoarchitecture of the olfactory and vomeronasal systems in reptiles is still poorly understood. Few reptiles have been studied, and at present it seems best to list the findings as they appear in the literature. The monoaminergic systems in reptiles have been summarized (Parent, 1979; Smeets, 1988b).

Dopaminergic cells occur in the olfactory bulb of *Trachemys scripta* (Halász et al., 1982a; Halász et al., 1982b; Smeets and Steinbusch, 1990); the cells lie in the periglomerular region, forming a dense network of fluorescent fibers. Immunofluorescent cells also occur in the external plexiform layer and in the outer part of the granule layer. Their occurrence has been confirmed using tritiated dopamine and L-dopa uptake (Halász et al., 1983). In the brain of *T. scripta elegans*, dopamine occurs in fibers and terminals of the olfactory bulb, the olfactory tubercle, anterior olfactory nucleus, and lateral cortex (Smeets et al., 1987).

Gekko gecko has dopamine immunoreactive cells in the glomerular layer of the olfactory bulb (Smeets and Steinbusch, 1990). A few dopaminergic cells also occur in the external plexiform layer. Immunoreactive fibers lie in the internal granular layer of the olfactory bulb, internal plexiform layer, olfactory tubercle, and amygdaloid complex, including nucleus sphericus (Smeets et al., 1986). Noradrenergic fibers occur in the internal granular and plexiform layers of the olfactory bulb and, to a lesser extent, in the mitral cell layer and surrounding glomeruli. Noradrenergic fibers also occur in the olfactory peduncle and lateral cortex (Smeets and Steinbusch, 1989).

In *Python regius,* dopamine-containing cell bodies occur around the glomeruli (probably as periglomerular cells) and in the external plexiform layer for both main and accessory olfactory bulbs. In addition, dopaminergic axons occur in the olfactory tubercle, lateral cortex, nucleus of the accessory olfactory tract, and nucleus sphericus (Smeets, 1988a).

The olfactory bulbs of *Thamnophis sirtalis* have been reported to contain the highest concentrations of extrahypothalamic thyrotropin-releasing hormone in the brain (Jackson and Reichlin, 1974). It is unclear whether the main and accessory olfactory bulbs differ in the distribution of thyrotropin-releasing hormone.

Muscarinic binding sites and benzodiazepine binding sites have been described in the anterior olfactory nucleus of *Trachemys scripta.* The lateral cortex also contains benzodiazepine binding sites (Schlegel and Kriegstein, 1987). Immunoreactive enkephalin cell bodies, fibers, and terminals have been reported in the nucleus of the lateral olfactory tract, bed nucleus of the stria terminalis, and anterior amygdaloid area of *Anolis carolinensis* (Naik et al., 1981); however, the role of these areas in chemoreception remains unknown.

The Timm method has allowed description of zinc concentrations in the brains of *Gallotia galloti* and *Podarcis hispanica,* specifically in their lateral cortex, nucleus sphericus, and the bed nucleus of the accessory olfactory tract (Molowny et al., 1987; Perez-Clausell, 1988; Perez-Clausell and Fredens, 1988).

Estradiol and testosterone concentrating neurons are preferentially found in the telencephalic structures of *Thamnophis,* such as the nucleus sphericus, that receive accessory olfactory efferents (Halpern et al., 1982). Numerous such neurons are also found in structures to which the nucleus sphericus projects; examples are the preoptic area and the ventromedial nucleus of the hypothalamus. Estradiol concentrating neurons are found in the nucleus ventromedialis of the telencephalon (a presumed homologue of the mammalian amygdala), the nucleus interstitialis striae terminalis, the preoptic area, and the ventromedial hypothalamus of *Anolis carolinensis* (Martinez-Vargas et al., 1978). These results were replicated by Morrell et al. (1979), who also report that these structures concentrate dihydrotestosterone and testosterone.

The significance of the variety of neurotransmitters, hormone-concentrating neurons, and other chemicals found in the reptilian brain is still unknown. Some information concerning the roles of neurotransmitters in the olfactory bulbs has been obtained using electrophysiological techniques (see below). As many of the hormone-concentrating regions are involved in chemosensitive pathways, as well as in reproductive behaviors, one may assume that these represent the morphological basis for hormonal control of reproduction in the reptiles studied.

K. Development

A comprehensive survey and review of the adult anatomy of the reptilian nose is presented by Parsons (1959b). This monograph also treats the embryology and offers a systematic discussion of the development of the peripheral portions of the nasal chemical senses.

A recent series of papers by Slabý (1979a, b, c, 1981, 1982, 1984) describes the embryonic development of the nasal cavities in *Archaeolacerta saxicola, Chalcides chalcides, Gekko gecko,* and members of the Varanidae, Agamidae, and Chamaeleonidae. As one would expect, the relative development of the main and accessory olfactory systems is reflected in their embryological development. For example, the olfactory and vomeronasal epithelia develop early in the Varanidae and can be identified in the 15-mm embryo (Slabý, 1979c). In Agamidae, the vomeronasal organ becomes relatively large during development. The main olfactory region is reduced in adults but passes through a fully developed stage during early embryological development (Slabý, 1981). In contrast, according to Slabý (1984), the vomeronasal organ of Chamaeleonidae, which is absent in adults, never develops embryologically. However, Haas (1947) reported that *Chamaeleo chamaeleon* has a vomeronasal organ in embryos as well as in neonates.

In the development of telencephalic olfactory receiving areas of *Podarcis sicula,* the main and accessory olfactory bulbs differentiate in parallel (Rudin, 1975). The olfactory bulbs grow out between 5 and 6 days of incubation. One to 2 days later, the accessory olfactory bulb appears as a distinct structure and then laminates. At about the 10th day of incubation, an inner plexiform layer separates the mitral cell layer from the inner cell layers, and by the 12th day of incubation an inner granular layer and ependymal layer can be distinguished. Glomeruli are recognizable by the 15th day of incubation, at which time, according to Rudin, the olfactory fila may be seen to contact the dendritic tips of mitral cells. The anterior olfactory nucleus and the olfactory tubercles begin to form on the 5th day of incubation. By the 10th day of incubation, the nucleus of the lateral olfactory tract and nucleus sphericus can be distinguished, although they do not develop their adult, laminated appearance until after the 16th day of incubation. The lateral pallium becomes distinguishable from the rest of the pallial mantle after the 13th day of incubation.

The olfactory and vomeronasal systems of *Thamnophis* (Figs. 7.10 and 7.11) are the subject of current investigation (Holtzman and Halpern, 1987, 1989, 1990). An in vitro technique has facilitated the selection of appropriate embryonic stages for study and experimental manipulation, permitting stimulation with odorants and injections with tritiated thymidine (Holtzman and Halpern, 1989). As one would expect, the thickness of the olfactory and vomeronasal epithelia changes as a function of embryonic age. The olfactory epithelium increases in thickness from embryonic stages 23 to 28, remains the same

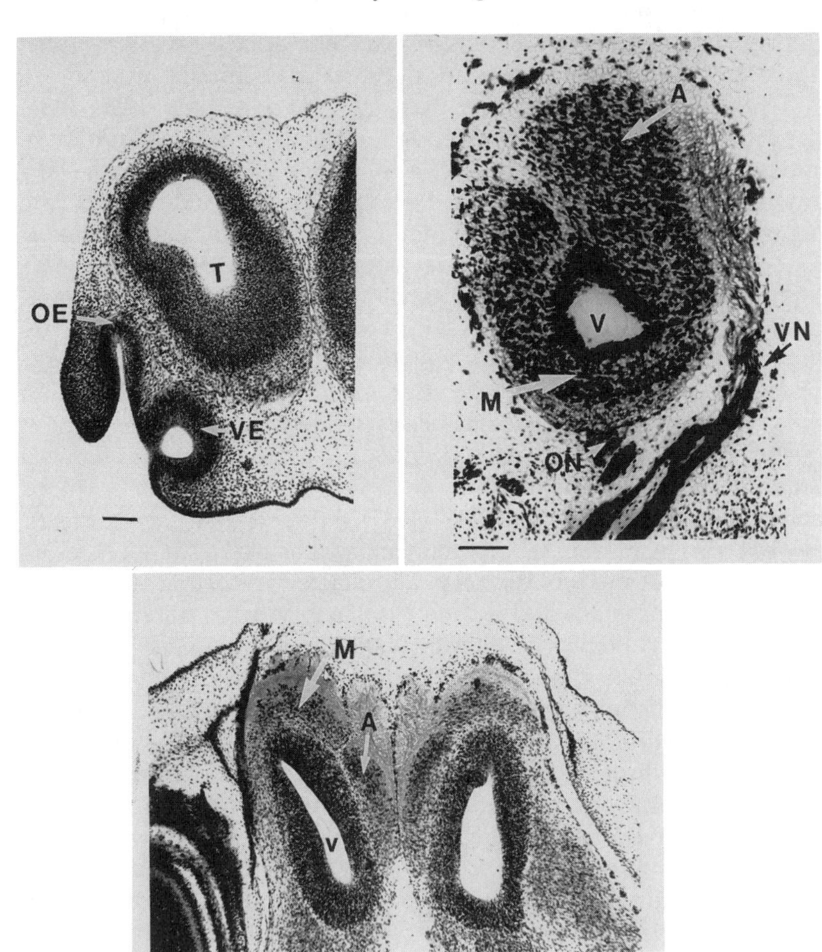

Fig. 7.10. *Thamnophis* sp. Top left: Zehr Stage 23 embryo head showing the relationship between the developing vomeronasal epithelium (VE), olfactory epithelium (OE), and anterior telencephalon. Coronal section; hematoxylin-eosin. Bar = 100 μm. Top right: Presumptive accessory (A) and main (M) olfactory bulbs in a Zehr Stage 28 embryo. Both the vomeronasal (VN) and olfactory (ON) nerves can be seen entering the dorsomedial accessory olfactory bulb and ventral main olfactory bulb, respectively. V, Olfactory ventricle; coronal section; Bodian's method. Bar = 50 μm. Bottom: Accessory olfactory bulb (A) and main olfactory bulb (M) of a Zehr Stage 32 embryo. v, Olfactory ventricle. Horizontal section; Bodian's method. Bar = 100 μm. (From Holtzman and Halpern, 1990.)

thickness to stage 32, decreases in thickness from stages 32 to 36, and then does not undergo a change in thickness through the 12 days following birth. The vomeronasal epithelium increases in thickness throughout embryonic development. Interestingly, the thickness of the supporting cell layer of the vomeronasal epithelium decreases as

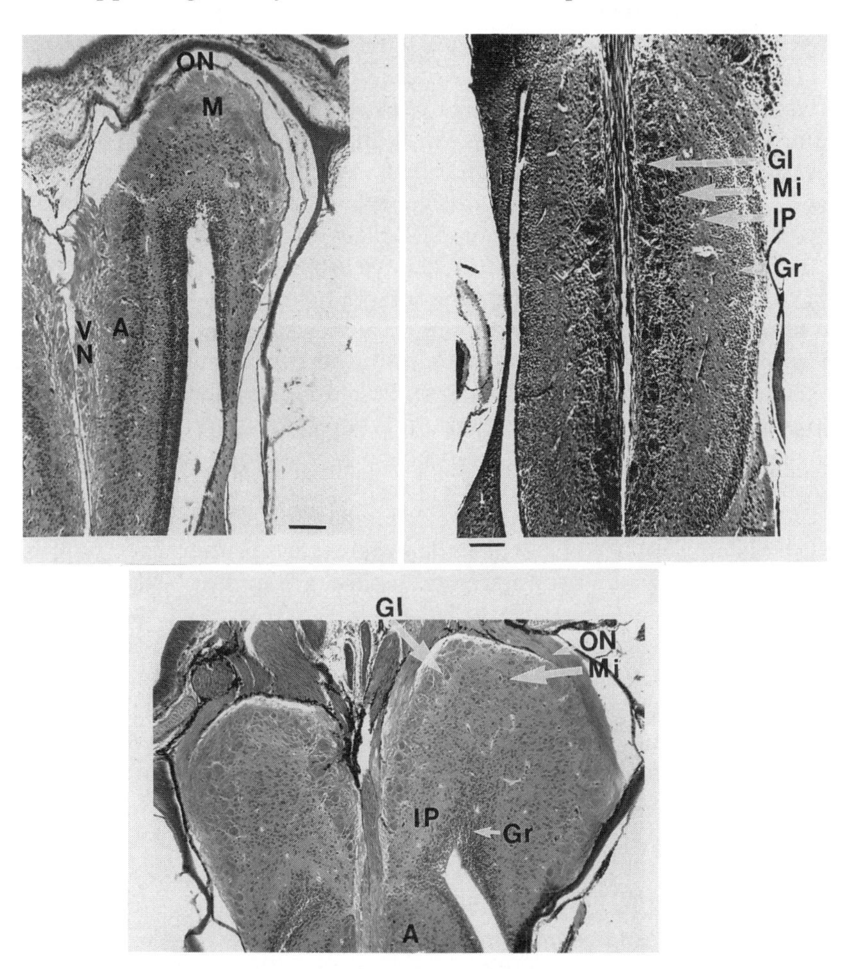

Fig. 7.11. *Thamnophis* sp. Top left: Accessory olfactory bulb (A) and main olfactory bulb (M) of a Zehr Stage 36 embryo. Olfactory nerve (ON) is seen entering main olfactory bulb rostrally and vomeronasal nerve (VN) entering the accessory bulb rostrally and vomeronasal nerve (VN) entering the accessory bulb medially. Horizontal section; hematoxylin-eosin. Bar = 100 μm. Top right: Accessory olfactory bulb of a 0- to 2-day-old neonate. Bar = 100 μm. Bottom: Main olfactory bulb of a 0- to 2-day-old neonate. Bar = 100 μm. A, Accessory olfactory bulb; Gl, glomerular layer; Gr, granular cell layer; IP, internal plexiform layer; Mi, mitral cell layer; ON, olfactory nerve. (From Holtzman and Halpern, 1990.)

that of the bipolar undifferentiated cell layer increases. The olfactory bipolar axons arrive at the main olfactory bulb before the vomeronasal axons arrive at the accessory olfactory bulb. However, as in *Podarcis sicula,* the main and accessory bulbs and the tertiary structures of the two systems develop in parallel. The nucleus sphericus and the lateral cortex first show their adult structure at Zehr stage 36 (just before birth).

Postnatal neurogenesis has been reported in the olfactory bulbs of *Podarcis hispanica* (Garcia-Verdugo et al., 1989). Use of tritiated thymidine autoradiography displays labeled neurons in the main and accessory olfactory bulbs of lizards that were injected during the perinatal period (24 to 25 mm snout-vent length), as juveniles (30 to 35 mm), and as adults (40 to 50 mm). The labeling occurs in the granular layer of lizards killed 7, 18, and 28 days after injection. The main olfactory bulb contains a spatiotemporal gradient with labeled cells progressing further from the ventricular layer as survival time increases. However, no such gradient occurs in the accessory bulb. The cells in the mitral cell layer of the accessory bulb are only labeled in animals injected during the perinatal period, whereas those of the mitral cell layer of the main olfactory bulb become labeled in lizards injected during the perinatal period and as juveniles.

A thorough analysis of the development of the nasal chemical sense systems in reptiles is particularly important, as members of this group are known to use their vomeronasal systems at birth (see discussion below on response to prey odors). At present we lack an understanding of when these systems become functional during ontogeny. There are, as well, few systematic studies using similar methods in several genera or different methods of developmental analysis in a single genus or species. Completion of such studies should provide a more comprehensive understanding of the developmental timetables of the nasal chemical senses in reptiles.

IV. ELECTROPHYSIOLOGY

A. Main Olfactory System

Electrophysiological analysis of chemical senses is valuable in determining the course of information processing throughout these systems. Such studies can also be used as tools for evaluating the excitatory or inhibitory actions of chemical stimuli and their interactions with each other.

In general, electrophysiological techniques for an examination of the nasal chemical senses of reptiles have lagged far behind behavioral and anatomical approaches. Most electrophysiological studies on the olfactory and accessory olfactory systems of reptiles used turtles. An excellent review summarizes this literature and offers an introduction to the interpretation of these studies (Scott, 1979).

Several types of electrical recordings may be made in the olfactory system. In the periphery, summed unit activity may be recorded as an electro-olfactogram (EOG) from the surface of the epithelium, a DC shift may be recorded from just below the epithelial surface, and single units may be recorded extracellularly or intracellularly. Intracellular recordings are rarely attempted, as the cells are usually considered too small to be penetrated without irreversible injury. Between the epithelium and the olfactory bulbs, recordings may be made of the whole nerve, parts of the nerve, or single axons. Such recordings are all extracellular; because the unmyelinated axons of the olfactory and vomeronasal nerves have very small diameters, they are not normally impaled individually. Within the telencephalon, summed activity—the so-called slow potential—can be recorded either from the surface or from deep in the olfactory bulb. Single unit activity can be recorded extracellularly or intracellularly in any of the structures that comprise the chemical sense systems.

The EOG in response to amyl acetate has been recorded in *Gopherus polyphemus* (Shibuya and Shibuya, 1963). Removal of the mucus from the surface of the epithelium decreases the amplitude of the EOG but not the neural response recorded from the olfactory nerve, suggesting that the EOG is not the generator potential (Shibuya, 1964). Although other perireceptor events such as mucus secretion could contribute to the EOG, it is generally accepted that the generator potential is the principal component contributing to the EOG (Getchell and Getchell, 1987). Single unit spikes in the olfactory epithelium have been recorded in *Gopherus polyphemus* and *Chinemys reevesii* (Shibuya and Shibuya, 1963; Shibuya, 1969). The number of spikes in a response train, the frequency of spikes, and the height of the spikes varies with odor (amyl acetate) concentration (Shibuya and Shibuya, 1963). A DC shift occurs near the mucosal surface and may be evoked by chemical or electrical stimulation. The amplitude and duration of the DC shift increases with increasing stimulus concentration (Shibuya, 1969).

Single unit responses in the olfactory epithelium and olfactory bulb of *Gopherus polyphemus* were recorded to odor stimulation (Mathews and Tucker, 1966; Mathews, 1972). All responses of the olfactory mucosa are facilitatory. Very low spontaneous firing rates have been reported: 0 to 1 spike per minute, with maximum stimulated firing rates of 10 spikes per second (Mathews and Tucker, 1966) and 16 spikes per second (Mathews, 1972). Nineteen of 40 sampled mucosal cells respond to at least one odorant, with cells being sensitive to 1 to 15 odorants. All olfactory bulb units sampled respond to at least one odorant. Bulbar units respond to odorants either by increased firing rates, or their firing rates increase when the units are stimulated by some odorants and decrease when they are stimulated by others.

Olfactory nerve twig responses occur to aliphatic *n*-fatty acids and aliphatic *n*-acetates in *Terrapene carolina* (Tonosaki and Tucker, 1982). Response magnitude for aliphatic *n*-acetates is greatest at C_4 and decreases with increasing carbon chain length. For aliphatic *n*-fatty acids, response magnitude increases up to C_7 and then decreases with increasing carbon chain length. The responses to *n*-fatty acids are always smaller than the responses to *n*-pentyl acetate (the standard stimulus).

Electrophysiological studies of the olfactory bulbs of turtles have provided a detailed understanding of their synaptic organization (Beuerman, 1975, 1977; Greer et al., 1981; Jahr and Nicoll, 1980; Mathews, 1972; Mori and Shepherd, 1979; Mori et al., 1981a, b, c, d, e, 1984; Nowycky et al., 1978, 1981a, b, 1983; Orrego, 1961a, b, 1962; Orrego and Lisenby, 1962; Shibuya et al., 1977, 1978; Waldow et al., 1981). Electrophysiological studies of turtles demonstrate that the upper parts of the olfactory epithelium project to the dorsal parts of the olfactory bulb, whereas the lower parts of the epithelium project to the ventral aspects of the bulb. Surface negative responses are evoked in the dorsal olfactory bulb by stimulation of the dorsal aspect of the olfactory nerve; surface positive responses are evoked in the ventral olfactory bulb by stimulation of the dorsal aspect of the nerve. The opposite responses are evoked by stimulation of the ventral aspect of the olfactory nerve; this supports the finding of topographic projections (Orrego, 1961a). Synaptic conduction through the glomerular layer is slow, with a delay of 8 to 10 msec. A three- to four-component, long-lasting response is evoked in the olfactory bulb following electrical stimulation of the olfactory nerve. The response consists initially of (1) a short latency, sharp diphasic potential lasting about 10 msec followed by (2) a low-amplitude surface negative wave and followed again by (3) a higher-voltage wave of 50 to 60 msec duration. A fourth component of variable amplitude is variably present. Component 1 appears to represent integrated spike activity from the incoming nerve, as it can be recorded best from the anterior bulb and its latency is directly proportional to the distance from the bulb of the stimulating electrode. Component 2 appears to arise from interglomerular presynaptic fibers, and component 3 is thought to reflect mitral cell activity (Orrego, 1962; Scott, 1979).

Slow potentials have been recorded in *Terrapene carolina* and *Gopherus polyphemus* (Beuerman, 1975, 1977) in response to odor stimuli directed into the nose. Three waves appear in the olfactory bulb: (1) a short-duration wave with a latency of 150 to 200 msec, which is positive deep in the bulb and negative at the surface; (2) a longer-duration wave, which is also positive deep and negative at the surface; and (3) an oscillating potential superimposed on wave 2. All three potentials reach their greatest magnitude in the deeper layers of the bulb. Olfactory nerve section blocks bulbar response to odor stimulation. Partial nerve section confirms the topographic projection from mucosa to bulb: lateral aspects of the nerve innervate lateral and ventrolateral aspects of the bulb and ventromedial nerve fibers innervate ventromedial aspects of the bulb. Amyl acetate is the most effective stimulus and geraniol is the least effective.

Single unit activity in response to odorants delivered to the nasal cavity has been recorded intracellularly from mitral cells of *Testudo horsfieldii, Gekko gecko,* and *Caiman crocodilus* (Shibuya et al., 1977, 1978). Two types of responses occur: excitation (increases in firing with increased concentration of odorant to some asymptote) and suppression (decreased responding of spontaneously active neurons during odor stimulation). Some neurons demonstrate excitatory response to some odorants and suppression to others. Some topographic correlations exist with respect to the position of mitral cells excited or suppressed by various odorants. Mitral cells excited by amyl acetate tend to lie in central areas of the bulb, whereas those suppressed by amyl acetate tend to lie at the fringes of the bulb. Excitation by d-limonene is more likely to occur anteriorly in the bulb, whereas suppression by d-limonene mainly occurs posteriorly.

The brain of *Trachemys scripta* has been particularly useful in elucidating neuronal circuitry and synaptic mechanisms in the olfactory bulb. The ability to study the electrical activity of the isolated olfactory bulb in vitro has permitted manipulations of the system that would be very difficult in vivo. Experimental studies indicate that the in vitro preparation maintains the properties of the in vivo preparation (Waldow et al., 1981). Mitral cells respond to olfactory nerve stimulation with a spike (1), followed by small afterhyperpolarization (2), followed by gradual depolarization (3), followed by slow, long-lasting hyperpolarization. The initial spike represents the excitatory postsynaptic potential generated by the axon terminal of the olfactory bipolar cell synapsing on the mitral cell dendrite. The early inhibitory (afterhyperpolarization) potential is thought to be mediated by the dendrodendritic pathway from granule cell to mitral cell, and the later potentials are thought to arise from intrinsic bulbar circuits (Mori and Shepherd, 1979; Mori et al., 1981b). Inhibitory postsynaptic potentials can be elicited by antidromic activation of mitral cells (Mori et al., 1981d; Nowycky et al., 1981a). This inhibition may be blocked with bicuculline, suggesting that at least some of the inhibition is GA-

BAergic and represents inhibition of mitral cells by dendrodendritic synapses of granule cells (Nowycky et al., 1981a).

The effects of some drugs on odor-evoked responses of mitral cells in the olfactory bulb of *Gekko gecko* have been recorded with extracellularly placed microelectrodes during stimulation with *n*-amyl acetate (Tonosaki and Shibuya, 1985). Firing rate of mitral cells in response to odor stimulation increases, decreases, or does not change. Norepinephrine applied directly to the glomerular layer or iontophoretically to the external plexiform layer decreases spontaneous activity and the odor-evoked response. γ-Aminobutyric acid (GABA) applied to the external plexiform layer decreases spontaneous activity and the excitatory odor-evoked response. Carnosine has a complex effect on both spontaneous and odor-evoked responses.

B. Vomeronasal System

As with the olfactory system, most studies using electrophysiological recording from vomeronasal system structures derive from turtles. This is particularly surprising, because of the earlier controversy about the existence of a vomeronasal system in turtles.

Early electrophysiological studies of the reptilian vomeronasal system used "typical" olfactory stimuli to test the responsiveness of the system. In *Gopherus polyphemus,* recording from the vomeronasal nerve yielded response to lower members of a homologous series of alcohols and fatty acids (Tucker, 1963). Similar results were obtained in *Lacerta agilis* and *Podarcis muralis* (Müller, 1971). However, the EOG recorded from the olfactory mucosa of lizards was considerably higher than the EOG elicited by the same substances from the vomeronasal epithelium.

The electrical activity of the dorsal (olfactory) and ventral (vomeronasal) nasal sensory epithelia has been compared in *Testudo graeca* (Kruzhalov and Boĭko, 1987). A series of laboratory chemicals, as well as conspecific chemicals, has been applied to the epithelia. For all stimuli the dorsal epithelium is more sensitive than the ventral one. The response of both male and female subjects is higher to female cloacal odor than to male odor.

The electrical response of the olfactory and vomeronasal epithelia of *Thamnophis sirtalis parietalis* for odors of amyl acetate, butanol, and prey (earthworm) wash has been compared (Inouchi and Halpern, 1988). The response magnitude recorded from the olfactory epithelium increases exponentially with logarithmic increase in stimulus concentration. Amyl acetate and butanol produce a response in the vomeronasal epithelium that also increases with increasing concentration of the stimulus, but earthworm vapor does not produce a detectable response. Response amplitudes recorded from the vomeronasal epithelium are 10 times lower than EOGs recorded from the olfactory epithelium, and thresholds are 10 times higher in the vomeronasal system than in the olfactory system.

Electrical activity in the form of an induced wave can be recorded from the surface of the accessory olfactory bulb in response to airborne or liquid delivery of chemical stimuli (Fig. 7.12) to the ventral epithelium of *Chinemys reevesii* (Fig. 7.13), *Trachemys scripta* (Fig. 7.14), and *Mauremys japonica* (Hatanaka and Hanada, 1986; 1987; Hatanaka et al., 1988). Several types of responses have been reported: on excitation, off excitation, long-lasting excitation, as well as no response to some odorants. A variety of stimuli are effective excitants; these include standard odorants, salts, acids, and amino acids. Single units in the accessory olfactory bulb of *Chinemys reevesii* respond to vapor and liquid stimuli applied to the ventral nasal epithelium (Hatanaka and Shibuya, 1989). Both excitatory and inhibitory responses occur but have differing time courses. Complex responses occur, such as excitation followed by suppression, and last as long as 10 seconds. Electrical stimulation of the vomeronasal nerve produces similar re-

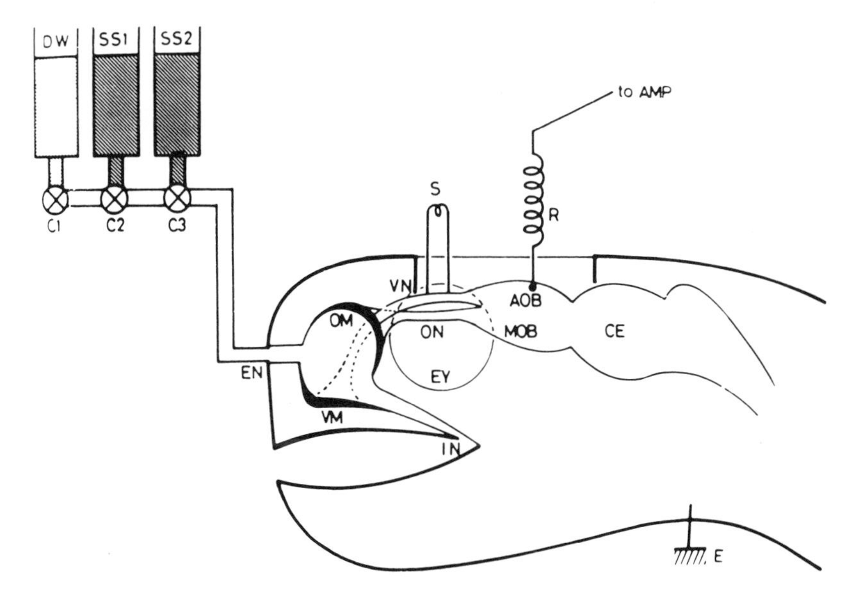

Fig. 7.12. Experimental arrangement for recording of induced wave responses of turtle accessory olfactory bulb and applying stimulants to the vomeronasal mucosa. AOB, Accessory olfactory bulb; C, three-way valve; CE, cerebrum; DW, distilled water; E, earth; EN, external naris; EY, eye; IN, internal naris; MOB, main olfactory bulb; OM, olfactory mucosa; ON, olfactory nerve; R, recording electrode; S, stimulating electrode; SS, stimulating solution; VM, vomeronasal mucosa; VN, vomeronasal nerve. (From Hatanaka et al., 1988, © 1988, Pergamon Press, plc.)

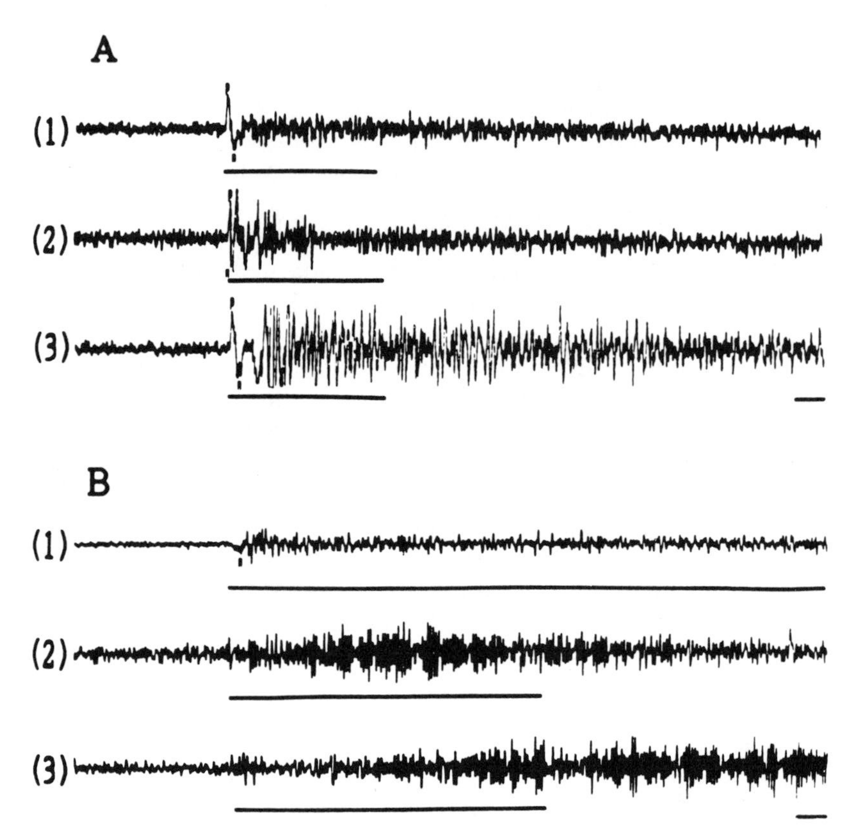

Fig. 7.13. *Chinemys reevesii.* Induced wave responses of the accessory olfactory bulb. (A) Responses to vapor stimuli of geraniol (1) and formic acid (2) and (3). (B) Responses to distilled water (1) and aqueous stimuli of 10^{-1}M ammonium chloride (2) and 10^{-1}M hydrochloric acid (3). The horizontal lines below each record show the stimulation. Calibration: 1 sec. (From Hatanaka et al., 1988, © 1988, Pergamon Press plc.)

sponses. Effective stimuli include standard odorants, such as amyl acetate and geraniole, salts, acids, and amino acids.

Odorants delivered to the vomeronasal organ of *Thamnophis radix* (Meredith and Burghardt, 1978) and *T. sirtalis parietalis* (Inouchi et al., 1988, 1989) alter unit activity in the accessory olfactory bulb. Prey washes (Fig. 7.15), amino acids, and standard odorants are effective stimuli. The response may involve an increase or a decrease in unit firing, and the changes in firing rate are concentration-dependent. Interestingly, the vomeronasal epithelium shows no EOG changes in response to the airborne vapor of prey washes, whereas liquid delivery of the same prey washes to the vomeronasal epithelium is an effective stimulus for altering activity in the accessory olfactory bulb (Inouchi et al., 1988, 1989).

Reptiles have proved to be especially valuable in the analysis of chemical sense systems, particularly in differentiating between functional domains of the vomeronasal and olfactory systems. Turtles and snakes are the only animals in which natural stimuli have been successfully used to activate central targets of the vomeronasal system. These recent studies, however, need careful follow-up. For example, species of turtles differ with respect to their normal environment, some being aquatic, others terrestrial. It is important to determine whether these differences are reflected in the responsiveness of the testudinian vomeronasal and olfactory systems to odorants delivered in airstreams or as liquids. Issues that must be clarified include the importance of the medium of odor delivery to driving the olfactory

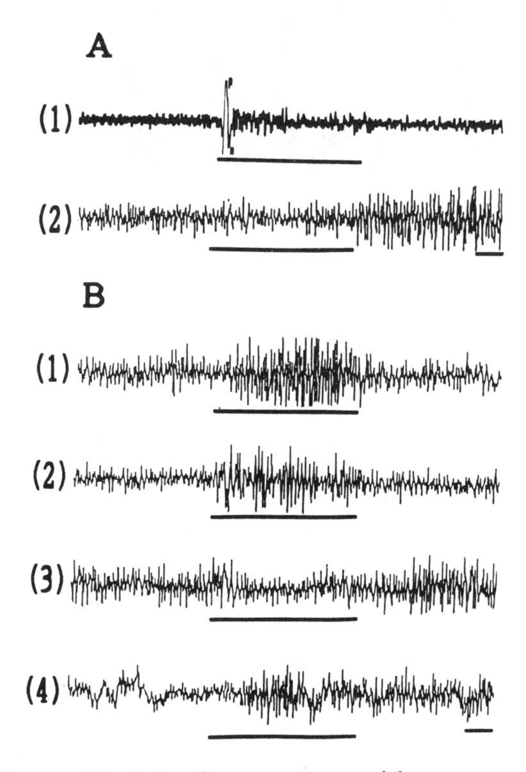

Fig. 7.14. *Trachemys scripta*. Induced wave responses of the accessory olfactory bulb. (A) Responses to vapor stimuli of caproic acid (1) and formic acid (2). (B) Responses to aqueous stimuli of 10^{-1}M potassium chloride (1), 10^{-1}M sodium citrate (2), 10^{-1}M acetic acid (3) and 10^{-1}M L-arginine (4). Symbols are as in Figure 13. (From Hatanaka et al., 1988, © 1988, Pergamon Press plc.)

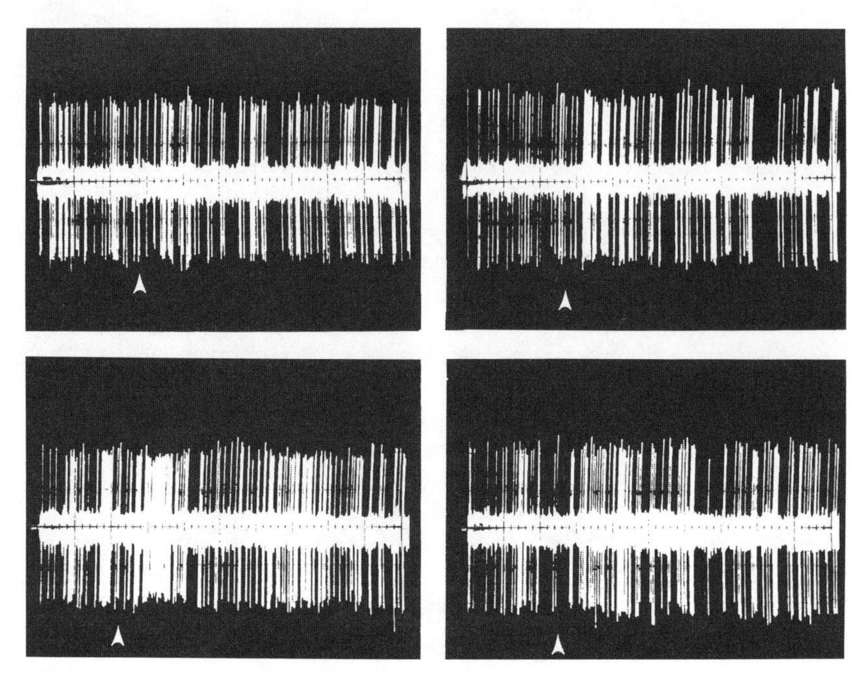

Fig. 7.15. *Thamnophis sirtalis.* Single-unit responses obtained from the accessory olfactory bulb in response to liquid delivery of stimuli to the vomeronasal epithelium. Left top: Ringer's solution. Left bottom: Earthworm wash (6 g earthworm to 20 ml Ringer's). Top right: Earthworm wash diluted 10 times. Bottom right: Earthworm wash diluted 100 times. Scale: total tracing = 50 secs; one vertical subdivision = 0.5 mV; arrow marks stimulus delivery. (Courtesy, Jun Inouchi.)

and vomeronasal systems, respectively, and the relationship of the peripherally recorded electro-olfactogram or electrovomeronasogram to the activity recorded in the main and accessory olfactory bulbs.

In a single study, electrical stimulation has been applied to olfactory or vomeronasal recipient zones in the telencephalon (Distel, 1978). Tongue-flicking, a response known to deliver odorants to the vomeronasal organ (see below), has been elicited by stimulation of the nucleus sphericus of *Iguana iguana.* However, caution must be taken in interpreting such findings. After all, tongue-flicking responses have been reported to be elicited fairly nonspecifically during stimulation of structures with no known connection to the vomeronasal system. Furthermore, odor delivery to the vomeronasal organ of *I. iguana* via a tongue-flicking mechanism has never been demonstrated, and tongue-flicking could be related to activation of a taste-sampling mechanism or to general activation of the animal.

V. ODORANT ACCESS TO VOMERONASAL ORGAN

It is well established that terrestrial vertebrates deliver gaseous odorants to the olfactory mucosa by inhaling or sniffing with the nose. The molecules that stimulate the main olfactory apparatus must be both volatile and adsorbent to the olfactory mucosa. The mechanism of odorant transport to the vomeronasal system is poorly understood and is not consistent across species. With few exceptions, the vomeronasal organs of vertebrates are sequestered from the nasal and oral cavities, and the ducts leading to them are exceedingly narrow. It appears that animals must actively enlarge the openings to the organs, activate a pumping suction mechanism, or deliver odorants directly into the organ.

With the exception of turtles, the vomeronasal organs of reptiles are separated by narrow ducts from the oral or nasal cavities or both. The tongue-flicking behavior of snakes and lizards is commonly thought to be a mechanism for odorant delivery to the vomeronasal organ (Bellairs, 1970). Until recently, the evidence for this idea was largely circumstantial, based on the common observation that squamates tongue-flick during chemically mediated, species-typical behaviors that were thought to depend on a vomeronasal system. Prior to 1980, the only direct evidence that the tongue of a snake is capable of delivering particles to the vomeronasal organ was the demonstration by Kahmann (1932) that a snake sacrificed shortly after tongue-flicking on a piece of carbon had a concentration of soot in the lumen of the vomeronasal organ. Since then the role of the tongue in delivery of nonvolatile substances to the vomeronasal organ has been demonstrated in the snake genus *Thamnophis* (Halpern and Kubie, 1980) and the lizard *Chalcides ocellatus* (Graves and Halpern, 1989). Both of these animals were made to tongue-flick cotton swabs soaked in prey extracts mixed with tritiated proline. Several hours later the animals were killed and their heads processed for autoradiography. In those snakes and lizards which the animals tongue-flick the swabs, large accumulations of radioactive material appear in the vomeronasal organs but not in the olfactory epithelium. Vomeronasal duct closure (by suturing in the snakes and by sealing the duct with tissue adhesive in the lizards) prevents odorant access to the vomeronasal organs. However, removal of tongues followed by touching the snouts of animals with the radioactive cotton swabs results in accumulation of radioactivity in the vomeronasal organs of both snakes and lizards, although to a much reduced extent in the former.

Although the studies described above indicate that the tongue is part of the normal mechanism for delivery of odorants to the vomeronasal organ, the transfer of those odorants from the tongue to the vomeronasal duct and through the duct to the lumen of the organ is still at issue. Use of radio-dense materials and fluoroscopy does not disclose entry of the tongue of *Varanus exanthematicus* (Oelofsen and Van Den Heever, 1979) or *Boa c. constrictor* (Young, 1990) into the vomeronasal organ; however, the tongue does move past the ductal opening. Use of lamp black to determine if the tongues of *Crotaphopeltis hotamboeia* and *Duberria lutrix* contact the opening of the vomeronasal ducts disclosed soot particles in the vomeronasal organs (Oelofsen and Van Den Heever, 1979).

If the tongue were used to pick up odorants from the substrate and transfer them to the mouth, one might expect morphological specializations that would aid in this process. Histological studies, including scanning electron microscopy of snake tongues, do suggest the occurrence of such specializations. The tongues of *Trimeresurus elegans* and *T. tokarensis* contain minute papillalike processes of various sizes on the surface and on the lips of the grooves that run along the lateral sides of the tongue (Kikuchi et al., 1981). The surface of the tongue furthermore contains many spoon-shaped, shallow depressions. The authors conclude that the tongue appears to be differentiated for picking up and transporting substances. The forked regions of *Nerodia rhombifera* and *Elaphe obsoleta* have modified epithelial cells that are covered by numerous projections (microfacets) (Morgans and Heidt, 1978). In *Elaphe* these microfacets are circumscribed by pores. These structural differentiations may also be responsible for transport of chemical molecules into the mouth and from there to the vomeronasal organ. The epithelium of the preforked (proximal) region of the tongue of *Thamnophis radix* is covered with ridges that appear to be made up of projections resembling microvilli (Ridlon, 1985). These ridges disappear just posterior to the bifurcation; microfacets then begin to appear, becoming more numerous distally, and completely covering the apical surface of epithelial cells from bifucation to tongue tip. The microfacets are circumscribed by small pores. The microfacets described in *Thamnophis* are considerably larger (0.7 to 1.5 μm) than those described by Morgans and Heidt (1978) for *Elaphe obsoleta* (10 nm).

Gillingham and Clark (1981b) have suggested that the anteriormost portions of the sublingual plicae (anterior processes)—specializations in the floor of the oral cavity—effect the transfer of molecules from the tongue to the vomeronasal organ (Figs. 7.16 and 7.17). Cinematographic analysis of *Elaphe obsoleta* (Gillingham and Clark, 1981b) and *Python molurus* (Young, 1990) has demonstrated that as the tongue is retracted into the mouth after a tongue-flick, the anterior processes, which lie just ventral to the opening of the vomeronasal organs, are elevated and directly contact the tongue. The tongue-flicks filmed (by

Gillingham and Clark) were open-mouthed, noted during the last phase of swallowing food or during aggressive displays in response to a gloved hand. Open-mouthed tongue-flicks and tongue-flicks during agonistic displays may be unrepresentative of chemosensory tongue-flicks, which almost always occur with the mouth closed, so some caution must be suggested in interpreting these observations. However, in the study by Young (1990), closed-mouth tongue-flicks were recorded, but these were not stimulated by chemical cues. Although it is not established that the anterior processes effect the transfer of particles from tongue to vomeronasal ducts, there are structural specializations that could facilitate such transfer. The epithelia of the anterior processes of several colubrid species contain goblet cells and mucous glands; these could provide lubrication and particle adherence for transfer to the vomeronasal duct (Ten Eyck and Gillingham, 1985). Scanning electron microscopy of the anterior processes of the rat snake *Elaphe obsoleta* (Clark, 1981) suggests that their surface anatomy, transverse folds, and sagittal and parasagittal pores, which may be openings to mucus glands, is well suited for this function. Smith (1986) also discusses this possibility for *Varanus*.

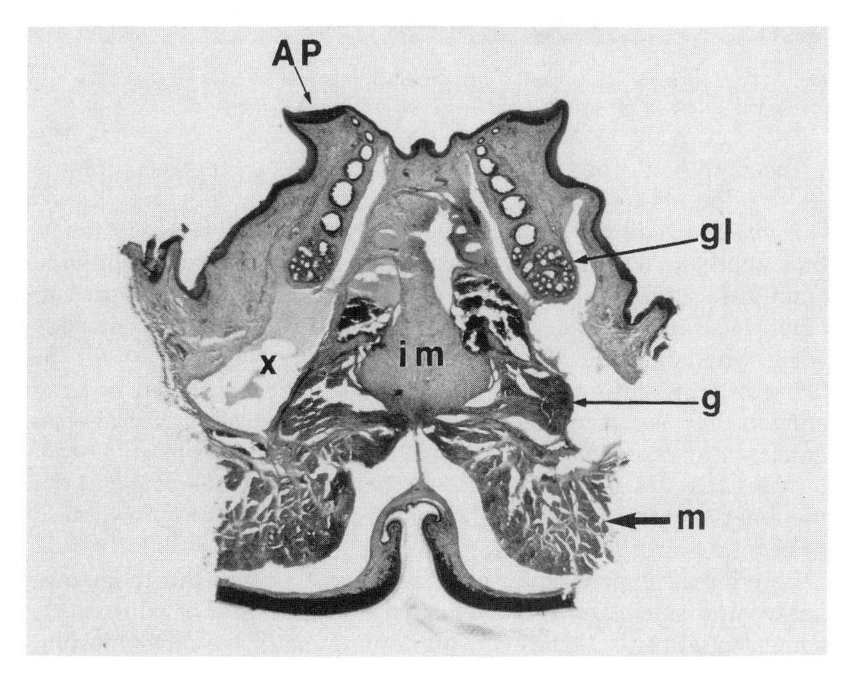

Fig. 7.16. *Elaphe guttata*. Section through lower jaw showing anterior processes (AP), intermaxillary ligament (im), lateral sublingual gland (gl), genioglossus muscle (g), geniomyoideus muscle (m), and unidentified cavity (x). (Courtesy, Gary Ten Eyck and James Gillingham.)

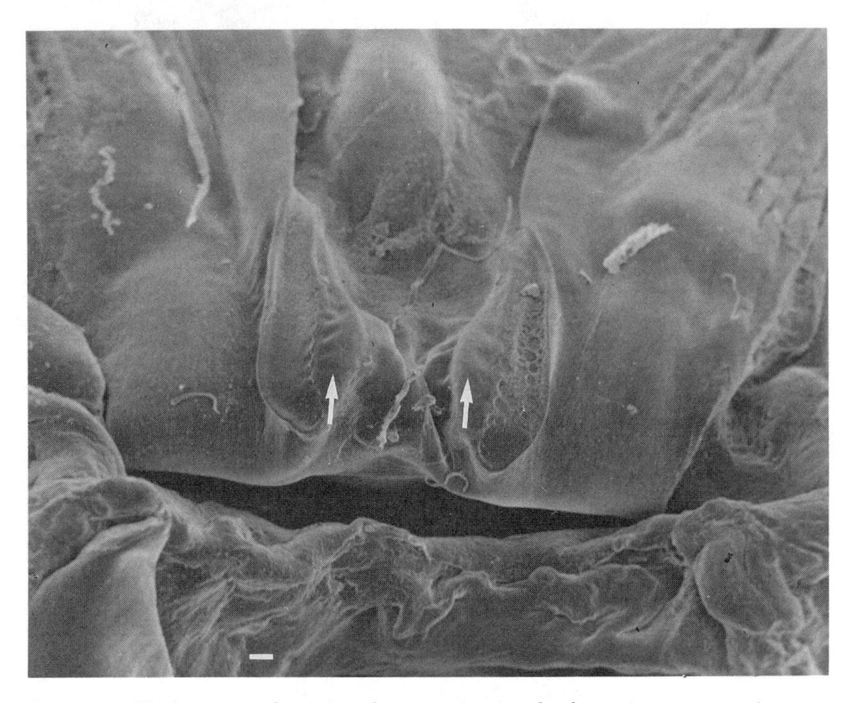

Fig. 7.17. *Elaphe guttata*. Scanning electron micrograph of anterior processes (arrows). Bar = 1 mm. (Courtesy, Gary Ten Eyck and James Gillingham.)

To test the concept that the anterior processes are involved in transfer of odorants to the vomeronasal organ, the anterior processes of *Thamnophis sirtalis* were surgically removed, and the snakes tested for their ability to detect fish odor in an open field apparatus (Gillingham and Clark, 1981b). Experimental animals spend less time in areas containing fish extract than do control animals. However, the critical issue is whether a functional vomeronasal system is necessary for this behavioral task. If not, then the apparent deficit exhibited by these animals may be due to some other factor. Alternatively, removal of the anterior processes may make it difficult for the snakes to guide their tongues into the area near the vomeronasal ducts. The proposal that the anterior processes effect the transfer of odorants to the vomeronasal duct is intriguing and worthy of further examination.

Convincing experimental evidence suggests that the tongues of snakes and some lizards are involved in the transfer of odors to the vomeronasal organ. However, the tongue may have other morphological features related to chemical senses (Schwenk, 1988), such as taste buds, as in the tuatara, *Sphenodon punctatus* (Schwenk, 1986), and lizards (Schwenk, 1985). Taste buds have also been observed in the oral epithelium of the blind snake, *Leptotyphlops dulcis*, on the dermal folds around the paired openings of the vomeronasal organ (Kroll, 1973). There are numerous other reports of taste buds in reptiles; mention is intended only as a reminder that the use of the tongue may relate to gustation as well as the vomeronasal sense.

Another behavior that may be related to delivery of odorants to the vomeronasal organ is gaping of the mouth (Graves and Duvall, 1983b, 1985a). *Crotalus v. viridis* mouth gape significantly more often after their snouts are coated with conspecific skin lipids than when they are coated with vehicle alone. Mouth gapes are more frequently followed by tongue-flicks than the other way around. It has been suggested that mouth-gaping may facilitate vomeronasal olfaction; this hypothesis deserves a more direct test. *C. v. viridis* have also been placed in a novel environment with their vomeronasal ducts sutured closed (Graves and Duvall, 1985b). These snakes mouth-gaped more and exhibited more head shakes than did sham sutured controls. The number of tongue-flicks emitted by the two groups did not differ. Mouth-gaping and head-shaking may aid in delivery of odorants to the vomeronasal organ; however, gaping and head-shaking could reflect the discomfort of having the lateral ridges of the fenestra vomeronasalis sutured closed. There is no independent evidence that the vomeronasal system is involved in the mouth-gaping and head-shaking response in a novel environment; better evidence will be needed before such behaviors are added to those involved in odorant access to the vomeronasal system.

VI. TONGUE-FLICK BEHAVIOR

Because of the close association between tongue-flicking and vomeronasal function, it has become common to assume that any observed increase in tongue-flicking in squamates signifies vomeronasal use. As stated several times below, this assumption is based on faulty logic. For example, although no deafferentation studies have ever demonstrated the use of the vomeronasal system in response to chemical stimuli in *Sceloporus jarrovi,* the following statements appear in the literature. "*S. jarrovi* was chosen as the test animal for several reasons. It has been shown to use the vomeronasal system to determine the presence of conspecifics. . . . In addition this species uses the system at a moderate level that allows for the correlation of tongue extrusions with external stimuli. In one study, for example, some teiids extruded their tongues over 300 times in 30 min, while *S. jarrovi* tongues were extruded only eight times in 30 min (Bissinger and Simon, 1979). Consequently, the tongue extrusions of these teiids are correlated with all behaviors and the vomeronasal system seems to be functioning constantly" (Simon and Moakley, 1985). The meaning of changes in tongue-flick rate must be interpreted cautiously.

Tongue-flick rates may covary with locomotion and are affected by environmental and experiential factors. Elicitation of tongue-flicking by chemical cues may depend on appropriate visual or contextual cues that occur in conjunction with the chemical ones. Finally, disturbance may artificially elevate or depress tongue-flick rates. Nevertheless, this chapter includes a review of recent literature that reports increases in tongue-flick rate in response to chemical and nonchemical stimuli. The manner in which tongue-flick interest scores are calculated and the effects of varying formulas on the interpretation of experimental outcomes are also of importance (Cooper and Burghardt, 1990b).

Various lizards and snakes differ in their patterns of tongue-flicking (Bissinger and Simon, 1979; Gove, 1979). Baseline tongue-flick rates range from a low of 3.4 long extrusions per 30 minutes for the Cordylidae to a high of 299.5 per 30 minutes for the Teiidae; indeed, teiids extrude their tongues significantly more often than do any other lizards (Bissinger and Simon, 1979). There are three types of tongue-flick: (1) simple downward extension, (2) single oscillation, and (3) multiple oscillation (Gove, 1979). The first type of tongue-flick occurs in all lizards and snakes studied; the second was observed in the more advanced lizards and all snakes; and the third occurs in all but one species of snake and in all but three species of lizards.

Tongue-flick patterns may differ with context. Duration, tongue-flick oscillations, oscillation number, and relative area circumscribed have been analyzed under different testing conditions in a comparative survey of 4 species of lizards and 10 of snakes (Gove and Burghardt, 1983). Exploratory tongue-flicks are consistently shorter and have smaller relative circumscribed areas than defensive tongue-flicks. Tongue-flicks used in feeding, exploration, social context, and defense all have different characteristics.

A common finding is an increase in tongue-flick rates in novel environments (Chiszar et al., 1976, 1978; DeFazio et al., 1977; Duvall, 1979; Simon et al., 1981; Burghardt et al., 1986). In lizards, the tongue touches of the substrate and air licks frequently differ. Locomotion may differentially correlate with tongue-touching and air-licking (Burghardt et al., 1986). In *Anolis carolinensis* air movement increases air-licking but not substrate licking. Tongue-flick rates in novel environments are not significantly different from those in handling control conditions and stress reduces tongue-flick rates.

Sceloporus jarrovi exhibits significantly higher tongue-flicking in the cage of a conspecific than in its home cage and significantly more tongue-flicks in a clean cage than in its previously occupied home cage (DeFazio et al., 1977).

Dipsosaurus dorsalis that inhabit kangaroo rat burrows exhibit significantly more tongue-flicks to kangaroo rat bedding than to clean sand, home cage sand, conspecific odor, or heterospecific odor (Pedersen, 1988). Although this finding has been interpreted in terms of the use of chemical information to select appropriate burrows, control with odor of a nonsympatric mammalian species would have been appropriate.

Tongue-flick behavior of *Crotalus viridis* and *Agkistrodon piscivorus* has been tested under the following conditions: home cage after brief handling, in the presence of hognose snake odor, in the presence of king snake (predator) odor, and in a clean open field (Chiszar et al., 1978). Tongue-flick rates are highest in the open field and in the cage of a hognose snake and lowest in the home cage. The intermediate tongue-flick rate in the presence of predator odor has been interpreted as indicating inhibition of exploration in potential prey. However, it points out the difficulty of expecting an isomorphic relationship between tongue-flick rate and vomeronasal stimulation.

VII. CHEMICAL SIGNALS: TURTLES

Chemoreception in turtles has been studied for sexual and nonsexual social odors, food odors, and habitat odors. In general, studies agree that turtles respond to social odors, particularly to those contained in feces and other cloacal excretions (Harless, 1979). A T maze discrimination task permitted demonstration of individual recognition based on visual and olfactory cues in *Pseudemys floridana peninsularis* and *P. nelsoni* (Kramer, 1986).

Chemoreceptive cues for spring aggregation in *Chrysemys picta* has been assayed by baiting traps with male and female turtles and thereafter covering the traps with porous canvas sacking to prevent the use of visual cues (Vogt, 1979). Interestingly, in the absence of visual and tactile cues, traps with males are equally likely to attract males as are traps with females. These results have been interpreted to suggest that female pheromone, if present, is not capable of attracting males at a distance but requires direct contact. Alternatively, other sensory cues such as visual and tactile stimuli may be important in sexual attraction.

Olfactory imprinting has been the subject of a number of studies, recently reviewed by Owens et al. (1982, 1986). The concept of olfactory imprinting deals with a developmental process, thought to involve learning, that modifies subsequent odor preferences. These odorants may arise from food, physical environments, or conspecifics.

Olfactory imprinting to conspecific odors has been studied in *Trachemys scripta* and *Chelydra serpentina* by pairing hatchlings in the same holding box (Evans et al., 1973) or exposing individually housed animals to water conditioned by a paired turtle (Evans et al., 1974). After a period of exposure to the other individual or its water, animals are tested to determine if they prefer proximity with the animal to which they have been previously exposed. During testing, turtles more frequently occupy the area close to "familiar" individuals; however, it is unclear whether the animals are responding to odor. Only the *Chrysemys* were visually isolated during "imprinting." *Chelydra* were never visually isolated, and both species were tested under conditions in which visual and other cues could be used.

Caretta caretta develops preferences for the food available to them during the interval beginning 5 days after hatching and continuing for 14 days thereafter (Grassman and Owens, 1982; Steele et al., 1989). *C. caretta* will also show a preference for nonbiological odors (e.g., morpholine) to which they are exposed in the nest and during the posthatch period (Grassman and Owens, 1981). Similar results were obtained for *Chelonia mydas* exposed to morpholine and 2-phenylethanol in the nest and during the posthatching period (Grassman and Owens, 1987). Perinatal olfactory imprinting may be responsible for olfactory homing by migratory turtles. *Lepidochelys kempii* artificially imprinted to Padre Island, Texas, were found to respond differentially to water obtained from Padre Island and to that from Galveston Island (Grassman et al., 1984).

Olfactory learning has been demonstrated in *Chelonia mydas* (Manton et al., 1972a,b). Turtles are trained to press a key in the presence of an odor to receive food reward. The turtles are able to discriminate among a number of chemicals including β-phenethylalcohol, isopentyl acetate, triethylamine, and cinnamaldehyde. They are unable to discriminate L-serine and glycine. Discriminations are disrupted for 1 to 5 days following zinc sulfate irrigation of the olfactory mucosa. Such findings suggest that the coagulation necrosis of the olfactory (and vomeronasal) epithelium caused by the zinc sulfate treatment impairs olfactory recognition of the discriminative stimuli. However, the short duration of the effect indicates that only part of the epithelium could have been destroyed; regeneration and subsequent return of function would have required several weeks.

The zinc sulfate method has been used to assess the role of the chemical senses in homing behavior of *Testudo hermanni* (Chelazzi and Delfino, 1986). Following destruction of the olfactory epithelium with zinc sulfate, adult tortoises collected along the coast of southern Tuscany were equipped with small radiotransmitters and released 500 to 1,000 m from the capture site. Control animals headed toward their home area. The responses of zinc sulfate–treated animals (lesions were histologically verified) were drastically impaired. Their orientation did not differ from chance, and only 1 of 26 treated animals was recovered near the capture site.

During courtship and mating, extensive male sniffing, head-bobbing, and ramming and lifting of the female carapace have been reported in apparent response to female cloacal and chin gland secretions in a number of turtles, for example, *Rhinoclemmys pulcherrima incisa* (Hidalgo, 1982), *Gopherus berlandieri* (Rose, 1970), *Geochelone radiata* (Auffenberg, 1978a), *Testudo graeca* (Boiko, 1984b). The roles of the nasal chemical sense systems in these reproductive behaviors have been tested by cutting the olfactory and vomeronasal nerves of *Emys orbicularis* (Boiko, 1984a). Transection of the olfactory or the vomeronasal nerves leads to a 60% to 70% reduction in reproductive behavior. Cutting of both nerves eliminates reproductive behavior. Olfactory and vomeronasal nerve section also impairs food searching behavior of *E. orbicularis;* the effect is particularly pronounced if vision is also compromised (Boiko, 1983).

There appears to be good evidence that the nasal chemical senses are involved in a number of social and nonsocial behaviors of turtles. A problem with the current state of knowledge about the role of the nasal chemical senses in behavior is the absence of correlated anatomical, behavioral, and physiological data for any single species.

VIII. CHEMICAL SIGNALS: CROCODILIANS

Alligator mississippiensis may detect chemical cues emanating from prey and orient its searching by using these cues. Alligators presented with perforated bags filled with meat (beef, nutria, or rattlesnake) or soaked in meat extract in a water-filled channel will contact the meat-filled bags first, contact the meat-filled bags more frequently, and are more likely to remove the bags with meat odors than control bags containing paper (Scott and Weldon, 1990; Weldon et al., 1990). Alligators on land detect meat and raccoon carcasses placed under baskets (Weldon et al., 1990). The findings from these seminatural studies support the idea that chemoreception plays a role in prey localization by alligators. The absence of a vomeronasal organ in these animals leaves olfaction and taste the most likely candidates to play a role in this behavior. Future studies would benefit from individual testing and appropriate use of odor controls.

Two studies in *Alligator mississippiensis* (Johnsen and Wellington, 1982; Weldon et al., 1990) report that yearling alligators change their gular pumping rate differentially to a variety of odors delivered in an air stream. Amyl acetate, citronellol, androstenol, mandibular gland and cloacal gland secretion from one alligator, and cloacal gland secretion from another alligator have been tested. Presentation of all stimuli, except amyl acetate and citronellol, significantly increases the rate of gular pumping over that of blank controls. Response to mandibular and cloacal gland secretions is significantly higher than all other stimuli (Johnsen and Wellington, 1982). Gular pumping also is increased significantly in response to airborne beef odor as compared to control (distilled water) odor (Weldon et al., 1990).

IX. CHEMICAL SIGNALS: LIZARDS

A. General

An extensive literature now exists on the response of lizards to a variety of chemical substances including odors from prey, predators, and conspecifics. This literature has been reviewed recently (Simon, 1983). Mason (this volume) reviews the pheromonal literature; therefore the present review does not deal with pheromones in detail.

The roles of the chemical senses in social and feeding behavior in *Varanus komodoensis* are discussed as they relate to individual and sexual recognition, carrion location, and prey ambush and stalking (Auffenberg, 1978b). Chemosensory behavior of the Komodo monitor includes considerable investigation of fecal pellets, extensive tongue-touching, and body surface investigation. In addition, fecal pellets are deposited in response to chemical signals. However, although much of this behavior is strongly suggestive of chemosensory communication, we lack definitive studies identifying the chemosensory systems mediating these behaviors.

B. Prey Odors

The ability of lizards to detect prey odors and discriminate them from nonprey odors has been examined by Cooper (1990). Adult *Tupinambis rufescens* respond to cotton swabs coated with mouse odors with significantly more tongue-flicks and/or attacks than to cotton swabs dipped in cologne or distilled water. Similarly, adult *Podarcis hispanica,* when presented with cotton swabs coated with cricket odors, respond with significantly greater numbers of tongue-flicks and/or attacks as compared to control odors (Cooper, 1990).

Newly hatched, ingestively naive *Eumeces fasciatus* respond differentially to extracts of two prey—mealworms (*Tenebrio molitor*) and mice—but do not respond to extracts of nightcrawler (*Lumbricus terrestris*) or water controls (Burghardt, 1973). These findings suggest, as do those with snakes (see below), that responsiveness to prey chemicals are innate. However, Loop and Scoville (1972) found that *Eumeces inexpectatus* poorly discriminates prey chemicals. No overt attacks are displayed, tongue-flicks to prey chemicals do not differ significantly, and only differences in an orientation time measure reach statistical significance.

Although some lizards are clearly able to detect prey using their chemical sense systems, other sensory systems such as vision and vibratory sense may be more important in the decision process. For example, the readiness of *Anolis lineatopus* to attack four prey species is unaffected by the presence or absence of odor (Curio and Möbious, 1978). *Scincus scincus* appear to be guided primarily by vibratory cues within the sand in detecting and locating prey (Hetherington, 1989). They spend most of the time buried in sand but can detect crickets and mealworms moving over the surface at distances up to 15 cm. Freshly dead insects placed on top of the sand, and presumably retaining their full complement of odors, are less likely to be detected or to elicit a response and are only responded to within 5 cm.

Interspecific differences clearly exist in response to prey odors. *Eumeces laticeps* significantly increase the rate of tongue-flicks and attack cotton swabs containing cricket odors (Cooper and Vitt, 1989). In contrast, *Anolis carolinensis, Sceloporus malachiticus,* and *Calotes mystaceus* do not differentiate between cotton applicators containing prey odors and those lacking them (Cooper, 1989). The investigator providing the last observations concluded that the vomeronasal system does not need to be stimulated to elicit feeding in these species and that olfaction is not important for initiation of attack. Such conclusions may in fact be valid; however, the study reported does not test them. First, it would be important to demonstrate that the cotton swab testing technique yields differential responses to some stimulus in these species. Second, cotton swab testing can only reveal the efficacy of a stimulus on the swab in eliciting a response to that swab. The swab technique would not yield the critical information about stimulus sampling, attack, and ingestion if response to prey involves a complex interaction of stimuli. Furthermore, one needs to know whether response to prey will be "normal" in the absence of a functional vomeronasal or olfactory system or both. Finally, it might be interesting to check whether the rate of tongue-flick responses correlates with other physiological rates, such as cardiac or pulmonary ones.

Scincella lateralis respond to prey with increased rate of tongue-flicking, followed by orientation and attack (Nicoletto, 1985a, b). Visual cues alone will elicit these responses, as will visual and chemical cues associated with the prey. However, supranormal concentrations of prey odor are required to elicit a response in the absence of visual cues. As these observations involved presentation of odor by pumping air over the prey, the chemical cues were airborne and preferentially stimulated the olfactory system. Chemical cues appear to be of less importance whenever visual cues are present; however, removal of the visual stimulus associated with the prey will increase the tongue-flick rate in response to prey odor.

C. Predator Odors

The ability of *Lacerta vivipara* to detect chemical stimuli deposited by predatory snakes has been tested by observing lizards placed in cages previously inhabited by snakes of the species *Vipera berus* and *Coronella austriaca* (lizard predators) or *Natrix natrix,* which rarely takes lizards (Thoen et al., 1986). Lizards have been tested in a clean cage and in one sprayed with civet cat odor. A significant increase in tongue-flick rate occurs in the presence of predator odor. This increase in tongue-flick rate is accompanied by a disruption of locomotor patterns, an increase in "slow motion" locomotion, tail vibrations, and starts. Tongue-flick rate is also increased over controls by the odor of *Natrix* but not as much as by the odor of predators. The results of this study are quite provocative. Whereas *Vipera berus* is a sympatric predator, *Coronella austriaca* is an allopatric one, yet both produce elevated tongue-flick rates. However, several design problems may compromise the usefulness of these data. The order of odor presentation does not appear to have been randomized. There is no indication that the cages were cleaned following each trial, no intertrial interval is stated, and no time was provided for the animals to recover from the stress of being handled.

D. Conspecific Odors (Pheromones)

There is an extensive survey of what is currently known about pheromonal communication in the genus *Eumeces* (Cooper and Vitt, 1986b,c). Techniques that minimize visual, somesthetic, and other nonchemosensory cues have provided considerable evidence that *Eumeces* are able to discriminate conspecific odors from congeneric odors and those of males from those of females. In general most of the studies utilize a similar behavioral paradigm. Skinks are presented with cotton swabs that contain the odors obtained from stimulus animals, and tongue-flicks are counted. Water or cologne on the swabs is used to control for texture and pungency. It is assumed that tongue-flick rate reflects chemosensory investigation of a stimulus; however, as discussed earlier in this review, tongue-flick rates may vary independent of chemosensory environment and therefore this assumption cannot be accepted uncritically.

Courting male *Eumeces laticeps* repeatedly tongue-flick females, particularly in the cloacal region. Postreproductive male *E. laticeps* respond with elevated tongue-flick rates to cloacal odors of postreproductive male and female conspecifics in comparison to distilled water (Cooper and Vitt, 1984a). Female *Eumeces laticeps* show higher tongue-flick rates in response to male or female cloacal odors than to distilled water (Cooper and Vitt, 1984b).

Adult *Eumeces laticeps* flick their tongues at higher rates to conspecific cloacal odors of the opposite sex than to those of other lizards of the same genus (*E. fasciatus* and *E. inexpectatus*) (Cooper and Vitt, 1986b). Male *E. fasciatus* and *E. inexpectatus* respond differentially to cloacal odors of female conspecifics and congenerics (Cooper and Vitt, 1986a). Males display higher tongue-flick rates to the cloacal odors of conspecific females than to odors of congeners. Thus, adults of the genus *Eumeces* appear able to discriminate cloacal odors obtained from conspecifics from other odors including those obtained from congeners. The ability of males to recognize conspecific odors obviously acts as a premating isolating mechanism. Evidence for such a mechanism is provided by the observation that male *E. inexpectatus* will court female *E. fasciatus* that have been coated with odors of estrogen-treated female *E. inexpectatus* (Cooper and Vitt, 1986c). Interestingly, male *E. fasciatus* will not court female *E. inexpectatus* primed with odor of estrogen-treated female *E. fasciatus*. However, although male *E. inexpectatus* will court female *E. fasciatus* under the conditions described, the female *E. fasciatus* will not permit mating to occur.

Further evidence that *Eumeces* uses chemical cues in intraspecific communication is the finding that male *Eumeces laticeps* will follow trails produced by dragging the ventral side of their adult females along a substrate (Cooper and Vitt, 1986d). Direct tongue contact with the trail occurs whenever males follow female trails. In contrast, females do not follow scent trails of males or females.

One of the sources of the female pheromone to which male *Eumeces laticeps* respond is the urodaeal gland (Cooper et al., 1986), a gland found only in females, that becomes highly secretory during the breeding season (Trauth et al., 1987). Postreproductive females bearing extracts of urodaeal gland obtained from estradiol-treated females are courted by males with greater intensity than control females (Cooper et al., 1986). Extracts of urodaeal gland elicit higher tongue-flick rates whenever presented on cotton swabs than will cloacal odors, and both of these elicit higher tongue-flick rates than water. Chemical fractionation of the urodaeal gland and subsequent assay indicates that the male attractant component is found in the neutral lipid fraction of the gland homogenate (Cooper and Garstka, 1987).

These studies demonstrate that male *Eumeces* can discriminate male from female odors and conspecific from congeneric odors. Testosterone increases the response of males to female odors, and the odors used in species and sexual identification derive from the urodaeal gland.

The pheromones permitting species identification have an obvious role in copulatory behavior. These pheromones also could be important in identifying appropriate targets for agonistic encounters (Cooper and Vitt, 1987). Male *Eumeces* direct most of their agonistic behavior to conspecific males. No fights or agonistic displays would normally occur during encounters between two males of different species. Mutual investigation by the males would typically involve tongue-flicking. As males are similar in appearance, chemical cues could act as discriminative stimuli in initiating aggressive encounters. Male *E. inexpectatus* do not normally show aggression toward male *E. fasciatus*. However, if skin and cloacal odors of male *E. inexpectatus* are transferred to male *E. fasciatus*, then male *E. inexpectatus* behave aggressively to the congeneric males, suggesting that some substance in the transferred chemical conveys information about the species of the donor that is then used to assess the appropriate social response.

Male *Sceloporus occidentalis* respond with higher rates of substrate licks or tongue-flicks to surfaces marked by conspecifics than to control surfaces (Duvall, 1979, 1981a). Immediately after licks of marked substrates, but not unmarked substrates, males exhibit push-ups, which are species-typical visual displays. Males show a greater increase in flick rate to male-marked surfaces than to female-marked surfaces.

Sceloporus jarrovi initiate tongue-flicking to odors of male conspecifics earlier than to odors from female conspecifics, to those of a male sympatric congener (*S. virgatus*), or to cologne or water. No other measure of response to the odorants differs significantly between stimulus conditions (Simon and Moakley, 1985).

The importance of fecal boli as the source of visual and chemical signals for conspecific interaction has been analyzed in *Sceloporus occidentalis* (Duvall et al., 1987). Lizards approach papers with fecal boli more readily, explore them more, exhibit more bolus licks, more substrate licks, and more tongue flicks to them than to control papers. Fifty percent of male and female lizards exhibit species-typical displays when uncovered fecal boli are licked; however, in the absence of chemical cues, only one of nine males and three of eight females exhibited the display after licking plastic-covered boli (Duvall, 1979). Apparently, visual cues provide important information to fence lizards during investigatory behavior. Species-typical displays are elicited more readily by the addition of chemical cues.

The situation is similar in *Dipsosaurus dorsalis* (Alberts, 1989). Femoral gland secretions of these desert iguanas strongly absorb long-wavelength ultraviolet light, are relatively nonvolatile, and remain in the environment for long periods. Electrophoretic and proton nuclear magnetic resonance analysis of the femoral gland secretion of *Dipsosaurus dorsalis* indicates that the secretions are composed of approximately 80% protein and 10% lipid. The electrophoretic patterns of femoral gland proteins differ among individuals (Alberts, 1990). Although blindfolded lizards respond to femoral gland secretions with more tongue touches than to control substrates, under conditions where vision is possible, the iguanas detect femoral gland deposits better if the light source contains ultraviolet than if it is incandescent (Alberts, 1989). More tongue touches are emitted to femoral gland secretions compared to control stimuli whenever the animals are tested with a light source, the ultraviolet emission of which is congruent with the absorption of femoral gland secretion. Unless the snout is touched with the test sample, the rate of tongue touches to femoral gland secretion does not differ significantly from that to controls with the marking illuminated by incandescent tungsten light (no ultraviolet emission). This study strongly supports the interaction between visual and chemical cues from femoral gland secretions in guiding lizards to locations that may contain pheromones to be investigated.

Male and female *Sceloporus occidentalis* in natural breeding condition respond differentially (with tongue-flicks and push-up displays) to surfaces marked with exudates collected from male and female conspecifics (Duvall, 1981a, b). Conspecific body licking is a species-typical behavior of this fence lizard (Duvall, 1982), and both males and females lick males more than they lick females. The areas licked may contain structures producing or releasing pheromones that are detected by the tongue-vomeronasal system (Duvall, 1982). The Harderian gland may also provide some of the pheromones present in the head region. These hypotheses are sufficiently provocative to warrant further investigation.

Sniffing and licking of the vent region of conspecifics by *Phrynosoma platyrhinos* and *P. coronatum* before and after courtship and during male-male encounters have been reported (Tollestrup, 1981). Again this behavior is suggestive of chemical communication of sexual status; the roles of the various chemical sense systems in this behavior should be investigated.

Male *Eublepharis macularius* respond to female conspecifics with courtship and to males with agonistic behavior. However, when females shed their skins, males respond with aggressive behavior typical of their response to conspecific males (Mason and Gutzke, 1990). This finding is interpreted as suggesting that the differential response of males to males and females is based on detection of chemicals normally present in the skin of females that are unavailable while the female is shedding.

Eumeces fasciatus and *Scincella lateralis* respond with approach or avoidance, increased snout dipping, and elevated tongue-flick rates to conspecific airborne odors delivered through an olfactometer (Duvall et al., 1980a). A recent report notes inability to replicate these findings (Cooper and Vitt, 1985). The discrepancy in the results of these two studies could be accounted for by the physical parameters of odor delivery: air speed (10 km/hr versus 180 km/hr) and temperature during testing (30°C versus 22°C) in Cooper and Vitt versus Duvall et al., respectively. The air speed differences are great and in the right direction, but the temperature differences are in the wrong direction if one assumes that higher temperature would increase volatilization and volatiles are what is being detected. It would have been helpful if the later study, having obtained discrepant results, had subsequently tried to replicate the Duvall study exactly. Cooper and Vitt suggest that the failure to demonstrate significant differences results from the absence of direct contact with the odor source. This hypothesis is reasonable and could be tested easily in the same apparatus by allowing the stimulus animal to make a trail and testing the ability of an experimental animal, with direct access to the trail, to approach or avoid the arm of the maze with the trail.

Kinship recognition and aggregation of hatchling *Iguana iguana* has been studied in a series of fairly complicated experiments (Werner et al., 1987). The clutch of eggs from one female has been subdivided and half placed in each of two different incubators. After emergence, hatchlings from one group were placed into an enclosure with hatchlings from another female. Kin pairs formed more frequently than would be expected by chance. Eight kin groups were studied in a second experiment. Hatchlings separated from kin do not show a preference for the side of a chamber closer to that of their kin group than for that of a nonkin group. However, once posthatching experience has taken place, hatchlings prefer to be in proximity to kin. Recognition of kin apparently occurs through odors contained in feces. This interesting study is difficult to review because of the complexity of design and the manner in which the data were analyzed, for example, disregarding groups larger than three. In several instances the treatments were confounded, such as isolation of animals that did not have experience with kin odors and group housing with adult animals. However, the results are suggestive of chemical communication at a very early stage of development.

X. CHEMICAL SIGNALS: SNAKES

A. General

Of all reptiles, snakes appear to have been the most broadly studied in terms of their responses to chemical stimuli and the chemical senses that mediate those responses. This literature was most recently reviewed by Halpern and Kubie (1984). Snakes respond to nonbiological odors, predator odors, prey odors, and conspecific odors. As indicated earlier, Burghardt (1970a) provides an exhaustive critical review of this literature; therefore, with few exceptions, the following review deals with the literature since 1970.

B. Nonbiological Odors

Garter snakes, *Thamnophis radix* and *T. sirtalis sirtalis*, can be trained to respond differentially to airborne odorants (Begun et al., 1988). Individual *T. radix* have been trained to move into a compartment scented with lemon for food reward. After reaching criterion performance, they were trained to move away from the scented compartment. In a second experiment individuals of *T. s. sirtalis* were trained to traverse a Y maze toward or away from an amyl acetate odor. Five of seven snakes achieved criterion performance (two training sessions with cumulative correct responding above the .05 confidence level) within 85 trails. These studies demonstrate that snakes can be trained to respond to an arbitrary airborne odorant. It remains to be seen if this response is mediated by the olfactory, vomeronasal, or both systems.

Adult and young *Thamnophis radix* avoid a fumigant containing pyrethrum (Secoy, 1979). The snakes also have been tested on a number of other common chemicals and commercial repellants but do not appear to respond to them. However, the original description of the behavior of the snakes indicates that they were trying to escape from the apparatus, a behavior that could have compromised the results.

C. Predator Odors

The classic experiments of Bogert (1941), Cowles (1938), and Cowles and Phelan (1958) established the ability of snakes to detect predator odors and the importance of the chemical senses in that detection. In the latter study these authors enunciated what has become known as the Cowles and Phelan hypothesis. It states: "rattlesnakes possess an olfactory chemoreceptive alerting mechanism, possibly of low discrimination but high sensitivity, which operates independently of the tongue and Jacobson's organ and probably serves as a trigger mechanism for the lingual air sampling response. This implies that alerting may be initiated from a resting nonalert condition by the arrival of vagrant airborne odors such as that of putrefying protein or the presence of a dangerous predator, and that following such general stimulation a more explicit analysis by the use of the tongue and Jacobson's organ may follow." Although there are several reports in the literature that claim to have tested this hypothesis, in fact none of the tests has been satisfactory.

Response to predator odors has been assayed in 21 taxa of crotaline pit vipers exposed to skin substances from colubrid snakes known to feed on lizards, snakes, or both. The vipers responded by elevating the middle portion of the body in a defensive posture called body bridging. Newborn snakes from two of the three species tested gave the response, suggesting an innate ability to sense potential predators (Marchisin, 1980). Ten species of pit vipers representing the genera *Agkistrodon, Crotalus,* and *Sistrurus* were tested for response to predator odors using the ophiophagous Eastern king snake, *Lampropeltis getulus getulus,* and the Eastern indigo snake, *Drymarchon corais couperi* (Marchisin, 1980). Pit vipers, using chemical cues alone or in conjunction with visual or tactile stimuli, could distinguish ophiophagous snakes. The ability to recognize snake predators and utilize the appropriate defense response was shown to be innate by tests with predator-naive, newborn crotalines. The nuchodorsal regions of *Lampropeltis* and *Drymarchon* are sources of the odorants to which the pit vipers respond with defensive postures.

Predator-naive garter snakes, *Thamnophis elegans vagrans* and *T. e. terrestris,* respond differentially to skin chemicals derived from ophiophagous snakes (Weldon, 1982). Tongue-flick rates of garter snakes rise after the snakes are presented with cotton swabs rubbed against skins of *Lampropeltis getulus* but not for swabs from nonophiophagous snakes. Significantly more tongue-flicks are emitted to cotton swabs with odor of another ophiophagous snake, *Coluber constrictor,* than to those of a control snake or blank swabs. Snakes tested in an olfactometer emitted more tongue-flicks in the presence of *Lampropeltis* odor than in the presence of odor from the nonophiophagous snake, *Elaphe obsoleta.* However, there was a significant increase over baseline in tongue-flicks in the presence of *Elaphe* odors, suggesting that *Thamnophis* is capable of responding to the odors of *Elaphe.*

Young *Elaphe guttata* respond to cotton swabs rubbed against the skin of the ophiophagous *Lampropeltis getulus* with significantly increased tongue flicks as compared to their response to blank swabs. Tongue-flicks to swabs rubbed against skin of *Thamnophis radix haydeni* did not differ from response to blank swabs or to *L. getulus.* Tongue flick responses to chloroform extracts of skin from *E. obsoleta, L. getulus,* and *Masticophis flagellum* are all significantly elevated compared to solvent alone, and although response rates to skin extracts of *M. flagellum* are significantly lower than those to *E. obsoleta* and *L. getulus,* the response to the latter two does not differ. These results do not support the hypothesis that *E. obsoleta* discriminates between ophiophagous and nonophiophagous snakes on the basis of chemicals contained in their skin.

D. Alarm Pheromones

Prairie rattlesnakes have been studied to determine whether they respond to alarm pheromones (Graves and Duvall, 1984, 1988). Heart rates were monitored in animals exposed to odors from alarmed snakes either after removal of cloacal sacs or after sham surgery and to control odor. Heart rate does not change before and after exposure to any of the odors. However, heart rates do increase in individuals exposed to odors from sham-operated conspecifics that were subsequently stimulated with a threatening stimulus. Odors from alarmed conspecifics with their cloacal sacs surgically removed and control odors do not represent a threatening stimulus nor significantly increase heart rate. Caution must be used in interpreting the results of this experiment. One cannot conclude (and the authors do not) that cloacal gland secretions increase heart rate or that the secretion is an alarm substance, although the authors suggest the latter. The observations that the heart rate of snakes exposed to odor of alarmed snakes with intact cloacal glands significantly increases whenever the exposed snakes are subsequently threatened is based on a subject group of four individuals. The prestimulation heart rate of these animals was lower (although not significantly so) than that of the other groups and therefore left more room for elevation of heart rate. The heart rates during the threat condition were in fact equivalent for the three groups.

E. Aggregation Pheromones

The demonstration that snakes aggregate and use visual and chemical cues to locate habitats containing other snakes had been demonstrated early in *Storeria dekayi*, *Thamnophis sirtalis*, *T. butleri* (Noble and Clausen, 1936), and *Diadophis punctatus arnyi* (Dundee and Miller, 1968).

More recent aggregation studies have expanded the number of species in which grouping behavior is observed, examined such behavior in neonates, investigated conspecific, nonsexual trailing, determined the location of the deposited conditioning cues used to locate favored habitats, and attempted to identify the source of the chemical cues used in trailing and habitat location.

The number of species for which aggregation behavior in the laboratory has been confirmed could recently be increased. Evidence of grouping behavior is now available for *Thamnophis marcianus*, *T. proximus*, *T. radix*, and *Elaphe guttata* (Ten Eyck and Halpern, 1988, 1989). The strength of aggregation differs interspecifically. Young *E. guttata* consistently show very strong tendencies to group. Unless the test apparatus is cleaned between trials, certain locations are consistently preferred over others, suggesting that the substrate is being conditioned by the animals. Snakes (*Elaphe*) tested individually are more likely to return to shelters preferred during group testing.

Newborn *Storeria dekayi* and *Thamnophis sirtalis* aggregate and are more likely to be found in clumps involving conspecifics, although mixed species aggregates are also found (Burghardt, 1983). *Thamno-*

phis show stronger tendencies to aggregate than do *Storeria*. Substrate conditioning and development of preferred shelters also have been observed.

Thamnophis sirtalis are more likely to be found below shelters than in the open field. They repeatedly return to preferred shelters whether or not tested in groups or as individuals (Heller and Halpern, 1982a). The cues used by the snakes in detecting preferred shelters derive from substances deposited by the snakes onto the substrates of the shelters. Snakes tested in aquaria, the contents (shelters, rocks, water dishes, and flooring) of which are changed on each trial, continue to aggregate but fail to establish preferred shelter location.

The basic findings of these aggregation studies strongly suggest that some substance deposited by the snakes is then used by them or others to return to certain locations. Neonate water snakes (*Nerodia s. sipedon*) tested in a two-choice preference task with unsoiled substrates or those soiled by conspecifics spend more time exploring areas soiled by conspecifics (Scudder et al., 1980). Snakes show higher tongue-flick rates and snout-rubbing in the presence of conspecific odors.

However, conspecific odors are not necessarily attractants. *Crotalus v. viridus* exhibit no preference for the side of an arena marked by conspecifics and after 36 hours in the apparatus are more likely to be found on the clean, rather than the marked, side (King et al., 1983).

Nerodia r. rhombifera display a significant preference for the clean side of an aquarium compared to an area soiled by a conspecific (Porter and Czaplicki, 1974). *Thamnophis radix* prefer areas soiled either by themselves or by conspecifics over clean areas but do not demonstrate preferences when one side of the arena is soiled by a sympatric water snake. These species differences could reflect natural differences in territoriality or aggregation. Species differences and substrate preference differences also are shown in a comparative study of four sympatric species of water snakes (Allen et al., 1984). In addition, under different conditions tongue-flick types and rates will vary among the different species. Interestingly, tongue-flick rates tend to be lower whenever snakes are on the preferred substrate.

Sensitivity to conspecific odors could develop in the perinatal period. Newborn *Crotalus viridis* during the immediate postpartum period exhibit several behaviors that are suggestive of chemical sense stimulation (Graves et al., 1987). These behaviors include tongue-flicking, mouth-gaping, head-shaking, and face-wiping. However, we lack independent verification of the chemosensory control of these behaviors.

How do snakes find each other prior to aggregation? No studies published thus far have actually observed animals during shelter selection. Trailing may be the mechanism used, as snakes will trail conspecifics, even those sexually inactive (Heller and Halpern, 1981).

Snakes of several species of *Thamnophis* and of varying ages (neonate, infant, and adult) have been tested in a Y maze (Heller and Halpern, 1981). Significantly greater numbers of snakes prefer the side of the maze previously conditioned by a large number of snakes. Unless large numbers of snakes have previously conditioned one side, individually tested snakes follow the immediately preceding snake.

It has been suggested that young snakes trail older snakes to locate hibernacula (e.g., Costanzo, 1986). Newborn *Crotalus horridus* trail conspecific adult males, nonreproductive females, and other newborns (Brown and MacLean, 1983). During trailing, newborns exhibit a distinctive behavior, observed in 86% of the stimulus trials but never on no-trail trials, which the authors call "the trail contact response." The behavior is characterized by a pause upon first encounter with the scent trail, lowering of the head toward the trail, and a series of tongue-flicks directed to the trail. Short rapid tongue-flicks are typical during the trail contact response, and tongue tips then contact the substrate.

Thamnophis sirtalis, tested in the laboratory during October, have been reported to follow scent trails laid by a large female conspecific (Costanzo, 1989). However, the data are presented in a manner that precludes analysis of the results or evaluation of the method. Furthermore, the conclusion that this study demonstrates the ability of snakes to follow each other to dens is erroneous, as the only similarity between this laboratory study and navigation to the hibernaculum is that both occurred in the fall of the year.

Leptotyphlops dulcis and *Typhlops pusillus* not only follow conspecific trails but their individual trails as well (Gehlbach et al., 1971).

The source of chemical signals used by snakes to respond to conspecifics has been investigated in neonate *Thamnophis radix* (Graves and Halpern, 1988), *T. proximus* (Graves et al., 1991), and *Crotalus v. viridis* (Graves et al., 1986). In these studies, snakes are reported to respond differentially to conspecific skin lipids but not to skin lipids derived from a congener or unrelated species.

The chemistry and/or anatomy of several suspected glandular sources of chemosignals have been examined, although behavioral data for their chemosensory signaling functions is not always present. The morphology, histology, and histochemistry of squamate scent glands has been described by Whiting (1969), who includes an excellent discussion of glandular structure and the nature and possible functions of the secretions. Oldak (1976) has compared the scent glands of 25 snakes using thin-layer chromatography; he considered the active odor components to be lipoidal. The major lipids of the scent gland secretions of *Acrantophis dumerili* have been analyzed by gas chromatography–mass spectrometry (Simpson et al., 1988). The cloacal sac secretion of *Leptotyphlops dulcis* has been biochemically analyzed. Nearly all of the odor-containing material consists of free fatty acids (Blum et al., 1971).

F. Sex-related Pheromones

It has long been known that the dorsal surface of the skin of female snakes contains a substance that attracts males at the time of mating (reviews by Burghardt, 1970a; Crews, 1980; Garstka and Crews, 1986; Mason, this volume). Several aspects of the nature of this substance and the mechanism by which it fosters gender and species recognition and initiates courtship have been studied in recent years.

The sex-attractant female pheromone appears to be responsible, at least in part, for the preference of males for conspecific female *Thamnophis butleri* and *T. sirtalis* (Devine, 1977a). Male *Thamnophis butleri* and *T. sirtalis* have been tested in a Y maze apparatus for preference in following the trails of female conspecifics and heterospecifics (Devine, 1977a). Males can discriminate between conspecific male and conspecific female trails and can discriminate trails of conspecific females from heterospecific females. Males are attracted only by the chemical substances of conspecific females.

Estradiol-injected female *Thamnophis radix* become sexually attractive to males immediately after shedding (Kubie et al., 1978a). The substance released by the shed is capable of being transferred to females that are group housed with the treated female. Males that prefer one estradiol-treated female to another treated female generalize that preference to the penmates of the preferred treated female. This suggests that the sex attractant contains individual-specific information as well as species-specific information.

Males may find sexually attractive females by trailing (Lillywhite, 1985; Ford, 1986). An elegant series of experiments has dealt with sexual trailing and species recognition in a number of species of *Thamnophis*. Male garter snakes have been tested in a Y maze for their ability to discriminate pheromone trails of conspecific and congeneric females (Ford, 1982). Male *Thamnophis sirtalis* discriminate female conspecific trails from no trails and female conspecific trails from female congeneric (*T. butleri*) trails but do not discriminate female congeneric trails from no trail. *T. butleri* discriminate conspecific trails from no trail, congeneric (*T. sirtalis*) trails from no trail, trails of two congeners (*T. sirtalis, T. radix*) from each other, but surprisingly do not discriminate conspecific trails from one of the two congeners tested (*T. radix*).

Similar studies to those described above indicate that male *Thamnophis radix radix* and *T. r. haydeni* prefer trails of female conspecifics over trails of females of the closely related but allopatric *T. marcianus* (Ford and Schofield, 1984). Male *Thamnophis marcianus* also discriminate trails of female conspecifics from trails of sympatric female congeners (*Thamnophis cyrtopsis*) (Ford and O'Bleness, 1984; Ford and O'Bleness, 1986). Whenever the males follow female conspecific trails, the behavior exhibited is virtually identical to the trail contact response described by Brown and MacLean (1983) (see above). Although requiring further confirmation, these studies suggest that snakes are less likely to follow trails of sympatric than of allopatric congeners; the mechanism for this differential response remains to be elucidated. Clearly, males of these species of *Thamnophis* are able to discriminate odor trails of female conspecifics.

While trailing females, male *Thamnophis radix* make use of positional cues on the substrate to determine the direction that a female has taken (Ford, 1986; Ford and Low, 1982, 1984). Females traversing an environment push against objects as they move. The pheromone used by males to trail females is deposited during contact of the skin of the female with objects she passes by. When pegs are placed in an arena, thus providing a surface onto which females can deposit the pheromone, males subsequently follow the trail left by the female (Ford and Low, 1982). Rotating the pegs 180° causes the males to reverse direction. Whenever the floor remains, but the pheromone deposited on the pegs is removed, males fail to trail accurately, suggesting that the information deposited on the floor is insufficient and that the source of the pheromone is likely to be the lateral surface of the females (Ford and Low, 1984; Ford, 1986).

Thamnophis sirtalis is a seasonal breeder, copulating in late spring and fall. The pheromone trailing by male snakes is best in April, May, and June, least good in late July, and intermediate in the fall (Ford, 1981). Male *T. proximus*, the seasonal breeding pattern of which is less well understood, show no differential trailing ability as a function of season. However, the absence of controls limits the usefulness of these observations.

Most breeding activity of *Thamnophis sirtalis parietalis* occurs on emergence from hibernation (e.g., Crews, 1980; Crews and Garstka, 1982). Male snakes appear to become more discriminating with time out of dormancy (Garstka et al., 1982). Males show no preference for estradiol-treated females 2 days after their emergence, but 10 days later all prefer estradiol-treated females. This result raises the intriguing question of whether the sensory system detecting the pheromone is becoming less sensitive with time, requiring more pheromone, whether it is becoming more discriminating, or whether central neural mechanisms are changing in response to a constant stimulation.

Considerable evidence now exists to support the idea that sex pheromones are produced by the skin of female *Thamnophis butleri* and *T. sirtalis* and not in cloacal scent glands (Devine, 1977a). Female skin–derived sex attractants from *T. butleri* and *T. sirtalis* consist of nonvolatile, relatively nonpolar neutral lipids (Devine, 1977a) that might be wax esters.

That the skin is the source of the female pheromone has been further demonstrated in a clever experiment in which tubes of female skin were placed on male *Thamnophis sirtalis sirtalis* (Gillingham and Dickinson, 1980). The probability of a male being courted by a conspecific male increased from 0% for males with male skin tubes to 33.3% for males with female skin tubes. Females with male skin tubes were less likely to be courted (20.2%) than females with female skin tubes (56.2%).

A vitellogeninlike sex pheromone has been identified on the skin of female garter snakes, *Thamnophis sirtalis parietalis* (Garstka and Crews, 1981a, 1986; Garstka et al., 1982). A second female sex pheromone of *T. s. parietalis* consists of a series of saturated and monounsaturated long-chain methyl ketones (Mason and Crews, 1985b; Mason et al., 1987a, b, 1989). When applied to the backs of males or onto paper towels these substances result in male courtship of the animal or substrate containing the purified or isolated pheromone.

The female sex attractant for males of *Lampropeltis getulus splendida* does not appear to derive from the scent glands (Price and LaPointe, 1981). If males and females are exposed to cloacal gland secretions of males, females, and controls, the tongue-flicks show no significant differences. The tongue-flick responses to secretions are significantly depressed compared to those for controls for females, after all secretions are grouped, and the rates for male secretions are depressed compared to controls for males in the month of February only. However, conduction of multiple T tests for each comparison compromise the assumptions of such a statistic and the p values derived from it.

Courtship-inhibiting pheromones have been reported for *Thamnophis butleri*, *T. sirtalis* (Devine, 1977b), and *T. radix* (Ross and Crews, 1977). Successful mating generates a mating plug in the cloaca of the mated female. Male garter snakes will not court a recently mated female, but detection (by the male) of whatever substance is deterring courtship seems to require close contact with the female (Devine, 1977b). After removal of the mating plugs, females again become sexually attractive. If the expressed plug is placed into sexually attractive females, these are rendered sexually uninteresting to males (Ross and Crews, 1977). These studies suggest that during mating some material is deposited or secreted into the cloacal region of the female and that this material contains a chemosignal that acts as a sexual deterrent.

Male garter snakes may produce a pheromone that signals other males that they are inappropriate objects for courtship. The evidence for such a pheromone is still somewhat tenuous, but some intriguing findings suggest that it probably does exist. Hibernated males have been observed to court nonhibernated males but not to court other hibernated males (Vagvolgyi and Halpern, 1983). Whenever hexane extracts of males are added to extracts of female sex pheromone, male courtship behavior ceases (Mason et al., 1989); this adds support to the idea that male odors may inhibit male-male courtship. Although the presence of a courtship-inhibiting substance on males has not been demonstrated experimentally, the occurrence of such a substance is suggested.

Some male *Thamnophis sirtalis parietalis,* so-called "she-males," are courted by males under natural conditions (Mason and Crews, 1985a, 1986). The attractiveness of these males is not caused by a transfer of female pheromone, as rubbing of males against recently emerged females does not cause them to be courted by other males. Toluene-extracted samples of skin surface of females and she-males elicits courtship by males; extracts from males do not. Chemical analysis indicates that the methyl ketones of she-males are proportionately different from those of males and females.

Females may initiate courtship behavior in *Thamnophis melanogaster* (Garstka and Crews, 1981b). Release of estrogen-treated females into a presumably nonbreeding natural population of garter snakes initiates a breeding period during which untreated females are also courted. This period is followed by increases of androgen levels, testes size, and spermatogenic activity, suggesting that the treated females have initiated a reproductive cycle that would not otherwise have occurred. The absence of a control group makes interpretation of this result difficult. An unanswered question remains whether the female chemical signal alone would have initiated the reproductive cycle.

G. Prey Odors

The response of snakes to prey chemicals has been thoroughly reviewed (Burghardt, 1970a). One of the most interesting phenomena associated with detection of prey by snakes is the occurrence of species-specific prey preferences in newborn, ingestively naive snakes. The newborn respond to the aqueous surface washes of prey that constitute the normal diet of their species (Burghardt, 1966, 1967a, b, 1968a, 1969, 1975; von Achen and Rakestraw, 1984; Mushinsky and Lotz, 1980). These studies developed a standard paradigm in which cotton swabs are dipped in prey washes or control substances (water), the swab is placed in front of the snake, and tongue-flicks and latency to attack by the snake of the swab are monitored. Geographic differences have been observed in newborn litters of *Thamnophis sirtalis* (Burghardt, 1970b). Newborn *Nerodia sipedon* exhibit prey preferences congruent with diverse ecological backgrounds, preferring the extracts of sympatric prey species (Gove and Burghardt, 1975). Geographic (Arnold, 1980) and genetic factors involved in these neonatal preferences have been studied in *T. elegans* (Arnold, 1981a, b). Factor analysis of genetic correlations between two conspecific populations, representing coastal and inland races, suggests that at least three groups of genes underlie the innate prey preferences exhibited at birth. The three groups affect responsiveness to each of three types of prey: one group affects response to some amphibians (anurans and some salamanders), a second group affects response to slugs and leeches, and a third group affects response to the salamander *Taricha*. Geographic variation in the slug-eating behavior of *Thamnophis elegans* has a genetic basis (Arnold, 1981b). Populations that are sympatric with slugs are most likely to eat slugs, whereas populations allopatric with slugs are less likely to eat slugs. Crosses between populations indicate at least partial dominance for slug refusal. Maternal effects are not observed, and the incidence of slug eating in the F1 generation is equivalent for sire or maternal donor.

Neonate garter snake prey preferences are unaffected by the diet of the mother during gestation (Burghardt, 1971). Pregnant female *Thamnophis sirtalis* were fed exclusively a diet of fish or earthworms. Ingestively naive newborns tested 6 or 7 days after birth respond similarly to extracts of fish and worm regardless of the diet of the mother. This study strongly supports the concept that neonate prey preferences have a genetic basis. It suggests that such preferences do not develop as a result of chemicals transferred from mother to offspring during gestation.

Although an absence of differentiated response to chemicals derived from different prey objects has been reported for newborn *Lampropeltis getulus* (Brock and Myers, 1979), *Elaphe obsoleta quadrivittata*, and *E. o. obsoleta* (Morris and Loop, 1969), a subsequent study demonstrated significant differential responses among newborn, ingestively naive *E. obsoleta*, *E. vulpina*, and *E. guttata* (Burghardt and Abeshaheen, 1971). The conflicting results suggest that the cotton swab technique for testing chemical prey preference may not be equally appropriate for all species. Before one can conclude that chemical stimuli are not discriminated one must demonstrate a reliable response. For example, if a chemical stimulus must be coupled with an appropriate visual (or physiological) cue to elicit tongue-flick, attack, or both, then it is incumbent on the experimenter to demonstrate that the response being used as a measure of discriminability of the stimuli can be elicited. Variation of the chemical parameters should not differentially affect the response.

Neonate, ingestively naive *Uromacer frenatus* respond differentially to water extracts of the two lizards *Anolis carolinensis* and *Cnemidophorus sexlineatus* (Henderson et al., 1983). These lizard-eating snakes tongue-flick significantly more to prey extracts than to water but do not attack the swabs containing the extracts. It may be that the swabs do not provide an appropriate visual search image normally required to elicit an attack.

Innate prey preferences can, within limits, be modified through feeding experience (Fuchs and Burghardt, 1971). Exposure to prey chemicals in the neonatal period can alter responsiveness to prey chemicals (Burghardt, 1968b). Inexperienced newborn *Thamnophis sirtalis* that habituate to one of two prey odors during testing are more responsive if they are subsequently tested with the unfamiliar odor. The effects of experience on neonate prey preference depend, in part, on the kind of prey available to the young and normally preferred (Arnold, 1978). The response of neonate *Thamnophis sirtalis* to dead, motionless fish can be enhanced by experience with living fish. Snakes that normally do not respond to fish extracts will attack fish extracts after feeding on living fish; however, snakes fed frog tadpoles during the neonatal period do not respond to fish extract. In contrast, snakes need not have had early experience with specific prey to respond subsequently to extracts of earthworm, bullfrog, or salamander. Also, responses to extracts of preferred prey can be drastically altered by pairing exposure to the extract with illness induced by injections of lithium chloride (Burghardt et al., 1973; Czaplicki et al., 1975).

Experience is not necessarily sufficient to explain changes in responsiveness to prey extracts. Newborn, ingestively naive *Nerodia erythrogaster,* a species for which anurans comprise 85% of the diet, show little selectivity among prey species extracts. For 2 to 6 months after birth, *N. erythrogaster* prefer fish extracts to other prey species extracts, regardless of whether they are fed exclusively on fish (*Gambusia affinis*) or frogs (*Rana clamitans*) (Mushinsky and Lotz, 1980). However, between 8 and 9 months of age, *N. erythrogaster,* reared exclusively on fish or on frogs, demonstrate an increased response to prey chemicals with a switch to a preference for frog extracts. *N. fasciata* feeds on fish (78%) and frogs (22%). Newborn, ingestively naive snakes demonstrate a preference for extracts of fish compared to extracts of frog skin. This preference remains at 2 to 6 months of age regardless of whether the snakes are reared on a diet exclusively of fish or frogs. *N. s. sipedon* respond to airborne odorants from prey with increased tongue-flicking (Dunbar, 1979). Interestingly, regardless of postnatal feeding experience, they respond to frog (*Rana sylvatica*) more than to mosquito fish (*Gambusia affinis*).

Captive-reared *Lampropeltis getulus* fed only laboratory mice after birth were tested with seven stimuli, including prey odors (mouse, snake [*Elaphe*], and chicken), human scent, and controls (distilled water, mammalian saline, and amphibian saline) (Williams and Brisbin, 1978). The greatest number of tongue-flicks are elicited by prey odors, and these responses do not significantly differ from each other. Responses to human scent and the controls were lower and do not significantly differ from each other. However, stimuli eliciting intermediate rates of tongue-flicking, snake, chicken, and human odor also did not differ significantly from each other, nor did response to chicken, human scent, distilled water, or amphibian saline odors. If the snakes, despite their restricted diet, retain their ability to recognize several different prey odors, as the authors conclude, then one would expected the responses to snake, chicken, and human odor to differ significantly from each other, which they do not.

Neonate *Lampropeltis getulus* presented with cotton swabs coated with the skin chemicals of snakes that could be appropriate prey, tongue-flick, attack, and attempt to ingest the swabs (Weldon and Schell, 1984). King snakes also respond differentially to airborne snake odors.

Responsiveness of snakes to prey odors may be affected by a variety of factors, including mode of presentation, environment in which the testing occurs, and prior history of the animal (Loop, 1970). Adult specimens of *Crotalus viridis* and *C. enyo* were repeatedly tested in clean cages and cages containing mouse odors over an 8-week period, during which they were not fed (Chiszar et al., 1981b). During the first 4 weeks tongue-flick rates were indistinguishable between clean cages and those containing mouse odors. During the last 3 weeks mean tongue-flick rate in the mouse odor condition increased significantly over the clean cage condition. This finding is suggestive of an effect of food deprivation on sensitivity to or responsiveness to potential food odors; however, this experiment needs to be repeated with a control group.

Repeated presentation (at 90- and 360-second intervals) of chemical extracts of normal prey to adult *Nerodia r. rhombifera* causes habituation of the response (Czaplicki, 1975). Dishabituation occurs if the snakes are presented with a novel chemical stimulus.

Adult *Thamnophis radix* and *T. sirtalis sirtalis* will follow trails of prey odor (aqueous earthworm wash) for food rewards in a two- or four-choice maze (Kubie and Halpern, 1975, 1978). During trailing the tongue touches the substrate on virtually every tongue-flick, and tongue-flick rates are high, stable from trial to trial, and show individual variation. Similar observations of trailing prey have been made for *Crotalus adamanteus* (Brock and Means, 1977), *Natrix natrix* (Kahmann, 1932; Wiedemann, 1931), and *Vipera aspis* (Naulleau, 1965). Tongue-flick rates of *Thamnophis* are highest in animals that traverse the maze rapidly (Kubie and Halpern, 1978; Halpern and Kubie, 1983). *Thamnophis* follows highly concentrated earthworm wash trails more accurately than diluted trails, and tongue-flick rates are higher on intense trails than on diluted trails (Kubie and Halpern, 1978). Dried trails are also followed accurately; however, the snakes must incur tongue contact with the odor source. Film analysis shows that a decreased concentration of earthworm wash effects the major change of increased tongue-flick duration. Tongue-flicks of longer duration are characterized by longer extensions of the tongue. Keiser (1975) hypothesized that long-extrusion tongue-flicks serve to accumulate detectable amounts of odorant on the tongue in a stimulus-poor environment.

Leptotyphlops dulcis follow trails deposited by army ants (*Neivamyrmex nigrescens*) as interspecific chemosignals (Watkins et al., 1967, 1969). *Thamnophis sirtalis* respond preferentially to electric shock secretions of earthworms, which contain an alarm pheromone (Halpern et al., 1987). In both of these cases, the predators are using secretions that presumably have survival value for the prey.

The mechanism by which snakes find prey under natural conditions is unknown. *Thamnophis* may sense a chemical signal emitted by earthworm castings (Gillingham et al., 1991). Direct contact with the castings is not necessary, suggesting that volatiles are involved. Also, fresh castings are more effective stimuli than castings 24 hours old.

Snakes can detect airborne odorants from prey and may attack an airstream containing such odors (Burghardt, 1977b, 1980). Tongue-flick rates of *Thamnophis* increase significantly during presentation of air blown over goldfish or over earthworms rather than distilled water (Halpern and Kubie, 1983; J. Halpern et al., 1985). Scavenging behavior has rarely been reported in snakes. However, *Crotalus atrox* can locate putrescent mice buried in gravel but fail to find freshly killed animals (Gillingham and Baker, 1981). In the absence of other cues, the mice would seem to be located by chemical signals.

Prey chemicals may direct locomoting *Crotalus v. viridis* to areas with high concentrations of rodent prey. Colorado *C. v. viridis* prefer areas of test arenas that contain shavings soiled by house mice (Duvall et al., 1990). Interestingly, Wyoming *C. v. viridis* do not demonstrate as strong a response, exhibiting an increase in only one dependent measure, ambush posture, when, and only when, they are permitted direct contact with the prey-soiled substrate. These results are interpreted in terms of differences between Wyoming and Colorado snakes in habitat and in recent evolutionary histories that may have resulted in different foraging strategies.

Although considerable evidence exists that chemical cues are important for prey detection, visual cues may also be used under natural conditions. *Thamnophis sirtalis* selectively attack rotating rather than stationary earthworm sections (Burghardt and Denny, 1983). In the absence of odor cues and at relatively high rotation speeds, attacks decrease. In the absence of earthworm odor, the tongue-flick rate gradually declines. Whenever earthworm odor permeates the test arena, the tongue-flick rates remain high.

Visual stimuli alone can elicit out-of-water approaches, orientations, and attacks in *Nerodia s. sipedon* (Drummond, 1979). Prey odor in the water results in faster orientation and approach and increased frequency of attack. The snakes will attack and attempt to seize and ingest pebbles coated with fish odor. *N. sipedon* and several species of *Thamnophis* increase orientations and attacks to fish models whenever the water contains fish odor (Drummond, 1985). Fish odor also increases aquatic searching behavior.

Rattlesnakes typically consume prey large enough to harm them if not released immediately after the strike. However, release of the envenomated animal frequently leads to disappearance of the prey as it wanders from the ambush site. To retrieve the prey in the absence of visual, auditory, or tactile signals, snakes explore the chemical environment for cues to the location of the dying or dead prey. Strike-induced chemosensory searching (SICS) follows successful strikes and subsequent release of prey by venomous snakes (reviewed by Chiszar and Scudder, 1980; Chiszar et al., 1983a). The behavior is characterized by locomotion and frequent tongue-flicking and usually culminates in localization, seizing, and ingestion of the envenomated prey (Baumann, 1927, 1928, 1929; Dullemeijer, 1961; Naulleau, 1965). The prevailing view is that the initial envenomation event is mediated by visual and thermal cues and that trailing during SICS is mediated by the vomeronasal system (Burghardt, 1970). Experiments showed that snakes with their tongues removed fail to trail prey (Dullemeijer, 1961; Naulleau, 1965); however, congenitally alingual *Vipera russellii* display chemosensory searching (Carr et al., 1982).

Chemosensory searching or prey trailing in rattlesnakes is initiated by the strike of prey. Within 2 to 3 minutes after the strike, tongue-flick rates begin to rise, and this increase continues until swallowing starts. The dependence of SICS on a strike has been demonstrated in a large number of species, including members of *Agkistrodon, Bitis, Crotalus, Eristicophis, Naja, Sistrurus,* and *Vipera* (Chiszar et al., 1977, 1982a, 1983c, Chiszar and Scudder, 1980).

Recent research has attempted to determine the importance of several variables in the expression of SICS. These include the role of striking on the elevation of tongue-flick rate after encounters with prey, the role of chemosensory stimuli available during the strike in creating a search image, the importance after envenomation takes place of cues in the environment in initiating and directing movement and tongue-flicking, the generality of SICS among snake species and various prey parameters on the pre-envenomation and poststrike behavior of snakes.

Chemical cues in the absence of predatory strikes appear to have a relatively weak effect on chemosensory searching (Chiszar et al., 1981c). *Crotalus e. enyo, C. v. viridis,* and *Sistrurus catenatus tergeminus* were tested for elevated tongue-flick rates after presentations of chemical, visual, and thermal stimuli, alone or in combination. Visual and thermal cues together elevate tongue flick rates during the 5 minutes following presentation, but chemical cues alone do not do so. In contrast, *Thamnophis radix haydeni* exhibit elevated tongue-flick rates to chemical or visual stimulation.

Crotalus viridis and *C. enyo* have been tested under four experimental conditions to determine the effects of various stimuli on the duration of SICS (Chiszar et al., 1982b). In Condition 1 snakes see, smell, and detect thermal cues from living mice but are not permitted to strike. In Condition 2, strikes are permitted but the envenomated mouse is then removed. In Condition 3, a strike is permitted, but the envenomated mouse remains in the cage. In Condition 4, the envenomated mouse remains in the cage and a second mouse is introduced after the first is consumed. High tongue-flick rates are emitted by snakes in all conditions except if strike is precluded. When envenomated mice are removed, SICS lasts for 150 minutes following the strike. SICS terminates after ingestion but may be reinstituted by presentation of a living mouse as long as 2.5 hours after striking.

Adult rodents envenomated by snakes are typically released by the snake and allowed to wander as the venom takes effect. The trail followed by snakes may originate from urine, integument, or other excretions or exudates of the prey. In laboratory tests, *Crotalus viridis* follows integumentary trails of rodent prey (*Mus musculus*), but not urine trails (Chiszar et al., 1990). This finding strongly supports the idea that urine trails of envenomated prey are insufficient to support poststrike trailing; however, the source of the integumentary trail has yet to be determined.

Rattlesnakes discriminate and prefer mice they or a conspecific have envenomated in contrast to control mice killed by some other means (Duvall et al., 1978, 1980b). Tongue-flicks during the investigation of envenomated prey are preferentially directed to the oral-nasal area, and the snakes are capable of discriminating oral-nasal tissue from anogenital tissue of envenomated mice but do not respond differentially to those of nonenvenomated mice. These data suggest that envenomation changes the chemical signals emitted by dead or dying prey, and the critical stimuli are found in the head region of the prey. However, multiple cues may be involved in head-first prey ingestion, some unrelated to changes occurring following envenomation (Klein and Loop, 1975).

In *Agkistrodon piscivorus*, strike-induced chemosensory searching persists for about 70 minutes (Chiszar et al., 1985). As chemical cues derived from the strike would not remain available for that long a period, it has been hypothesized that a chemical search may sustain SICS and aid in identification of the struck prey. Snakes that struck either rodent or fish prey were later tested with either. Presentation of prey for ingestion of the same type as that struck initially causes snakes to grasp the prey quickly. When the prey offered for ingestion differs from the prey struck initially, only a few snakes grasp the prey quickly; others do so with very long latencies. This supports the idea that information obtained during envenomation remains available through the poststrike period.

Evidence that chemical learning occurs during the strike and is subsequently used to guide poststrike rattlesnake behavior is presented in studies by Melcer and Chiszar (1989a, b). *Crotalus viridis* have been permitted to strike mice misted with dilute perfume or control. In subsequent tests, the snakes prefer carcasses of nonenvenomated mice scented with perfume to unscented carcasses of nonenvenomated mice. A similar experiment using mice fed on two different diets yielded similar results, suggesting that chemical cues picked up during the strike direct subsequent prey preferences.

The importance of chemical cues in the poststrike environment may depend on the nature of prey typically eaten by the snake (Cruz et al., 1987). After a predatory strike, *Crotalus viridis*, a rodent specialist, maintains high tongue-flick rates regardless of the presence or absence of rodent odors. In contrast, *C. pricei*, a lizard specialist, only maintains high tongue-flick rates in the poststrike period in the presence of prey odors. Although SICS supposedly requires vomeronasal feedback in *C. pricei*, no evidence is presented that the prey odor effect on this behavior is vomeronasally mediated.

The nature of preferred prey may have other effects on SICS. For example, snakes that release their prey after striking may trail envenomated prey more effectively than snakes that typically do not release their prey (Chiszar et al., 1986b). *Crotalus lepidus klauberi* primarily feed on lizards, tongue-flick less, and spend less time on prey trails than *C. v. viridis,* a rodent specialist that typically releases its prey. As in the previous experiment, tongue-flick rates of *C. v. viridis* were high after striking, regardless of the presence of prey odor, whereas the tongue-flick rates of *C. l. klauberi* were high only in the presence of a prey odor. However, before one can accept the proposition that species that do not release prey are poorer at trailing, more species belonging to each category need to be tested.

Agkistrodon piscivorus preys on rodents and fish, releasing rodents but retaining fish (Chiszar et al., 1986a). Both prey may be trailed, and tongue-flick rates do not differ for trails of fish and mice. Tongue-flick rates are higher following strikes and decrease when strikes are blocked, but striking does not affect trailing behavior. At least in this species, two elements of SICS can be dissociated: tongue-flicking and trail following. The former is clearly affected by the presence or absence of a strike; the latter is affected primarily by the chemical environment present at the time of trailing.

Tongue-flick rate following presentation of prey to rattlesnakes is consistently higher after striking than in the absence of a strike (Chiszar and Scudder, 1980). The nature of the strike, predatory versus defensive, affects the level of tongue-flicking; although both types of strikes increase tongue-flick rate, predatory strikes increase tongue-flick rate more than defensive strikes (O'Connell et al., 1982). Visual and thermal cues from living mice increase tongue-flick rates above baseline levels without a strike. This increase is sustained longer in the presence of odor cues than in their absence (Chiszar et al., 1983a).

Although SICS is clearly facilitated by striking, tongue-flick rates may be elevated following presentation of visual and thermal cues from appropriate prey that are then removed from the field of the snake predator (Gillingham and Clark, 1981a). *Crotalus atrox* have been tested for responses to mice that initially were in the field of view and then removed from it. Tongue-flick rates were depressed while the snakes could see the prey; rates increased significantly if the mouse moved away and decreased again when the mouse was again visible. This observation suggests that searching, and the increased tongue-flick rate that accompanies it, may be triggered in the absence of a strike. An interesting aspect of the observation here noted is that the stimulus mouse was free moving (but tethered so as to keep it from getting too close to the snake) and therefore may have provided a more realistic prey image than a mouse dangling from forceps.

Until recently, SICS has been reported primarily in venomous snakes. Recently two colubrids, *Elaphe guttata* and *Thamnophis sirtalis*, were studied to determine whether they also exhibited it (Cooper et al., 1989). The paradigm involves presenting prey, permitting a strike, then forcibly removing the prey and counting the number of tongue-flicks following the manipulation. A variety of chemical, visual, or tactile cues that represent components of the normal prey stimulus are presented, and appropriate controls are included to compensate for the effects of handling and nonspecific stimulation. Tactile stimulation of the mouth is used as a control for chemical stimulation of the oral mucosa; pulling the snake backward, rather than removing the prey, is used to control for handling during prey removal. In both snakes tongue-flick rates increase after prey attack and prey removal. An issue in such experiments is that the behavior of release of prey may not be part of the normal behavioral repertoire of the animal. The basis of SICS is that animals that strike and release prey can search to find them, as there is expectation that envenomation leads to death. Search for SICS in animals that normally do not envenomate and release prey creates a laboratory paradigm by interfering with normal behavior. Furthermore, chemosensory searching is more than increase in tongue-flick rate. It involves directed locomotion and trail following. Movement and tongue-flicks need to be monitored together, and there must be some independent observation that SICS is under chemosensory control.

Graves and Duvall (1985b) reported that rattlesnakes (*Crotalus viridis*) with their vomeronasal ducts sutured closed fail to strike rodent prey. This finding is particularly interesting considering the general acceptance of the idea that the vomeronasal system is critical for the poststrike chemosensory searching portion of prey capture. However, before the finding can be accepted as demonstrating that the vomeronasal system mediates prey attack in *Crotalus*, the chemosensory control of prey attack and the specificity of the duct suture effect must be demonstrated.

H. Chemical Analysis of Prey Substances

Prey chemicals that are detected by a functional vomeronasal system (see below) have been the subject of chemical analysis in recent years. Earthworm surface washes and secretions have now yielded some purified substances to which *Thamnophis* respond with attack characteristic of their response to prey. Warm water washes of *Lumbricus terrestris* contain prey chemicals that are nonvolatile—highly stable macromolecules in excess of 5,000 daltons that can be lyophilized without losing activity (Sheffield et al., 1968; Reformato et al., 1983; Halpern et al., 1984, 1986; Kirschenbaum et al., 1985).

Earthworm- and fish- (fathead minnows, *Pimephales promelas*) derived prey chemicals have been extracted from aqueous prey washes with chloroform:methanol (Burghardt et al., 1988; Schell et al., 1990). The water-soluble material results from partition of the chloroform:methanol phase passed through a gel filtration column and yields two active fractions of high and low molecular weights. The high-molecular-weight fraction from earthworms is estimated to be 432 kilodaltons (kDa), that from minnow, 519 kDa. The low-molecular-weight component is estimated to be between 1.35 and 1.8 kDa for both prey types. Biologically active material from minnows or earthworms exhibits similar gel filtration profiles, carbohydrate and amino acid contents, and NMR spectra. However, these analyses were performed on nonpurified fractions of the prey washes.

The active substances in earthworm wash that were erroneously identified as earthworm cuticle collagen (Kirschenbaum et al., 1986) have since been identified as glycoproteins (Wang et al., 1988; Schell et al., 1990; Burghardt et al., 1988). At least one such protein has been isolated and purified from earthworm wash using covalent chromatography (Wang et al., 1988). The proteinaceous attractant contains free sulfhydryl groups and has an apparent mass of 20 kDa. The activity of the purified protein can be destroyed by heating, as well as by proteolysis. The activity of the protein can also be reversibly blocked by mixed disulfide formation. The free sulfhydryls may be essential in letting the snake recognize this material; this hypothesis must be tested further.

A substance that snakes attack has been isolated from electric shock–induced secretions of earthworms and purified by permeation chromatography and semipreparative nondenaturing polyacrylamide gel electrophoresis (PAGE) (Jiang et al., 1990). The purified glycoprotein is more active on a unit protein basis than is the crude secretion. It loses activity after proteolytic digestion. The glycoprotein, which consists of a single polypeptide chain with an *N*-terminal alanine, has a minimum molecular weight of 14.4 kDa calculated from amino acid and carbohydrate contents and an apparent molecular weight of about 20 kDa estimated from SDS PAGE. The sequence of its *N*-terminal 15 amino acid residues has been determined. The purified glycoprotein binds to snake vomeronasal epithelial membrane fractions but not to membrane extracts of the nonsensory mushroom body (Jiang et al., 1990). The binding is saturable and reversible, suggesting that receptors in the epithelium specifically bind this glycoprotein. The substance yields a behavioral response whenever delivered to the vomeronasal organ. If it is applied directly to the exposed vomeronasal epithelium of a snake prepared for electrophysiological recording, it modifies the firing of single units in the accessory olfactory bulb (Fig. 7.15) (Inouchi et al., 1990; Jiang et al., 1990).

XI. CHEMICAL SENSES: ROLES IN BEHAVIOR

The chemical senses used in reptilian responses to chemosignals have been investigated primarily by inactivating one of the two nasal chemical sense systems. Inactivation may occur (1) at the periphery, for example by covering the external nares, zinc sulfate irrigation of the olfactory epithelium, cutting the tongue (presumably to prevent odorant delivery to the vomeronasal organ), suturally or chemically closing the vomeronasal ducts; (2) in the pathway from periphery to central nervous system, for example by cutting the nerves; or (3) centrally, by placing lesions in one of the target regions of the olfactory or vomeronasal system.

Few recent studies have inactivated the olfactory system and then assessed reptilian behavior. The homing behavior of *Testudo hermanni* has been assessed after they were rendered anosmic by zinc sulfate (Chelazzi et al., 1981). Such turtles were displaced 800 m from the capture point and allowed to travel for 7 and 14 days. Control animals return to the home area, but anosmic turtles remain more or less randomly dispersed around the release site.

The role of the olfactory sense in capture of fruit flies has been investigated in *Hemidactylus frenatus* and *Cosymbotus platyurus* (Chou et al., 1988). After the geckos are made anosmic by severing their olfactory nerves, only *Cosymbotus* show a significant impairment in their capacity for capturing fruit flies. In both species, auditory and visual cues seem to be more important than olfactory cues.

The role of the olfactory system in several snake behaviors has been investigated as part of experiments designed to determine the relative contributions of the olfactory and vomeronasal systems in detecting chemical stimuli. Covering the nostrils of newborn, ingestively naive *Thamnophis* with collodion does not significantly affect response to prey extracts (Burghardt and Hess, 1968; Burghardt and Pruitt, 1975). Olfactory nerve section does not impair trailing for prey by *Thamnophis* (Kubie and Halpern, 1979), prey attack and feeding (Halpern and Frumin, 1979), response to earthworm electric shock secretions (Schulman et al., 1987), male courtship behavior (Kubie et al., 1978b), or aggregation behavior (Heller and Halpern, 1982b). However, attack on prey extract (Halpern and Frumin, 1979) and return of individual snakes to preferred shelters (Heller and Halpern, 1982b) significantly improve following section of the olfactory nerve. This suggests that volatiles detected by the main olfactory system may provide cues that interfere with discrimination of prey and conspecific odors.

Olfactory nerve lesions severely compromise lingual air sampling by *Thamnophis* in response to airborne odorants (J. Halpern et al.,

1985). Snakes tested in an olfactometer show increased tongue-flick rates in response to amyl acetate, limonene, earthworm, and fish odors. However, section of the olfactory nerve does not lead to significant elevation in tongue-flick rate during or after delivery of these odorants. Interpretation of the effect is complicated because vomeronasal nerve lesions also compromise the response to airborne odorants, although less obviously. It is possible, as suggested by the Cowles and Phelan hypothesis (1958), that airborne odorants perceived by the olfactory system initiate tongue-flicking and that maintained elevated tongue-flick rates depend on a functional vomeronasal system. This idea has yet to be put to the critical test.

All recent studies using vomeronasal deafferentation in reptiles have used snakes or lizards. *Dipsosaurus dorsalis,* with their vomeronasal ducts sealed, do not respond differentially to cotton swabs containing food odors (Cooper and Alberts, 1991). *Chalcides ocellatus,* with their vomeronasal ducts sealed closed, tongue-flick less than do normal skinks and do not elevate their tongue-flick rate in a novel environment as do unoperated skinks (Graves and Halpern, 1990). Prey consumption, but not prey attack, is severely affected by closure of the vomeronasal duct. It is possible that prey attack is mediated by multiple cues, and is therefore not compromised by the duct closure technique.

The detection of female sex pheromones by male snakes depends on a functional vomeronasal system. Male *Vipera berus,* with tape containing lidocaine (Xylocaine) placed over the vomeronasal organs, do not court sexually attractive females (Andrén, 1982). Male *Thamnophis radix* do not court estradiol-injected sexually attractive females if tongue-flicking is prevented by the injection of Xylocaine into the chin (Halpern and Kubie, 1983) or their vomeronasal nerves are sectioned (Kubie et al., 1978b). Interestingly, *Thamnophis sirtalis parietalis* that sustain lesions of the nucleus sphericus (the major target of accessory olfactory bulb efferents) before hibernation initiate courtship significantly sooner and exhibit courtship longer than control animals (Krohmer and Crews, 1987). In addition to the behavioral effect, snakes with such lesions had a significantly greater quantity of sexual granules in the epithelium of the renal sex segment. This study is particularly provocative because it suggests that nucleus sphericus has an inhibitory function in reproductive behavior. However, it does not include negative surgical controls because septal lesions also facilitate courtship.

In *Thamnophis,* aggregation and shelter selection behavior is also affected by vomeronasal deafferentation. Tongueless newborn garter snakes do not aggregate as strongly as do normal snakes (Burghardt,

1980). Adult garter snakes with vomeronasal nerve lesions do not return to previously preferred shelter locations when tested individually (Heller and Halpern, 1982b). However, if these animals are tested in a group that includes normal snakes or those with olfactory nerve lesions, they return to previously preferred shelters. Groups of *Thamnophis* with their vomeronasal ducts sutured closed are significantly depressed in aggregation compared to control animals or to those blindfolded or olfactory nerve sectioned.

As mentioned above, adult *Thamnophis proximus* discriminate between skin lipid extracts from conspecifics and those from sympatric heterospecifics (*Elaphe guttata*). Whenever the vomeronasal ducts of *T. proximus* are sealed closed, these animals do not respond differentially to conspecific skin lipids (Graves and Halpern, 1991).

Thamnophis respond to odor-laden airstreams with increased tongue-flicking (J. Halpern et al., 1985). Although snakes with bilateral vomeronasal nerve lesions show increased tongue-flick rate during odor delivery over the values before odor delivery, the tongue-flick rate decreases significantly during odor delivery compared both to preoperative levels and the rate of control animals. The deficit is considerably less than that displayed by snakes with their olfactory nerves sectioned. It is possible that the deficit following vomeronasal nerve lesions results not from a detection deficit, but from interference in the vomeronasal, tongue-flick feedback loop.

The role of the vomeronasal organ in prey recognition has been reviewed (Burghardt, 1970a; Halpern, 1980a, 1983, 1987; Halpern and Kubie, 1984). Removal of the tongues of newborn, ingestively naive *Thamnophis* results in a loss of discrimination response to aqueous prey extracts (Burghardt and Pruitt, 1975). Attacks on prey extract and unconditioned prey bite do not occur after the vomeronasal nerves are lesioned in adults (Halpern and Frumin, 1979). Garter snakes with vomeronasal nerve lesions cannot accurately follow earthworm extract trails in a maze and extinguish their earthworm reward eating behavior in time (Kubie and Halpern, 1979). Vomeronasal duct sutures also result in loss of trailing behavior and the gradual loss of prey attack behavior. Discriminated response to electric shock–induced earthworm secretions and earthworm wash also ceases following vomeronasal nerve lesions (Schulman et al., 1987).

Consequently, appetitive behaviors are extinguished following vomeronasal deafferentation, and the vomeronasal system is important in a number of critical species-typical behaviors. These observations have led to the hypothesis that the vomeronasal system transmits to the central nervous system information on the reinforcing value of unconditioned stimuli (Halpern, 1988). Support for this proposal has been obtained from studies demonstrating that experience before vomeronasal disruption attenuates the effects of deafferentation and that vomeronasal stimuli can be used as unconditioned or reinforcing stimuli to condition a response to an arbitrary stimulus.

XII. CONCLUSION

Although most of the relatively recent research on nasal chemical senses in reptiles has been conducted on snakes, lizards are now the subjects of an increasing number of studies. Squamates have been particularly useful for these studies because their tongue-flick is an observable, quantifiable behavioral correlate to vomeronasal stimulation; this has been very useful for investigating vomeronasal function in a noninvasive manner. However, the literature is replete with statements uncritically linking increased tongue-flick rates with vomeronasal stimulation. For each behavior, it is necessary to demonstrate that chemical stimulation enhances tongue-flick rates and that tongue-flick rates fail to be elevated in the absence of a functional vomeronasal system.

Another advantage of using squamates for such studies is that, at least in snakes, the vomeronasal and main olfactory systems can be denervated independently. Our inability to do this in mammals has been a major stumbling block to contrasting the two systems functionally.

The facility for investigation of prenatal and postnatal development and perinatal use of the nasal chemical senses represents another advantage of reptilian subjects. However, surprisingly little recent work has been published on the developmental morphology of the olfactory and vomeronasal systems in reptiles.

The paucity of studies on nonsquamate reptiles is a constraint on our ability to generalize about the role of the nasal chemical senses in the behavior of reptiles and to discuss with any generality the relative contributions of the olfactory and vomeronasal senses. A recent resurgence of interest in Testudines, particularly for use in vomeronasal electrophysiology, may change that situation. However, we need more well-designed experiments identifying the behavioral domains of the vomeronasal and olfactory systems in turtles. The sophistication of the electrophysiology will be ill used unless biologically relevant odorants are used as stimuli.

Broadly, future research on reptilian nasal chemical sense systems should aim at increasing our understanding of the roles of these systems in naturally occurring behaviors, their interactions with other sensory systems, and the neural and molecular mechanisms by which significant chemicals evoke, guide, and direct behaviors.

ACKNOWLEDGMENTS

The author's research is supported by NIH grant No. NS11713. I am grateful to my collaborators and colleagues who permitted me to use illustrations from their publications and unpublished work. Drs. David Holtzman and Brent Graves made many helpful suggestions on a previous version of this manuscript.

REFERENCES

von Achen, P. H., and Rakestraw, J. L. (1984). The role of chemoreception in the prey selection of neonate reptiles. In *Vertebrate Ecology and Systematics: a Tribute to Henry S. Fitch* (R. A. Seigel, L. E. Hunt, J. L. Knight, L. Malaret, and N. L. Zuschlag, eds.). University of Kansas Press, Lawrence, pp. 163–172.

Agarwal, P. N., and Sharma, S. (1979). On the brain and the cranial nerves of Indian cobra, *Naja naja naja* (Linn.). Fam: Elapidae. *Ind. J. Zool.* 7, 23–24.

Alberts, A. C. (1989). Ultraviolet visual sensitivity in desert iguanas: implications for pheromone detection. *Anim. Behav.* 38, 129–137.

Alberts, A. C. (1990). Chemical properties of femoral gland secretions in the desert iguana, *Dipsosaurus dorsalis. J. Chem. Ecol.* 16, 13–25.

Allen, B. A., Burghardt, G. M., and York, D. S. (1984). Species and sex differences in substrate preference and tongue flick rate in three sympatric species of water snakes (*Nerodia*). *J. Comp. Psychol.* 98, 358–367.

Allison, A. C. (1953). The morphology of the olfactory system in the vertebrates. *Biol. Rev.* 28, 195–244.

Altner, H., and Muller, W. (1968). Elektrophysiologische und elektronenmikroskopische Untersuchungen an der Riechschleimhaut des jacobsonschen Organs von Eidechsen (*Lacerta*). *Z. Verg. Physiol.* 60, 151–155.

Altner, H., Muller, W., and Brachner, I. (1970). The ultrastructure of the vomero-nasal organ in Reptilia. *Z. Zellforsch.* 105, 107–122.

Amiguet, M., Martinez-Garcia, F., Olucha, F., and Poch, L. (1986). Olfactory bulbs connections in the lizard *Podarcis hispanica. Neurosc. Lett.*, Suppl., 26, S379.

Andrén, C. (1982). The role of the vomeronasal organs in the reproductive behavior of the adder *Vipera berus. Copeia* 1982, 148–157.

Arnold, S. J. (1978). Some effects of early experience on feeding responses in the common garter snake, *Thamnophis sirtalis. Anim. Behav.* 26, 455–462.

Arnold, S. J. (1980). The microevolution of feeding behavior. In *Foraging Behavior: Ecological, Ethological and Psychological Approaches* (A. C. Kamil and T. D. Sargent, eds.). Garland STPM Press, New York, 409–453.

Arnold, S. J. (1981a). Behavioral variation in natural populations. I. Phenotypic, genetic and environmental correlations between chemoreceptive responses to prey in the garter snake, *Thamnophis elegans. Evolution* 35, 489–509.

Arnold, S. J. (1981b). Behavioral variation in natural populations. II. The inheritance of a feeding response in crosses between geographic races of the garter snake, *Thamnophis elegans*. *Evolution* 35, 510–515.

Auen, E. L., and Langebartel, D. A. (1977). Cranial nerves of the colubrid snakes *Elaphe* and *Thamnophis*. *J. Morphol.* 154, 205–222.

Auffenberg, W. (1978a). Courtship and breeding behavior in *Geochelone radiata* (Testudines: Testudinidae). *Herpetologica* 34, 277–287.

Auffenberg, W. (1978b). Social and feeding behavior in *Varanus komodoensis*. In *Behavior and Neurology of Lizards* (N. Greenberg and P. D. MacLean, eds.). National Institutes of Mental Health, Bethesda, Md., pp. 301–331.

Balaban, C. D. (1977). Olfactory projections in emydid turtles (*Pseudemys scripta elegans* and *Graptemys pseudogeographica*). *Am. Zool.* 17, 887.

Bannister, L. H. (1968). Fine structure of the sensory endings in the vomeronasal organ of the slow-worm *Anguis fragilis*. *Nature* (London) 217, 275–276.

Baumann, F. (1927). Experimente über den Geruchssinn der Viper. *Rev. Suisse Zool.* 34, 173–184.

Baumann, F. (1928). Ueber die Bedeutung des Bisses und des Geruchssinnes für den Nahrungserwerb der Viper. *Rev. Suisse Zool.* 35, 233–239.

Baumann, F. (1929). Experimente uber den Geruchssinn und den Beuterwerb der Viper (*Vipera aspis* L.). *Z. Vergl. Physiol.* 41, 329–343.

Begun, D., Kubie, J. L., O'Keefe, M. P., and Halpern, M. (1988). Conditioned discrimination of airborne odorants by garter snakes (*Thamnophis radix* and *T. sirtalis sirtalis*). *J. Comp. Psychol.* 102, 35–43.

Belekhova, M. G., and Kenigfest, N. B. (1983). A study of the hippocampal mediodorsal cortex connections in lizards by means of horseradish peroxidase axonal transport. *Neurofiziologiya* 15, 145–152. [In Russian, English abstract.]

Bellairs, A. D'A. (1970). *The Life of Reptiles*. Universe Books, New York.

Bellairs, A. D'A., and Boyd, J. D. (1950). The lachrymal apparatus in lizards and snakes. II. The anterior part of the lachrymal duct and its relationship with the palate and with the nasal and vomeronasal organs. *Proc. Zool. Soc. Lond.* 120, 269–309.

Bertmar, G. (1981). Evolution of vomeronasal organs in vertebrates. *Evolution* 35, 359–366.

Beuerman, R. W. (1975). Slow potentials of the turtle olfactory bulb in response to odor stimulation of the nose. *Brain Res.* 97, 61–78.

Beuerman, R. W. (1977). Slow potentials in the turtle olfactory bulb in response to odor stimulation of the nose and electrical stimulation of the olfactory nerves. *Brain Res.* 128, 429–445.

Bissinger, B. E., and Simon, C. A. (1979). Comparison of tongue extrusions in representatives of six families of lizards. *J. Herpetol.* 13, 133–139.

Blum, M. S., Byrd, J. B., Travis, J. R., Watkins, J. F., II., and Gehlbach, F. R. (1971). Chemistry of the cloacal sac secretion of the blind snake *Leptotyphlops dulcis*. *Comp. Biochem. Physiol.* 38(1B), 103–107.

Bogert, C. M. (1941). Sensory cues used by rattlesnakes in their recognition of ophidian enemies. *Ann. N.Y. Acad. Sci.* 41, 329–344.

Boĭko, V. P. (1983). The jaw testing movements and role of sensory systems

in food searching behaviour of *Emys orbicularis* (Testudines, Emydidae). *Zool. Zh.* 62, 1528–1532. (In Russian, English abstract.)
Boĭko, V. P. (1984a). The participation of chemoreception in the organization of reproductive behavior in *Emys orbicularis* (Testudines, Emydidae). *Zool. Zh.* 63, 584–589. (In Russian, English abstract.)
Boĭko, V. P. (1984b). Reproductive behavior of the Mediterranean tortoise *Testudo graeca. Zool. Zh.* 63, 228–232. (In Russian, English abstract.)
Brock, O. G., and Means, D. B. (1977). Preliminary observations on the prey trailing behavior of the eastern diamondback rattlesnake, *Crotalus adamanteus. American Society of Ichthyology and Herpetology,* Gainesville, Fla. (Abstract.)
Brock, O. G., and Myers, S. N. (1979). Responses of ingestively naive *Lampropeltis getulus* (Reptilia, Serpentes, Colubridae) to prey extracts. *J. Herpetol.* 13, 209–212.
Brown, W. S., and MacLean, F. M. (1983). Conspecific scent-trailing by newborn timber rattlesnakes, *Crotalus horridus. Herpetologica* 39, 430–436.
Bruce, L. L., and Butler, A. B. (1984). Telencephalic connections in lizards. I. Projections to cortex. *J. Comp. Neurol.* 229, 585–601.
Burghardt, G. M. (1966). Stimulus control of the prey attack response of naive garter snakes. *Psychon. Sci.* 4, 37–38.
Burghardt, G. M. (1967a). Chemical perception in newborn snakes. *Psychol. Today* 1, 50–59.
Burghardt, G. M. (1967b). Chemical-cue preferences of inexperienced snakes: comparative aspects. *Science* (N.Y.) 157, 718–721.
Burghardt, G. M. (1968a). Chemical preference studies on newborn snakes of three sympatric species of *Natrix. Copeia* 1968, 732–737.
Burghardt, G. M. (1968b). The reinforcement potential of the consummatory act in the modification of chemical-cue preferences. *Am. Zool.* 8, 745.
Burghardt, G. M. (1969). Comparative prey-attack studies in newborn snakes of the genus *Thamnophis. Behaviour* 33, 77–114.
Burghardt, G. M. (1970a). Chemical perception in reptiles. In *Advances in Chemoreception. I. Communication by Chemical Signals* (J. W. Johnston, D. G. Moulton, A. Turk, eds.). Appleton-Century-Crofts, New York, pp. 241–308.
Burghardt, G. M. (1970b). Intraspecific geographical variation in chemical food cue preferences of newborn garter snakes (*Thamnophis sirtalis*). *Behaviour* 36, 246–257.
Burghardt, G. M. (1971). Chemical-cue preferences of newborn snakes: Influence of prenatal maternal experience. *Science* (N.Y.) 171, 921–923.
Burghardt, G. M. (1973). Chemical release of prey attack: Extension to naive newly hatched lizards, *Eumeces fasciatus. Copeia* 1973, 178–181.
Burghardt, G. M. (1975). Chemical prey preference polymorphism in newborn garter snakes *Thamnophis sirtalis. Behaviour* 52, 202–225.
Burghardt, G. M. (1977a). Learning processes in reptiles. In *Biology of the Reptilia,* Vol. 7 (C. Gans and D. W. Tinkle, eds.). Academic Press, New York, pp. 555–681.
Burghardt, G. M. (1977b). The ontogeny, evolution and stimulus control of

feeding in humans and reptiles. In *The Chemical Senses and Nutrition* (M. R. Kare and O. Maller, eds.). Academic Press, New York, pp. 253–275.

Burghardt, G. M. (1978). Behavioral ontogeny in reptiles: Whence, whither, and why? In *The Development of Behavior: Comparative and Evolutionary Aspects* (G. M. Burghardt and M. Bekoff, eds.). Garland STPM Press, New York, pp. 149–174.

Burghardt, G. M. (1980). Behavioral and stimulus correlates of vomeronasal functioning in reptiles: feeding, grouping, sex, and tongue use. In *Chemical Signals of Vertebrates and Aquatic Invertebrates* (D. Müller-Schwarze and R. M. Silverstein, eds.). Plenum Press, New York, pp. 275–301.

Burghardt, G. M. (1983). Aggregation and species discrimination in newborn snakes. *Z. Tierpsychol.* 61, 89–101.

Burghardt, G. M., and Abeshaheen, J. P. (1971). Responses to chemical stimuli of prey in newly hatched snakes of the genus *Elaphe*. *Anim. Behav.* 19, 486–489.

Burghardt, G. M., Allen, B. A., and Frank, H. (1986). Exploratory tongue flicking by green iguanas in laboratory and field. In *Chemical Signals in Vertebrates. 4. Ecology, Evolution and Comparative Biology* (D. Duvall, D. Müller-Schwarze, and R. M. Silverstein, eds.). Plenum Press, New York, pp. 305–321.

Burghardt, G. M., and Denny, D. (1983). Effects of prey movement and prey odor on feeding in garter snakes. *Z. Tierpsychol.* 62, 329–347.

Burghardt, G. M., Goss, S. E., and Schell, F. M. (1988). Comparison of earthworm and fish derived chemicals eliciting prey attack by garter snakes (*Thamnophis*). *J. Chem. Ecol.* 14, 855–881.

Burghardt, G. M., and Hess, E. H. (1968). Factors influencing the chemical release of prey attack in newborn snakes. *J. Comp. Physiol. Psychol.* 66, 289–295.

Burghardt, G. M., and Pruitt, C. H. (1975). Role of the tongue and senses in feeding of naive and experienced garter snakes. *Physiol. Behav.* 14, 185–194.

Burghardt, G. M., Wilcoxon, H. C., and Czaplicki, J. A. (1973). Conditioning in garter snakes: aversion to palatable prey induced by delayed illness. *Anim. Learn. Behav.* 1, 317–320.

Carey, J. H. (1967). The nuclear pattern of the telencephalon of the blacksnake, *Coluber constrictor constrictor.* In *Evolution of the Forebrain. Phylogenesis and Ontogenesis of the Forebrain* (R. Hassler and H. Stephan, eds.). Plenum Press, New York, pp. 73–80.

Carpenter, C. C., and Ferguson, G. W. (1977). Variation and evolution of stereotyped behavior in reptiles. In *Biology of the Reptilia,* Vol. 7 (C. Gans and D. W. Tinkle, eds.). Academic Press, New York, pp. 335–554.

Carr, J., Maxion, R., Sharps, M., Weiss, D., O'Connell, B., and Chiszar, D. (1982). Predatory behavior in a congenitally alingual Russell's viper (*Viper russelli*). 1. Strike-induced chemosensory searching. *Bull. Maryland Herp. Soc.* 18, 196–204.

Chelazzi, G., Calfuni, P., Graninetti, A., Carla, M., Delfino, G., and Callen, C. (1981). Modification of homing behavior in *Testudo hermanii* Gmelin

(Reptilia: Testudinidae) after intranasal irrigation with zinc sulfate solution. *Monit. Zool. Ital.* 15, 306–307.

Chelazzi, G., and Delfino, G. (1986). A field test on the use of olfaction in homing by *Testudo hermanni* (Reptilia: Testudinidae). *J. Herpetol.* 20, 451–455.

Chiszar, D., Andren, C., Nilson, G., O'Connell, B., Mestas, Jr., J. S., Smith, H. M., and Radcliffe, C. W. (1982a). Strike-induced chemosensory searching in Old World vipers and New World pit vipers. *Anim. Learn. Behav.* 10, 121–125.

Chiszar, D., Carter, T., Knight, L., Simonsen, L., and Taylor, S. (1976). Investigatory behavior in the plains garter snake (*Thamnophis radix*) and several additional species. *Anim. Learn. Behav.* 4, 273–278.

Chiszar, D., Melcer, T., Lee, R., Radcliffe, C. W., and Duvall, D. (1990). Chemical cues used by prairie rattlesnakes (*Crotalus viridis*) to follow trails of rodent prey. *J. Chem. Ecol.* 16, 79–86.

Chiszar, D., Radcliffe, C., Boyd, R., Radcliffe, A., Yun, H., Smith, H. M., Boyer, T., Atkins, B., and Feiler, F. (1986a). Trailing behavior in cottonmouths (*Agkistrodon piscivorus*). *J. Herpetol.* 20, 269–272.

Chiszar, D., Radcliffe, C., and Feiler, F. (1986b). Trailing behavior in banded rock rattlesnakes (*Crotalus lepidus klauberi*) and prairie rattlesnakes (*C. viridis viridis*). *J. Comp. Psychol.* 100, 368–371.

Chiszar, D., Radcliffe, C. W., O'Connell, B., and Smith, H. M. (1981a). Strike-induced chemosensory searching in rattlesnakes (*Crotalus viridis*) as a function of a disturbance prior to presentation of rodent prey. *Psychol. Rec.* 31, 57–62.

Chiszar, D., Radcliffe, C. W., O'Connell, B., and Smith, H. M. (1982b). Analysis of the behavioral sequence emitted by rattlesnakes during feeding episodes. II. Duration of strike-induced chemosensory searching in rattlesnakes (*Crotalus viridis, C. enyo*). *Behav. Neural Biol.* 34, 261–270.

Chiszar, D., Radcliffe, C. W., Overstreet, R., Poole, T., and Byers, T. (1985). Duration of strike-induced chemosensory searching in cottonmouths (*Agkistrodon piscivorus*) and a test of the hypothesis that striking prey creates a specific search image. *Can. J. Zool.* 63, 1057–1061.

Chiszar, D., Radcliffe, C. W., and Scudder, K. M. (1977). Analysis of the behavioral sequence emitted by rattlesnakes during feeding episodes. 1. Striking and chemosensory searching. *Behav. Biol.* 21, 418–425.

Chiszar, D., Radcliffe, C. W., Scudder, K. M., and Duvall, D. (1983a). Strike-induced chemosensory searching by rattlesnakes: the role of envenomation-related chemical cues in the post-strike environment. In *Chemical Signals in Vertebrates 3* (D. Müller-Schwarze and R. M. Silverstein, eds.). Plenum Press, New York, pp. 1–24.

Chiszar, D., Radcliffe, C. W., Smith, H. M., and Bashinski, H. (1981b). Effect of prolonged food deprivation on response to prey odors by rattlesnakes. *Herpetologica* 37, 237–243.

Chiszar, D., and Scudder, K. M. (1980). Chemosensory searching by rattlesnakes during predatory episodes. In *Chemical Signals. Vertebrates and Aquatic Invertebrates* (D. Müller-Schwarze and R. M. Silverstein, eds.). Plenum Press, New York, pp. 125–139.

Chiszar, D., Scudder, K., Knight, L., and Smith, H. M. (1978). Exploratory behavior in prairie rattlesnakes (*Crotalus viridis*) and water moccasins (*Agkistrodon piscivorus*). *Psychol. Rec.* 28, 363–368.

Chiszar, D., Stimac, K., and Boyer, T. (1983b). Effect of mouse odors on visually-induced and strike-induced chemosensory searching in prairie rattlesnakes (*Crotalus viridis*). *Chem. Senses* 7, 301–308.

Chiszar, D., Stimac, K., Poole, T., Miller, T., Radcliffe, C. W., and Smith, H. M. (1983c). Strike induced chemosensory searching in cobras (*Naja naja kaouthia, N. mossambica pallida*). *Z. Tierpsychol.* 63, 51–62.

Chiszar, D., Taylor, S. V., Radcliffe, C. W., Smith, H. M., and O'Connell, B. (1981c). Effects of chemical and visual stimuli upon chemosensory searching by garter snakes and rattlesnakes. *J. Herpetol.* 15, 415–424.

Chou, L. M., Leong, C. F., and Choo, B. L. (1988). The role of optic, auditory and olfactory senses in prey hunting by two species of geckos. *J. Herpetol.* 22, 349–351.

Clark, D. L. (1981). The tongue-vomeronasal transfer mechanism in snakes. *Micron* 12, 299–300.

Cooper, W. E., Jr. (1989). Absence of prey odor discrimination by iguanid and agamid lizards in applicator tests. *Copeia.* 472–478.

Cooper, W. E., Jr. (1990). Prey odor detection by teiid and lacertid lizards and the relationship of prey odor detection to foraging mode in lizard families. *Copeia.* 237–242.

Cooper, W. E., Jr., and Alberts, A. C. (1991). Tongue-flicking and biting in response to chemical food stimuli by an iguanid lizard (*Dipsosaurus dorsalis*) having sealed vomeronasal ducts: vomerolfaction may mediate these behavioral responses. *J. Chem. Ecol.* 17, 135–146.

Cooper, W. E., Jr., and Burghardt, G. M. (1990a). Vomerolfaction and vomodor. *J. Chem. Ecol.* 16, 103–105.

Cooper, W. E., Jr., and Burghardt, G. M. (1990b). A comparative analysis of scoring methods for chemical discrimination of prey by squamate reptiles. *J. Chem. Ecol.* 16, 45–65.

Cooper, W. E., Jr., and Garstka, W. R. (1987). Lingual responses to chemical fractions of urodaeal glandular pheromone of the skink *Eumeces laticeps*. *J. Exp. Zool.* 242, 249–253.

Cooper, W. E., Jr., Garstka, W. R., and Vitt, L. J. (1986). Female sex pheromone in the lizard *Eumeces laticeps*. *Herpetologica* 42, 361–366.

Cooper, W. E., Jr., McDowell, S. G., and Ruffer, J. (1989). Strike-induced chemosensory searching in the colubrid snakes *Elaphe g. guttata* and *Thamnophis sirtalis*. *Ethology* 81, 19–28.

Cooper, W. E., Jr., and Vitt, L. J. (1984a). Conspecific odor detection by the male broad-headed skink, *Eumeces laticeps:* effects of sex and site of odor source and of male reproductive condition. *J. Exp. Zool.* 230, 199–209.

Cooper, W. E., Jr., and Vitt, L. J. (1984b). Detection of conspecific odors by the female broad-headed skink, *Eumeces laticeps*. *J. Exp. Zool.* 229, 49–54.

Cooper, W. E., Jr., and Vitt, L. J. (1985). Responses of the skinks, *Eumeces fasciatus* and *E. laticeps,* to airborne conspecific odors: further appraisal. *J. Herpetol.* 19, 481–486.

Cooper, W. E., Jr., and Vitt, L. J. (1986a). Interspecific odour discriminations among syntopic congeners in scincid lizards (genus *Eumeces*). *Behaviour* 97, 1–9.

Cooper, W. E., Jr., and Vitt, L. J. (1986b). Interspecific odour discrimination by a lizard (*Eumeces laticeps*). *Anim. Behav.* 34, 367–376.

Cooper, W. E., Jr., and Vitt, L. J. (1986c). Lizard pheromones: behavioral responses and adaptive significance in skinks of the genus *Eumeces*. In *Chemical Signals in Vertebrates. 4. Ecology, Evolution and Comparative Biology* (D. Duvall, D. Müller-Schwarze, and R. M. Silverstein, eds.). Plenum Press, New York, pp. 323–340.

Cooper, W. E., Jr., and Vitt, L. J. (1986d). Tracking of female conspecific odor trails by broad-headed skinks (*Eumeces laticeps*). *Ethology* 71, 242–248.

Cooper, W. E., Jr., and Vitt, L. J. (1987). Intraspecific and interspecific aggression in lizards of the scincid genus *Eumeces:* chemical detection of conspecific sexual competitors. *Herpetologica* 43, 7–14.

Cooper, W. E., Jr., and Vitt, L. J. (1989). Prey odor discrimination by the broad-headed skink (*Eumeces laticeps*). *J. Exp. Zool.* 249, 11–16.

Costanzo, J. P. (1986). Influences of hibernaculum microenvironment on the winter life history of the garter snake (*Thamnophis sirtalis*). *Ohio J. Sci.* 86, 199–204.

Costanzo, J. P. (1989). Conspecific scent trailing by garter snakes (*Thamnophis sirtalis*) during autumn. Further evidence for use of pheromones in den location. *J. Chem. Ecol.* 15, 2531–2538.

Cowles, R. B. (1938). Unusual defense postures assumed by rattlesnakes. *Copeia* 1938, 13–16.

Cowles, R. B., and Phelan, R. L. (1958). Olfaction in rattlesnakes. *Copeia* 1958, 77–83.

Crews, D. (1980). Studies in squamate sexuality. *Bioscience* 30, 835–838.

Crews, D., and Garstka, W. R. (1982). The ecological physiology of a garter snake. *Sci. Am.* 247, 158–168.

Crosby, E. C. (1917). The forebrain of *Alligator mississippiensis*. *J. Comp. Neurol.* 27, 325–402.

Crosby, E. C., and Humphrey, T. (1939). Studies of the vertebrate telencephalon. I. The nuclear configuration of the olfactory and accessory olfactory formations and of the nucleus olfactorius anterior of certain reptiles, birds and mammals. *J. Comp. Neurol.* 71, 121–213.

Cruz, E., Gibson, S., Kandler, K., Sanchez, G., and Chiszar, D. (1987). Strike-induced chemosensory searching in rattlesnakes: a rodent specialist (*Crotalus viridis*) differs from a lizard specialist (*Crotalus pricei*). *Bull. Psychonom. Soc.* 25, 136–138.

Curio, V. E., and Möbious, H. (1978). Versuche zum Nachweis eines Riechvermögens von *Anolis l. lineatopus* (Rept., Iguanidae). *Z. Tierpsychol.* 47, 281–292.

Czaplicki, J. (1975). Habituation of the chemically elicited prey-attack response in the diamond-backed water snake, *Natrix rhombifera rhombifera*. *Herpetologica* 31, 403–409.

Czaplicki, J. A., Porter, R. H., and Wilcoxon, H. C. (1975). Olfactory mimicry involving garter snakes and artificial models and mimics. *Behaviour* 54, 60–71.

DeFazio, A., Simon, C. A., Middendorf, G. A., and Romano, D. (1977). Iguanid substrate licking: a response to novel situations in *Sceloporus jarrovi*. *Copeia* 1977, 706–709.

Delfino, G., Bigazzi, M., and Chelazzi, G. (1986). Olfactory mucosa of *Testudo hermanni* (Gmelin) (Reptilia: Testudinidae): occurrence of paracrystalline inclusions in supporting cells. *Z. Mikrosk.-anat. Forsch.* (Leipzig) 100, 867–880.

Devine, M. C. (1977a). "Chemistry and Source of Sex-attractant Pheromones and their Role in Mate Discrimination by Garter Snakes." Ph.D. dissertation, University of Michigan, Ann Arbor. University Microfilms No. 77–26, 227.

Devine, M. C. (1977b). Copulatory plugs, restricted mating opportunities and reproductive competition among male garter snakes. *Nature* (London) 267, 345–346.

Distel, H. (1978). Behavioral responses to the electrical stimulation of the brain in the green iguana. In *Behaviour and Neurology of Lizards* (N. Greenberg and P. D. MacLean, eds.). National Institutes of Mental Health, Bethesda, Md., pp. 135–147.

Drummond, H. M. (1979). Stimulus control of amphibious predation in the northern water snake (*Nerodia s. sipedon*). *Z. Tierpsychol.* 50, 18–44.

Drummond, H. (1985). The role of vision in the predatory behaviour of natricine snakes. *Anim. Behav.* 33, 206–215.

Dullemeijer, P. (1961). Some remarks on the feeding behavior of rattlesnakes. *Konnkl. Ned. Acad. Wetenschapp. Proc. Ser. C.* 64, 383–396.

Dunbar, G. L. (1979). Effects of early feeding experience on chemical preference of the Northern water snake, *Natrix s. sipedon* (Reptilia, Serpentes, Colubridae). *J. Herpetol.* 13, 165–169.

Dundee, H. A., and Miller, M. C., III. (1968). Aggregative behavior and habitat conditioning by the prairie ringneck snake, *Diadophis punctatus arnyi*. *Tulane Stud. Zool. Bot.* 15, 41–58.

Duvall, D. (1979). Western fence lizard (*Sceloporus occidentalis*) chemical signals. I. Conspecific discriminations and release of a species-typical visual display. *J. Exp. Zool.* 210, 321–326.

Duvall, D. J. (1981a). "Pheromonal Mechanisms in the Social Behavior and Communication of the Western Fence Lizard, *Sceloporus occidentalis biseriatus*." Dissertation Abstracts International, 41, #8103092.

Duvall, D. (1981b). Western fence lizard (*Sceloporus occidentalis*) chemical signals. II. A replication with naturally breeding adults and a test of the Cowles and Phelan hypothesis of rattlesnake olfaction. *J. Exp. Zool.* 218, 351–361.

Duvall, D. (1982). Western fence lizard (*Sceloporus occidentalis*) chemical signals. III. An experimental ethogram of conspecific body licking. *J. Exp. Zool.* 221, 23–26.

Duvall, D. (1986). A new question of pheromones: aspects of possible chemical signaling and reception in the mammal-like reptiles. In *The Ecology and Biology of Mammal-like Reptiles* (N. Hotton, III., P. D. MacLean, J. J. Roth, and E. C. Roth, eds.). Smithsonian Institution Press, Washington, D.C., pp. 219–238.

Duvall, D., Chiszar, D., Hayes, W. K., Leonhardt, J. K., and Goode, M. J. (1990). Chemical and behavioral ecology of foraging in prairie rattlesnakes (*Crotalus viridis viridis*). *J. Chem. Ecol.* 16, 87–101.

Duvall, D., Chiszar, D., Trupiano, J., and Radcliffe, C. W. (1978). Preference for envenomated rodent prey by rattlesnakes. *Bull. Psychonom. Soc.* 11, 7–8.

Duvall, D., Graves, B. M., and Carpenter, G. C. (1987). Visual and chemical composite signaling effects of *Sceloporus* lizard fecal boli. *Copeia* 1987, 1028–1031.

Duvall, D., Herskowitz, R., and Trupiano-Duvall, J. (1980a). Responses of five-lined skinks (*Eumeces fasciatus*) and ground skinks (*Scincella lateralis*) to conspecific and interspecific chemical cues. *J. Herpetol.* 14, 121–127.

Duvall, D., Scudder, K. M., and Chiszar, D. (1980b). Rattlesnake predatory behavior: mediation of prey discrimination and release of swallowing by cues arising from envenomated mice. *Anim. Behav.* 28, 674–683.

Ebbeson, S. O. E., and Voneida, T. J. (1969). The cytoarchitecture of the pallium in the tegu lizard (*Tupinambus nigropunctatus*). *Brain Behav. Evol.* 2, 431–466.

Evans, L. T., Preston, W., and Hardy, E. (1973). Imprinting in *Chrysemys* turtles. *Am. Zool.* 13, 1269.

Evans, L. T., Preston, W. C., and McGeary, M. E. (1974). Imprinting in turtles. *Am. Zool.* 14, 1278.

Ferri, D., Liquori, G. E., and Labate, M. (1982). La mucosa del *cavum nasi proprium* della lucertola campestre (*Podarcis sicula campestris* de betta). *Atti Soc. Pelorit. Sc. Fis. Mat. Natur.* 28, 75–81.

Ford, N. B. (1981). Seasonality of pheromone trailing behavior in two species of garter snake, *Thamnophis* (Colubridae). *Southwest. Nat.* 26, 385–388.

Ford, N. B. (1982). Species specificity of sex pheromone trails of sympatric and allopatric garter snakes (*Thamnophis*). *Copeia* 1982, 10–13.

Ford, N. B. (1986). The role of pheromone trails in the sociobiology of snakes. In *Chemical Signals in Vertebrates. 4. Ecology, Evolution and Comparative Biology* (D. Duvall, D. Müller-Schwarze, and R. M. Silverstein, eds.). Plenum Press, New York, pp. 261–278.

Ford, N. B., and Low, J. R., Jr. (1982). A biological mechanism for determining direction of pheromone trails having low volatility. *Am. Zool.* 22, 852.

Ford, N. B., and Low, J. R., Jr. (1984). Sex pheromone source location by garter snakes: a mechanism for detection of direction in nonvolatile trails. *J. Chem. Ecol.* 10, 1193–1199.

Ford, N. B., and O'Bleness, M. L. (1984). Species and sexual specificity of pheromone trails of the checkered garter snake, *Thamnophis marcianus*. *Am. Zool.* 24, 16A.

Ford, N. B., and O'Bleness, M. L. (1986). Species and sexual specificity of pheromone trails of the garter snake, *Thamnophis marcianus*. *J. Herpetol.* 20, 259–262.

Ford, N. B., and Schofield, C. W. (1984). Species specificity of sex pheromone trails in the plains garter snake, *Thamnophis radix*. *Herpetologica* 40, 51–55.

Fuchs, J. L., and Burghardt, G. M. (1971). Effects of early feeding experience on the responses of garter snakes to food chemicals. *Learn. Motiv.* 2, 271–279.

Gabe, M., and Saint Girons, H. (1976). Contribution a la morphologie comparée des fosses nasales et de leur annexes chez les lépidosoriens. *Mem. Mus. Nat. d'Hist. Nat. Paris,* 98 Serie A, 1–87.

Gamble, H. J. (1956). An experimental study of the secondary olfactory connexions in *Testudo graeca. J. Anat.* (London) 90, 15–29.

Gans, C. (1980). Allometric changes in the skull and brain of *Caiman crocodilus. J. Herpetol.* 14, 297–301.

García-Verdugo, J. M., Llahi, S., Farinas, I., and Martin, V. (1986). Laminar organization of the main olfactory bulb of *Podarcis hispanica:* an electron microscopic and Golgi study. *J. Hirnforsch.* 27, 87–100.

García-Verdugo, J. M., Llahi, S., Ferrer, I., and Lopez-García, C. (1989). Postnatal neurogenesis in the olfactory bulbs of a lizard. A tritiated thymidine autoradiographic study. *Neurosci. Lett.* 98, 247–252.

Garstka, W. R., Camazine, B., and Crews, D. (1982). Interactions of behavior and physiology during the annual reproductive cycle of the red-sided garter snake (*Thamnophis sirtalis partietalis*). *Herpetologica* 38, 104–123.

Garstka, W. R., and Crews, D. (1981a). Female sex pheromone in the skin and circulation of a garter snake. *Science* (N.Y.) 214, 681–683.

Garstka, W. R., and Crews, D. (1981b). The role of the female in the initiation of the reproductive cycle of the snake, *Thamnophis melanogaster. Am. Zool.* 21, 960.

Garstka, W. R., and Crews, D. (1986). Pheromones and reproduction in garter snakes. In *Chemical Signals in Vertebrates.* 4. *Ecology, Evolution and Comparative Biology* (D. Duvall, D. Müller-Schwarze, and R. M. Silverstein, eds.). Plenum Press, New York, pp. 243–260.

Gehlbach, F. R., Watkins, J. F., II, and Kroll, J. C. (1971). Pheromone trail-following studies of typhlopid, leptotyphlopid, and colubrid snakes. *Behaviour* 40, 282–294.

Getchell, T. V., and Getchell, M. L. (1987). Peripheral mechanisms of olfaction: biochemistry and neurophysiology. In *Neurobiology of Taste and Smell* (T. E. Finger and W. L. Silver, eds.). John Wiley & Sons, New York, pp. 91–123.

Gillingham, J. C. (1987). Social behavior. In *Snakes: Ecology and Evolutionary Biology* (R. A. Seigel, J. T. Collins, and S. S. Novak, eds.). Macmillan, New York, pp. 184–209.

Gillingham, J. C., and Baker, R. E. (1981). Evidence for scavenging behavior in the Western diamondback rattlesnake (*Crotalus atrox*). *Z. Tierpsychol.* 55, 217–227.

Gillingham, J. C., and Clark, D. L. (1981a). An analysis of prey-searching behavior in the Western diamondback rattlesnake, *Crotalus atrox. Behav. Neur. Biol.* 32, 235–240.

Gillingham, J. C., and Clark, D. L. (1981b). Snake tongue-flicking: Transfer mechanics to Jacobson's organ. *Can. J. Zool.* 59, 1651–1657.

Gillingham, J. C., and Dickinson, J. A. (1980). Postural orientation during courtship in the eastern garter snake, *Thamnophis s. sirtalis. Behav. Neur. Biol.* 28, 211–217.

Gillingham, J. C., Rowe, J., and Weins, M. A. (1991). Chemosensory orientation and earthworm location by foraging eastern garter snakes, *Thamnophis s. sirtalis.* In *Chemical Signals in Vertebrates* V (D. McDonald, D. Müller-Schwarze and S. Natynczyk, eds.). Oxford University Press, New York, pp. 522–532.

Goldby, F., and Gamble, H. J. (1957). The reptilian cerebral hemispheres. *Biol. Rev.* 32, 383–420.

Gove, D. (1979). A comparative study of snake and lizard tongue-flicking, with an evolutionary hypothesis. *Z. Tierpsychol.* 51, 58–76.

Gove, D., and Burghardt, G. M. (1975). Responses of ecologically dissimilar populations of the water snake *Natrix s. sipedon* to chemical cues from prey. *J. Chem. Ecol.* 1, 25–40.

Gove, D., and Burghardt, G. M. (1983). Context-correlated parameters of snake and lizard tongue-flicking. *Anim. Behav.* 31, 718–723.

Grassman, M. A., and Owens, D. W. (1981). Olfactory imprinting in loggerhead turtles (*Caretta caretta*). *Am. Zool.* 21, 924.

Grassman, M. A., and Owens, D. W. (1982). Development and extinction of food preferences in loggerhead sea turtle, *Caretta caretta. Copeia* 1982, 965–969.

Grassman, M. A., and Owens, D. (1987). Chemosensory imprinting in juvenile green sea turtles, *Chelonia mydas. Anim. Behav.* 35, 929–931.

Grassman, M. A., Owens, D. W., McVey, J. P., and Marquez, M. R. (1984). Olfactory-based orientation in artificially imprinted sea turtles. *Science* (N.Y.) 224, 83–84.

Graves, B. M., Carpenter, G. C., and Duvall, D. (1987). Chemosensory behaviors of neonate prairie rattlesnakes, *Crotalus viridis. Southwest. Nat.* 32, 515–517.

Graves, B. M., and Duvall, D. (1983a). A role for aggregation pheromones in the evolution of mammal like reptile lactation. *Am. Nat.* 122, 835–839.

Graves, B. M., and Duvall, D. (1983b). Occurrence and function of prairie rattlesnake mouth gaping in a non-feeding context. *J. Exp. Zool.* 227, 471–474.

Graves, B. M., and Duvall, D. (1984). An alarm pheromone from the cloacal sacs of prairie rattlesnakes. *Am. Zool.* 24, 17A.

Graves, B. M., and Duvall, D. (1985a). Mouth gaping and head shaking by prairie rattlesnakes are associated with vomeronasal organ olfaction. *Copeia* 1985, 496–497.

Graves, B. M., and Duvall, D. (1985b). Avomic prairie rattlesnakes (*Crotalus viridis*) fail to attack rodent prey. *Z. Tierpsychol.* 67, 161–166.

Graves, B. M., and Duvall, D. (1988). Evidence of an alarm pheromone from the cloacal sacs of prairie rattlesnakes. *Southwest. Nat.* 33, 339–345.

Graves, B. M., Duvall, D., King, M. B., Lindstedt, S. L., and Gern, W. A. (1986). Initial den location by neonatal prairie rattlesnakes: functions, causes, and natural history in chemical ecology. In *Chemical Signals in Vertebrates. 4. Ecology, Evolution and Comparative Biology* (D. Duvall, D. Müller-Schwarze, and R. M. Silverstein, eds.). Plenum Press, New York, pp. 285–304.

Graves, B. M., and Halpern, M. (1988). Neonate plains garter snakes (*Thamnophis radix*) are attracted to conspecific skin extracts. *J. Comp. Psychol.* 102, 251–253.

Graves, B. M., and Halpern, M. (1989). Chemical access to the vomeronasal organs of the lizard *Chalcides ocellatus*. *J. Exp. Zool.* 249, 150–157.

Graves, B. M., and Halpern, M. (1990). Roles of vomeronasal organ chemoreception in tongue flicking, exploratory and feeding behaviour of the lizard, *Chalcides ocellatus*. *Anim. Behav.* 39, 692–698.

Graves, B. M., and Halpern, M. (1991). Snake aggregation pheromones: source and chemosensory mediation in Western Ribbon Snakes (*Thamnophis proximus*). *J. Comp. Psychol.* In press.

Graziadei, P. P. C., and Tucker, D. (1970). Vomeronasal receptors in turtles. *Z. Zellforsch.* 105, 498–514.

Greenberg, N. (1985). Exploratory behavior and stress in the lizard, *Anolis carolinensis*. *Z. Tierpsychol.* 70, 89–102.

Greer, C. A., Mori, K., and Shepherd, G. M. (1981). Localization of synaptic responses in the in vitro turtle olfactory bulb using the [^{14}C]2-deoxyglucose method. *Brain Res.* 217, 295–303.

Gregory, P. T., Macartney, J. M., and Larsen, K. W. (1987). Spatial patterns and movements. In *Snakes: Ecology and Evolutionary Biology* (R. A. Seigel, J. T. Collins, and S. S. Novak, eds.). Macmillan, New York, pp. 366–395.

Haas, G. (1937). The structure of the nasal cavity in *Chamaeleo chameleon* (Linnaeus). *J. Morphol.* 61, 433–451.

Haas, G. (1947). Jacobson's organ in the chameleon. *J. Morphol.* 81, 195–207.

Halász, N., Hökfelt, T., Nowycky, M. C., Shepherd, G. M., and Goldstein, M. (1982a). Comparison of the rat and turtle olfactory bulb catecholamine systems. *Neurosc. Lett.*, Suppl. 10, S227.

Halász, N., Nowycky, M., Hökfelt, T., Shepherd, G. M., Markey, K., and Goldstein, M. (1982b). Dopaminergic periglomerular cells in the turtle olfactory bulb. *Brain Res. Bull.* 9, 383–389.

Halász, N., Nowycky, M. C., and Shepherd, G. M. (1983). Autoradiographic analysis of [^{3}H]dopamine and [^{3}H]dopa uptake in the turtle olfactory bulb. *Neuroscience* 8, 705–715.

Halpern, J., Erichsen, E., and Halpern, M. (1985). Role of olfactory and vomeronasal senses on garter snake response to airborne odorants. *Soc. Neurosci. Abstracts.* 11 (2), 1221.

Halpern, J., Schulman, N., and Halpern, M. (1987). Earthworm alarm pheromone is a garter snake chemoattractant. In *Olfaction and Taste IX* (S. D. Roper and J. Atema, eds.). New York Academy of Sciences, New York, pp. 328–329.

Halpern, M. (1976). The efferent connections of the olfactory bulb and accessory olfactory bulb in the snakes, *Thamnophis sirtalis* and *Thamnophis radix*. *J. Morphol.* 150, 553–578.

Halpern, M. (1980a). Chemical ecology of terrestrial vertebrates. In *Animals and Environmental Fitness* (R. Gilles, ed.). Pergamon Press, Oxford, pp. 263–282.

Halpern, M. (1980b). The telencephalon of snakes. In *Comparative Neurology of the Telencephalon* (S. O. E. Ebbeson, ed.). Plenum Press, New York, pp. 257–295.

Halpern, M. (1983). Nasal chemical senses in snakes. In *Advances of Vertebrate Neuroethology* (J.-P. Ewert, R. R. Capranica, and D. J. Ingle, eds.). Plenum Publishing, New York, pp. 141–176.

Halpern, M. (1987). The organization and function of the vomeronasal system. *Ann. Rev. Neurosci.* 10, 325–362.

Halpern, M. (1988). Vomeronasal system functions: role in mediating the reinforcing properties of chemical stimuli. In *The Forebrain of Reptiles: Current Concepts of Structure and Function* (W. K. Schwerdtfeger and W. J. A. J. Smeets, eds.). Karger, Basel, Switzerland, pp. 142–150.

Halpern, M., and Frumin, N. (1979). Roles of the vomeronasal and olfactory systems in prey attack and feeding in adult garter snakes. *Physiol. Behav.* 22, 1183–1189.

Halpern, M., and Kubie, J. L. (1980). Chemical access to the vomeronasal organs of garter snakes. *Physiol. Behav.* 24, 367–371.

Halpern, M., and Kubie, J. L. (1983). Snake tongue flicking behavior: clues to vomeronasal system functions. In *Chemical Signals in Vertebrates,* III (R. M. Silverstein and D. Müller-Schwarze, eds.). Plenum Publishing, New York, pp. 45–72.

Halpern, M., and Kubie, J. L. (1984). The role of the ophidian vomeronasal system in species-typical behavior. *Trends Neurosci.* 7, 472–477.

Halpern, M., Morrell, J. I., and Pfaff, D. W. (1982). Cellular [^{3}H]estradiol and [^{3}H]testosterone localization in the brains of garter snakes: an autoradiographic study. *Gen. Comp. Endocrinol.* 46, 211–224.

Halpern, M., Schulman, N., and Kirschenbaum, D. M. (1986). Characteristics of earthworm washings detected by the vomeronasal system of snakes. In *Chemical Signals in Vertebrates.* 4. *Ecology, Evolution and Comparative Biology* (D. Duvall, D. Müller-Schwarze, and R. M. Silverstein, eds.). Plenum Press, New York, pp. 63–77.

Halpern, M., Schulman, N., Scribani, L., and Kirschenbaum, D. M. (1984). Characterization of vomeronasally-mediated response-eliciting components of earthworm wash, II. *Pharm. Biochem. Behav.* 21, 655–662.

Harless, M. (1979). Social behavior. In *Turtles: Perspectives and Research* (M. Harless and H. Morlock, eds.). John Wiley & Sons, New York, pp. 475–492.

Hatanaka, T., and Hanada, T. (1986). Structure of the vomeronasal system and induced wave in the accessory olfactory bulb of red eared turtle. In *Proceedings of the 20th Japanese Symposium on Taste and Smell* (T. Shibuya and S. Saito, eds.). JASTS, Gifu, Japan, pp. 183–186.

Hatanaka, T., and Hanada, T. (1987). Activity of the accessory olfactory bulb to chemical stimuli delivered to the vomeronasal mucosa in turtles. *Zool. Sci.* 4, 964.

Hatanaka, T., Matsuzaki, O., and Shibuya, T. (1982). Fine structure of vomeronasal receptor cells in the Reeve's turtle, *Geoclemys reevesii. Zool. Mag.* (Tokyo) 91, 190–193.

Hatanaka, T., and Shibuya, T. (1989). Odor response patterns of single units in the accessory olfactory bulb of turtle, *Geoclemys reevesii. Comp. Biochem. Physiol.* 92A, 505–512.

Hatanaka, T., Shibuya, T., and Inouchi, J. (1988). Induced wave responses of the accessory olfactory bulb to odorants in two species of turtle, *Pseudemys scripta* and *Geoclemys reevesii*. *Comp. Biochem. Physiol.* 91A, 377–385.

Heimer, L. (1969). The secondary olfactory connections in mammals, reptiles and sharks. *Ann. N.Y. Acad. Sci.* 167, 129–146.

Heller, S., and Halpern, M. (1981). Laboratory observations on conspecific and congeneric scent trailing in garter snakes (*Thamnophis*). *Behav. Neural. Biol.* 33, 372–377.

Heller, S. B., and Halpern, M. (1982a). Laboratory observations of aggregative behavior of garter snakes, *Thamnophis sirtalis*. *J. Comp. Physiol. Psychol.* 96, 967–983.

Heller, S. B., and Halpern, M. (1982b). Laboratory observations of aggregative behavior of garter snakes, *Thamnophis sirtalis:* roles of the visual, olfactory, and vomeronasal senses. *J. Comp. Physiol. Psychol.* 96, 984–999.

Henderson, R. W., Binder, M. H., and Burghardt, G. M. (1983). Responses of neonate Hispaniolan vine snakes (*Uromacer frenatus*) to prey extracts. *Herpetologica* 39, 75–77.

Hetherington, T. E. (1989). Use of vibratory cues for detection of insect prey by the sandswimming lizard *Scincus scincus*. *Anim. Behav.* 37, 290–297.

Hidalgo, H. (1982). Courtship and mating behavior in *Rhinoclemmys pulcherrima incisa* (Testudines: Emydidae: Batagurinae). *Trans. Kansas Acad. Sci.* 85, 82–95.

Holtzman, D. A., and Halpern, M. (1987). Development of olfactory and vomeronasal systems in the red-sided garter snake, *Thamnophis sirtalis parietalis*. In *Olfaction and Taste,* IX (S. D. Roper and J. Atema, eds.). New York Academy of Sciences, New York, pp. 373–374.

Holtzman, D. A., and Halpern, M. (1989). In vitro technique for studying garter snake (*Thamnophis* sp.) development. *J. Exp. Zool.* 250, 283–288.

Holtzman, D. A., and Halpern, M. (1990). Embryonic and neonatal development of the vomeronasal and olfactory systems in garter snakes (*Thamnophis* spp.). *J. Morphol.* 203, 123–140.

Hoogland, P. V., and Stoll, C. J. (1984). Laminar organization of cortical afferents to the medial cortex in the lizard *Gekko gecko*. *Neurosc. Lett.*, Suppl. 18, S390.

Hoogland, P. V., and Vermeulen-Van der Zee, E. (1988). Intrinsic and extrinsic connections of the cerebral cortex of lizards. In *The Forebrain of Reptiles. Current Concepts of Structure and Function* (W. K. Schwerdtfeger and W. J. A. J. Smeets, eds.). Karger, Basel, Switzerland, pp. 20–29.

Hoogland, P. V., and Vermeulen-Van der Zee, E. (1989). Efferent connections of the dorsal cortex of the lizard *Gekko gecko* studied with *Phasiolus vulgaris*-leucoagglutinin. *J. Comp. Neurol.* 285, 289–303.

Hoogland, P. V., Witjes, R., and Lohman, A. H. M. (1985). Laminar organization of secondary and tertiary olfactory projections to the lateral cortex in the lizard *Gekko gecko*. *Neurosci. Lett.*, Supp. 22, S459.

Inouchi, J., and Halpern, M. (1988). Responses of olfactory and vomeronasal epithelia to airborne odors in garter snakes. *Chem. Senses* 13, 700.

Inouchi, J., Kubie, J. L., and Halpern, M. (1988). Liquid odorants delivered to the vomeronasal organ of garter snakes increase firing of cells in the accessory olfactory bulb. *Soc. Neurosci. Abstracts* 14 (2), 1167.

Inouchi, J., Kubie, J. L., and Halpern, M. (1989). Accessory olfactory bulb neurons respond to liquid delivery of odorants to vomeronasal organ. *Chem. Senses* 14, 712.

Inouchi, J., Jiang, X.-C., Wang, D., Kubie, J., and Halpern, M. (1990). Garter snake accessory olfactory bulb neurons respond to a chemoattractive protein purified from earthworm secretions. *Chem. Senses* 15, 650.

Iwahori, N. (1984). The vomeronasal and olfactory epithelium in the snake (*Elaphe quadrivirgata*): a Golgi and Nissl study. *Neurosci. Lett.*, Suppl. 17., S136.

Iwahori, N., Kiyota, E., and Nakamura, K. (1987). Olfactory and respiratory epithelia in the snake, *Elaphe quadrivirgata. Okajimas Folia Anat. Japn.* 64, 183–192.

Iwahori, N., Nakamura, K., and Mameya, C. (1989a). A Golgi study on the main olfactory bulb in the snake *Elaphe quadrivirgata. Neurosci. Res.* 6, 411–425.

Iwahori, N., Nakamura, K., and Mameya, C. (1989b). A Golgi study on the accessory olfactory bulb in the snake *Elaphe quadrivirgata. Neurosci. Res.* 7, 55–70.

Jackson, I. M. D., and Reichlin, S. (1974). Thyrotropin-releasing hormone (TRH): distribution in hypothalamic and extrahypothalamic brain tissues of mammalian and submammalian chordates. *Endocrinology* 95, 854–862.

Jahr, C. E., and Nicoll, R. A. (1980). Dendrodendritic inhibition: demonstration with intracellular recording. *Science* (N.Y.) 207, 1473–1475.

Jiang, X.-C., Inouchi, J., Wang, D., and Halpern, M. (1990). Purification and characterization of a chemoattractant from earthworm electric shock-induced secretion, its receptor binding and signal transduction through vomeronasal system of garter snakes. *J. Biol. Chem.* 265, 8736–8744.

Johnsen, P. B., and Wellington, J. L. (1982). Detection of glandular secretions by yearling alligators. *Copeia* 1982, 705–708.

Kahmann, H. (1932). Sinnesphysiologische Studien an Reptilien: I. Experimentelle Untersuchungen über das jakobsonsche Organ der Eidechsen und Schlangen. *Zool. Jhb. Zool. Physiol.* 51, 173–238.

Keiser, E. D. (1975). Observation on tongue extension in vine snakes (genus *Oxybelis*) with suggested behavioral hypothesis. *Herpetologica* 31, 131–133.

Kikuchi, S., Sawai, Y., Toshioka, S., Okuyama, Y., Matsui, T., and Numata, T. (1981). Scanning electronmicroscopy of the scale and tongue of Japanese snakes. 2. *Trimeresurus elegans* (Gray, 1849) and *T. tokarensis* (Nagai, 1928). *Snake* 13, 79–88.

King, M., McCarron, D., Duvall, D., Baxter, G., and Gern, W. (1983). Group avoidance of conspecific but not interspecific chemical cues by prairie rattlesnakes (*Crotalus viridis*). *J. Herpetol.* 17, 196–198.

Kirschenbaum, D. M., Schulman, N., Yao, P., and Halpern, M. (1985). Chemo-attractant for the garter snake: Characterization of vomeronasally-mediated response-eliciting components of earthworm wash, III. *Comp. Biochem. Physiol.* 82B, 447–453.

Kirschenbaum, D. M., Schulman, N., and Halpern, M. (1986). Earthworms produce a collagen-like substance detected by the garter snake vomeronasal system. *Proc. Natl. Acad. Sci. USA* 83, 1213–1216.

Klein, J., and Loop, M. (1975). Headfirst prey ingestion by newborn *Elaphe* and *Lampropeltis*. *Copeia* 1975, 366.

Kowell, A. P. (1974). "The Olfactory and Accessory Olfactory Bulbs in Constricting Snakes: Neuroanatomic and Behavioral Considerations." Ph.D. dissertation, University of Pennsylvania, Philadelphia, University Microfilms #74–22, 865.

Kramer, M. (1986). Individual discrimination in *Pseudemys* turtles. *Am. Zool.* 26, 79A.

Kratzing, J. E. (1975). The fine structure of the olfactory and vomeronasal organs in lizard (*Tiliqua scincoides scincoides*). *Cell Tiss. Res.* 156, 239–252.

Krohmer, R. W., and Crews, D. (1987). Facilitation of courtship behavior in the red-sided garter snake (*Thamnophis sirtalis parietalis*) following lesions of the septum or nucleus sphericus. *Physiol. Behav.* 40, 759–765.

Kroll, J. C. (1973). Taste buds in the oral epithelium of the blind snake, *Leptotyphlops dulcis* (Reptilia: Leptotyphlopidae). *Southwest. Nat.* 17, 365–370.

Kruzhalov, N. B., and Boïko, V. P. (1987). Electrical reactions of the olfactory epithelium and olfactory bulb of the tortoise *Testudo graeca* to some natural and artificial odours. *J. Evol. Biochem. Physiol.* 23, 617–623.

Kubie, J. L., Cohen, J., and Halpern, M. (1978a). Shedding enhances the sexual attractiveness of oestradiol treated garter snakes and their penmates. *Anim. Behav.* 26, 562–570.

Kubie, J. L., and Halpern, M. (1975). Laboratory observations of trailing behavior in garter snakes. *J. Comp. Physiol. Psychol.* 89, 667–674.

Kubie, J. L., and Halpern, M. (1978). Garter snake trailing behavior: effects of varying prey-extract concentration and mode of prey-extract presentation. *J. Comp. Physiol. Psychol.* 92, 362–373.

Kubie, J. L., and Halpern, M. (1979). Chemical senses involved in garter snake prey trailing. *J. Comp. Physiol. Psychol.* 93, 648–667.

Kubie, J. L., Vagvolgyi, A., and Halpern, M. (1978b). Roles of vomeronasal and olfactory systems in courtship behavior of male garter snakes. *J. Comp. Physiol. Psychol.* 92, 627–641.

Labate, M., Liquori, G. E., and Ferri, D. (1982). Il naso di *Podarcis sicula campestris* de betta (Reptilia, Lacertidae). (Studio anatomo topografico). *Atti Soc. Pelorit. Sc. Fis. Mat. Natur.* 28, 57–74.

Lillywhite, H. B. (1985). Trailing movements and sexual behavior in *Coluber constrictor*. *J. Herpetol.* 19, 306–308.

Liquori, G. E., Ferri, D., and Labate, M. (1982). Ultrastutturali aspetti dell'epitelio olfattorio della lucertola campestre (*Podarcis sicula campestris* de betta). *Atti Soc. Pelorit. Sc. Fis. Mat. Natur.* 28, 83–105. (In Italian, English abstract.)

Llahi, S., and García-Verdugo, J. M. (1986). Laminar organization of the accessory olfactory bulb of the lizard *Podarcis hispanica:* a light microscropic study. *Neurosci. Lett.*, Suppl. 26, S443.

Llahi, S., and García-Verdugo, J. M. (1989a). Ultrastructural organization of the accessory olfactory bulb of the lizard *Podarcis hispanica*. *J. Morphol.* 202, 1–11.

Llahi, S., and García-Verdugo, J. M. (1989b). Neuronal organization of the accessory olfactory bulb of the lizard *Podarcis hispanica:* Golgi study. *J. Morphol.* 202, 13–28.

Lohman, A. H. M., Hoogland, P. V., and Witjes, R. J. G. M. (1988). Projections from the main and accessory olfactory bulbs to the amygdaloid complex in the lizard *Gekko gecko.* In *The Forebrain of Reptiles: Current Concepts of Structure and Function* (W. K. Schwerdtfeger and W. J. A. J. Smeets, eds.). Karger, Basel, Switzerland, pp. 41–49.

Lohman, A. H. M., and Mentink, G. M. (1972). Some cortical connections of the tegu lizard (*Tupinambis teguixin*). *Brain Res.* 45, 325–344.

Lohman, A. H. M., and Van Woerden-Verkley, I. (1976). Further studies on the cortical connections of the tegu lizard. *Brain Res.* 103, 9–28.

Loop, M. S. (1970). The effects of feeding experience on the response to prey-object extracts in rat snakes. *Psychonomic Sci. Sect. Anim. Physiol. Psychol.* 21, 189–190.

Loop, M. S., and Scoville, S. A. (1972). Response of newborn *Eumeces inexpectatus* to prey object extracts. *Herpetologica* 28, 254–256.

Madison, D. M. (1977). Chemical communication in amphibians and reptiles. In *Chemical Signals in Vertebrates* (D. Müller-Schwarze and M. M. Mozell, eds.). Plenum Press, New York, pp. 135–168.

Manton, M. L. (1979). Olfaction and behavior. In *Turtles. Perspectives and Research* (M. Harless and H. Morlock, eds.). John Wiley & Sons, New York, pp. 289–301.

Manton, M. L., Karr, A., and Ehrenfeld, D. W. (1972a). Chemoreception in the migratory sea turtle, *Chelonia mydas. Biol. Bull.* 143, 184–195.

Manton, M. L., Karr, A., and Ehrenfeld, D. W. (1972b). An operant method for the study of chemoreception in the green turtle, *Chelonia mydas. Brain Behav. Evol.* 5, 188–201.

Marchisin, A. (1980). "Predatory-prey Interactions between Snake-eating Snakes and Pit Vipers." Ph.D. dissertation, Rutgers University, The State University of New Jersey, Newark, University Microfilms Number ADG80–20143.

Martinez-García, F., Amiguet, M., Olucha, F., Perez-Clausell, J., and Lopez-Garcia, C. (1985). Afferent connections of the lateral telencephalic cortex of the lizard *Podarcis hispanica:* an HRP study. *Neurosci. Lett.*, Suppl. 22, S459.

Martinez-García, F., Amiguet, M., Olucha, F., and Lopez-Garcia, C. (1986a). Connections of the lateral cortex in the lizard *Podarcis hispanica. Neurosci. Lett.* 63, 39–44.

Martinez-García, F., Amiguet, M., Poch, L., and Olucha, F. (1986b). Commissural connections of the cerebral cortex of the lizard *Podarcis hispanica. Neurosci. Lett.*, Suppl. 26, S443.

Martinez-García, F., and Olucha, F. E. (1988). Afferent projections to the Timm-positive cortical areas of the telencephalon of lizards. In *The Forebrain of Reptiles: Current Concepts of Structure and Function* (W. K. Schwerdtfeger and W. J. A. J. Smeets, eds.). Karger, Basel, Switzerland, pp. 30–40.

Martinez-Vargas, M. D., Keefer, D. A., and Stumpf, W. E. (1978). Estrogen localization in the brain of the lizard, *Anolis carolinensis. J. Exp. Zool.* 205, 141–147.

Mason, R. T., Chinn, J. W., and Crews, D. (1987a). Sex and seasonal differences in the skin lipids of garter snakes. *Comp. Biochem. Physiol.* 87B, 999–1003.

Mason, R. T., Chinn, J. W., and Crews, D. (1987b). Skin lipids of garter snakes serve as semiochemicals. In *Olfaction and Taste,* IX (S. D. Roper and J. Atema, eds.). New York Academy of Sciences, New York, pp. 472–474.

Mason, R. T., and Crews, D. (1985a). Female mimicry in garter snakes. *Nature* (London) 316, 59–60.

Mason, R. T., and Crews, D. (1985b). Analysis of sex attractant pheromone in garter snakes. *Am. Zool.* 25, 76A.

Mason, R. T., and Crews, D. (1986). Pheromone mimicry in garter snakes. In *Chemical Signals in Vertebrates. 4. Ecology, Evolution and Comparative Biology* (D. Duvall, D. Müller-Schwarze, and R. M. Silverstein, eds.). Plenum Press, New York, pp. 279–283.

Mason, R. T., Fales, H. M., Jones, T. H., Chinn, J. W., Pannell, L. K., and Crews, D. (1989). Sex pheromones in snakes. *Science* (N.Y.) 245, 290–293.

Mason, R. T., and Gutzke, W. H. N. (1990). Sex recognition in the leopard gecko, *Eublepharis macularius* (Sauria: Gekkonidae): possible mediation by skin-derived semiochemicals. *J. Chem. Ecol.* 16, 27–36.

Mathews, D. F. (1972). Response patterns of single neurons in the tortoise olfactory epithelium and olfactory bulb. *J. Gen. Physiol.* 60, 166–180.

Mathews, D. F., and Tucker, D. (1966). Single unit activity in the tortoise olfactory mucosa. *Fed. Proc.* 25, 329.

Matsuzaki, O., and Shibuya, T. (1978). Ultrastructure of olfactory nerves and number of olfactory cells in some reptiles and birds. *Zool. Mag.* (Tokyo) 87, 402. (In Japanese.)

Matsuzaki, O., Shibuya, T., and Hatanaka, T. (1980). The number of olfactory receptor cells and fine structure of the olfactory nerve fibers in several species of vertebrate. *Proc. Jap. Symp. Taste Smell* 14, 173–176.

Melcer, T., and Chiszar, D. (1989a). Striking prey creates a specific chemical search image in rattlesnakes. *Anim. Behav.* 37, 477–486.

Melcer, T., and Chiszar, D. (1989b). Strike-induced chemical preferences in prairie rattlesnakes (*Crotalus viridis*). *Anim. Learn Behav.* 17, 368–372.

Meredith, M., and Burghardt, G. M. (1978). Electrophysiological studies of the tongue and accessory olfactory bulb in garter snakes. *Physiol. Behav.* 21, 1001–1008.

Molowny, A., Martinez-Calatayud, J., Juan, M. J., Martinez-Guijarro, F. J., and Lopez-Garcia, C. (1987). Zinc accumulation in the telencephalon of lizards. *Histochemistry* 86, 311–314.

Morgans, L. F., and Heidt, G. A. (1978). Comparative tongue histology and scanning electron microscopy of the diamondback water snake (*Natrix rhombifera*) and black rat snake (*Elaphe obsoleta*) (Reptilia, Serpentes, Colubridae). *J. Herpetol.* 12, 275–280.

Mori, K., Nowycky, M. C., and Shepherd, G. M. (1981a). Dendritic spikes in mitral cells of the isolated turtle olfactory bulb. *J. Physiol.* (London) 320, 89P.

Mori, K., Nowycky, M. C., and Shepherd, G. M. (1981b). An intracellular analysis of the synaptic responses of mitral cells to olfactory nerve volleys in the *in vitro* turtle olfactory bulb. *Neurosci. Lett.*, Suppl. 6., S103.

Mori, K., Nowycky, M. C., and Shepherd, G. M. (1981c). Electrophysiological analysis of mitral cells in the isolated turtle olfactory bulb. *J. Physiol.* (London) 314, 281–294.

Mori, K., Nowycky, M. C., and Shepherd, G. M. (1981d). Analysis of synaptic potentials in mitral cells in the isolated turtle olfactory bulb. *J. Physiol.* (London) 314, 295–309.

Mori, K., Nowycky, M. C., and Shepherd, G. M. (1981e). Analysis of a long-duration inhibitory potential in mitral cells in the isolated turtle olfactory bulb. *J. Physiol.* (London) 314, 311–320.

Mori, K., Nowycky, M. C., and Shepherd, G. M. (1984). Synaptic excitability and inhibitory interactions at distal dendritic sites on mitral cells in the isolated turtle olfactory bulb. *J. Neurosci.* 4, 2291–2296.

Mori, K., and Shepherd, G. M. (1979). Synaptic excitation and long-lasting inhibition of mitral cells in the in vitro turtle olfactory bulb. *Brain Res.* 172, 155–159.

Morrell, J. I., Crews, D., Ballin, A., Morgentaler, A., and Pfaff, D. W. (1979). H^3-estradiol, H^3-testosterone and H^3-dihydrotestosterone localization in the brain of the lizard *Anolis carolinensis:* an autoradiographic study. *J. Comp. Neurol.* 188, 201–224.

Morris, D. D., and Loop, M. S. (1969). Stimulus control of prey attack in naive rat snakes: a species duplication. *Psychon. Sci.* 15, 141–142.

Moulton, D. G., and Beidler, L. M. (1967). Structure and function in the peripheral olfactory system. *Physiol. Rev.* 47, 1–52.

Müller, W. (1971). Vergleichende elektrophysiologische Untersuchungen an den Sinnesepithelien des jacobsonschen Organs und der Nase von Amphibien (*Rana*), Reptilien (*Lacerta*) und Säugetieren (*Mus*). *Z. vergl. Physiol.* 72, 370–385.

Mushinsky, H. R. (1987). Foraging ecology. In *Snakes: Ecology and Evolutionary Biology* (R. A. Seigel, J. T. Collins, and S. S. Novak, eds.). Macmillan, New York, pp. 302–334.

Mushinsky, H. R., and Lotz, K. H. (1980). Chemoreceptive responses of two sympatric water snakes to extracts of commonly ingested prey species. Ontogenetic and ecological considerations. *J. Chem. Ecol.* 6, 523–535.

Naik, D. R., Sar, M., and Stumpf, W. E. (1981). Immunohistochemical localization of enkephalin in the central nervous system and pituitary of the lizard, *Anolis carolinensis. J. Comp. Neurol.* 198, 583–602.

Naulleau, G. (1965). La biologie et le comportement predateur de *Vipera aspis* au laboratoire et dans la nature. *Bull. Biol. France Belgique* 99, 395–524.

Negus, V. E. (1958). *The Comparative Anatomy and Physiology of the Nose and Paranasal Sinuses.* F & S Livingston, Edinburgh.

Nicoletto, P. (1983). The role of vision and olfaction in prey selection by the skink *Scincella lateralis. Virginia J. Sci.* 34, 131.

Nicoletto, P. F. (1985a). The relative roles of vision and olfaction in prey detection by the ground skink, *Scincella lateralis. J. Herpetol.* 19, 411–415.

Nicoletto, P. F. (1985b). The roles of vision and the chemical senses in predatory behavior of the skink, *Scincella lateralis. J. Herpetol.* 19, 487–491.

Noble, G. K., and Clausen, H. J. (1936). The aggregation behavior of *Storeria dekayi* and other snakes, with especial reference to the sense organs involved. *Ecol. Monogr.* 6, 269–316.

Northcutt, R. G. (1967). Architectonic studies of the telencephalon of *Iguana iguana*. *J. Comp. Neurol.* 130, 109–148.

Northcutt, R. G. (1970). The telencephalon of the western painted turtle (*Chrysemys picta belli*). *Ill. Biol. Monogr.* 43, 1–113.

Northcutt, R. G. (1978). Forebrain and midbrain organization in lizards and its phylogenetic significance. In *Behavior and Neurology of Lizards* (N. Greenberg and P. D. MacLean, eds.). National Institutes of Mental Health, Bethesda, Md. pp. 11–64.

Nowycky, M. C., Halász, N., and Shepherd, G. M. (1983). Evoked field potential analysis of dopaminergic mechanisms in the isolated turtle olfactory bulb. *Neuroscience* 8, 717–722.

Nowycky, M. C., Mori, K., and Shepherd, G. M. (1981a). GABAergic mechanisms of dendrodendritic synapses in isolated turtle olfactory bulb. *J. Neurophysiol.* 46, 639–648.

Nowycky, M. C., Mori, K., and Shepherd, G. M. (1981b). Blockade of synaptic inhibition reveals long-lasting synaptic excitation in isolated turtle olfactory bulb. *J. Neurophysiol.* 46, 649–658.

Nowycky, M. C., Waldow, U., and Shepherd, G. M. (1978). Electrophysiological studies in the isolated turtle brain. *Neurosci. Abstr.* 4, 583.

O'Connell, B., Poole, T., Nelson, P., Smith, H. M., and Chiszar, D. (1982). Strike-induced chemosensory searching by prairie rattlesnakes (*Crotalus v. viridis*) after predatory and defensive strikes which made contact with mice (*Mus musculus*). *Bull. Maryland Herpetol. Soc.* 18, 152–160.

Oelofsen, B. W., and Van Den Heever, J. A. (1979). Role of the tongue during olfaction in varanids and snakes. *South African J. Sci.* 75, 365–366.

Oldak, P. D. (1976). Comparison of the scent gland secretion lipids of twenty-five snakes: implications for biochemical systematics. *Copeia* 1976, 320–326.

Olucha, F., Poch, L., Amiguet, M., and Martinez-García, F. (1986). Differential connections of the dorsal cortex subregions in lizards. *Neurosci. Lett.*, suppl. 26, S444.

Orrego, F. (1961a). The reptilian forebrain. I. The olfactory pathways and cortical areas in the turtle. *Arch. Ital. Biol.* 99, 425–445.

Orrego, F. (1961b). The reptilian forebrain. II. Electrical activity in the olfactory bulb. *Arch. ital. Biol.* 99, 446–465.

Orrego, F. (1962). The reptilian forebrain. III. Cross connection between the olfactory bulbs and the cortical areas in the turtle. *Arch ital. Biol.* 100, 1–16.

Orrego, F., and Lisenby, D. (1962). The reptilian forebrain. IV. Electrical activity in the turtle cortex. *Arch. ital. Biol.* 100, 17–30.

Owens, D., Comuzzie, D. C., and Grassman, M. (1986). Chemoreception in the homing and orientation behavior of amphibians and reptiles, with special reference to sea turtles. In *Chemical Signals in Vertebrates. 4. Ecology, Evolution and Comparative Biology* (D. Duvall, D. Müller-Schwarze, and R. M. Silverstein, eds.). Plenum Press, New York, pp. 341–355.

Owens, D. W., Grassman, M. A., and Hendrickson, J. R. (1982). The imprinting hypothesis and sea turtle reproduction. *Herpetologica* 38, 124–135.

Parent, A. (1979). Monoaminergic systems of the brain. In *Biology of the Reptilia,* Vol. 10 (C. Gans, R. G. Northcutt, and P. Ulinski, eds.). Academic Press, New York, pp. 247–285.
Parsons, T. S. (1959a). Nasal anatomy and phylogeny of reptiles. *Evolution* 13, 175–187.
Parsons, T. S. (1959b). Studies on the comparative embryology of the reptilian nose. *Bull. Mus. Comp. Zool.* 120, 103–277.
Parsons, T. S. (1967). Evolution of the nasal structure in the lower tetrapods. *Am. Zool.* 7, 397–413.
Parsons, T. S. (1970). The nose and Jacobson's organ. In *Biology of the Reptilia,* Vol. 2 (C. Gans and T. S. Parsons, eds.). Academic Press, New York, pp. 99–191.
Pedersen, J. M. (1988). Laboratory observations on the function of tongue extrusion in the desert iguana (*Dipsosaurus dorsalis*). *J. Comp. Psychol.* 102, 193–196.
Perez-Clausell, J. (1988). Organization of zinc-containing terminal fields in the brain of the lizard *Podarcis hispanica:* a histochemical study. *J. Comp. Neurol.* 267, 153–171.
Perez-Clausell, J., and Fredens, K. (1988). Chemoarchitectonics in the telencephalon of the lizard *Podarcis hispanica.* In *The Forebrain of Reptiles: Current Concepts of Structure and Function* (W. K. Schwerdtfeger and W. J. A. J. Smeets, eds.). Karger, Basel, Switzerland, pp. 85–96.
Porter, R. H., and Czaplicki, J. A. (1974). Responses of water snakes (*Natrix r. rhombifera*) and garter snakes (*Thamnophis sirtalis*) to chemical cues. *Anim. Learn. Behav.* 2, 129–132.
Price, A. H., and La Pointe, J. L. (1981). Structure-functional aspects of the scent gland in *Lampropeltis getulus splendida*. *Copeia* 1981, 138–146.
Reformato, L. S., Kirschenbaum, D. M., and Halpern, M. (1983). Preliminary characterization of response-eliciting components of earthworm extract. *Pharm. Biochem. Behav.* 18, 247–254.
Reiner, A., and Karten, H. J. (1985). Comparison of olfactory bulb projections in pigeons and turtles. *Brain Behav. Evol.* 27, 11–27.
Ridlon, R. W., Jr. (1985). Scanning electron microscopy of the tongue of the snake, *Thamnophis radix*. *J. Herpetol.* 19, 536–538.
Rolon, R. R., and Skeen, L. C. (1980). Afferent and efferent connections of the olfactory bulb in the soft-shell turtle (*Trionyx spinifer spinifer*). *Abstr. Soc. Neurosci.* 6, 305.
Rose, F. L. (1970). Tortoise chin gland fatty acid composition: behavioral significance. *Comp. Biochem. Physiol.* 32, 577–580.
Ross, P., Jr., and Crews, D. (1977). Influence of the seminal plug on mating behaviour in the garter snake. *Nature* (London) 267, 344–345.
Rudin, W. (1974). Untersuchungen am olfaktorischen System der Reptilien. III. Differenzierungsformen einiger olfaktorischen Zentren bei Reptilien. *Acta Anat.* 89, 481–515.
Rudin, W. (1975). Untersuchungen am olfaktorischen System der Reptilien. I. Bau und Ontogenese der telencephalon olfaktorischen Zentren von *Lacerta sicula* (Rafinesque). *Verh. naturf. Ges. Basel* 85, 101–134.
Saint Girons, M. H. (1975). Développement respectif de l'épithélium senso-

riel du cavum et de l'organe de Jacobson chez les Lépidosauriens. *C. R. Seanc. hedb. Acad. Sc. Paris* 280, 721–724.

Scalia, F., Halpern, M., and Riss, W. (1969). Olfactory bulb projections in the South American caiman. *Brain Behav. Evol.* 2, 238–262.

Schell, F. M., Burghardt, G. M., Johnston, A., and Coholich, C. (1990). Analysis of chemicals from earthworms and fish that elicit prey attack by ingestively naive garter snakes (*Thamnophis*). *J. Chem. Ecol.* 16, 67–77.

Schlegel, J. R., and Kriegstein, A. R. (1987). Quantitative autoradiography of muscarinic and benzodiazepine receptors in the forebrain of the turtle, *Pseudemys scripta*. *J. Comp. Neurol.* 265, 521–529.

Schulman, N., Erichsen, E., and Halpern, M. (1987). Garter snake response to the chemoattractant in earthworm alarm pheromone is mediated by the vomeronasal system. In *Olfaction and Taste, IX* (S. D. Roper and J. Atema, eds.). New York Academy of Sciences, New York, pp. 330–331.

Schwenk, K. (1985). Occurrence, distribution and functional significance of taste buds in lizards. *Copeia* 1985, 91–101.

Schwenk, K. (1986). Morphology of the tongue in the tuatara, *Sphenodon punctatus* (Reptilia: Lepidosauria), with comments on function and phylogeny. *J. Morphol.* 188, 129–156.

Schwenk, K. (1988). Comparative morphology of the lepidosaur tongue and its relevance to squamate phylogeny. In *Phylogenetic Relationships of the Lizard Families. Essays Commemorating Charles L. Camp* (R. Estes and G. Pregill, eds.). Stanford University Press, Stanford, Calif., pp. 569–598.

Scott, T. P., and Weldon, P. J. (1990). Chemoreception in the feeding behavior of adult American alligators, *Alligator mississippiensis*. *Anim. Behav.* 39, 398–405.

Scott, T. R., Jr. (1979). The chemical senses. In *Turtles: Perspectives and Research* (M. Harless and H. Morlock, eds.). John Wiley & Sons, New York, pp. 267–287.

Scudder, K. M., Stewart, N. J., and Smith, H. M. (1980). Response of neonate water snakes (*Nerodia sipedon sipedon*) to conspecific chemical cues. *J. Herpetol.* 14, 196–198.

Secoy, D. M. (1979). Investigatory behavior of plains garter snakes, *Thamnophis radix* (Reptilia: Colubridae), in tests of repellant chemicals. *Can. J. Zool.* 57, 691–693.

Sheffield, L. P., Law, J. H., and Burghardt, G. M. (1968). On the nature of chemical food sign stimuli for newborn snakes. *Commun. Behav. Biol.* 2, 7–12.

Shibuya, T. (1964). Dissociation of olfactory neural response and mucosal potential. *Science* (N.Y.) 143, 1338–1340.

Shibuya, T. (1969). Activities of single olfactory receptor cells. In *Olfaction and Taste,* III (C. Pfaffmann, ed.). Rockefeller University Press, New York, pp. 109–116.

Shibuya, T., Aihara, Y., and Tonosaki, K. (1977). Single cell responses to odors in the reptilian olfactory bulb. In *Food Intake and Chemical Senses* (Y. Katsuki, ed.) University of Tokyo Press, Tokyo, pp. 23–32.

Shibuya, T., and Shibuya, S. (1963). Olfactory epithelium: unitary responses in the tortoise. *Science* (N.Y.) 140, 495–496.

Shibuya, T., Tonosaki, K., and Matsuzaki, O. (1978). Intracellular responses and morphological types of the mitral cell in the olfactory bulb. *Zool. Mag.* (Tokyo) 87, 403. (In Japanese.)

Simon, C. A. (1983). A review of lizard chemoreception. In *Lizard Ecology: Studies of a Model Organism* (R. B. Huey, E. R. Pianka, and T. W. Schoener, eds.). Harvard University Press, Cambridge, Mass., pp. 119–133.

Simon, C. A., Gravelle, K., Bissinger, B. E., Eiss, I., and Ruibal, R. (1981). The role of chemoreception in the iguanid lizard *Sceloporus jarrovi*. *Anim. Behav.* 29, 46–54.

Simon, C. A., and Moakley, G. P. (1985). Chemoreception in *Sceloporus jarrovi:* Does olfaction activate the vomeronasal system? *Copeia* 1985, 239–242.

Simpson, J. T., Weldon, P. J., and Sharp, T. R. (1988). Identification of major lipids from the scent gland secretions of Dumeril's ground boa (*Acrantophis dumerili* Jan) by gas chromatography-mass spectrometry. *Z. Naturforsch.* 43c, 914–917.

Skeen, L. C., Pindzola, R. R., and Schofield, B. R. (1984). Tangential organization of olfactory, association, and commissural projections to olfactory cortex in a species of reptile (*Trionyx spiniferus*), bird (*Aix sponsa*), and mammal (*Tupaia glis*). *Brain Behav. Evol.* 25, 206–216.

Skeen, L. C., and Rolon, R. R. (1982). Olfactory bulb interconnections in soft shell turtle. *Neurosci. Lett.* 33, 223–228.

Slabý, O. (1979a). Morphogenesis of the nasal capsule, the nasal epithelial tube and the organ of Jacobson in Sauropsida. I. Introduction and morphogenesis of the nasal apparatus in members of the families Lacertidae and Scincidae. *Fol. Morphol.* 27, 245–258.

Slabý, O. (1979b). Morphogenesis of the nasal apparatus in *Gecko verticillatus* Laur (Family Geckonidae). II. Morphogenesis of the nasal capsule, the nasal epithelial tube and the organ of Jacobson in Sauropsida. *Fol. Morphol.* 27, 259–269.

Slabý, O. (1979c). Morphogenesis of the nasal apparatus in a member of the family Varanidae. III. Morphogenesis of the nasal capsule, the nasal epithelial tube and the organ of Jacobson in Sauropsida. *Fol. Morphol.* 27, 270–281.

Slabý, O. (1981). Morphogenesis of the nasal apparatus in Sauropsida. IV. Morphogenesis of the nasal capsule, the epithelial nasal tube and the organ of Jacobson in a member of the family Agamidae. *Fol. Morphol.* 29, 305–317.

Slabý, O. (1982). Morphogenesis of the nasal capsule, the epithelial nasal tube and the organ of Jacobson in Sauropsida. VI. Morphogenesis of the nasal apparatus in *Iguana iguana Shaw* and morphological interpretation of the individual structures. *Fol. Morphol.* 30, 75–85.

Slabý, O. (1984). Morphogenesis of the nasal apparatus in a member of the genus *Chamaeleon* L. VIII. Morphogenesis of the nasal capsule, the epithelial nasal tube and the organ of Jacobson in Sauropsida. *Fol. Morphol.* 32, 225–246.

Smeets, W. J. A. J. (1988a). Distribution of dopamine immunoreactivity in the forebrain and midbrain of the snake *Python regius:* a study with antibodies against dopamine. *J. Comp. Neurol.* 271, 115–129.

Smeets, W. J. A. J. (1988b). The monoaminergic systems of reptiles investigated with specific antibodies against serotonin, dopamine, and noradren-

aline. In *The Forebrain of Reptiles: Current Concepts of Structure and Function* (W. K. Schwerdtfeger and W. J. A. J. Smeets, eds.). Karger, Basel, Switzerland, pp. 97–109.
Smeets, W. J. A. J., Hoogland, P. V., and Voorn, P. (1986). The distribution of dopamine immunoreactivity in the forebrain and midbrain of the lizard *Gekko gecko:* An immunohistochemical study with antibodies against dopamine. *J. Comp. Neurol.* 253, 46–60.
Smeets, W. J. A. J., Jonker, A. J., and Hoogland, P. V. (1987). Distribution of dopamine in the forebrain and midbrain of the red-eared turtle, *Pseudemys scripta elegans,* reinvestigated using antibodies against dopamine. *Brain Behav. Evol.* 30, 121–142.
Smeets, W. J. A. J., and Steinbusch, H. W. M. (1989). Distribution of noradrenaline immunoreactivity in the forebrain and midbrain of the lizard *Gekko gecko. J. Comp. Neurol.* 285, 453–466.
Smeets, W. J. A. J., and Steinbusch, H. W. M. (1990). New insights into the reptilian catecholaminergic systems as revealed by antibodies against the neurotransmitters and their synthetic enzymes. *J. Chem. Neuroanat.* 3, 25–43.
Smith, K. K. (1986). Morphology and function of the tongue and hyoid apparatus in *Varanus* (Varanidae, Lacertilia). *J. Morphol.* 187, 261–287.
Smith, M., and Bellairs, A. D'A. (1947). The head glands of snakes, with remarks on the evolution of the parotid gland and teeth of the Opisthoglypha. *J. Linn. Soc.* (London) 41, 351–368.
Soliman, M. A. (1974). The main special sensory nerves; nervus olfactorius, nervus opticus and nervus octavus in lizards. *Bull. Fac. Sci. Cairo Univ.* 47, 127–135.
Steele, C. W., Grassman, M. A., Owens, D. W., and Matis, J. H. (1989). Application of decision theory in understanding food choice behavior of hatchling loggerhead sea turtles and chemosensory imprinting in juvenile loggerhead sea turtles. *Experientia* 45, 202–205.
Takami, S., and Hirosawa, K. (1987). Light microscopic observations of the vomeronasal organ of habu, *Trimeresurus flavoviridis. Japan J. Exp. Med.* 57, 163–174.
Takami, S., and Hirosawa, K. (1990). Electron microscopic observations on the vomeronasal sensory epithelium of a crotaline snake, *Trimeresurus flavoviridis. J. Morphol.* 205, 45–61.
Ten Eyck, G. R., and Gillingham, J. C. (1985). Comparative tongue and anterior process histology in five colubrid snakes. *Am. Zool.* 25, 44a.
Ten Eyck, G., and Halpern, M. (1988). Aggregation in infant corn snakes (*Elaphe guttata*) and garter snakes (*Thamnophis radix*). *Chemical Senses* 13, 740–741.
Ten Eyck, G., and Halpern, M. (1989). Laboratory studies on garter snake (*Thamnophis* sp.). and corn snake (*Elaphe guttata*) aggregation. Animal Behavior Society, Highland Heights, Ky. (Abstract.)
Thoen, C., Bauwens, D., and Verheyen, R. F. (1986). Chemoreceptive and behavioral responses of the common lizard *Lacerta vivipara* to snake chemical deposits. *Anim. Behav.* 34, 1805–1813.
Tollestrup, K. (1981). The social behavior and displays of two species of

horned lizards, *Phrynosoma platyrhinos* and *Phrynosoma coronatum*. *Herpetologica* 37, 130–141.
Tonosaki, K., and Shibuya, T. (1985). The effect of some drugs on the mitral cell odor-evoked responses in the gecko olfactory bulb. *Comp. Biochem. Physiol.* 80C, 361–370.
Tonosaki, K., and Tucker, D. (1982). Olfactory receptor cell responses of dog and box turtle to aliphatic *n*-acetates and aliphatic *n*-fatty acids. *Behav. Neural Biol.* 35, 187–199.
Trauth, S. E., Cooper, W. E., Jr., Vitt, L. J., and Perrill, S. A. (1987). Cloacal anatomy of the broad-headed skink, *Eumeces laticeps,* with a description of a female pheromonal gland. *Herpetologica* 43, 458–466.
Tucker, D. (1963). Olfactory, vomeronasal and trigeminal receptor responses to odorants. In *Olfaction and Taste* (Y. Zotterman, ed.). Macmillan, New York, pp. 45–69.
Ulinski, P. S. (1976). Intracortical connections in the snakes *Natrix sipedon* and *Thamnophis sirtalis*. *J. Morphol.* 150, 463–484.
Ulinski, P. S. (1983). *Dorsal Ventricular Ridge: a Treatise on Forebrain Organization in Reptiles and Birds*. John Wiley & Sons, New York.
Ulinski, P. S., and Kanarek, D. A. (1973). Cytoarchitecture of nucleus sphericus in the common boa, *Constrictor constrictor. J. Comp. Neurol.* 151, 159–174.
Ulinski, P. S., and Peterson, E. H. (1981). Patterns of olfactory projections in the desert iguana, *Dipsosaurus dorsalis. J. Morphol.* 168, 189–227.
Ulinski, P. S., and Rainey, W. T. (1980). Intrinsic organization of snake lateral cortex. *J. Morphol.* 165, 85–116.
Vagvolgyi, A., and Halpern, M. (1983). Courtship behavior in garter snakes: effects of artificial hibernation. *Can J. Zool.* 61, 1171–1174.
Vogt, R. C. (1979). Spring aggregating behavior of painted turtles, *Chrysemys picta* (Reptilia, Testudines, Testudinidae). *J. Herpetol.* 13, 363–365.
Voneida, T. J., and Sligar, C. M. (1979). Efferent projections of the dorsal ventricular ridge and the striatum in the tegu lizard, *Tupinambis nigropunctatus. J. Comp. Neurol.* 186, 43–64.
Waldow, U., Nowycky, M. C., and Shepherd, G. M. (1981). Evoked potential and single unit responses to olfactory nerve volleys in the isolated turtle olfactory bulb. *Brain Res.* 211, 267–283.
Wang, D., Chen, P., Jiang, X. C., and Halpern, M. (1988). Isolation from earthworms of a proteinaceous chemoattractant to garter snakes. *Arch. Biochem. Biophys.* 267, 459–466.
Wang, R. T., and Halpern, M. (1980a). Light and electron microscopic observations on the normal structure of the vomeronasal organ of garter snakes. *J. Morphol.* 164, 47–67.
Wang, R. T., and Halpern, M. (1980b). Scanning electron microscopic studies of the surface morphology of the vomeronasal epithelium and olfactory epithelium of garter snakes. *Am. J. Anat.* 157, 399–428.
Wang, R. T., and Halpern, M. (1982a). Neurogenesis in the vomeronasal epithelium of adult garter snakes. 1. Degeneration of bipolar neurons and proliferation of undifferentiated cells following experimental vomeronasal axotomy. *Brain Res.* 237, 23–39.
Wang, R. T., and Halpern, M. (1982b). Neurogenesis in the vomeronasal epi-

thelium of adult garter snakes. 2. Reconstitution of the bipolar neuron layer following experimental vomeronasal axotomy. *Brain Res.* 237, 41–59.

Wang, R. T., and Halpern, M. (1983). Surface modifications of the olfactory cilia in some reptiles. *Anat. Rec.* 205, 209A-210A.

Wang, R. T., and Halpern, M. (1988). Neurogenesis in the vomeronasal epithelium of adult garter snakes. 3. Use of H^3-thymidine autoradiography to trace the genesis and migration of bipolar neurons. *Am. J. Anat.* 183, 178–185.

Watkins, J., II, Gehlbach, F., and Baldridge, R. S. (1967). Ability of the blind snake, *Leptotyphlops dulcis* to follow pheromone trails of army ants, *Neivamyrmex nigrescens* and *N. opacithorax*. *Southwest. Nat.* 12, 455–462.

Watkins, J., II, Gehlbach, F., and Kroll, J. (1969). Attractant-repellent secretions in the intra- and interspecific relations of blind snakes (*Leptotyphlops dulcis*) and army ants (*Neivamyrmex nigrescens*). *Ecology* 50, 1098–1104.

Weldon, P. J. (1982). Responses to ophiophagous snakes by snakes of the genus *Thamnophis*. *Copeia* 1982, 788–794.

Weldon, P. J., Ford, B. B., and Perry-Richardson, J. J. (1990). Responses by corn snakes (*Elaphe guttata*) to chemicals from heterospecific snakes. *J. Chem. Ecol.* 16, 37–44.

Weldon, P. J., and Schell, F.M. (1984). Responses by king snakes (*Lampropeltis getulus*) to chemicals from colubrid and crotaline snakes. *J. Chem. Ecol.* 10, 1509–1520.

Weldon, P. J., Swenson, D. J., Olson, J. K., and Brinkmeier, W. G. (1990). The American alligator detects food chemicals in aquatic and terrestrial environments. *Ethology* 85, 191–198.

Werner, D. I., Baker, E. M., Gonzalez, E. del C., and Sosa, I. R. (1987). Kinship recognition and grouping in hatchling green iguanas. *Behav. Ecol. Sociobiol.* 21, 83–89.

Whiting, A. M. (1969). "Squamate Cloacal Glands: Morphology, Histology and Histochemistry." Ph.D. dissertation, Pennsylvania State University, University Park. (Unpublished.)

Wiedemann, E. (1931). Zur Biologie der Nahrungs-aufnahme europäischer Schlangen. *Zool. Jhrb. Syst.* 61, 621–636.

Williams, P. R., Jr., and Brisbin, I. L., Jr. (1978). Responses of captive-reared eastern kingsnakes (*Lampropeltis getulus*) to several prey odor stimuli. *Herpetologica* 34, 79–83.

Wysocki, C. J., and Meredith, M. (1987). The vomeronasal system. In *Neurobiology of Taste and Smell* (T. E. Finger and W. L. Solver, eds.). John Wiley & Sons, New York, pp. 125–150.

Young, B. A. (1987). The cranial nerves of three species of sea snakes. *Can. J. Zool.* 65, 2236–2240.

Young, B. A. (1990). Is there a direct link between the ophidian tongue and Jacobson's organ? *Amphibia-Reptilia* 11, 263–276.

Contributors

William E. Cooper, Jr.
Department of Biological Sciences
Indiana University—Purdue University at Fort Wayne
Fort Wayne, IN 46805-1899

David Crews
Department of Zoology
University of Texas at Austin
Austin, TX 78712

Carl Gans
Department of Biology
University of Michigan
Ann Arbor, MI 48109-1048

Neil Greenberg
Department of Physiological Ethology and Department of Zoology
University of Tennessee
Knoxville, TN 37996-0810

Mimi Halpern
Department of Anatomy and Cell Biology
Downstate Medical Center
Brooklyn, NY 11203-2098

Jonathan Lindzey
Department of Zoology
University of Texas at Austin
Austin, TX 78712

Robert T. Mason
Department of Zoology
Oregon State University
Corvallis, OR 97331

Michael C. Moore
Department of Zoology
Arizona State University
Tempe, AZ 85287

Richard R. Tokarz
Department of Biology
University of Miami
Coral Gables, FL 33124

Herbert Underwood
Department of Zoology
North Carolina State University
Raleigh, NC 27695-7616

Joan M. Whittier
Department of Anatomy
University of Queensland
St. Lucia, Brisbane
Queensland 4067
Australia

Author Index

Subject Index